International Scientific and Practical Conference "Modern Problems of Ecology, Transport and Agricultural Technologies" (MPETAT' 2020)

IOP Conference Series: Materials Science and Engineering
Volume 941

Barnaul, Russia
26-27 June 2020

ISBN: 978-1-7138-4776-2
ISSN: 1757-8981

Printed from e-media with permission by:

Curran Associates, Inc.
57 Morehouse Lane
Red Hook, NY 12571

Some format issues inherent in the e-media version may also appear in this print version.

Institute of Physics
Dirac House, Temple Back
Bristol BS1 6BE UK

Phone: 44 1 17 929 7481
Fax: 44 1 17 920 0979

techtracking@iop.org

Additional copies of this publication are available from:

Curran Associates, Inc.
57 Morehouse Lane
Red Hook, NY 12571 USA
Phone: 845-758-0400
Fax: 845-758-2633
Email: curran@proceedings.com
Web: www.proceedings.com

TABLE OF CONTENTS

3. CURRENT PROBLEMS AND PROSPECTS FOR THE DEVELOPMENT OF GROUND TRANSPORTATION AND TRANSPORT-TECHNOLOGICAL MACHINES

Abstract

Conference Modern problems of ecology, transport and agricultural technologies (MPETAT' 2020)

PREFACE

The conference serves as a platform for discussion of advanced and innovative technical solutions in various fields of ecology, transport and agricultural technologies. The key topics of the conference cover priority areas of research in the field of technical and related Sciences and take into account global development trends.

MPETAT ' 2020 IS based on sustainable development with a close focus on all forms of engineering professions and research tools, including the following topics:

1. Technosphere and environmental safety, Agroecology

2. Modern problems and prospects of agricultural technologies development

3. Modern problems and prospects of development of land transport and transport and technological machines

IOP Conf. Series: Materials Science and Engineering **941** (2020) 011002 doi:10.1088/1757-899X/941/1/011002

Peer review declaration

All papers published in this volume of *IOP Conference Series: Materials Science and Engineering* have been peer reviewed through processes administered by the Editors. Reviews were conducted by expert referees to the professional and scientific standards expected of a proceedings journal published by IOP Publishing.

- **Type of peer review:** Single-blind
  We had a blind review. 2 reviewers were assigned to one paper.
- **Conference submission management system:**
  https://www.altstu.ru/structure/unit/noo/scienceevent/4190/
- **Number of submissions received:** 108
- **Number of submissions sent for review:** 85
- **Number of submissions accepted:** 74
- **Acceptance Rate (Number of Submissions Accepted / Number of Submissions Received X 100:** 68.5
- **Average number of reviews per paper:** 2 reviewers
- **Total number of reviewers involved:** 8 reviewers
- **Any additional info on review process:** During review process, all papers were checked for plagiarism, compliance with the topic of the journal, acceptable English language, and good quality of graphic images.
- **Contact person for queries:** Alexander Saprykin (mpetat2020_astu@mail.ru)

IOP Conf. Series: Materials Science and Engineering **941** (2020) 012001 doi:10.1088/1757-899X/941/1/012001

Zinc in the body and somatic health of industrial workers

G Slepchenkol, E Moiseeva[1] * and I Khlusov[1,2]

[1]National Research Tomsk Polytechnic University, Lenina avenue 30 634050 Tomsk Russia
[2]Siberian State Medical University, 2 Moskovsky tract Tomsk 634050 Russia
E-mail: microlab@tpu.ru

Abstract 30% of industrial workers have a chronic (at least 3-4 months) zinc deficiency as a vital trace element necessary for the functioning of a significant amount of metalloproteins. The cause of zinc deficiency may be copper intoxication noted earlier, which together is important for the development of structural pathology of internal organs.

1. Introduction

Dysmicroelementoses are common in the Russian Federation as a whole and in the Siberian region, which is also characterized by a low level of population health [1]. For industrial workers, the amount of metals in the body can be increased 20-50 times according to the methodological recommendation of the USSR Ministry of Health for professional activities (1988). However, it is not known in which concentration range in the body the vital elements become toxic. Analysis of human hair and nails is considered an adequate method for mass non-invasive assessment of elemental status [1] within 3-4 months prior to the examination. However, the official boundaries of the "norm" for the content of chemical elements in human hair have not been established [2], in particular, due to their dependence on the region of residence [3]. Due to the wide range of values, poorly regulated definitions of "recommended levels" and "biologically acceptable levels" are proposed [2-4].

In certain quantities, zinc is an essential element in the absence of fluctuations in its concentration in the environment. The paper presents the results of a study of the zinc content in the body of workers of an industrial enterprise in Tomsk and some indicators of their somatic health.

2. Materials and methods

A survey of employees of the enterprise producing methanol was carried out as part of a preventive medical examination. The examination of healthy volunteers was carried out with the permission of the Ethics Committee (opinion No. 583 of March 19, 2007)

Siberian State Medical University (Tomsk). The subjects were informed about the features of the diagnostic procedures performed in accordance with ethical requirements. In the occipital region, closer to the neck, for men and women employed in the production of the Metanol OJSC enterprise (Tomsk), hair was cut at 3-5 points in strands 2-3 mm2 thick. The total sample volume was about 100 mg. Quantitative chemical analysis of hair samples for zinc and copper is based on the inversion voltammetric method for determining the mass concentrations of elements in the solution of the prepared sample [5]. Semi-quantitative assessment of the state of parenchymal organs was carried out in points according to ultrasound data (ultrasound). The absence of structural changes in the parenchyma was taken for 1 point. In the case of detected organic changes in each organ (liver, kidney, pancreas, thyroid, prostate), an additional score was added.

IOP Conf. Series: Materials Science and Engineering **941** (2020) 012001 doi:10.1088/1757-899X/941/1/012001

The results were expressed as mean value (X) and standard deviation (SD), median (Me), 1st (Q1) and 3rd (Q3) percentiles. Statistical processing of the results was carried out by methods of variation statistics using the Mann-Whitney test and the Spearman rank correlation coefficient.

3. Results

An analysis of the type of distribution (universal criteria of Lilliefors and Shapiro-Wilk) of the quantitative contents of zinc in the hair of men and women showed a distribution close to normal (Fig. 1a, b).

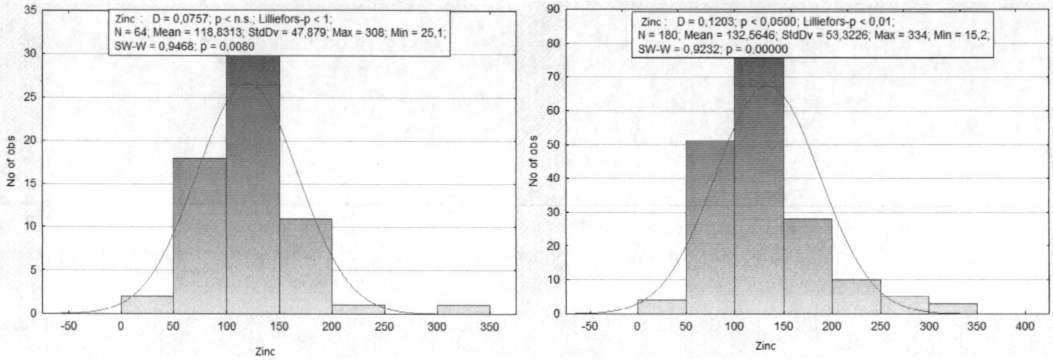

Figure 1. A histogram of the distribution of zinc in the hair of employees: a) in women; b) in men

Given the nature of the distribution, the values of zinc in the hair of the employees of the enterprise were described both as an average value (X ± SD) and were presented as values of centile intervals of concentrations (Me (Q1-Q3).

A physical examination of the enterprise employees did not show a statistically significant change in the average zinc content in comparison with both the recommended levels (Table 1) and with the regional indicator in Tomsk [6]. Gender differences in zinc accumulation were not detected (Table 1); therefore, the groups were combined. Only in 5 workers (2% of the examined) did the zinc concentration in the hair exceed the extreme limit of 280 kg / kg, which removes the question of possible mass zinc intoxication in methanol production using metal catalysts. More common was reduced (less than 100 mg / kg) zinc levels in 73 people (30%). According to ultrasound data, only 11% of the examined employees of the enterprise, mainly young people (less than 33 years), did not reveal structural changes in internal organs. In other cases, polyorgan (in 2-3 internal organs) structural changes were found (Table 2), primarily of the gall bladder, kidneys, liver, and pancreas.

Table 1. The level of trace elements in the hair of the examined employees of the enterprise, X ± SD, Me (Q1-Q3)

Group	Age (years)	Zinc (mg / kg)
Recommended levels for men and women [1,3,4]	-	140-280 100-250
Male workers, n=180	43.7±11.4	124.5 (97.4-151.0)
Female workers, n=64	42.0±8.5	121.0 (91.8-147.0)

Employees of the enterprise (men and women),n=244	43.3+10.8	124.0 (96.2-150.0)

However, correlation analysis did not reveal statistically significant correlations between the Zn content in the hair, passport age (Fig. 2a, b) and somatic pathology (Table 2, 3).

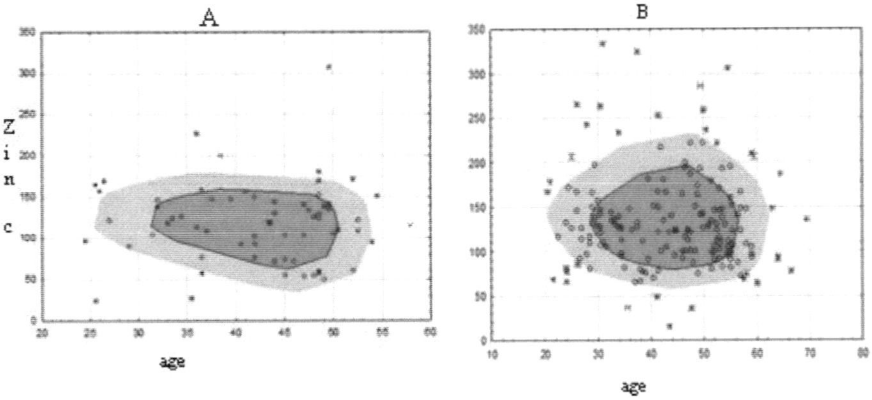

Figure 2. Dependence of the distribution of zinc in the hair of employees on the age: a) in women; b) in men.

Table 2. Results of correlation analysis of the dependence of emerging pathologies on the content of zinc in the hair, at P=0.95 (according to Spearman)

Group	Spearman rank order correlation	Significance of the coefficient, t(N-2)
Female workers, n=63	0.039	0.302
Male workers, n=168	0.026	0.342

Table 3. Data of ultrasound investigation of internal organs in employees, X±SD.

% of examined employees					Organ pathology, points
Renal system	Liver	Thyroid gland	Pancreas gland	Prostate	
41	36	14	6	3	2.81±1.15* n=237

4. Discussion

IOP Conf. Series: Materials Science and Engineering **941** (2020) 012001 doi:10.1088/1757-899X/941/1/012001

The reduced level of zinc in the hair of the company's employees testifies, first of all, in favor of the chronic pathology of the liver, pancreas and kidneys. Nevertheless, the nature of organ pathology among the employees of the enterprise as a whole did not differ from that among the population of the Tomsk region of comparable age. Apparently, the main reason was occupational hazards associated with the constant use of copper (Cu) catalyst in the methanol production process, which is accompanied by excessive intake of heavy metal in the body of workers of this enterprise [6]. Cu and Zn are biological competitors [4].

5. Conclusion

30% of industrial workers have a chronic (at least 3-4 month) zinc deficiency as a vital trace element necessary for the functioning of metalloproteins that regulate metabolism, cell function and antioxidant defense of the body [7]. The cause of zinc deficiency may be copper intoxication noted earlier, which together is important for the development of structural pathology of internal organs.

6. Acknowledgments

The researches and experimental calculations are carried out within the Framework of Tomsk Polytechnic University Competitiveness Enhancement Program grant. The researches was supported by State task "Science" (basic) No. 20.0025. Gzb.2020

References

[1] A V Skalniy, A T Bykov 2003 Environmental and physiological aspects of the use of macro- and microelements in regenerative medicine (Orenburg: RIK GOU OGU).

[2] A V Skalny, A L Gorbachev, M V Veldanova 2004 Elemental status of children in the North-East of Russia (Orenburg: RIK GOU OGU).

[3] A V Skalny, A T Bykov, E P Serebryansky, M G Rocky 2003 Medico-ecological assessment of the risk of hypermicroelementoses in the metropolis population (Orenburg: RIK GOU OGU).

[4] V M Lifshits, V I Sidelnikova 1998 Biochemical analyzes in a clinic: Reference book (Moscow: MIA).

[5] Hair. Inversion voltammetric method for measuring the mass concentrations of zinc, cadmium, lead, copper, iron, arsenic, manganese, nickel, selenium (FR.1.31.2006.02273)

[6] I Khlusov, G Slepchenko, E Moiseeva 2019 An anthropogenic load of copper and somatic health *IOP Conf. Ser .: Earth Environ. Sci. 395 012008*. doi: 10.1088 / 1755-1315 / 395/1/012008

[7] V G Rebrov, O A Gromova 2008 Vitamins, macro- and microelements (Moscow, Russia: GAOTAR-Media)

IOP Conf. Series: Materials Science and Engineering **941** (2020) 012002 doi:10.1088/1757-899X/941/1/012002

The effectiveness to cleanup oil spills from the water environment of the combined sorbent on the basic of PUF and chitin

T Y D Trang, L A Zenitova

Kazan National Research Technological University, 68 Karl Marx Street, Kazan, Tatarstan, 420015, Russia

E-mail: tydtrang@gmail.com

Abstract. This study provides information about evaluating the parameters (density, buoyancy, sorption capacity) of a combined oil sorbent based on polyurethane foam filled with chitin (PPU10M). In comparison with the original chitin, the developed sorbent has better buoyancy, and commensurate with the pure polyurethane foam (PUF). The effect of the contact time, the type of the water, the thickness of the oil layer and the adsorbent dosage are studied. The results show that the optimal contact time of the sorption process is 60 minutes. In which, the oil capacity of the sorbent PPU10M in artificial seawater reaches 13.6 g/g and in artificial river water 13.2 g/g. The sorbent dose depends on the thickness of the oil layer on the water surface. For a thin oil layer, the use of 1g of adsorbent reaches higher oil capacity but the oil removal percentage lower in comparison to the dose of 2g sorbent. For an oil layer thickness equal to or greater than the height of the sorbent cube, the efficiency of oil removal is observed better when using 2g of sorbents. Thus, it is considered that the use of 2g sorbents is optimal. A generalization of the research results indicates the prospects of the developed combined sorbent based on chitin and polyurethane for oil spill treatment.

1. Introduction

Nowadays, because of the quick development of industry and transport, leading to an increase in oil demand [1]. Mostly oil is transported by waterways such as through rivers, seas, and oceans. It is one of the main reasons that lead to accidents and the spill of oil into the sea and river environment. These accidents can not only cause serious pollution in the water environment but also negatively affect the surrounding environment, and these are also one of the causes of the significant loss of valuable resources [2]. Therefore, it is urgent to remove oil spills from the water environment [3]. To treatment the oil spills problem, commonly used methods include on-site burning, mechanics method, chemistry method and the use of sorbents [4]. Each method has its advantages and disadvantages.

The sorbent method can be considered one of the most effective methods for the cleanup of oil spills and regeneration of oil after the process adsorption [5]. A good sorbent should include criteria such as high sorption capacity, selectivity for oil adsorption, the possible reusability, good oil regeneration, high buoyancy and a moderate cost [6, 7]. There are three main types of oil adsorbents: inorganic mineral absorbents, organic natural absorbents, and synthetic absorbents. Among them, the most potential and promising method is the use of synthetic absorbents, because their absorption capacity is significantly higher than other groups. However, they have a fairly high cost and decompose very slowly.

IOP Conf. Series: Materials Science and Engineering **941** (2020) 012002 doi:10.1088/1757-899X/941/1/012002

Furthermore, the effects on the environment of these sorbents are still not clear. Therefore, it is important to develop environmentally-friendly oil absorbents with high adsorption capacity, reasonable cost and high ability to recycle process and reuse [2].

Recently, polyurethane foam (PUF) is one of the synthetic sorbents that has attracted the attention of many scientists, because of its high oil capacity, the reusability, and high buoyancy. However, its cost is quite high. Reducing the cost of this synthetic adsorbent by the use of fillers from natural materials with low density is considered a viable method [8]. For this aim, it is recommended to use chitin - an available natural material that has many advantages, such as low cost, low density, environmental friendliness, and biodegradability. However, chitin has a not high adsorption capacity and hydrophobicity in comparison to the synthetic materials [8]; more its reusability is impossible. To resolve these disadvantages of both materials, in this work, the polyurethane foam is combined with a natural sorption material - chitin.

The adsorption capacity of the sorbent PPU10M for oil spilled on the surface of seawater and river water is studied.

2. Materials and methods

2.1. Materials

To produce the PUF, components A and B from the company "Dau Izolan" (Vladimir City, Russia) are used. Component A has the density - 1.02 g/cm^3 and the viscosity - 1200 MPa.s; component B has the density - 1.20 g/cm^3 and the viscosity - 50 MPa.s. The filler is chitin obtained from shrimp shells with a particle size of 1-3 mm, with moisture content - 2.54%, ash content <0.1%. Chitin is purchased from the company "Chitosan Vietnam" (Ho Chi Minh City, Vietnam).

As a pollutant, oil is purchased from the Vereisky region (Tatarstan republic, Russia) with a density at 20°C - 0.984 g/cm^3, and a kinematic viscosity at 20°C - 55.7 MPa.s. Artificial seawater has been used according to ASTM D1141-98 [9] and artificial river water according to the method described in the work [10] for simulating oil spills in river water and seawater.

2.2. Synthesis of the combined sorbent PPU10M

For the synthesis of the combined sorbent PPU10M, component A is mixed with chitin by using the high-speed agitator, then component B is added to this chitin-component A mixture, and continue to mix them. The ratio between the components A and B by weight is 10:6. In the previous study, it is showed that the sorbent containing 10% chitin with the crushed size of 1-3 mm (PPU10M) has the highest oil capacity [11]. After the preparation process, the material is left in 24 hours at room temperature. Next, it is cut into particles with a size of 10×10×10 mm.

2.3. Characteristics of the sorbent

To determine the apparent density of materials, method GOST 409-77 is used [12]. The buoyancy of sorbents is evaluated by method ASTM F 726-99 [13].

2.4. Sorption capacity of the sorbent

The experiments are carried out in the oil-water system. To simulate accident spilled oil on the water, 250 ml of the artificial seawater or artificial river water is placed in a glass with a volume 500 ml, and then an amount of oil is added until it forms a layer on the surface of the water. Next, an amount of the sorbent is weighed and placed in this system. The sorption process is carried out within 120 minutes. Then the sorbents are removed and drained for 2 min and the total mass of the sorbents after sorption process is determined on the analytical balance. The remaining oil-water after adsorption are separated by 100 ml toluene. This obtaining mixture is separated by the separatory funnel to obtain the remaining water after adsorption. It is weighed and the result is recorded to calculate the mass of absorbed water. The oil capacity of the sorbent – (q, g/g) calculate by the formula (1).

$$q = \frac{m_s - (m_w + m_o)}{m_o} \quad (1)$$

where: m_s – the mass of saturated sorbent (water, oil and initial sorbent), g; m_w – the mass of the adsorbed water, g; m_0 – the initial mass of the dry sorbent, g.

The degree of oil removal – (P, %) is determined according to the formula (2).

$$P = \frac{m_r}{m_n}.100 \quad (2)$$

where m_r – the mass of the adsorbed oil, g; m_n – the initial mass of oil, g.

The efficiency of oil adsorption by the sorbent PPU10M – (E, %) is determined by the ratio between the mass of adsorbed oil (m_r, g) and the total mass of the sorbent after sorption process (m_s, g) (formula 3).

$$E = \frac{m_r}{m_s}.100\% \quad (3)$$

2.4.1. Effect of the contact time on the sorption capacity

About 50 g of oil is added to 250 ml of the water (artificial seawater or artificial river water) in the glass beaker, then 2 g of the sorbent PPU10M is placed into this oil-water system. Sorption capacity is measured after 5, 15, 30, 60, 90 and 120 min, respectively.

2.4.2. Effect of oil layer thickness and sorbent dose on adsorption capacity

Oil is added to a glass beaker with 250 ml of water until it creates a layer with a thickness of 3, 5, 7, 9, 11, and 13 mm, respectively in the surface of the water. Next, 1, 2 or 3 g of sorbent, respectively, is placed into these simulate oil spills system. The contact time of the adsorption process is 60 minutes.

3. Results and discussions

3.1. Characteristics of the materials

Evaluation of the density of sorbents is necessary because it affects the buoyancy of the sorbents. As can be seen from the data in table 1, the density of chitin is lower than the density of the sorbents. Therefore, its presence in the sorbent PPU10M doesn't affect the density of the sorbent. But in view of the fact that foaming of PPU filling with chitin is somewhat difficult, the density of the combined sorbent with chitin is higher than the blank PUF.

Table 1. Density and buoyancy of the materials.

Characteristics	PUF	Chitin	PPU10M
Density, kg/m^3	59.10	13.12	69.91
Buoyancy in the saturated state, %	95%	55-60%	95%
Description of the buoyancy	Floats (100% mass)	> 70% of the mass submerged in water	Floats (100% mass)

The sorbent with good buoyancy leads to reducing the negative effects on the organisms living in the aquatic environment [7]. The research results about the buoyancy of materials show that the buoyancy of the combined sorbent PPU10M is not different from the original PUF, but it is significantly higher than in comparison to the chitin. The loss of buoyancy of chitin at the saturated state not only affects the oil recovery process and the removal process of the solid waste after adsorption, but it is also one of the causes of secondary pollution and negative influents on aquatic organisms. Due to the high buoyancy of the sorbent PPU10M, after the adsorption process, it is easily removed by using the simple and cheap mechanical methods, for example, the use of nets. This contributes to the significant reduction in the cost of the oil spill removal process.

3.2. Adsorption capacity of the sorbent PPU10M

3.2.1. Effect of the contact time

With the purpose to establish the optimal contact time for maximum oil absorption, the adsorption process is studied within 120 minutes. It is noticed that the oil sorption of the sorbent PPU10M increases with the increase of contact time (figure1a). For both types of water, the speed of oil sorption increases significantly in the first 30 minutes due to the high oil concentration. In the next 30 minutes, the oil concentration decreases, leading to the reduction gradually of the speed of the oil sorption. After 60 minutes the oil sorption reaches the equilibrium state, so the oil capacity hardly increases. The oil sorption on the sorbent PPU10M can be explained because of the large number of capillary pores in its structure, that provide a sufficient volume to absorb oil through intermolecular interactions [14-16]. Next, the oil continues to enter into the capillary pores of the PPU10M. It leads to a low speed of the oil adsorption process and after 60 minutes the equilibrium state is setup.

In the treatment of oil spills, the degree of oil removal and the efficiency of oil adsorption are two important parameters for evaluating the sorbent. Figure 1b shows that the oil removal percentage of the sorbent PPU10M increases significantly from 5 to 60 minutes, after 60 minutes there is almost no difference in the process of oil removal. Similarly, when the contact time is longer than 60 minutes, the efficiency of oil sorption also significantly reduces. Therefore, the optimal contact time of the oil adsorption by the sorbent PPU10M is 60 minutes with oil capacity reaches 13.6 g/g for artificial seawater and 13.2 g/g for artificial river water.

The difference in adsorption capacity between artificial river water and artificial seawater of the combined adsorbent PPU10M is explained by the presence of various salts in the composition of these types of water. It leads to the changes in the electrical double layer between the sorbent and the adsorbate, which affects the sorption ability of the sorbent in different aqueous environments [17, 18]. But generally, the difference in oil adsorption capacity of the sorbent PPU10M is small, it ranges 0.19-0.35 g/g. This indicates that the chemical composition of the aqueous phases does not significantly affect the capacity to remove oil spills by the combined adsorbent based on polyurethane foam and chitin. Therefore, the sorbent PPU10M can be used widely to treatment oil spills both on the rivers and in the sea.

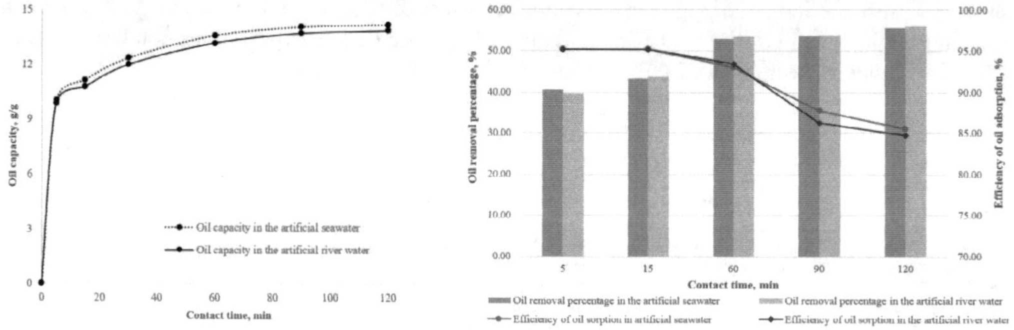

(a) (b)

Figure 1. The dependence of the oil capacity (a), the degree of oil removal and the efficiency of oil sorption (b) on the contact time of the PPU10M in the artificial seawater and the artificial river water.

3.2.2. Effect of oil layer thickness and adsorbent dose

The dose of sorbent and the thickness of the oil layer are two factors that significantly affect the effectiveness to clean up oil spills. Three different dosages of the sorbent (1, 2 and 3g, respectively) and the various thicknesses of the oil layer (3, 5, 7, 9, 11 and 13 mm, respectively) are investigated.

Due to an increase in the thickness of the oil layer leads to an increase in the contact area between the adsorbent and adsorbate, at the same time, it leads to the decrease of the contact between the absorbent particles and the water surface (figure 2). Therefore, with an increase in the thickness of the oil layer in the limit of this study, the oil capacity of the PPU10M increases for all three cases of the adsorbent dosage (figure 3a).

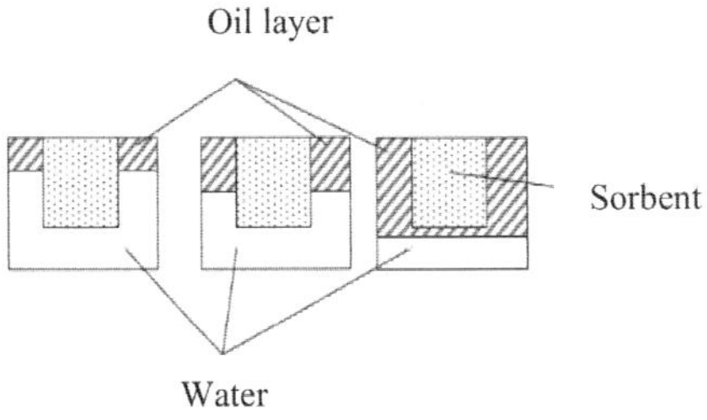

Figure 2. The scheme of the sorbent at different thicknesses of the oil layer.

The research results show that the use of adsorbent dosage depends on the thickness of the oil layer. With a thin layer of oil, the increased dose of the sorbent leads to a decrease in oil capacity and the efficiency of oil adsorption. Because a large amount of absorbent leads to contact of the adsorbent not only to the oil but also to the water. Therefore, the water capacity increases, while the oil capacity reduces. But the low dose of adsorbent (1g) reaches the lower significant degree of oil removal in comparison to the higher dose (2g and 3g) (figure 3b). With the oil layer thickness commensurate with the height of the sorbent cube (d_{oil} = 9-11 mm), the increase of adsorbent dosage from 1g to 2 g significantly increases oil capacity. When the use of the sorbent increases from 2g to 3g, the degree of oil removal and the efficiency of oil sorption are not significantly different. In the case the thickness of the oil layer exceeds the size of the cubes PPU10M ($d_{oil} \geq 11$ mm), the more the adsorbent dosage increases, the more the oil capacity and the degree of oil removal increases. In general, in comparing the parameters: oil capacity, removal percentage and efficiency of the oil sorption show that the optimal dose of the sorbent is 2g. At a dosage of 2g PPU10M, its oil capacity reaches about 5.2–13.4 g/g, the degree of oil removal is in the range 56–90%, and the efficiency of oil absorption is relatively high about 44–92%.

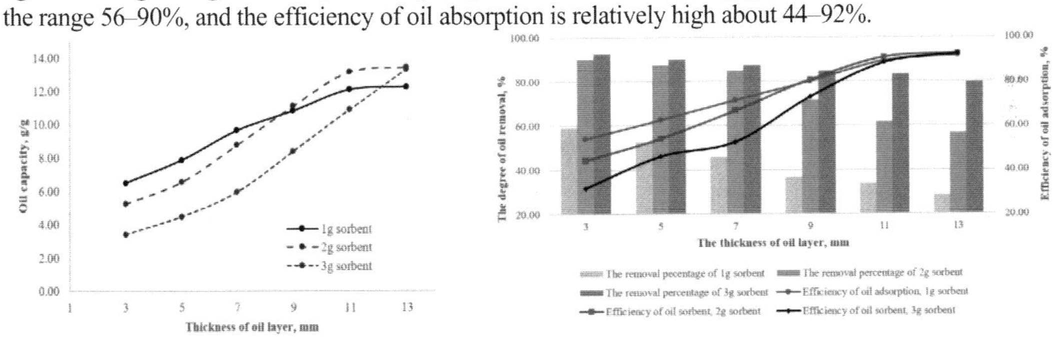

(a) (b)

Figure 3. The dependence of the oil capacity of the PPU10M (a), the degree of oil removal and the efficiency of oil sorption (b) on the thickness of the oil layer on the surface of water.

Conclusion

The oil adsorption capacity of the combined adsorbent based on chitin and polyurethane foam was studied in this paper. It is showed that the fill of chitin to the component PUF doesn't significantly change the density of the combined adsorbent PPU10M in comparison to blank PUF, it reaches 69.91 kg/m3. The buoyancy of the combined sorbent is also higher compared to the original chitin and corresponds to the buoyancy of the blank PUF. The study of the effect of sorption time on the efficiency of oil sorption on the adsorbent in artificial river water and artificial seawater shows that the optimal contact time of the oil sorption process is 60 minutes; the oil capacity at 60 minutes reaches 13.6 g/g for seawater and 13.2 g/g for river water. At the same time, the chemical composition of the aqueous environment does not significantly affect the efficiency to remove the oil spill of the material PPU10M. The initial thickness of the oil layer on the water surface affects the use of the adsorbent dosage. In detail, for the thin oil layer, the use of 1 g of adsorbent reaches better oil capacity but the oil removal percentage is not high as the dose of 2g; for the oil layer thickness equal to or exceeding to the size of the sorbent cube, then the use of 2g PPU10M is most effective. In general, the optimal adsorbent dosage is 2g.

References

IOP Conf. Series: Materials Science and Engineering **941** (2020) 012002 doi:10.1088/1757-899X/941/1/012002

[1] Wang J, Zeng Y, Wang A 2012 *Chemical Engineering Journal* **213** 1–7
[2] Wu Z Y, Li C, Liang H W, Zhang Y N, Wang X, Chen J F and Yu S H 2014 *Scientific Reports* **4(4049)** 1–6
[3] Dong X, Chen J, Ma Y, Wang J, Chan-Park M B, Liu X and Chen P 2012 *Chemical Communications* **48** 10660–10662
[4] Al-Majed A A, Adebajo A R, Hossain M E 2012 *Journal Enviromental Manage* **113** 213–217
[5] Banerjee S S, Josh M V, Jayaram R V 2006 *Chemosphere* **64** 1026–1031
[6] Alideza K, Hamid Z, Amir A, Ahmad A 2015 *Journal of environmental Manegement* **157** 279–286
[7] Trang Y D T, Zenitova L A 2019 IOP Study on the sorption capacity of the adsorbent based on polyurethane and chitin to remove oil spills // IOP *Conf. Ser.: Earth Environ. Sci.* **337** 012008
[8] Hoang P H, Hoang A T, Chung N H, Dien L Q, Nguyen X P and Pham X D 2017 *Energy sources, part A: Recovery, utilization, and environmental effects* **2017** 1–8
[9] *ASTM D1141-98* (ASTM International)
[10] Ruben A, Nobuyasu Y, Katsuji T and Masao N 2003 *Journal of Health Science* **49(5)** 379–385
[11] Trang T Y D, Zenitova L A 2019 *Bulletin of PNIPU. Chemical technology and biotechnology* **2** 33–47
[12] *GOST 409-77* (Moscow: Official publication, and IC Publishing Standards)
[13] *ASTM F726-99* (ASTM International. West Conshohocken PA)
[14] Tanobe V O A, Sydenstricker T H D, Amico S C, Vargas J V C and Zawadzki S F 2009 *Journal Applied Polymer Science* **111** 1842–1849
[15] Harikrishnan G, Patro T U and Khakhar D 2006 *Index Eng. Chemical Res.* **45** 7126–7134
[16] Wu D, Fang L, Qin Y, Wu W, Mao C and Zhu H 2014 *Mar Pollut Bull.* **84** 263–267
[17] Bjelopavlic M G et al 1999 *Journal and Interface Science* **210** 271–280
[18] Drikas M 1997 *Carbon* **35** 1239–1250

IOP Conf. Series: Materials Science and Engineering **941** (2020) 012003 doi:10.1088/1757-899X/941/1/012003

Resource-saving technologies in the processing ofnepheline ores by sintering

I I Shepelev[1], E N Eskova[1], E V Kirushin[1], O V Pilyaeva[2] and A Yu Sakhachev[3]

[1]Krasnoyarsk state agrarian university, 660049, Krasnoyarsk, Mira Ave, 90, Russia
[2]Achinsk branch of Krasnoyarsk state agrarian university, 662150, Achinsk, Kommunisticheskaya str., 49 Russia
[3] JSC "RUSAL Achinsk", 662153, Yuzhnaya promzona, kvartal XII, stroenie 1 Achinsk, Russia
[4]E-mail: Ekoing@mail.ru

Abstract. To solve ecological problems in alumina production, ecological measures were implemented to reduce the consumption of mineral natural resources and by involving secondary technogenicmaterials in the production. In the processing ofnepheline ores, a number of technologies were tested and implemented into production for the introduction of various additives of technogenic raw materials into the raw alumina charge: spent chamotte refractory bricks, ferrotitanium production slag, and combined heat and power plant ash. The introduction of technology for processing oftechnogenic raw materials provided the use of spent chamotte bricks of more than 70,000 tons, while saving natural raw materials: nepheline ore of about 42,000 tons. In addition to saving mineral resources, work was carried out to reduce waste water discharges into open water basin and reduce the water intake of treated water for production needs. For this purpose, the technological equipment for water supply of the combined heat and power plant was replaced. As a result of the introduction of new cooling towers, the waste water discharges into an open water basin was eliminated and the intake of treated water from the river was reduced by 40.07 millionm3per year.

Introduction

At the modern level and scale of natural resource consumption, the importance of full use factors, resource conservation and involvement in the production of secondary resources is of great importance. The wide involvement of secondary material resources in the production process, including technogenic industrial products, will ensure the preservation of natural raw materials and lead to the development and implementation of low-waste technologies. The limited and non-reproducible nature of mineral resources on the one hand and the high intensity of their exploitation on the other, have led to the need to search for new resource-saving technologies, including in the processing of nepheline ores [1]. The Kiya-Shaltyrsky nepheline mine located in Siberiaof the Russian Federation has almost exhausted its resource of unique nepheline ores. And it is searching for new aluminum-containing raw materials. However, at present, the Kiya-Shaltyrsky nepheline mine has accumulated more than 90 million tons of substandard nepheline ore, which is located in special dumps. The involvement in production using the existing technology is difficult to implement and not economically profitable. Processing of this substandard mineral raw material can be carried out with the introduction of aluminum enhancing additives. The development of technologies aimed at saving raw materials in the processing of substandard nepheline ores, with the involvement of production waste containing aluminum oxide, is an urgent problem today [2-5].

Results and discussion

The paper evaluates the possibility of using combined heat and power plant ash and ferrotitanium production slag as aluminum – raising additives in limestone-nepheline charge. The use of ash as an additive in the alumina raw material charge becomes possible due to the similarity of its chemical characteristics with nepheline ore and the increased content of alumina in it [6, 7]. The content of alumina in the ash of various combined heat and power plants (CHPP)ranges from 27 to 35%.

The sintering charges were based on nepheline ore from the Kiya-Shaltyrskymine, limestone from the Mazulskymine, and white slime formed during the desilting of aluminate solutions. The dosage of white slime was 10 % of the weight of the ore mixture for dry materials. As a corrective additive, chemically pure soda was used.

Ash of CHPPis alkali-free raw materials and in comparison, with nepheline ore is characterized by high content of silicon dioxide (55.6 %), and a content of iron (6.18 %) and lower content of calcium oxide (5.6 %). The results of the analysis showed that insertion of ash(3% mass fraction) in the charges leads to increasing content of aluminum oxide in sintering 0.21%, from 16.32 % (no ash) to 16.53% (with the addition of 3% ash) and increase the dosage of the ash in the charges up to 10% mass fraction reduces the content of aluminum oxide in the sinter to 16.23 %. The actual alkali content in the sintering is lower than the calculated one, which is due to the removal of alkalis during sintering.

To conduct pilot tests, the ash was delivered to the field of the material receiving and loading unit, then fed through a nourishing hopper for joint crushing with the ore to prepare an alumina raw charge. Involvement of ash in the technological process of raw charge preparation in the amount of 1-2% reduces the consumption of the main raw material component - nepheline ore by 6500-7000 tons per year. At the same time, stored waste from combined heat andpower plants is disposed of, the harmful impact of production on the natural environment is reduced, and valuable components are extracted from them.

As another additive in the preparation of nepheline-limestone-soda charge, it is possible to use alumina slags of ferrotitanium production of JSC "Klyuchevsky Ferroalloy plant", whose reserves are about 3 million tons [3, 8]. The slag of ferrotitanium production is extremely heterogeneous in phase and chemical composition, since it has a technogenic origin and is represented by a large number of thinly sprouted mineral phases [9]. The content of Al_2O_3 in slag samples is in the range of 59.4-64.5% [8]. Alumina slags even exceed bauxite ores in their content of aluminum oxide, but fine grinding is required to open them and involve them in processing. They can be considered as an additional raw material source for a number of alumina enterprises.

Table 1 - Results of tests chemical analysis fromferrotitanium slag in JSC "Klyuchevsky Ferroalloy plant”

Nameofthematerial	Content of components in the slag, % mass fraction									
	Al_2O_3	CaO	SiO_2	Mg0	Fe_2O_3	Cr_2O_3	TiO_2	Na_2O	K_2O	SO_3
Ferrotitaniumproductionslag	62.5	13.5	1.89	6.35	2.41	1.36	11.24	0.52	0.18	0.05

In accordance with the data of x-ray phase analysis, the main aluminum-containing minerals of ferrotitanium slag are analcime, montmorillonite and hercynite, to a lesser extent iron-substituted grossular. On the basis of these studies, given that currently enter the slag ferrotitanium production in the limestone-nepheline-soda mixture in the raw material plant of JSC "RUSAL Achinsk" from Russia is a joint grinding nepheline ore with the 4th stage shredding, it was recommended to change the layout of crushing slag and to grinding in a separate mill. From figure1 it can be seen that the introduction of 5-10 % of ferrotitanium slag crushed to 100% class - 0.074 mm into the charge increased the content of aluminum oxide in the sintering from 16 to 17.5-18.5% (at the sintering temperature of 1250°C) and from 16.3 to 17.7-18.9% (at the temperature of 1270°C).

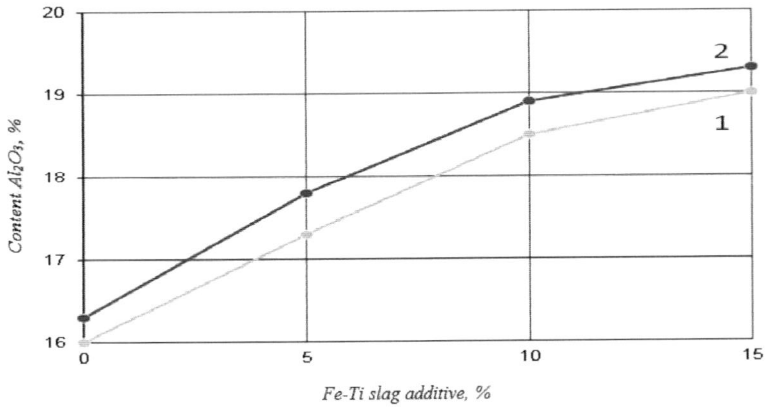

Figure 1. Dependence of the Al_2O_3 content in the sintering on the addition of slag to the charge: 1-sintering temperature 1250°C; 2-sintering temperature 1270°C.

Extraction of Al_2O_3 during the technological leaching of sintering with the introduction of 2-5% slag had a similar character, but with additions of more than 10% slag, there was a slight decrease in recovery in the range of sintering temperatures of 1225-1250 ° C. At the same time, an increase in alkali extraction was noted with the introduction of 2-5% ferrotitanium slag at a temperature 1250°C.

A resource-saving technology for processing of spent chamotte refractory bricks used for lining rotary sintering and other heat units was mastered and introduced in production. The chamotte refractories that had served their operational life included impurities of the raw materials of the alumina charge and alumina production products, which can be recovered by adding them as an additive to the raw alumina charge. Studies have shown that the technogenic addition of chamotte refractory bricks to alumina batch even in minimal amounts (0.11% mass fraction), allows to reduce the consumption of nepheline ore while maintaining the same technological parameters for the extraction of alumina. JSC "RUSAL Achinsk" involved more than 70,000 tons of spent chamotterefractory bricks into the charge preparation process and achieved the saving of natural raw materials: nepheline ore about 42,000 tons.

In addition to the work on the involvement of secondary technogenic resources in the production, ensuring the conservation of natural mineral raw materials, and leading to the implementation of low-waste technologies, JSC "RUSAL Achinsk" is carrying out work to reduce the impact of production on water basins and reduce water consumption [10]. As a result of the reconstruction of the power equipment in the CHPP, which is part of the alumina refinery, the waste water discharges into the open water was excluded. This became possible after replacing the cooling system and water supply technology for combined heat and power plant using modern cooling towers [11, 12]. To cool the waste water of the CHPP, two tower cooling towers (TCT) were reconstructed and a new five-section TCT-2000 fan-type cooling tower was built instead of the emergency-stopped TCT-1600, which had higher technical performance and its tests confirmed it. Tests have shown that the CHPPof the TCT - 2000 fan cooling tower is 2.5 times higher than the CHPPof the TCT -1600 tower. As a result of putting the tower into operation, the condensing capacity of the CHPP increased by 45 MW. The practical implementation of this technical solution made it possible to exclude the waste water discharges into an open basin and transfer the combined heat and power plant of JSC "RUSAL Achinsk" to a closed water circulation system. As a result of the commissioning from the constructed

IOP Conf. Series: Materials Science and Engineering **941** (2020) 012003 doi:10.1088/1757-899X/941/1/012003

cooling towers, the waste water discharge of the CHPP from the cooling water basin was eliminated, and the water basin was converted into a new slurry card to accommodate nepheline sludge and ash and slag waste from the plant. The new slurry map was built using modern technologies with the use of a waterproofing polymer membrane that excludes the effect of the sludge storage facility on groundwater. The elimination of the waste waterdischarge from the CHPP reduced the discharge of waste water into the Chulym river from 33.5 million m^3 in 2013 year up to 5.4 million m^3 in 2017 (figure2).

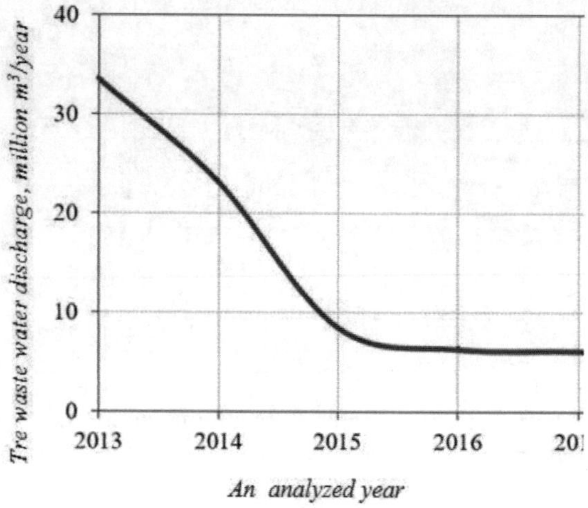

Figure 2. Volume of waste water discharge into an open basin after the transfer of the combined heat and power plant of JSC "RUSAL Achinsk" to a closed water circulation system

In order to reduce the waste water discharge into the open water, industrial sewage treatment plants were built at the same time, which also provided the return of treated water to a closed water cycle. The total intake of treated water from the Chulym river as a result of the introduction of water protection measures decreased by 40.07 millionm^3/year.

The solution of the above-mentioned ecological problems in JSC "RUSAL Achinsk" has led to a significant reduction in the impact of alumina production on natural water basins, virtually eliminating the waste water discharge into an open basin. Currently, the discharge into the basin is a one-time nature when cleaning the filters of the pumping and filtering station and the works continue to exclude all waste water discharges into the Chulym river.

Conclusion.

The research has allowed us to develop resource-saving technologies for processing nepheline ores, which have been successfully tested on an industrial scale. The introduction of various additives of technogenic raw materials into the alumina charge reduces irreplaceable losses of natural raw materials and reduces the impact of waste on the environment. In the course of research, the possibility of using combined heat and power plant ash, ferrotitanium production slag and waste of spent chamotte bricks as aluminum raising additives in the limestone – nepheline charge was studied. Involvement of ash in the technological process of raw chargepreparation in the amount of 1-2% reduces the consumption of the main raw material component - nepheline ore by 6500-7000 tons per year. At the same time, stored waste from combined heat and power plants is disposed of, the harmful impact of production on the environment is reduced, and valuable components are extracted from them.As an additive to substandard nepheline ore, alumina slags have been successfully tested, which are superior to bauxite ores in terms of aluminum oxide content. It is determined that their involvement in processing requires fine grinding. A resource-saving technology for processing of spent chamotte refractory bricks used for lining rotary sintering and other heat units has been developed and implemented in production. After serving their operational life, chamotte refractories were introduced as a raw material additive to the alumina charge. The introduction of technology for processing of waste chamotte bricks has provided savings of 42,000 tons of nepheline ore. In addition to saving mineral resources, work was carried out to reduce the consumption of treated water for production needs. To do this, the technological equipment was replaced and the industrial sewage treatment plants were built, which also provided the return of treated water to a closed water cycle. The total intake of treated water from the river as a result of the introduction of water protection measures decreased by 40.07 millionm^3/year. The implementation of ecological protection measures provided a solution to current ecological problems in JSC "RUSAL Achinsk" and led to a significant reduction in the impact of alumina production on the environment.

References:

[1] Shepelev I I, Bochkov N N, Golovnykh N V and others 2015 *Khimiya I khimicheskayatekhnologiy* **58** (1) pp 81-6
[2] Smirnov L A, Sorokin Yu V 2012 *Technogenous Waste Material Processing* (Ekaterinburg: UIPTs Ltd) p 607
[3] Perepelitsyn V A, Rytvin V M and Koroteev V A 2013 *Ural Technogenous Raw Materials* (Ekaterinburg: RIO UrO RAN) p 332
[4] Kitler I N, Liner Yu A 1962 *Nepelines are complex raw materials for aluminum industry (*Moscow: Metallurgizdat) p 238
[5] ShepelevI I, Sakhachev A Yu and Aleksandrov A V and others 2017 *Proceedings of XXII Intern.sci-techn. konf. "Scientifical backgrounds and practice of ore and technogenous raw materials processing",* 19-20 April 2017 (Ekaterinburg: UGGU Ekaterinburg:Tails Co) pp 244-49
[6] Ni L P, Rizman V L 1988 *Combined Methods of Low-grade Aluminum Raw Materials Processing (*Alma-Ata: Nauka) p 255
[7] Petrov V P 1982 *New Non-bauxite Types of Alumina Feed* (Moscow: Nauka) p 256
[8] Perepelitsyn V A, Rytvin V M and Gilvarg S I and others 2014 *Ferroalloy aluminothermy slags* (Ekaterinburg: Uralskiyrabochiy) p 368
[9] Shepelev I I, Sakhachev A Yu, Kozhevnikov V A and others 2014 *Proceedings of XIX Intern.sci-techn. konf. "Scientifical backgrounds and practice of ore and technogenous raw materials processing",* 23-24 April 2014 (Ekaterinburg: UGGU Ekaterinburg: Tails Co) pp 159-62
[10] Shepelev I I, Nemerov A M, Eskova E N, Zhukov E I, Sakhachev A Yu, Pilyaeva O V, Kiryushin E V and Potapova S O 2020 *Ecology and Industry of Russia* **24** (2) pp 4-9
[11] Vlasova E R 2016 Young scientist **24** pp 135-37
[12] Semenov I V 1998 Hydrotechnical construction **6** pp 33-40

IOP Conf. Series: Materials Science and Engineering **941** (2020) 012004 doi:10.1088/1757-899X/941/1/012004

Cultivation of biocultures for the purpose of their use for cleaning soils from oil pollution

S M Kurbatova[1a], L Yu Aisner[1], A G Rusakov[1] and A A Rusakov[2]

[1]Krasnoyarsk State Agrarian University, 90 Mira Ave., Krasnoyarsk, 660049, Russia
[2]Reshetnev Siberian State University of Science and Technology, 31 Krasnoyarskiy rabochiy Ave., Krasnoyarsk, 660037, Russia

[a] E-mail: sveta_kurbatova@mail.ru

Annotation. The article considers a number of aspects of the use of biocultures for cleaning the soil when it is contaminated with oil. Oil is represented as a multi-component hydrophobic and non-fermentable substrate. Heterotrophic microorganisms, when certain favourable conditions ensuring their activity are created, are able to completely utilize all the components of oil. An effective means of eliminating oil contamination of the soil is the activation of the regenerative ability of native biocenoses, or the introduction of certain selected and adapted cultures of microorganisms.

1. Introduction

Oil and oil products pollution affects the morphological, physical, chemical, biological (which determine its fertility) and ecological properties of the soil, its functions, processes of migration, accumulation and metabolism. The depth of penetration of petroleum products into the soil depends on various reasons, but the main amount is retained in the upper 15 cm layer [1]. Soil respiration reacts to oil pollution, especially in the first period, when the microflora is suppressed by a large amount of hydrocarbons. The complex of soil microorganisms after short-term inhibition responds to oil pollution by increasing the gross number and increasing activity. First of all, it refers to HCOM (hydrocarbon-oxidizing microorganisms), the number of which increases sharply relative to unpolluted soils [2].

Soil contamination with oil and petroleum products causes changes in its humus content, there is a decrease in the content of mobile forms of phosphorus and potassium, changes in soil acidity. Since the main element in the composition of oil is carbon (83-87 %), the content of organic matter per total carbon and humus in polluted soils increases due to the carbon of oil. Alongside the increase in the introduced carbon, there is a process of qualitative change in bituminous substances and the group composition of humus [3].

The impact of oil on the complex of soil microorganisms is ambiguous. On the one hand, oil pollution stimulates the growth of certain species, on the other hand, it inhibits. Any of the modern methods of collecting spilled oil and petroleum products (burning, mechanical collection, chemical and physical-chemical methods) can not do without the stage of biological purification. Thin films of oil remaining after physical and chemical methods can be removed only with the use of sorbents and biosorbents.

2. Cleaning of oil-contaminated soils

To clean oil-contaminated soils, various technologies are used that differ in the method of destruction of petroleum products (mechanical, physical-chemical, biological), implementation, duration and efficiency of the process, and economic costs.

A common technology for cleaning oil-contaminated soils is bioremediation technology, which is characterized by the duration of the process of biodegradation of oil hydrocarbons (several growing seasons) and the dependence of the process efficiency on climatic conditions, so that the intensification of this process is an important environmental and technological task.

One of the main and priority directions in the field of environmental protection and rational use of natural resources is the development and implementation of improved technologies for soil restoration by bioremediation method using biosorbents, which is based on the principle of self-cleaning of oil-polluted territories with the participation of native or introduced microflora.

Bioremediation is a method of cleaning and restoring soils by biological decomposition of pollutants by microorganisms (bacteria, microscopic fungi, actinomycetes), as a result of various biochemical reactions and physical and chemical processes carried out with the participation of biota [4].

Activation of the native microflora is aimed at creating an optimal environment for the development of certain groups of microorganisms that use oil as a food source, which is achieved by applying mineral fertilizers to the soil, maintaining the optimal temperature, humidity and acidity of the environment using agrotechnical techniques [5].

Soil application of various biologically-based products stemming from certain cultures of micro-organisms improves the efficiency of the process of cleaning soil from oil and oil products by increasing the number of HCOM [6].

Biological products can be used in the form of aqueous suspensions of microorganisms or dehydrated microbial biomass (immobilized on a solid carrier of microbial cells) [7].

When using biologics, the biological balance in nature is not disturbed, since they effectively decompose pollutants without the formation of toxic degradation products, gradually restoring soil fertility.

More promising is the area of work, in which the cleaning of the soil from oil contamination is carried out on site without removing the soil. Thus, a suspension of microorganisms in the nutrient medium is introduced into the treated soil using a mobile unit.

To prevent the leaching of HCOM cells on oil-contaminated soil, they are immobilized on a porous solid material (polymer sorbent). This process creates a microenvironment for the cells of HCOM, stabilizes their physiological condition, and increases the availability of nutrients, water, and oil.

The polymer sorbent has a number of unique properties: it is an aerating agent and a soil-aggregate stabilizer, it is non-toxic, it sorbs oil, preventing its spread, and it works in a wide pH range of the medium. During the treatment of the oil contamination site with a sorbent, it gradually changes and loses its structure, and does not require collection and disposal. These characteristics allow using it in combination with biomass of HCOM.

The use of such complex biologics allows creating favorable microenvironment conditions for HCOM cells, structuring the soil and ensuring their growth and reproduction due to the process of oxidative enzymatic activity and decomposition of hydrocarbons.

It is possible to stimulate biodegradation with the help of oxygen, using special equipment, or by introducing nutrients into the soil (nitrogen-and phosphorus-containing fertilizers) [8].

The decomposition of oil in the soil under the action of biologics is caused by the enzyme activity of living microorganisms in its composition, the ability of biologics to contribute favourably to the native microbial community of the soil, increasing its ability to utilize oil with the formation of organic compounds of a humus-like nature. The use of biologics together with the seeding of oil-tolerant grasses at all the stages of soil cleaning significantly reduces the cleaning time and contributes to the early restoration of soil fertility.

The development of bioremediation technology for soil contaminated with a chemical substance is a long multi-stage complex process that requires the collaboration of specialists in various fields: microbiologists, analytical chemists, biochemists, biotechnologists and toxicologists. This work allows us to develop effective, environmentally safe and inexpensive technologies for bioremediation of soils. [9]

Complete degradation of petroleum products in nature is possible only with the use of oxidative HCOM enzymes, and only in cases when the system has a set of conditions for their development. The microbiological synthesis of proteins and vitamins from oil is based on the process of enzymatic oxidation of hydrocarbons and their derivatives [10].

An important factor in the cultivation of microorganisms on hydrocarbon substrates is access to oxygen. Oxygen is required to build biomass, to oxidize carbon, and to meet the energy needs of bacteria.

Microbiological oxidation of petroleum hydrocarbons can be carried out in two directions: to grow microorganisms to be used in the form of protein and vitamin concentrates or to extract valuable products produced by microorganisms from the aquatic environment. The mechanism of oxidation of hydrocarbons by microorganisms, as shown by numerous experiments, differs from the mechanism of chemical oxidation. During the oxidation of hydrocarbons in the liquid phase, the primary product of oxidation is hydroperoxide, which is responsible for the further degenerate branching of complex chain oxidation.

The absorption of hydrocarbons by microorganisms depends on the following conditions: the penetration of hydrocarbons into the cell, the ability of microorganisms to adapt to hydrocarbons, and the presence of enzymes necessary for the primary oxidation of hydrocarbons. Microorganisms that absorb hydrocarbons must adapt to the nutrient medium, which undergoes great changes in natural conditions. Some microorganisms that fail to adapt to the new conditions die. Others, gradually adapting, begin to assimilate new substrates.

Microbiological oxidation of petroleum hydrocarbons occurs in a complex system of several phases. In the water environment, the hydrocarbon is dispersed, and there are also microbial cells. As a result of the growth of microorganisms, surfactants that cause additional emulsification of hydrocarbons are formed. The chemical processes here are influenced by factors related to the life of microorganisms (limiting and inhibiting), as well as the colloidal-chemical nature of the system [11].

One of the solutions to this problem is the use of a polymer sorbent with HCOM. The substrate-specific and activated actinomycete microflora introduced with the sorbent makes a significant contribution to the enzymatic degradation of hydrocarbon and provides a deeper reduction of the pollutant. The modified polymer sorbent significantly accelerates the processes of oxidative degradation of hydrocarbons. [12].

Microorganisms that can absorb oil hydrocarbons can be used to solve environmental problems, so when searching for destructive microorganisms, it is necessary to take into account that the microbial biomass introduced into the soil should not be foreign to the soil microflora. Another important condition is their non-pathogenicity. Destructive microorganisms must be highly resilient, since they can be exposed to adverse environmental factors, such as temperature fluctuations, high and low humidity, changes in the pH of the environment, and lack of nutrients.

To eliminate oil pollution in the soil, a number of biologics containing an active strain-destructor or community of microorganisms with catabolic activity against petroleum hydrocarbons, and mineral additives have been developed [13].

Pseudomonas bacteria have an extremely diverse metabolism. Some strains of the genus Pseudomonas are used to convert complex organic substances into biodegradable organic compounds. There are studies showing that strains of the genus Pseudomonas living in nature can be used to clean soils contaminated with oil and its products. [14

are widely distributed in nature, quickly grow and multiply, utilizing oil in the soil and water. Yeast fungi introduced at a dose of 108 KL / ml actively decompose crude oil, aromatic compounds with a concentration of 1 % within 180 days, and are also able to utilize kerosene and gas.

The preference for using Pseudomonas bacteria and yeast fungi of the genus Candida and Rhodotorula is due to their lower hazard to humans and the environment [15].

3. Use of biologics to clean the soil from oil pollution

Biological products on the basis of the UVM strains are divided into two main groups: mono - and polybacterial. Polybacterial biologics have wider adaptive and ecological possibilities for use.

As with any technological process, the use of biosorbents in the bioremediation technology of oil-contaminated soil has its advantages and disadvantages. The main advantages of using immobilized cells on the surface of the carrier can include [16]: the possibility of longer operation of cell properties in the immobilized condition; increased productivity as a result of increasing the concentration of microbial biomass per unit of working volume of the carrier; reduced energy consumption, since working environments often contain fewer soluble impurities, which makes it easier to isolate and clean the final products; cell preservation and partial deactivation of microorganisms; resistance of cells to various adverse inactivating external factors (temperature, acidity, concentration of electrolytes, etc.) as a result of immobilization; additional protection of the culture from the effects of pathogenic microflora in cases of the sterility failure of the biotechnological system.

In addition, when using a biosorbent, natural self-purification is activated due to natural mechanisms that are inhibited without the drug by the action of spilled petroleum products. Practical application has shown that the biosorbent simplifies the mechanical collection of oil by 50-60 %, if the hydrological and meteorological conditions are favourable for this. Treatment of an oil slick with a biosorbent prevents its spread (the effect of physical and chemical booms), which allows collecting more than 90% of oil in favourable hydrological and meteorological conditions [17].

When applying a biological product to oil-contaminated soil, the number of microorganisms begins to grow, using oil hydrocarbons as food. In natural conditions, the strains of these microorganisms retain their activity and population size for the time of biodegradation (elimination) of hydrocarbons, and when its quantity decreases, their number also decreases. As a result of complete absorption of petroleum products, bacteria lose their nutrient medium and die, creating humus on the ground, and silt in the water. Microorganisms also die if they are located separately from the carrier. Thus, the possibility of such a phenomenon as introduction - microbial contamination of the natural environment is excluded and makes the storage and use of biosorbent safe [18].

4. Experiments with oil-oxidizing strains

Bacteria of the genus Pseudomonas and yeast fungi of the genus Candida and Rhodotorula are prototrophs, so they are able to accumulate biomass in semi-synthetic environments where the source of carbon and energy is crude oil.

Experiments with oil-oxidizing strains, other things being equal, have shown that microorganisms are able to grow on crude oil and are resistant to it. The studied strains can be used for obtaining biologics, since only microorganisms that have enzymes for oxidative destruction are able to reproduce in an aerobic liquid-phase process.

In liquid-phase tests, the sorbents quickly changed their structure, which indicates the gradual decomposition of sorbents into components and their inclusion in the medium, which makes them an additional source of nutrients for HCOM. Consequently, the influence of the added filler affects the quality of the sorbent.

Oil contamination in the experiment was 1 %, yeast cultures within three weeks reduced the concentration of oil to 90 % of the introduced amount due to their enzymes, food source, oil energy and nutrients contained in the sorbent and in the medium. The speed of oil utilization is related to the properties of sorbents, which convert oil into a film form, increase the availability of oxygen for HCOM, and create a favourable microenvironment, due to which microorganisms perform their physiological functions – nutrition and reproduction.

As HCOM the biomass of new native strains of bacteria and yeast fungi as carriers of immobilized microflora was used. Eight types of modified polymer sorbents were used, with a detailed assessment of their main characteristics. Thus, the introduction of biologics based on UVOM into the oil-contaminated environment provides an environmentally safe, relatively fast and economical method of cleaning, which does not cause additional harm to nature. The type of the sorbent modification determines its properties that affect the microenvironment of cells during immobilization, which is manifested in the speed of the utilization processes of crude oil hydrocarbons.

The high rate of oil utilization in a liquid medium without additional aeration (which is economically important) is associated with the properties of sorbents, which increase the availability of oxygen for HCOM, oil (in film form), and create a favourable microenvironment, due to which microorganisms perform physiological functions (nutrition and reproduction due to the enzymatic destruction of hydrocarbons) instead of survival.

Modification of sorbents reduces dusting, which is important when using plant waste for the production of sorbents and their utilization (in the soil structure), expands the nutritional base for heterotrophs, which is important given their self-destruction in the soil.

5. Conclusions.

Experiments with biologics have proved their high efficiency. Thus, when exposed to biological products, the processes of biodegradation and biotransformation of oil cause effective cleaning of industrially polluted soil from petroleum products and improving its environmental condition. Based on the advantages of biologics, their use for introduction into the oil-contaminated environment is the most promising and economic methods of oil elimination.

The mixed culture of yeast fungi, immobilized for sorbents, effectively disposed of oil, which allows us to use the selected biologics for cleaning oil contamination of the soil.

References

[1] Brief information about the chemical composition of oil and its fractions [Electronic resource] Mode of access: http://www.borpak.ru/neftvanaya-otrasl/kratkie-svedeniya-o-ximicheskom-sostave - nefti-i-ee-frakciie.htm

[2] Shamraev A V, Shorina T S 2009 Influence of oil and petroleum products on various components of the environment *Bulletin of OSU* **6** Pp. 642-643

[3] Kozhevin P A 1997 Biotic component of soil quality and the problem of stability *Soil Science* **4** Pp. 44-48

[4] Bioremediation of polluted soils and soils [Electronic resource] Mode of access: http://bibliofond.ru, Shlegel G G 1987General Microbiology P. 194-197

[5] Sang-Hwan Lee at al 2008 Effect of various amendments on heavy mineral oil bioreme-diation and soil microbial activity *Bioresource Technology* **99** P. 2578-2587

[6] Kaporina L N, Morshakova G N 2001 Method of cleaning environmental objects from oil and oil hydrocarbons *Environmental systems and devices* **7** 57 p

[7] Kamenshchikov F A, Bogomolny E I 2003 Oil sorbents 268 p

[8] Arens V Zh, Gridin O M and Kondrashenko V M 2000 Oil sorbents: advertising illusions and real prospects *Oil and Gas vertical* **9** 108

[9] Velkov V V 1995 Bioremediation; principles, problems, approaches *Biotechnology* **3-4** P. 20-27

[10] Karpenko E V, Vildanova R I, Shcheglova N S, Pirog T P and Voloshina I N 2006 Perspective of using bacteria of the genus Rhodococcus and microbial surfactants for degradation of oil pollution *Applied biochemistry and Microbiology* V 42 **2** P. 175-179

[11] Petroleum products 2005 All abstracts of the network [Electronic resource] Mode of access: http://etelien.ru/Collection/20/2000007.html

[12] Ryazanova T V 2006 Method for obtaining polymer sorbents

[13] Sorbent "Unisorb" [Electronic resource] Access mode: http://www.ecosorb.su; Microbial biotechnologies of soil remediation [Electronic resource] Mode of access: http://www.cbio.ru

[14] Bacteria of the genus Pseudomonas [Electronic resource] Mode of access: http://www.gastroscan.ru/handbook/118/8124]

[15] Study of catalytic processes of microbial oxidation of petroleum hydrocarbons [Electronic resource Mode of access: https://cyber-leninka.ru/article/v/izuchenie-kataliticheskih-

IOP Conf. Series: Materials Science and Engineering **941** (2020) 012004 doi:10.1088/1757-899X/941/1/012004

protsessov-mikrobnogo-okisleniya-neftya-nyh-uglevodorodov
[16] Method for obtaining polymer sorbent [Electronic resource] Mode of access: http://www.findpatent.ru/patent/260/2604370.html
[17] Physiology of microorganisms [Electronic resource] Mode of access: http://medbe.ru/materials/mikrobiologiya-i-biotekhnologii/fiziologiya-mikroorganizmov.html
[18] Methods of cultivation of microorganisms [Electronic resource] Mode of access: http://studopedia.org/3-132512.html

Dynamics of steppe arable land in zones of oil fields development (Orenburg region)

D A Grudinin, K V Myachina

Institute of Steppe, Ural Branch of the Russian Academy of Sciences
11, Pionerskaya st., Orenburg, 460000, Russia

E-mail: mavicsen@gmail.com

Abstract. The reported paper presents the analysis data on the arable land dynamics in oil extraction zones. The long-term development of oil fields leads inevitably to farmland losses for a variety of destructive processes and, as a consequence, to withdrawal of lands from a turnover, and land abandonment. Our study was carried out on two sample territories in the steppe zone of Orenburg region with two key land utilization sectors – agriculture and long-term oil extraction. To reveal the dynamics of agricultural lands and oil fields we adapted time series of satellite (Landsat) imagery. An analysis was divided into several phases dealing with quantity-related assessment of fields withdrawn from the farmland turnover and tracing the relationship between their position and objects of oil extraction infrastructure. A period of interest covered 40 years – from 1980 to 2020. The research has highlighted a thirteen and eightfold increase of zones with oil production facilities on the first and second territory, respectively. Recently, 24% and 26% oil fields are located on arable lands. Another outcome of the study is that the expansion of territories designated for subsurface use is the reason that both directly abandoned areas and neighboring lands are removed from the farmland turnover. Agricultural lands affected by oil fields are at a high risk of being withdrawn from the turnover – losses may range to the firth of arable lands.

1 Introduction

Agriculture is a major economy sector in regions of steppe lowlands; therefore, farmlands dominate in the steppe landscape. In the context of threatening food scarcity at the global level a priority is supposed to be a comprehensive assessment of a relationship between agriculture and parallel economic activities, such as oil production. Industrial facilities of oil extraction are incorporated into the existing agricultural environment of steppe regions and create additional powerful sources of man-induced effect and synergy of impact that cause special multifaceted transformations of landscapes, and frequently even a loss of efficient arable lands.

A geo-ecological component in any global large-scale manufacturing activity plays an important role for modification and social and economic situation in all territorial and geographical segments since environment, economy, and society represent coupled systems. Almost fifty oil fields are developed in steppe regions of the Russian Federation with a tremendous man-made effect on natural and agricultural landscapes. Therefore, one of the conflict aspects influencing the co-existence of agriculture and oil production is related to removal of efficient farmlands from agricultural land turnover in order to develop profitable oil fields. At the same time, oil is the most important natural recourse; undoubtedly, its production is needed and driven by economic forces, as a result, areas within any categories of farmland may be render suitable for this economic activity. In many cases long-term development of oil fields brings about farmland losses, e.g. because of chemical or thermal pollution of landscape components (Myachina, Chibilev, 2020), activated erosion of soils (Ermolaev et al, 2012), and other destructive processes, as a consequence, these lands are taken out from the agricultural land turnover and get abandoned. An exclusively output-oriented activity of companies involved in extraction of natural recourses, an aim to maximize profits at the expense of a geo-ecological state of a territory give rise to mentioned above instances.

This research had a purpose to analyze the dynamics of arable land areas located near oil fields being developed on a long-term basis.

2 Objects and Methods

Our study was carried out in the steppe zone of Orenburg region, where agriculture and oil production are key sectors in land utilization. Two sample territories were selected and met two basic requirements: 1 – an existing agricultural activity, 2 – running oil fields close to or within the boundaries of agricultural lands. Sample territory №1 (approximately 400 km^2) was chosen in the lowland water-parting, one zone is designated for agricultural production, the other one represents oil fields being developed since the early 1980-s. Sample territory №2 (approximately 800 km^2) is also characterized by a lowland landscape, therefore, numerous agriculturally used areas (ploughlands, pastures or hayfields) alternate with objects of oil infrastructure – a system of oil production facilities has been evolving here since the early 1990-s.

Our study focused on the arable land dynamics in areas adjacent to oil fields comprised several phases dealing with quantity-related assessment of fields withdrawn from the farmland turnover and tracing the relationship between their position and objects of oil extraction infrastructure. A period of interest covered 40 years – from 1980 to 2020.

For the purpose of research we used time series of satellite (Landsat) imagery, controversial issues were clarified by means of historical maps of the platform Google Earth. We selected cloud free photos of the territory under consideration, which were taken every year within the timespan of interest from the middle of April to the end of Mai. A mid-April to late Mai time interval was analyzed on purpose because a majority of fields are tilled at that time. As a result of plowing they contrast with the background on panchromatic images, NDVI imagery, and visual data resulting from the expert photointerpretation.

To trace the dynamics in use of arable lands and areas designated for oil fields infrastructure a period of interest was divided into three time intervals: 1980-1987; 1998-2005; and 2013-2020. Thus, each interval is seven years. Such a time span is necessary not to miss out territories which might be not identified as agricultural lands in the period from the middle of April to the end of Mai (for instance, they might be a part of the crop rotation process, but be planted with winter crops or left fallow during autumn or winter). In addition, within this approach it was possible to identify farmlands, which are tilled spontaneously, i.e. a land user plows up some of fields he owns whenever he wants. As a consequence, a field is classified as a farmland if tilled at least two times during a seven year period, and plowing in the last third of the interval was a strong requirement.

3 Results and Discussion

From the data obtained it is apparent that a maximal area of arable lands was in the interval from 1980 to 1987. An arable land on the first sample territory ranged to 36000 ha and to 17000 ha on the second one, respectively. At that time lands were allotted to oil companies and they started the development of oil fields. It is suggested to be an initial phase in the launching infrastructure facilities of oil production, which were located close to or within the boundaries of arable lands (Figure 1, Figure 2). In the period of interest territories occupied by oil production facilities were insignificant, about 360 ha on the first sample territory and 260 ha on the second one (1% and 1.5% of a total area size, respectively) and were located mainly beyond agriculturally used lands.

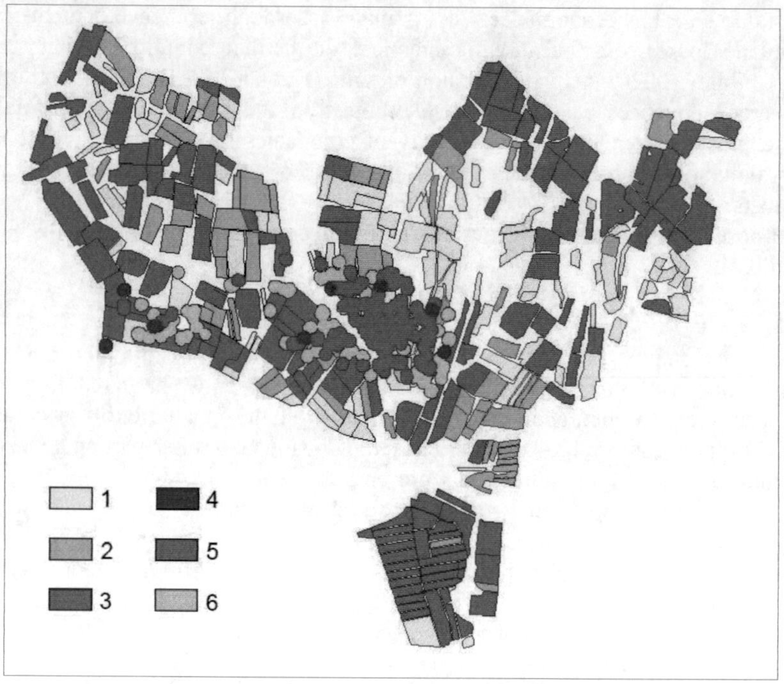

Figure 1. The dynamics of land utilization on sample territory №1: 1 – arable lands in 1980-1987, 2 – in 1998-2005, 3 – in 2013-2020; 4 – oil fields in 1980-1987; 5 – in 1998-2005; 6 – in 2013-2020.

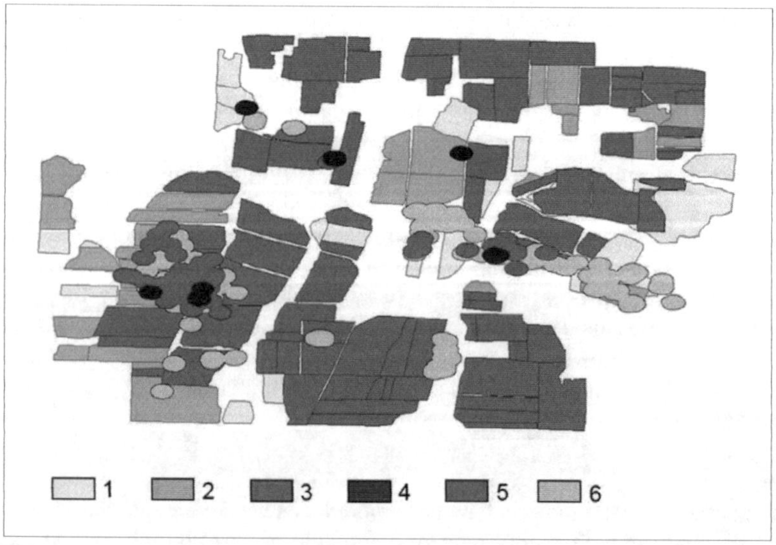

Figure 2. The dynamics of land utilization on sample territory №2: 1 – arable lands in 1980-1987, 2 – in 1998-2005, 3 – in 2013-2020; 4 – oil fields in 1980-1987, 5 – in 1998-2005, 6 – in 2013-2020.

Another time interval under consideration is from 1998 to 2005 and covers a post-crisis time in agricultural production with its less demanded lands (Levykin et al, 2012). We estimate an area of arable lands decreased by 23% on average in the zones under study; whereas territories with oil production infrastructure increased more than eightfold on the first sample territory and fourfold on the second sample territory (see Figure 1 and Figure 2). Importantly, a major part of oil extraction objects was located on arable lands. The objects of oil fields are both on arable and lay lands. For instance, shares of oil wells in the classes: arable lands / lay lands / other territories are 27%/ 42%/ 30% on sample territory №1.

On sample territory №2 a little over 5% zone with oil production objects overlapped with agricultural fields and approximately 7% of them – with lay lands.

Figure 3 and Figure 4 present the dynamics of a ratio between a total area of lay lands on sample territories and an area of lay lands with oil production facilities.

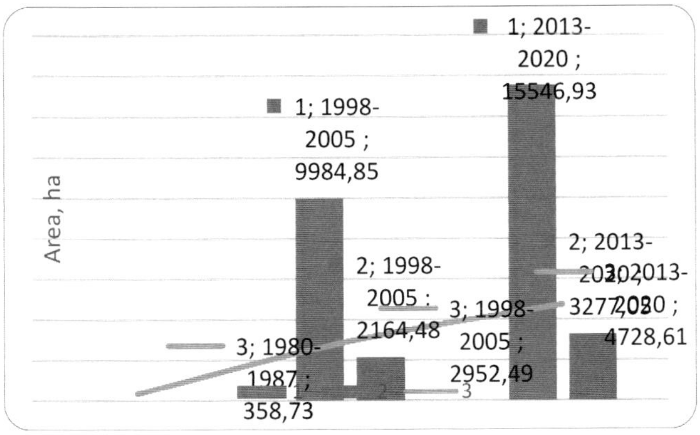

Figure 3. The dynamics of a lay lands area on sample territory №1 and oil production infrastructure from 1980 to 2020: 1 – total area of lay lands, 2 – area of lay lands with oil production facilities, 3 – total area of lands with oil production facilities.

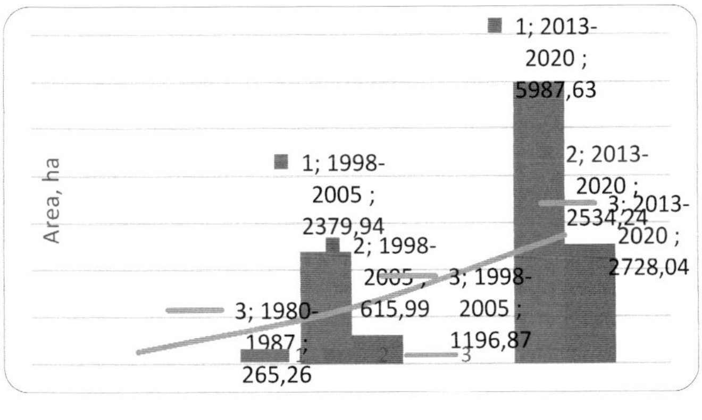

Figure 4. The dynamics of a lay lands area on sample territory №2 and oil production infrastructure from 1980 to 2020: 1 – total area of lay lands, 2 – area of lay lands with oil production facilities, 3 – total area of lands with oil production facilities.

The third time interval, from 2013 to 2020 is marked with the maximal expansion of lands removed from the agricultural turnover, as well as lands allotted to oil fields (see Figures 1, 2, 3, 4). The graphical data in Figure 3 and Figure 4 demonstrate a growth correlation between areas of oil fields and lay lands.

An area of arable lands in territories under consideration decreased by 40% on average by contrast to that in the 1980-s (the first time interval of interest), whereas an area of territories with oil production facilities increased thirteen and eightfold on the first and second sample territory, respectively. In the total area of territories with oil production infrastructure 24% and 26% of them are located on arable lands on the first and second sample territory, respectively, and 72% and 38% - on lay lands.

The study has revealed that the expansion of territories designated for subsurface use is the reason that both directly abandoned areas and neighboring lands are removed from the farmland turnover. Undoubtedly, the abandonment of farmlands may be a result of different factors (Meyfroidt et al., 2016), but sometimes an aggressive man-induced influence spreading far beyond areas where oil production facilities are located turn agricultural fields adjacent to oil fields into "useless areas" (Myachina, Chibilev, 2020). A land user should take these consequences into account when making a decision whether to withdraw an agriculturally used field from the farmland turnover or continue its targeted use.

It is an obvious fact that a spontaneous and often unreasonable removal of valuable agriculturally used lands from the farmland turnover never ceases unless companies engaged in use of subsurface resources change their attitude to a geo-ecological component of their activity. Such outcomes of oil production industry are in conflict with existing ideas to restrict legislatively the allotment of most efficient agricultural lands to nonfarm uses. The establishment of such untouchable "high-class land fund" can't be limited to soil rating only; it needs data of high-quality economic, landscape and land-improvement assessment (Chibilev, 2016). This point of view gains more importance in the context of threatening global food deficit and the necessity to meet present day social and environmental challenges.

An analysis of oil well pattern on arable and lay lands on sample territories suggests there should be a critical ratio between an arable land area and an area with oil production facilities, which plays a decisive role for ceasing the agricultural use of a particular field. In our opinion, the determination of this critical value is an essential step in the geo-environmental optimization of landscapes in oil-bearing steppe regions.

4 Conclusion

The proposed procedure of geo-informative analysis concerned with consequences of co-existing agricultural and oil production activities and based on a stepwise assessment of arable lands area takes into consideration the specifics of the farmland turnover on the territory of interest and is supportive when achieving the research aim.

The most important outcome to emerge from the analysis is maps showing the multiannual dynamics of land utilization on territories under study and comprising the tendencies of arable land areas and oil fields.

On territories affected by oil production arable lands are abandoned proportionally to a growing man-induced impact. Therefore, the development of methods to keep valuable agricultural fields bordering with oil fields in the farmland turnover represents a strong requirement for the support of agriculture and social and economic flourishing in oil-bearing regions.

Acknowledgements

The reported study was supported by RFBR grant No. 20-05-00122A (No. AAAA-A20-120011390069-6) and under the theme of state assignment of IS UrB RAS No. AAAA-A17-117012610022-5 (Goszadanie).

References

[1] Myachina K V., Chibilev A A 2020 *Doklady Earth Sciences* **492(1)** 366–9.
[2] Ermolaev O P, Rysin I I and Golosov V N 2016 *Geomorphology* 2 38-51 (*in Russian*)
[3] Levykin S V, Chibilyov A A and Kazachkov G V 2012 *Eurasian Steppes. Ecological Problems and Livelihoods in a Changing World* (Dordrecht: Springer) pp 491-505
[4] Meyfroidt P, Schierhorn F, Prishchepov A V, Müller D and Kuemmerle T 2016 *Global*

Environmental Change 37 1–15.
[5] Chibilev A A 1992 *Ecological optimization of steppe landscapes* (Sverdlovsk: Institute of plant and animal ecology) p 170 (*in Russian*)

An adsorption technique applied by plants and factories for purifying gas emissions using modified wastes available at power plants

E M Khusnutdinova, L A Nikolaeva and A N Khusnutdinov

Kazan State Power Engineering University
51 Krasnoselskaya street, Kazan 420061, Russia

E-mail: rr-088@mail.ru

Abstract: the paper proposes an adsorption-based method of how to remove sulfur dioxide from the gas emissions produced by industrial companies. A power plant waste – chemical water treatment sludge from Kazan CHPP-1 – was used as an adsorption material. The authors refere here to the chemical composition of the sludge and describe the method of its modification. A new sorption material based on power plant waste was tested for removal sulfur dioxide from the gases. This was resulted in kinetic dependence and adsorption isotherm. The mechanism of how the sulfur dioxide can be adsorbed by a new sorption material at different temperatures was studied. The tests revealed Gibbs free energy, differential heat, and adsorption activation energy. The economic and environmental effect of upgrading the process of removal of sulfur dioxide from gas emissions was calculated for the sodium bisulfite production line of L. Ya. Karpov Chemical Plant JSC

Introduction

Currently, new methods for reducing and limiting emissions release into the atmosphere generated by chemical, petrochemical, as well as fuel and energy industries are being actively developed. One of the toxic components of emissions are sulfur oxides and nitrogen oxides [1]. There are many methods for purification of gas emissions, one of the most effective methods is one that based on microporous sorbents adsorption. The performance indicators of adsorption treatment of gas emissions largely depend on the properties of adsorbents, the choice of which is determined by the energy and material costs spent for purification. Adsorbents must have a high sorption capacity, sufficient mechanical strength, be low-cost and made from affordable materials. Many materials of natural and artificial origin are known to be used as sorbents for purifying gas emissions: activated carbons, alumogels, silicates, silica gels, etc.[2]. Currently, the development of new sorption materials based on production waste is actively underway. The production of new cheap sorbents for purifying gas emissions is a challenge.

Results and discussion

Plants and factories are in the process of developing methods to reduce emissions release through the use of production waste. This production waste is the carbonate sludge of Water Treatment Plant (WTP) of Kazan CHPP-1. WTP sludge is a natural raw and stable mixture of a certain chemical composition, which depends on the chemical composition of raw water. The chemical composition of the WTP sludge produced by Kazan

CHPP-1 is as follows: calcite $CaCO_3$-72 %, brusite Mg $(OH)_2$ – 9 %, portlandite $Ca(OH)_2$ 1 %, quartz SiO_2 – 0.5 %, other compounds – 17.5 %. The WTP sludge produced by Kazan CHPP-1 has the following characteristics: bulk density – 560 kg/m^3; total pore volume – 0.375 cm^3/g; radius of effective size fraction ranged 0.05-1.40 mm; slime moisture – 3 %, ash content – 89 % [5]; organic carbon content – 11 %, humic compounds – 12 %. The number of the latter is determined from the total mass of the sample, which was detected by gas chromatography-mass spectroscopy [3, 4]. It is the significant specific surface of the sludge, the presence of a large number of active sites on it that determines its sorption properties.

The technologies for purifying gas emissions of plants and factories use pelletized adsorbents. Therefore, adsorbents granular in shape are used to reduce the resistance to flow in the bed through which gas emissions are passed. The paper proposes to modify the power plant waste – sludge – into pellets. To produce pellets, fine-dispersed sludge 0.01 to 0.09 mm in size is mixed with liquid sodium glass at a mass and volume ratio of 2:1, respectively. This ratio was determined experimentally [6, 7]. With a lower ratio, the sludge is partially impregnated with liquid sodium glass, the adsorption capacity drops and the pellets are shedded during subsequent firing. With a larger ratio, the adsorption capacity does not increase significantly and the binder is overspent. Then the mixture is brought to a homogeneous mass by rolling. The resulting pellets are kept in a muffle furnace for 3 hours [8, 9]. The processing temperature varies from 100 to 450 °C in increments of 50 °C. Then the pellets are cooled down to room temperature in the desiccator. The new sorption material was named pelletized sorption material and abbreviated as PSM. Pellets are 1-3.5 mm in size, have an abrasion strength of 78 % and average hydrophilicity.

Description of the pilot installation

A dummy installation with a fixed PSM bed made at the Department of Water and Fuel Technology of the Kazan State Power Engineering University was used for studying the sorption properties of PSM in a gas atmosphere that is close in composition to the outgoing gases (Figure 1).

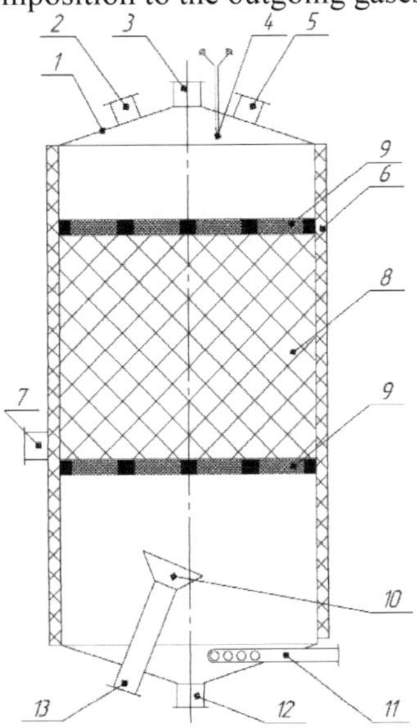

Figure 1. Pelletized sorption material (PSM) fixed bed reaction column.
1 – housing; 2 – adsorbent discharge pipe; 3 – purified gas discharge pipe; 4 – thermocouple; 5 – hatch for loading adsorbent; 6 – electric heater; 7 – hatch for unloading adsorbent; 8 – PSM bed; 9 – steel mesh with holes no more than 1 mm in diameter; 10 – sprayer, 11 – steam supply pipe; 12 – condensate discharge pipe; 13 – gas mixture supply pipe.

The reaction column is made of stainless steel, which ensures its resistance to corrosive environments. The gas mixture was supplied and discharged through pipes 3 and 13. The pipe 13 is threaded to the sprayer 10. This ensures a uniform distribution of gases over the PSM bed 8 on the steel mesh 9. The steel mesh 9 above the PSM bed prevents particles from carrying over. A thermostat with an electric heater 6 was used to heat up the column. The gas temperature was 50 °C, which was controlled by a chromel-drop type thermocouple.

The gas composition throughout the trial was as follows: O_2 – 4 to 5.7 %; N_2 – 75 to 78 %; H_2O to 3 %. The remaining volume of the gas mixture was CO_2. In the process of sorption, the concentration of sulfur dioxide varied in the range of 0–5,500 mg/m^3. The gas flow rate reduced to normal conditions was 4×10^{-4} m^3/s [10]. Sulfur dioxide was produced in the laboratory by reaction:

$$Na_2SO_3 + 2HCl \rightarrow 2NaCl + H_2O + SO_2\uparrow$$

The results of the studying the PSM sorption properties in terms of sulfur dioxide show that the adsorption capacity of PSM happen during the first minutes of contact and after 16 minutes reaches 140 mg/g for sulfur dioxide. The adsorption capacity is 14 wt%. Figure 2 illustrates how the sorption capacity of PSM kinetically changes with time.

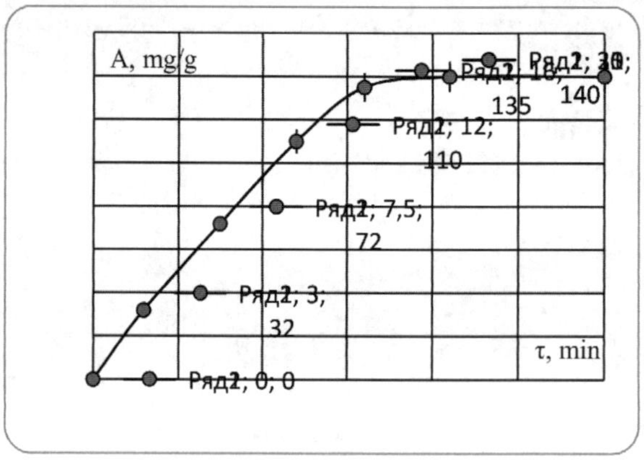

Figure 2. The PSM sulfur dioxide adsorption kinetic curve.

The optimal parameters of industrial treatment plants can be obtained using the sorption isotherm – the sorption capacity *A* vs. the change in the SO_2 concentration, as is shown in Figure 3a. The adsorption isotherm corresponds to the L-type Langmuir isotherm. The convex shape of the isotherm confirms effective physical adsorption. The adsorption isotherm is best described by the Freundlich equation $A=2.11C^{0.54}$(R_2>0.99) (Figure 3b).

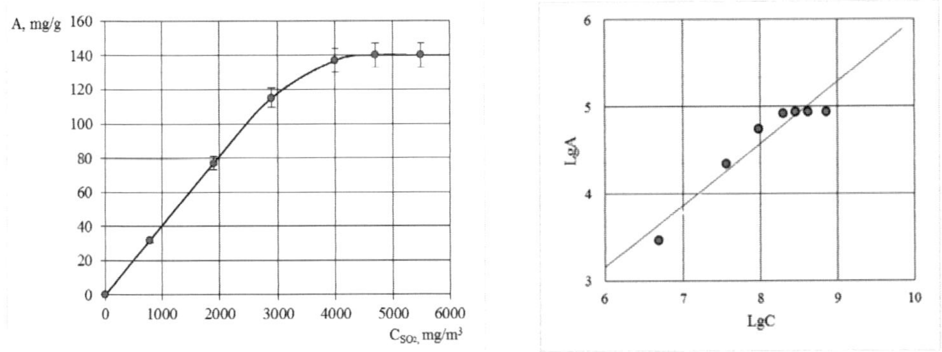

a) b)

Figure 3. Isotherm of SO_2 adsorption by pelletized sorption material (a) and its type in logarithmic coordinates (b).

For studying the mechanism of how the SO_2 is adsorbed by PSM, the process was studied under static conditions at different temperatures (293K, 313K, 323K, 333K). The resulting isotherm and isoster of adsorption on sulfur dioxide is shown in Figure 4 (during sorption of sulfur dioxide, the concentration varied in the range of 0–5500 mg/m^3).

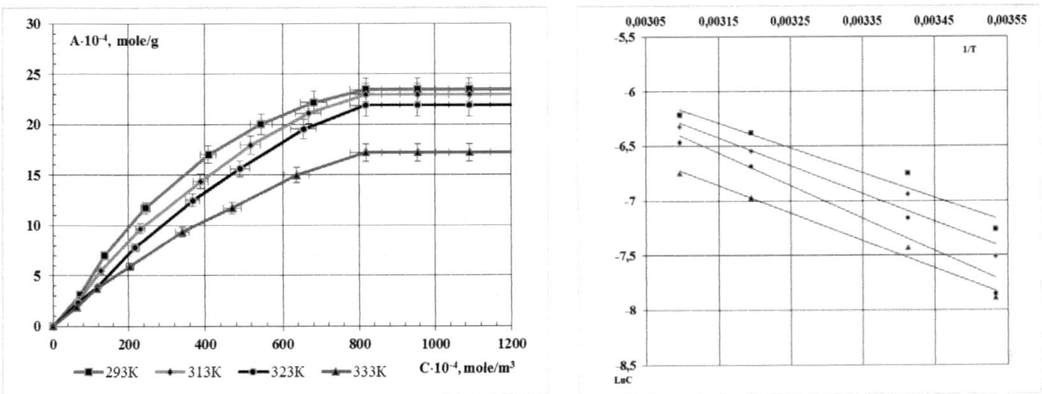

a) b)

Figure 4. Isotherms (a) and isosters (b) of SO_2 adsorption of PSM at different temperatures.

Isosters show the relationship between equilibrium temperatures and concentrations at a constant capacity of the sorbent.

An increase in temperature leads to a decrease in the SO_2 adsorption capacity, which is characteristic of the exothermic process and indicates the physical nature of the forces. The differential heat of adsorption was determined by the Clausius–Clapeyron relation (1):

$$\frac{\Delta lnC}{\Delta(1/T)} = -Q/R, \tag{1}$$

where C is the equilibrium concentration of sulfur dioxide in the gas, mol/m^3; T is the temperature, K; Q is the differential heat of the adsorption isoster, J/mol; R is the molar gas constant 8.341 J/mol·K.

IOP Conf. Series: Materials Science and Engineering **941** (2020) 012006 doi:10.1088/1757-899X/941/1/012006

The differential heat of the adsorption isoster is determined by the equation:

$$Q = -R\frac{\Delta lnC}{\Delta(1/T)}, \tag{2}$$

according to equation (2), the differential heat of adsorption of sulfur dioxide by PSM pellets was calculated according to the isoster angles of inclination. The calculation results are shown in Table1.

Table 1. Differential heat of adsorption of dissolved sulfur dioxide by PSM.

Adsorption capacity, $A \cdot 10^{-4}$, mol/g	Differential heat of adsorption Q, kJ/mol
17.19	9.31
21.88	12.17
22.97	13.64
23.44	14.22

Analysis of Table 1 shows that with the increase in adsorption of sulfur dioxide, the value of the differential heat of adsorption decreases.

The Gibbs energy of the adsorption process is determined by the equation (3):

$$\Delta G = -RT\ln K_л, \tag{3}$$

where ΔG is the Gibbs energy, J/mol; T is the temperature, K; K_l is the adsorption equilibrium constant.

The resulting Gibbs free energy values change slightly with increasing temperature and confirm spontaneous SO_2 adsorption by PSM.

The kinetics of the SO_2 adsorption by PSM was studied. Kinetic dependences were obtained at different temperatures (Figure 5). Kinetic equations were used to determine the adsorption rate constants. It is shown that with the increase in temperature, the adsorption rate constants decrease, which is typical of non-activated adsorption (Table 2).

Table 2. Influence of the temperature on the adsorption rate constant of sulfur dioxide by PSM.

Value	Temperature, K			
	293	313	323	333
$K \cdot 10^3$, s^{-1}	9.54	8.93	8.17	5.74

During adsorption, not all SO_2 molecules can penetrate the pores and be adsorbed, but only those that have an excess of energy – activation energy, so the rate of adsorption increases at high temperatures. The adsorption activation energy is calculated using the Arrhenius equation. The calculation results are shown in Figure 6. The value of the apparent activation energy (14.7 kJ/mol) indicates the course of physical adsorption.

IOP Conf. Series: Materials Science and Engineering **941** (2020) 012006 doi:10.1088/1757-899X/941/1/012006

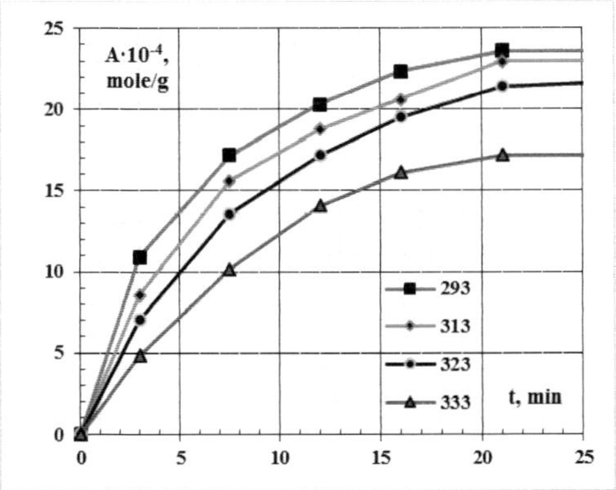

Figure 5. Kinetic dependences of SO_2 adsorption by PSM at different temperatures.

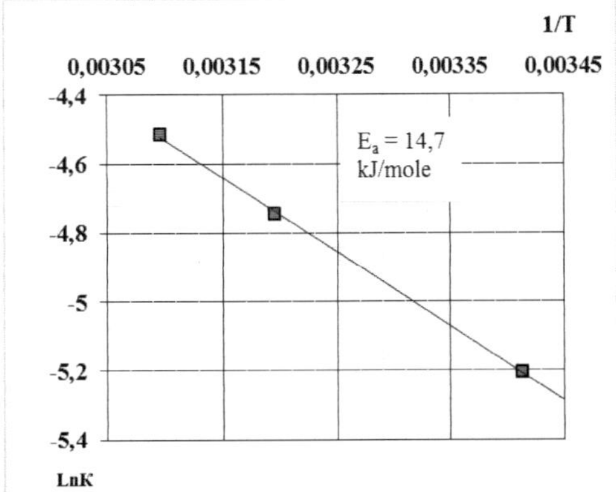

Figure 6. A graphic method for calculating the activation energy of the SO_2 adsorption by PSM.

The studies resulted in the following characteristics of the modified power plant waste – pelletized sorption material: the SO_2 adsorption capacity is 140 mg/g, the total pore volume is 0.450 cm^3/g, the specific surface area is 720 m^2/kg, and the abrasion strength is 78 %. The efficiency of PSM-based SO_2 removal was 99.9 %.

The pelletized sorption material has a high porosity, which is especially important when using its adsorption properties for gas passage. The low cost of sludge-based PSM, its availability, and the possibility of recovery, allow it to be used for purification of gas emissions produced by plants and factories with minimal costs and the greatest efficiency.

The obtained study results made it possible to propose upgrading the existing process of gas emissions purification for the sodium bisulfite production line at L. Ya. Karpov Chemical Plant JSC in the city of Mendeleyevsk. The main sources of sulfur dioxide emissions at the L. Ya. Karpov Chemical Plant are: Hyposulfite Salts Production Department, sulfite salts production, production of catalyst for the light paraffin hydrocarbon dehydrogenation, and production of extruded polystyrene plates [11]. According to [12], this chemical plant is classified as a hazard category III company.

The sodium bisulfite ($NaHSO_3$) production process has an adverse impact on the environment, since its production releases about 325 tons of SO_2 into the atmosphere annually. The output of sodium bisulfite at the L. Ya. Karpov Chemical Plant is about 1,750 t/y.

The sodium bisulfite process includes the following stages: production of sulfur dioxide, cooling of sulfur dioxide and its conveying, dissolution of soda ash, absorption of sulfur dioxide, preparation of sodium sulfite solutions and packaging. Figure 7 shows the current process of the sodium bisulfite production including the removal of sulfur dioxide from gas emissions.

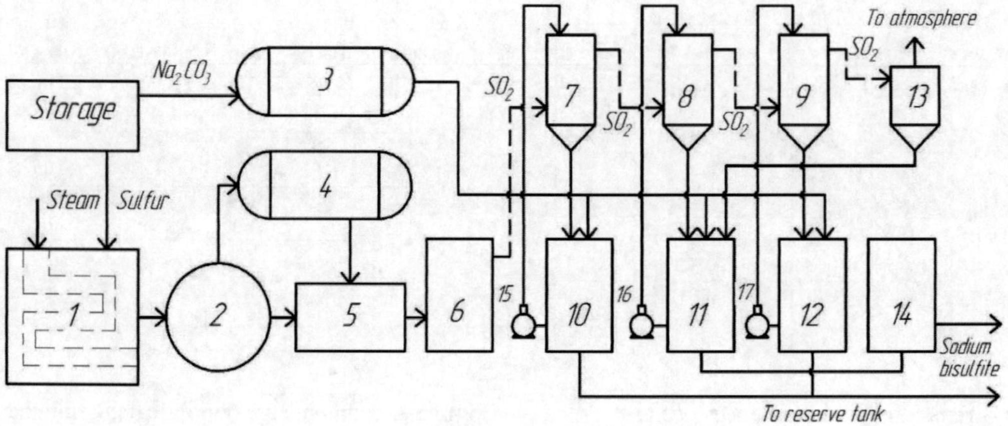

Figure 7. Sodium bisulfite production flow diagram.
1– sulfur smelter; 2 – settling tank; 3 – soda solution head tank; 4 – sulfur dioxide head tank; 5 – cyclone furnace; 6 – cooling tower; 7 – stage 1 absorber; 8 – stage 2 absorber; 9 – stage 3 absorber; 10 – circulation tank; 11 – bisulfite circulation tank; 12 – sanitary tank; 13 – spray trap; 14 – finished product tank; 15, 16, 17 – centrifugal pump.

For reduction of the sulfur dioxide content in the gas emissions in the process of sodium bisulfite production, it is proposed to convert the spray trap into an adsorber charged with PSM.

The optimal characteristics of the adsorber are calculated according to the isotherm of sulfur dioxide. The following initial parameters are accepted: pressure – 4.5 kPa; inlet gas flow rate – 300 m^3/h; sulfur dioxide content in the inlet gas – 1%. The duration of the adsorption and the amount of PSM per charge were calculated. The adsorber data are listed in Table 3.

Table 3. Adsorber data.

Parameter	Value
Diameter, m	0.85
Height, m	1.7
Specific surface area of PSM, m^2/kg	720.0
Amount of PSM for a single loading, kg	47.3
Height of PSM bed, m	0.91
Duration of adsorption, h	46.2

The continuity of purification is secured by two adsorbers, one of which is used for gas adsorption cleaning, the other for sequential recovery of PSM, while the adsorption is combined with drying and cooling of the absorber.

The modification to the waste gas treatment will allow to remove up to 99.9 % of SO_2 from the gas emissions by using the carbonate sludge.

The modified gas emissions purification train with the sulfur dioxide removal bed was calculated for environmental and economic effect. The amount of prevented environmental damage from soil and land degradation for Kazan CHPP-1 was 2.90 kilo rubles per year. The cost of PSM is 13 kilo rubles per ton. The cost of removal of sulfur dioxide from gas emissions when using PSM is 6.76 rubles/m^3.

Conclusion

1. A new sorption material based on power plant waste was tested for removal sulfur dioxide from the gases.

2. The process of adsorption of a dummy gas mixture containing SO_2 by a pelletized sorption material of water treatment was studied. The adsorption kinetics were analyzed. The adsorption isotherm corresponds to the L-type Langmuir isotherm. The convex shape of the isotherm confirms effective physical adsorption. The mechanism of how the sulfur dioxide can be adsorbed by the modified sorption material at different temperatures was studied. The tests revealed Gibbs free energy, differential heat, and adsorption activation energy.

3. The optimal data of a batch-operated adsorber with a fixed PSM bed are calculated: the diameter of the adsorber, the amount of PSM per charge, the height of the PSM bed, and the duration of adsorption.

4. Based on the study results, the gas emission purification train with the sulfur dioxide removal bed was modified at the L. Ya. Karpov Chemical Plant using the method of adsorption by pelletized sorption material.

5. The modified gas emissions purification train with the sulfur dioxide removal bed using the carbonate sludge-based PSM was calculated for environmental and economic effect. The prevented environmental damage will amount to 2.90 kilo rubles per year. The efficiency of removal of SO_2 from the gas emissions using PSM was 99.9 %.

References

[1] Ryzhkin V Ya 1987 *Thermal power plants: textbook for universities* 3rd ed., reprint. and add. Moscow: Energoatomizdat

[2] Ziganshin M G, Kolesnik A A and Posokhin V N 1998 *Design of dust and gas purification vessels*. Moscow: Ecopress-3M

[3] Nikolaeva L A and Khusnutdinov A N 2018 *A study of the absorption of nitrogen oxides from the boiler flue gases* Thermal Engineering Vol. 65. Iss. 8. pp. 575-579

[4] *Gas chromatograph/mass spectrometer-based determination of volatile organic substances in soil and waste production and consumption*: MUK 4.1.1062-2001: approved by the Chief State Sanitary Doctor of the Russian Federation - First Deputy Health Minister of the Russian Federation G. G. Onishchenko 18.07.2001: effective from 01.10.2001. Moscow: Federal Center of Gossanepidnadzor, 2001, p. 10

[5] Nikolaeva L A and Khusnutdinov A N 2018 *Purification of gas emissions of chemical industry enterprises by carbonaceous cutting*. Ecology and Industry of Russia. Vol. 22. Iss. 8. pp. 14-18

[6] Russian Federation Patent 117420. *Wastewater treatment system for removal petroleum products*. Declared on 10.01.2012. Published on 27.06.2012. Bul. № 3.

[7] Semakina O K, Yakusheva Yu S and Shevchenko A A 2013 *Selection of method of pelletizing adsorbent from waste products*. Fundamental research. No. 8-3. pp. 720-725

[8] Krylov I O, Lugovskaya I G, Korovushkin V V and Makeev D B 2008 *Influence of heat treatment on sorption and catalytic properties of shungite rocks*. Ecology and industry of Russia. No. 10. pp. 24-27

[9] Terekhova M V and Rusakova S M 2013 *Studying adsorption patterns of anions on the surface of red sludge*. Izvestiya MGTU MAMI. No. 3, Vol. 1. pp. 147-151

[10] Nikolaeva L A 2013 *Research of sorption processes using chemical water purification sludge for nitrogen and sulfur oxides contained in smoke gases emitted from a thermal power station*. Thermal Engineering. Vol. 60. No. 4. pp. 244-247

[11] Technological regulation No. 10 of production of sodium thiosulfate crystal photographic and technical, sodium sulfuric acid 5-water reagent, sodium bisulfite technical (aqueous solution) and sodium thiosulfate solution, 2015

[12] Methodological guide for calculating, regulating and controlling emissions of pollutants into the air, Saint Petersburg, 2012

IOP Conf. Series: Materials Science and Engineering **941** (2020) 012007 doi:10.1088/1757-899X/941/1/012007

Estimation of anti-chlorosis action of iron-lignosulfonate complex synthesized by anodic dissolution of iron

Yu G Khabarov, V A Veshnyakov and A E Shergin

Northern (Arctic) Federal University named after M.V. Lomonosov, 17 Northen Dvina Emb., Archangelsk, 163002, Russia

E-mail: khabarov.yu@mail.ru

Abstract. Iron-lignosulfonate complexes are used as trace element fertilizers to control plant chlorosis caused by iron deficiency. In this study, iron-lignosulfonate complexes were synthesized by anodic dissolution of iron in solutions of sodium lignosulfonates or lignosulfonic acids. The resulting solutions of the iron-lignosulfonate complex do not contain extraneous ions that are present in similar preparations synthesized chemically from salts. Electrosynthesis of iron-lignosulfonate complexes was carried out with two iron electrodes with alternating polarity, changing the period of switching the polarity of the electrode, temperature and the duration of electrolysis. The anti-chlorosis action of the synthesized iron-lignosulfonate complex was tested while growing lupine on calcareous soil. The plants were treated with solutions of the synthesized complex by the non-root method after 10 and 20 days of their cultivation. For comparison, lupine was also grown on normal soil and on calcareous soil without iron-lignosulfonate complex treatment. The content of chlorophyll, the content of carotenoids and the mass of the ground part of the plants treated with the synthesized complex were the same or higher than those of the plants grown on normal soil. In addition, all these parameters were 2-3 times higher than that of the plants grown on calcareous soil without treatment.

1. Introduction

The iron deficiency leads to the development of chlorosis in plants, which is manifested in the yellowing of the leaves due to a decrease in the content of chlorophyll. This decrease leads to the fall of the crop [1]. Chlorosis is usually observed in plants growing on calcareous soil that has a high concentration of bicarbonate anions [2, 3] and pH of 7.5–8.5 [4] or higher. Iron in this soil is in an insoluble form of hydroxycarbonate, therefore it is inaccessible to plants. Chlorosis control is a global problem as calcareous soils occupy about 1/3 of all agricultural areas. For this purpose, iron complexes with synthetic organic compounds that provide solubility are commonly used. However, synthetic complexones are poorly biodegradable and quite expensive.

The use of iron complexes with natural organic compounds is a more ecofriendly alternative. In particular, many studies have been devoted to the use of iron lignosulphonates for the control of chlorosis [5]. At the same time, iron lignosulfonates are usually obtained by mixing technical lignosulfonates with iron salts. As a result, these solutions contain extraneous ions. In this study, we used anodic dissolution of iron to synthesis iron-lignosulfonate complex without extraneous ions, and the resulting solutions with high iron concentrations were tested when lupine was grown on calcareous soils.

2. Materials and methods

2.1. Materials

The studies were carried out with industrial sodium lignosulfonates with the content of functional groups determined by known methods [6], %: total acid 5.51±0.07, sulfo groups 2.88±0.09, carboxyl 1.47±0.02, phenolic hydroxyl 1.16, carbonyl 1.43±0.04, methoxy 9.35±0.06. The solution of lignosulfonic acids was prepared by decationization of the sodium lignosulfonates solution with cation-exchange resin KU-2-8. The sodium lignosulfonates solution had pH 4.3 and contained, %: dry substances 62.5±0.4 including reducing substances 0.58±0.02 and ash content relative to dry substances 14.40±0.4. The lignosulfonic acids solution had pH 1.1 and contained, %: dry substances 50.9±0.2 and ash content relative to dry substances 0.33±0.03.

2.2. Electrochemical synthesis

Electrochemical synthesis of iron-lignosulfonate complexes was carried out in a 300 mL temperature-controlled flow cell with two iron electrodes (the iron content of 99.1%, the diameter of 5 mm, the length of 70 mm, the interelectrode gap was 5 mm) with continuous electrolyte circulation. The power supply served as an electronic unit, which allows to vary the constant voltage, control the current strength and change the polarity of the electrodes with a predetermined time interval. The concentration of the sodium lignosulfonate solution was 6%, the concentration of lignosulfonic acids was 5%. The iron concentration in resulting solutions was determined by photometry [7].

2.3. Estimation of anti-chlorosis action

Vegetation studies were carried out in greenhouse conditions on a gray forest soil (original soil). A calcareous soil was created artificially by adding 20% (by volume) of calcium carbonate to the original soil. Yellow lupine (*Lupinus luteus*, cultivar *Martin 2*, concultivar *Kopylovsky*) was used as an object of research. On the 10th day of cultivation, half of all plants growing on calcareous soil were treated with a non-root method with a solution of the synthesized iron-lignosulfonate complex with an iron concentration of 0.02%. The second similar treatment of the same plants was carried out on the 20th day of cultivation with a solution of the synthesized iron-lignosulfonate complex with an iron concentration of 0.03%. After 20 and 30 days of growing plants on the original and calcareous soils (without and with treatment), the contents of chlorophyll and carotenoids were determined after extraction with acetone [55]. The height and absolutely dry weight of the aerial parts of plants were measured after 30 days of cultivation.

3. Results and Discussion

The dissolution of the iron anode occurs directly under the action of the applied voltage. In the processes of electrochemical machining of metals, the voltage does not exceed 48 V. When using an alternating voltage, there was practically no dissolution of the anode, therefore, studies were carried out at a constant voltage. During the experiments, a rapid decrease in the magnitude of the anode current and the formation of a film on the anode were noted. This is apparently due to the fact that when an electric field is applied, the lignosulfonate polyanions move to the anode and settle on its surface to form a film that leads to an increase in resistance and thereby a decrease in the anode current density. The efficiency of anodic dissolution of a metal is proportional to the anodic current density.

To increase the rate of anodic dissolution, it is necessary to remove the resulting film from the surface of the anode. To solve this problem, a setup was used in which the polarity of the electrodes changes with a given period. The first experiment showed that the efficiency of electrochemical processing increases by 5–7 times. With an increase in voltage in the region of 4–30 V, the amount of iron transferred to the solution increases. The dependence of the concentration of iron on the switching period of the polarity of the electrodes (SPPE) passes through a maximum. At higher voltages, the maximum shifts towards lower SPPE. Obviously, as the voltage increases, the rate of formation of a lignosulfonate film at the anode increases, therefore, to maintain the process speed at the maximum level, a more frequent change in the polarity of the electrodes is necessary.

Subsequently, experiments were carried out at a voltage of 16 V. A full factorial experiment was implemented to systematically study the dependence of anodic dissolution of iron in solutions of lignosulfonic acid and lignosulfonates on temperature, process duration, and SPPE. Independent variables were changed in the following intervals: temperature was from 30 to 80°C, process duration was from 5 to 60 min, SPPE was from 1 to 41 s for lignosulfonic acid and from 1 to 170 s for lignosulfonate (in coded values from –1.683 to 1.683). Polynomials of 1–3 degrees were calculated from experimental data. The approximation error of the dependence of the iron concentration on variable factors was about 5% using third-order polynomials.

All selected independent factors have a significant impact on the output value. Moreover, their influence is not the same for solutions of lignosulfonates and lignosulfonic acids. SPPE has the greatest influence on the lignosulfonates solution, and in the case of lignosulfonic acids the duration of the process has the greatest effect. For both samples, there are coefficients with different signs for SPPE, which indicates the presence of an extremum on the dependence. An increase in both the temperature and the duration of the process leads to an increase in iron concentration. The optimal SPPE decreases with increasing electrolysis time. An increase in temperature also leads to a decrease in the optimal SPPE value for the lignosulfonates solution to 14–20 s at a temperature of 80°C, while for the lignosulfonic acids solution this parameter remains unchanged (7–8 s) with a process time of 20 min. The dissolution rate of the iron anode is significantly higher in lignosulfonic acids solutions than in lignosulfonates solutions. This is due to the lower pH value and high proton mobility.

The anti-chlorosis action of the synthesized iron-lignosulfonate complex was evaluated with lupine cultivation. Lupine was chosen as an object of vegetative research, since it is calciphobous, i.e. iron deficiency develops when it is grown on calcareous soil. For non-root treatment, we synthesized iron-lignosulfonate complexes with a maximum iron content (3 g/L), which were then diluted to the required concentration. The first treatment of plants growing on calcareous soil with the iron-lignosulfonate complex was 10 days after planting. At this point, chlorosis of plants growing on calcareous soil had already begun, and the chlorophyll content in plants was 30% lower than that of plants growing on the original soil. It was enough to carry out one foliar treatment so that the chlorophyll content in 3 days would increase by more than 2 times compared with control plants growing on calcareous soil without treatment. After 20 and 30 days of lupine cultivation, the total content of chlorophyll and carotenoids in the treated plants grown on calcareous soil was much higher than in the control plants grown on calcareous soil without treatment, and even exceeded those in plants grown on the original soil (figure 1).

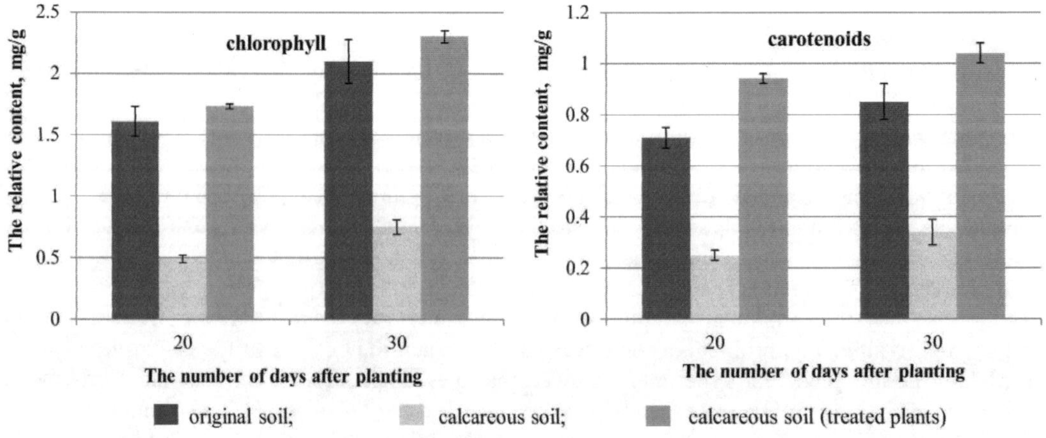

Figure 1. The relative content of pigments in lupine grown on original and calcareous soils without and with iron-lignosulfonates complex treatment of plant.

The weight of the ground part of the plants after treatment with iron-lignosulfonate complex increased 2.2 times compared with plants grown on carbonate soil and almost reached the mass of plants grown on the original soil (table 1).

Table 1. Height and weight of the ground part of plants grown on the original and calcareous soils without and with iron-lignosulfonates complex treatment of plant.

Type of soil	Height		Weight	
	cm	relative to plants grown on calcareous soil, %	g	relative to plants grown on calcareous soil, %
Original soil	13.9±0.9	125	2.5±0.3	227
Calcareous soils	11.1±1.1	100	1.1±0.2	100
Calcareous soils (treated plants)	12.1±0.6	109	2.4±0.5	218

4. Conclusion

Anodic dissolution of iron in a solution of lignosulfonic acids or lignosulfonates can be used to obtain iron-lignosulfonate complexes to avoid introducing additional anions, which occurs when using iron salts. The rate of anodic dissolution of iron is higher in a solution of lignosulfonic acids than in solutions of lignosulfonates. An increase in the concentration of iron in solution occurs with increasing temperature and the duration of electrolysis. It is necessary to use a direct current with periodic polarity reversal in order to avoid coating the electrodes with a lignosulfonate film. The dependence of the concentration of dissolved iron on the duration of the period of switching the polarity of the electrodes has a maximum, which shifts toward a shorter duration with increasing voltage. The synthesized iron-lignosulfonate complexes have a pronounced anti-chlorosis effect. The contents of pigments, the height and weight of the ground part of the treated plants grown on calcareous soil were close to those of plants grown on the original soil.

References

[1] Zuo Y and Zhang F 2011 *Plant and Soil* **339** 83-95.
[2] Briat J F, Dubos C and Gaymard F 2015 *Trends in Plant Science* **20** 33-40.
[3] Lucena J J and Hernandez-Apaolaza L 2017 *Plant and Soil* **418** 1-4.
[4] McCauley A, Jones C and Jacobsen J 2009 *Nutrient management module* **8** 1–12
[5] Rodríguez-Lucena P, Hernández-Apaolaza L and Lucena J J 2010 *J. Plant Nutr. Soil Sc.* **173** 120-6.
[6] Zakis G F 1994 *Functional analysis of lignins and their derivatives* (Atlanta: Tappi Press).
[7] Khabarov Yu G, Komarova G V and Mashyanova E A 1988 *Lesnoy Zhurnal-Forestry Journal* N5 124–5.
[8] Pochinok H N 1976 *Methods of biochemical analysis of plants* (Kiev: Naukova Dumka).

IOP Conf. Series: Materials Science and Engineering **941** (2020) 012008 doi:10.1088/1757-899X/941/1/012008

Environmental mapping of radiation-hazardous areas

M T Abisheva[1,2], V N Monayenko[1], E P Khlebnikova[2]

[1]Branch "Institute of Radiation Safety and Ecology" RSE NNC RK, 2 Beibyt Atom Str., Kurchatov, 071100, Kazakhstan
[2]Siberian State University of Geosystems and Technologies, 10 Plakhotnogo Str., Novosibirsk, 630108, Russia

E-mail: Abisheva@nnc.kz

Abstract. The article presents algorithmization of notation design process using computer technologies for environmental maps design. Cartographic models displaying natural resources are considered as an example. The results of model color design based on developed coloristic principles for the adjacent territory in the area of the "Sary-Uzen" test site of the former Semipalatinsk Nuclear Test Site are presented. As a solution of environmental mapping tasks, the authors have shown the possibilities of territories assessment and using the results to create and update thematic layers in the GIS.

1. Introduction

The Semipalatinsk Nuclear Test Site (SNTS) where 456 nuclear tests were conducted using 616 nuclear devices [1], [2], [3], has a large territory (18 300 km^2), characterized by relatively low population density and high radionuclide concentrations in selected sites.

Protection of the environment, especially humans, from the effects of residual and secondary radioactivity is the main point in solving the problems of the site. Additional work is required to assess the consequences of the SNTS contamination, followed by the development of recommendations on land use in terms of radiation safety. This needs scientific development and practical implementation of methods and ways to assess the impact of land cover contamination on the population living here.

Under these conditions, the role of radioecological mapping, which aims to develop environmental maps (radiation situation maps), in other words, cartographic works, designed to systematize the accumulated data on radiation situation of specific territories and to determine the most modified territories by people, areas of high risk to human life, etc., becomes relevant and obvious. Usually, these maps display the environmental situation in the studied area, areas of ecological disadvantage, areas of high risk to human life, etc.

The most common environmental maps are those that display:

- general radiation situation;
- soil contamination with both natural and technogenic radionuclides, and with other harmful substances;
- snow cover contamination;
- water quality assessment, etc.

Color scheme is applied to perceive effectively the radiation situation in the studied territory. The efficiency of applying color as the main graphic art is related to the visual perception of cartographic signs. Hence, color scheme for isolines of the isotopes areal distribution (cartographic method) has been developed, and the interval of specific activity of technogenic radionuclides corresponds to the established regulatory indicators in regulatory documents of the Republic of Kazakhstan.

2. Results and Discussion

According to the existing regulatory requirements, data about areas with the technogenic radionuclide activity of more than 300 Bq/kg are known as materials of restricted use (MRU), and the least significant specific activity (LSSA) is at: ^{241}Am – 1000 Bq/kg, ^{137}Cs – 1000 Bq/kg, ^{90}Sr – 100000 Bq/kg, $^{239+240}$Pu – 1000 Bq/kg [4].

Design and creation of isolines color scheme (color scales) for environmental maps of technogenic radionuclides starts after the analysis and assessment of radiation parameters of "background" territory of the former SNTS. Based on radioactive isotopes content in soil, the following level of background global fallout has been recorded: ^{241}Am – 1 Bq/kg, ^{137}Cs from 16 to 20 Bq/kg, $^{239+240}$Pu – 5.2 Bq/kg, ^{90}Sr from 0.9 to 16 Bq/kg [5].

The principle of color scheme is to choose a base color in such a way so that it will promote for a user a clear association of the given color with the essence of the phenomenon being mapped.

Color scheme from green to red (to dark red, in some cases) changing to the orange color, which is best distinguished by human eyes, is used for the best perception. This is especially important when creating environmental (radioecological) maps.

- green – the color of life, inspires calm;
- orange – the color of the warning;
- red – causes people to feel danger.

This color scheme is applied as follows: shades of green are used for up to fivefold of the global background value of the radioactive fallout, above fivefold of the global background value of the radioactive fallout to MRU begins to change from green to yellow-orange colors, values from MRU to LSSA and above changes from orange to red color (Table 1).

Table 1. Base color choice for the areal contamination with technogenic radionuclides

Radionuclides specific activity in soils, Bq/kg				Base color and its name	Coordinates of a base color in the RGB system
^{241}Am	$^{239+240}$Pu	^{137}Cs	^{90}Sr		
<1	<5.2	<20	<12	Pale green	127; 179; 127
1-2	5.2-10	20-40	12-25	Dark sea green 3	151; 201; 151
2-5	10-25	40-80	25-60	Dark sea green 2	184; 219; 164
5-10	25-50	80-120	60-90	Lemon chiffon	229; 242; 182
10-30	50-100	120-200	90-150	Yucca	255; 249; 179
30-100	100-200	200-250	150-200	Light goldenrod	255; 231; 150
100-300	200-300	200-300	200-300	Light salmon	255; 193; 112
300-1000	300-1000	300-1000	300-1000	Ochre	255; 122; 56
>1000	>1000	>1000	1000-45000	Mars red	255; 34; 0

	45000-100000	Red 3	181; 18; 0
	>100000	Red 4	115; 0; 0

Among all existing options for presenting data on a map showing the radiation situation in the studied area, isolines and signs are the two most popular mapping methods. The sign method is generally used to visualize single data (point data) and method of mapping with isolines is well suited to visualizing the areal distribution of any contaminants, e.g. radionuclides, both technogenic and natural.

In order to clarify and assess the overall radiation situation in the SNTS territory, a comprehensive survey is being carried out, including sampling of the surface soil layer using a regular net. In the course of laboratory testing of the selected samples, the content of contaminants such as ^{241}Am, ^{137}Cs, $^{239+240}$Pu, ^{90}Sr is determined. These radionuclides are one of the long-lived isotopes that can lead to irreversible consequences if exposed to long-term effects on humans and animals living in the studied area.

Based on the results of laboratory testing of selected samples, maps presenting the radiation situation in the studied area are developed. For example, for visual perception of the areal distribution of technogenic radionuclides ^{241}Am and ^{137}Cs in the upper soil layer, maps are designed using the proposed color scheme (Figure 1).

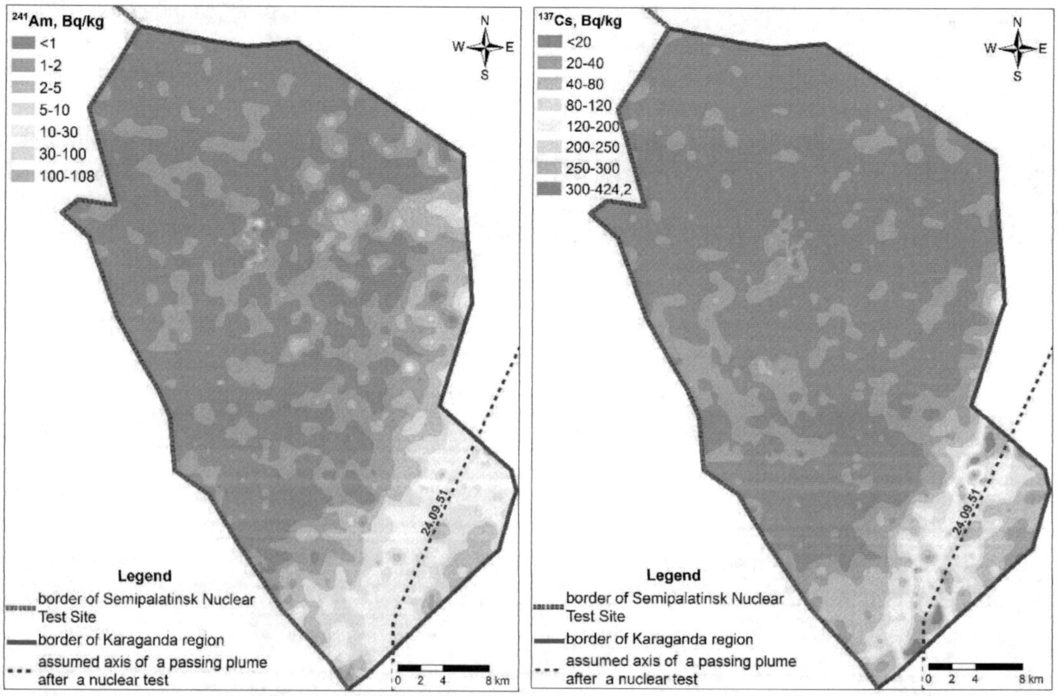

Figure 1. Areal distribution of technogenic radionuclides a) ^{241}Am, b) ^{137}Cs.

The applied color gradation allows to characterize visually the actual concentrations of the main contaminants in the soil surface layer. According to the data obtained, it can be seen that the main part of the contaminated soil with technogenic radionuclides is concentrated in the south-eastern part of the

IOP Conf. Series: Materials Science and Engineering **941** (2020) 012008 doi:10.1088/1757-899X/941/1/012008

territory of the studied area, in the place where the radioactive fallout plume of September 24, 1951 nuclear test occurred.

In addition to the creation of maps of the areal distribution of contaminants using the isoline method, environmental maps displaying the ecological situation using the sign method were also created. For example, maps showing the content of radioactive substances in the surface soil layer, such as $^{239+240}Pu$-and ^{90}Sr were created. This method does not display the general ecological situation in the studied area, but shows an accurate localization of these substances concentration.

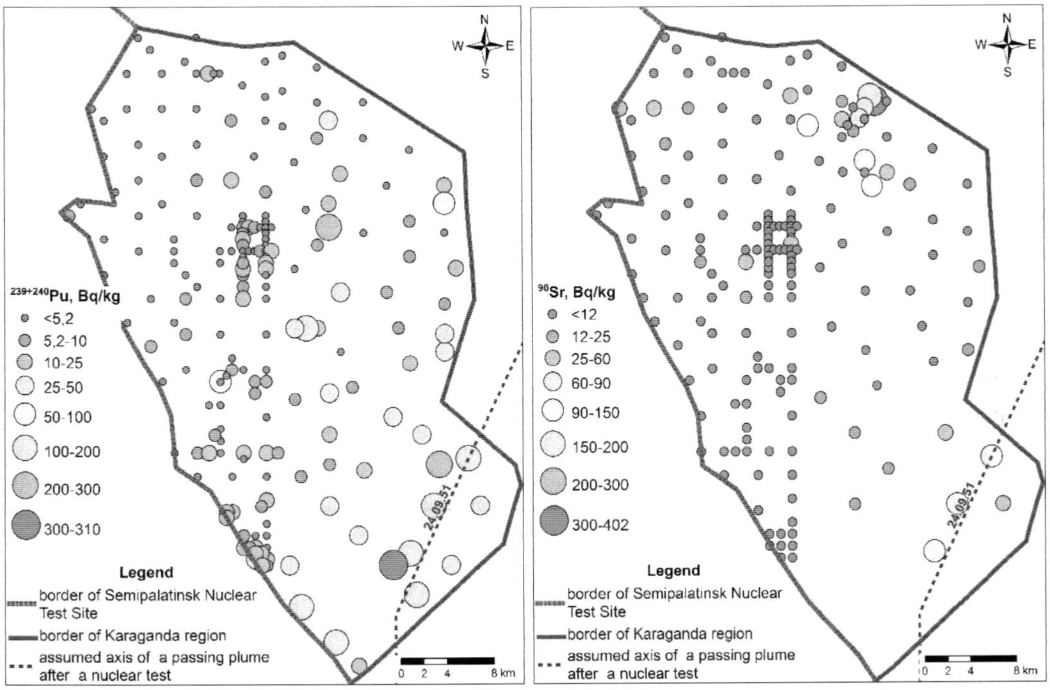

a) b)

Figure 2. Technogenic radionuclides distribution using sign method a) $^{239+240}Pu$, b) ^{90}Sr.

When creating these environmental maps, the method of color scheme presented was applied. On these maps, local areas with elevated values can be seen visually, which clearly indicates that the specific activity concentration of the studied radionuclides exceeds the background values. Accordingly, this provides ground for a more detailed survey of the areas with abnormally high values.

3. Conclusion

Accurate and logical design of environmental maps is one of the important stages of their creation. This allows to receive objective and reliable complex information on the general situation of environment of the studied area, and also on spatial differentiation of separate ecological problems in the studied territory and their combination.

The main advantages of forming environmental maps are:

- high information capacity;
- visibility and accessibility to perception;
- the possibility of spatial analysis;
- spatial assessment and prediction capability.

The high information capacity of environmental maps is achieved by improving the cartographic signage system. The application of GIS technologies in the design of cartographic products increases

IOP Conf. Series: Materials Science and Engineering **941** (2020) 012008 doi:10.1088/1757-899X/941/1/012008

the visibility, logic and accessibility of maps for direct perception and spatial analysis. For the best perception of the general environmental situation in the studied area, an important role is given to color scheme of this or that phenomenon display.

The development of color gradation was carried out in compliance with the existing regulatory requirements, which allows to characterize the overall radiation situation in the studied area more clearly.

References

[1] Nazarbayev N A, Shkolnik V S, Batyrbekov E G, Berezin S A, Lukashenko S N and Skakov M K 2016 *Conducting a Set of Scientific, Technical and Engineering Works to Bring the Former Semipalatinsk Test Site to a Safe State* vol 2 (Kurchatov: RSE NNC RK) p 448

[2] Logachev V A 1997 *Nuclear Tests of the USSR. Semipalatinsk Training Ground* (Moscow: RSE NNC RK) p 347

[3] Yakovenko Yu Yu, Monayenko V N and Abisheva M T 2017 Geoinformation of Identification of Sources of Radioactive Pollution on the Example of the Semipalatinsk Nuclear Test Site *Interexpo Geo-Siberia-2017* **8** pp 25-32

[4] *Hygiene Standards* n.d. Available at: https://online.zakon.kz/document/?doc_id=38534553 (Accessed 15 06 2020)

[5] Kashirsky V V, Lukashenko S N, Yakovenko Yu Yu and Romanenko E V 2013 *Actual Issues of Radioecology of Kazakhstan* vol 2 (Kurchatov: Institute of Radiation Safety and Ecology RSE NNC RK) pp 11–24

THE ALTERNATIVE METHOD OF CONDITIONING INDUSTRIAL WASTEWATER CONTAINING HEAVY METALS BASED ON THE HYDROTHERMODYNAMIC CAVITATION TECHNOLOGY

[1]Dubrovskaya O.G., [2]Kulagin V.A. , [3]Yao Limin

[1,2] Siberian Federal University,
79 Prospekt Svobodny, Krasnoyarsk, Russian Federation, 660041
[3] Institute of Advanced Technology of the Heilongjiang Academy of Sciences,
Harbin, PRC

Abstract. This article provides the results of studies of modifying the physicochemical properties of industrial wastewater when treated in a cavitation reactor. The authors performed an analysis of physicochemical changes in the makeup of industrial effluent and defined the efficiency of reducing the content of heavy metal ions under various modes of hydrothermodynamic action. The article also proposes the methods of mathematical modelling for determining the optimal parameters of cavitation treatment of wastewater and reveals the advantages of upgrading the wastewater process equipment with the inclusion of a unit of SC-reactors.

Keywords: industrial wastewater, heavy metals in effluents, cavitation, cavitation technology, recirculating water management, industrial wastewater treatment.

1. Introduction.

Currently, the environmental legislation, both of the Russian Federation and other countries of the world, imposes strict requirements on the methods of treatment and disposal of industrial wastewater. These requirements aim at reducing the negative environmental impact of industrial effluents on water bodies. Modern wastewater treatment technologies, both domestic and foreign, make it possible to achieve almost any degree of purification, however, the use of chemicals and the large volume of purification by-products (precipitates) make it impossible to consider the available technologies as environmentally friendly and economically feasible. In most countries, industrial enterprises are required to introduce recirculating water use, and, respectively, enterprises are revising low-efficient standard methods for conditioning industrial effluent and searching for alternative methods for wastewater treatment. The issue of the greatest relevance is treatment of effluents containing heavy metal ions. As a rule, such wastewater is typical for industrial enterprises of instrument engineering, mechanical engineering, metallurgy, chemical and pulp and paper industries. Heavy metals pose a serious environmental and toxicological threat, their negative impact on human physiology has been repeatedly noted.

At local wastewater treatment facilities, the above enterprises prevalently apply the chemical methods for purifying their effluents. Unfortunately, these methods do not allow reaching the efficiency of treatment which will comply with the standards permitting to discharge the treated effluent into fishery water bodies, or into centralised public sewers, but, moreover, the residual concentrations are too high to use this effluent as treated process water and form recirculating water use systems. It is important that the presence of heavy metal ions can affect pH and electrical conductivity of the solution, and chemical compounds negatively affect the work of the utility systems: undergoing hydrolysis, they can act as strong acids, intensify frothing of the effluent, and participate in scaling [1]. In this regard, it is expedient to develop and introduce into production alternative nonchemical innovative and highly effective methods of treatment of wastewater containing heavy metal ions. Such methods can include effluent treatment based on the effects of hydrothermodynamic cavitation.

The research objective is to identify the patterns of modifying the physicochemical parameters of water when exposed to hydrothermodynamic cavitation, to scientifically justify and develop a nonchemical technology for treating wastewater containing heavy metal ions using the methods of cavitation impact that occur on the phase boundaries in the medium being studied and to select optimal modes of cavitation treatment of wastewater in order to purify it effectively.
Method and methodology of work. The methodological base is the experimental methods of thermodynamic, kinetic and production research. Laboratory and bench semi-industrial cavitation installations have been used as an empirical base of research. When studying the effluent treatment efficiency, we used standard water quality assessment methods based on Russian state standards GOST 25150-82 and GOST 17.1.1.01-77, Natural Conservation Federal Statutory Document (CFSD) 1412.16.1-10, 14.1:2.105-97, and 14.1:2:4.214-06, as well as the methods of VBA-mathematical modelling and interpretation of the research results.

2. Results and discussions.

The change in the physicochemical properties of a liquid during hydrothermodynamic cavitation is based on high concentration (accumulation) of energy in a very small volume of a gaseous or liquid medium, followed by its release in a critically short period.
The physical characteristics of a liquid have a different effect on the intensity of cavitation impact, increasing or decreasing the velocity of cumulative jets near solid boundaries of the flow. This effect especially strongly recoils upon the last stage of the collapse of the bubbles, when their sizes are very small. A rise in viscosity and density reduces the intensity of cavitation action – surface tension forces accelerate the collapse of bubbles, while dissolved and undissolved gases in the liquid slow down this process, damping the collision of the walls of the bubble. Therefore, degassing of the liquid is one of the ways to intensify cavitation.
Changing the conditions of technological processes can also significantly affect the intensity of cavitation impact, and therefore the corresponding process rate. For example, lowering the temperature and pressure of saturated vapours increases the intensity of cavitation impact. A rise in pressure increases the velocity of the cumulative jet when the bubble collapses, however, with a significant increase in static pressure, it is difficult to obtain the modes of a developed cavitation.
The time of cavitation treatment also ambiguously affects the final result. Not always an increase in treatment time leads to an increase in the operational benefit. Therefore, applying hydrodynamic cavitation, it is necessary to take into account both the physical properties of liquids and the conditions of a specific technological process: pressure, temperature, treatment time, stream turbulence, solid particles, etc.
The object of the research was industrial wastewater, the core physical and chemical indicators of which were obtained on the basis of quantitative chemical analysis of wastewater from the workshops of an instrument-making plant. Initial concentrations of pollutants and the requirements for residual concentrations of treated effluent for the purpose of its reuse are shown in Table 1.

Table 1. Water quality requirements for the electronics industry according to Industrial Standard 11.029.003-80 for the Russian Federation, and according to ASTM D-5127-90 for the USA.

Index	Initial concentration mg/dm^3	Requirements for circulating process water of instrument-making enterprises						
		IST 11.029.003-80 TYPES			ASTM D-5127-90 TYPES			
		A	Б	B	E-1	E-2	E-3	E-4
Cu	**3.94**	0.002	0.002	0.003	0.001	0.001	0.002	0.5
Fe	**13.7**	0.015	0.02	0.03	-	-	-	-
Ni	**408.5**				0.1	1	2	500
Mn	**0.032**	-	-	-	-	-	-	-

As it has been mentioned above, hydrothermodynamic cavitation has been proposed as an alternative method of purifying water from heavy metal ions. At the same time, we used two independent research lines at different types of supercavitation reactors (SC-reactors). This formulation of the research problem concerning the effect of hydrothermodynamic cavitation on wastewater enabled obtaining the most adequate results on the reproducibility of the experimental conditions, and also made it possible to measure physical parameters in the cavitation area. Note, that the physicochemical effects are identical regardless of the method of excitation of cavitation, which is undoubtedly important when choosing SC-reactors for certain technological solutions.

It has been proved by [1–5] that complex physicochemical processes run under the conditions of hydrothermodynamic cavitation. These processes are divided into:

oxidation-reduction reactions involving organic and inorganic substances present in the aquatic medium due to the formation H2O2, O2, O3, O· and OH- in the solution;

chain reactions in the solution, which are initiated by the products of cleavage of impurities present in the solution;

destruction of macromolecules and initiation of depolymerization of polymer compounds;

reactions between dissolved gases inside cavitation bubbles.

So, to substantiate the direction of purification from heavy metals, we performed an analysis aimed at obtaining the dependences of changes in the physicochemical parameters of an aqueous solution (effluent) under certain cavitation treatment modes. These dependencies are given in Table 2, and pictorially shown in Figure 1.

Table 2. Change in physicochemical parameters of wastewater containing heavy metal ions under various modes of cavitation treatment and subsequent relaxation of the aqueous solution (effluent).

Experimental data		Cavitation mode at a constant treatment time of 10 minutes					
indicators of wastewater containing heavy metal ions	untreated wastewater	3000 rpm	5000 rpm	7000 rpm	10000 rpm	11000 rpm	12000 rpm
t ^{0}C	19	25	30	38	43	42	50
trel ^{0}C	21.5	24.5	26	28.5	32	30	34
pH	3.91	4.24	4.46	4.62	4.71	4.77	4.93
pHrel	3.83	4.26	4.49	4.68	4.68	4.86	4.97
REDOX	402	383	365	292	270	244	249
REDOXrel	330	342	334	331	316	282	265
O_2	7.64	8.56	8.89	9.95	10.37	9.89	8.57
O_2rel	7.75	8.23	7.93	8.02	8.06	8.06	7.12

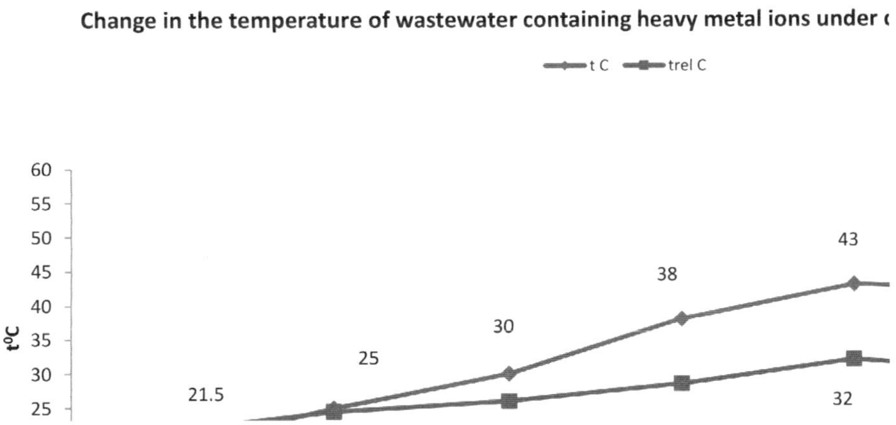

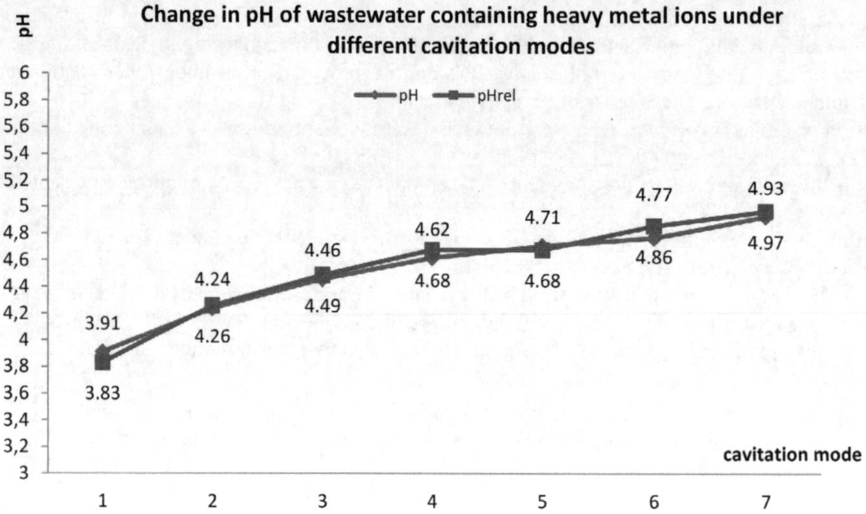
Change in pH of wastewater containing heavy metal ions under different cavitation modes
pH
pH
pHrel
cavitation mode
6
5,8
5,6
5,4
5,2
5
4,8
4,6
4,4
4,2
4
3,8
3,6
3,4
3,2
3
1
2
3
4
5
6
7
3.91
3.83
4.24
4.26
4.46
4.49
4.62
4.68
4.71
4.68
4.77
4.86
4.93
4.97

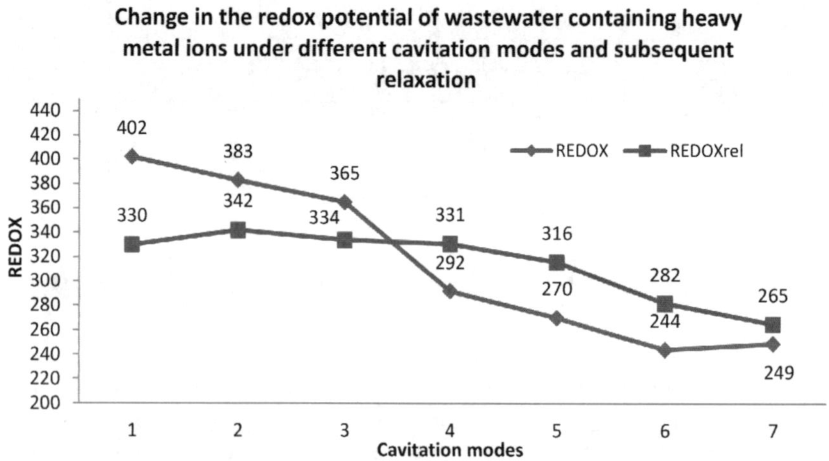
Change in the redox potential of wastewater containing heavy metal ions under different cavitation modes and subsequent relaxation
REDOX
REDOX
REDOXrel
Cavitation modes
440
420
400
380
360
340
320
300
280
260
240
220
200
1
2
3
4
5
6
7
402
383
365
330
342
334
331
292
316
270
282
244
265
249

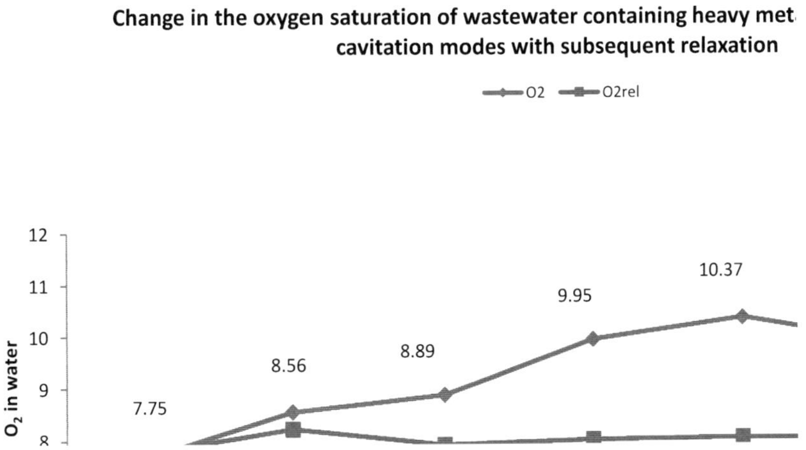

Fig. 1. The diagrams of the dependences of change in the physicochemical parameters of wastewater containing heavy metal ions and the cavitation mode.

In order to justify the environmental friendliness of the method and the economic feasibility of applying cavitation reactors while improving the process flows for industrial wastewater treatment, we performed a comparative analysis on the efficiency and energy consumption of the conventional chemical method for purifying wastewater from heavy metal ions and the proposed method of hydrothermal cavitation treatment. The results of comparing the purification efficiency using different methods of treatment using chemicals or hydrothermodynamic cavitation are given in Table 3.

Table 3. The efficiency of removing heavy metals for different methods of industrial effluent treatment

Ion	Initial concentration mg/dm^3	Concentration of ions after different methods of the effluent treatment						
		Chemical method				Cavitation at 10000 rpm		
		10 min	15 min	30 min	Effect, %	10 min	15 min	Effect, %
Cu	**3.94**	3.75	3.75	3.55	9.8	2.75	1.15	70.8
Fe	**13.7**	12.66	10.7	5.6	59.12	2.66	0.64	95.3
Ni	**408.5**	334.9	288.07	96.9	76.27	37.25	16.25	94.54
Mn	**0.032**	0.022	0.022	0.022	31.25	0.017	0.017	46.8

IOP Conf. Series: Materials Science and Engineering **941** (2020) 012009 doi:10.1088/1757-899X/941/1/012009

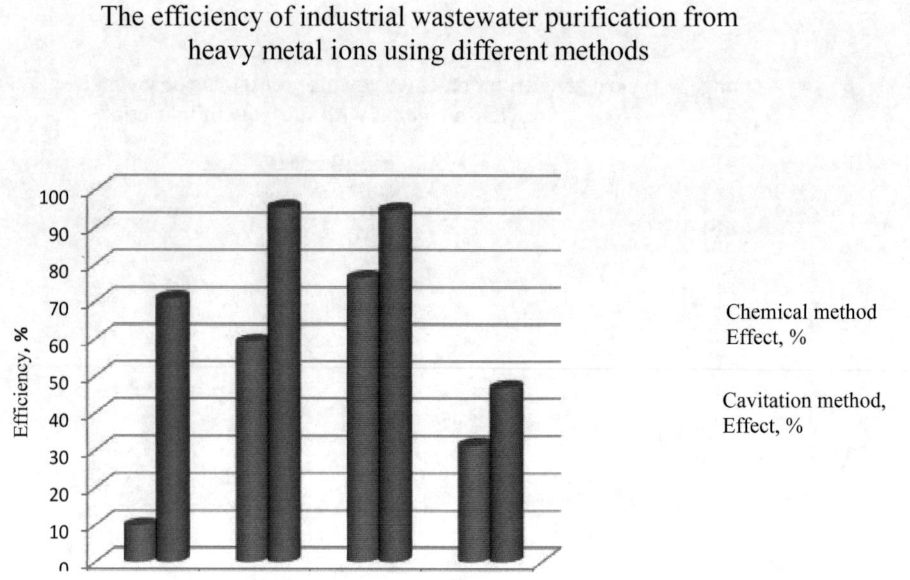

Fig. 2 Efficiency of purification of industrial wastewater from heavy metal ions using different methods of treatment.

Apart from the obvious advantage of cavitation technology for removing heavy metals from wastewater, it is worth noting that the chemical method of removing a complex of heavy metals requires bringing pH of the medium to 10.5-12.0, which surely entails an excessive use of chemicals. The insoluble precipitates which form in this process (as a rule, metal hydroxides) are subject to prolonged sedimentation and their water content is 60-75%, as a result, the technology of removing and treating such sediments of wastewater is becoming more complicated.

Analysing the obtained experimental data, we chose mathematical modelling with a set of VBA techniques, combined with the Excel computational capabilities (fig. 3) in order to determine the optimal mode of cavitation treatment of the industrial effluent.

Based on the matrices of change of the concentration of heavy metal ions in the wastewater being treated and of the obtained dependences on such factors as revolution speed of the cavitational impeller, treatment time, pH and temperature of the medium, second-order regression equations were obtained. These regression equations were used to build graphical interpretations in the form of surfaces (fig. 3), which allows to regulate the process of chemical-free conditioning of wastewater containing heavy metals and take into account technological and economic indicators later on. When calculating the data, we applied an algorithm for estimating the coefficients of a nonlinear regression model using the least-squares method. The evaluation criterion y1-2 is presented in the factor space of x1; x2; x3. Figure 3 shows an adjustment diagram of effluent treatment using copper ion as an example.

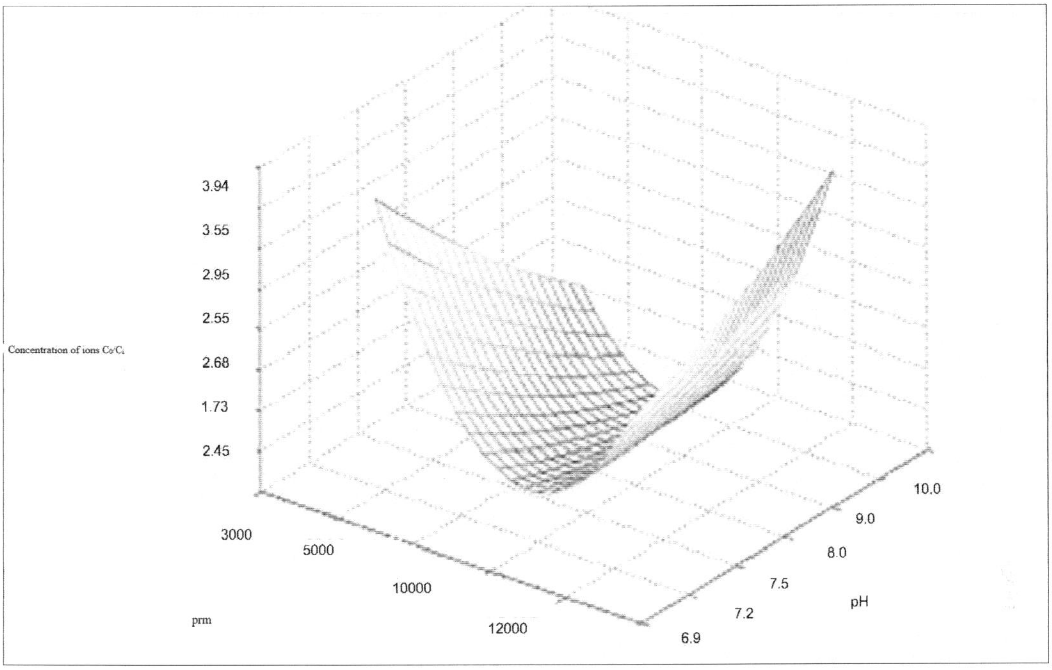

Fig. 3. The adjustment diagram of the treatment process for wastewater containing heavy metal ions (the case of Cu2+ ion) using cavitation treatment.

Interpreting the results of mathematical modelling and comparing them with the actual changes in the concentrations of heavy metal ions in various modes of hydrothermodynamic treatment of the wastewater model, we can conclude that the optimal mode for removing a complex of heavy metal ions is processing at 10.000 rpm for 60 seconds. Under a more intense cavitation mode, we observed the formation of intermediate products of metal oxidation and recombination of active ions in an aqueous medium with the formation of coagulated structures in the form of metal hydroxides, for example, Fe(OH)3, or hydroxo-aqua complexes such as [Cu(OH2)2(OH)2]0, [Zn(OH2)2(OH)2]0, [Ni(OH2)2(OH)2]0, [Ni(OH2)4(OH)2]0, [Fe(OH2)(OH)3]0, [Fe(OH2)3(OH)3]0. These substances, in turn, having electrostatic attraction due to a change in the charge of the complexion, from positive to negative (at pH above 10.0), act as coagulants. The resulting coagulated precipitate was also analyzed in the framework of this research, with determining the sizes of precipitate particles using the turbidimetric and sedimentation methods of the analysis. And the size of the precipitate flakes coagulated on the hydroxo-aqua complexes of Ni, Cu, Zn reaches 130-140 nm, while the precipitate flakes formed and sedimented on the Fe3+ ion complexes vary from 24 to 50 μm.

All the studies were performed by the Research Laboratory of the School of Engineering and Construction of Siberian Federal University (Testing Laboratory of Structural Materials and Chemical Analysis of Water) in collaboration with the Institute of Advanced Technology of the Heilongjiang Academy of Sciences (Harbin, PRC) and the International Laboratory of Cavitation Technologies (Russia-China).

Scope of application of the results. The results of this research can be applied in the upgrading of industrial local treatment facilities of enterprises with the introduction of a cavitation treatment unit in order to intensify the purification of wastewater from heavy metal ions.

Conclusion. These technical solutions allow achieving the water quality required for reuse. The main advantages of the proposed chemical-free method based on the effects of hydrothermodynamic cavitation are:

1. the opportunity to clean various effluents with significant fluctuation in the quantitative content of pollutants;
2. the opportunity of using the proposed technology when upgrading the existing equipment, as SC-reactors are easily integrated into the existing structures, for example, storage tanks, without significant alteration of their construction;
3. the opportunity to automate the purification process;
4. reduction of operational and economic costs;
5. reduction of negative impact on water bodies to which industrial wastewater is discharged.

The introduction of a cavitation treatment unit to intensify the purification of industrial effluent and to condition process water will permit forming a closed-loop water use cycle and significantly increasing the environmental sustainability of the region where industrial complexes are present.

This work was funded by RFBR and the Government of Krasnoyarsk Territory in the framework of research project 18-41-242008.

References

[1]. Kurilina T.A. The prospects of utilizing the modified sorption material to intensify purification of waste water from electroplating production/ Kurilina T.A., Dubrovskaya O.G., Kulagin V.A., Matyushenko A.I., Bobrik A.G.// Journal of Siberian Federal University. Engineering & Technologies. 2019. Vol. 12. No. 2. Pp. 182-191.

[2]. Dubrovskaya O.G. Reagent-free treatment of industrial wastewater containing heavy metals based on hydrothermodynamic cavitation technology /Dubrovskaya O.G., Kulagin V.A.// Journal of Siberian Federal University. Engineering & Technologies. 2019. Vol. 12. No 4. Pp. 460-467.

[3]. Dubrovskaya O.G. Resource-saving technologies for neutralizing and disposal of waste from enterprises of the thermal power complex of Krasnoyarsk Territory: monograph / O.G. Dubrovskaya, L.V. Priymak, I.V. Andrunyak. Krasnoyarsk: Sib. Feder. Univ., 2014. 164 p.

[4]. Dubrovskaya O.G. Conditioning of wastewater from power systems and complexes / O.G. Dubrovskaya, V.V. Evstigneev, V.A. Kulagin, Journal of Siberian Federal University. Engineering & Technologies, 6 (2011 4). Pp. 665-675.

[5]. Eldarzade E.A. Use of purified industrial and storm water as an alternative water source of the enterprise / E.A. Eldarzade // Building and Architecture – formation of the environment of vital activity. - Achinsk, 2016. Pp. 122-123.

[6]. Eldarzade E.A. Development of closed water treatment systems for oil and gas production enterprises as a basis for environmental security in the region / E.A. Eldarzade // Current State of Science and Technology. Sochi, 2017. Pp.73-79.

[7]. Kulikova N.I., Nozhevnikova A.N. Purification of municipal wastewater with the reuse of water and treated sediments: theory and practice / Eds Moscow: Logos, 2014. 400 p.

[8]. Dubrovskaya O.G. Current arrangements of the process flows for wastewater treatment using cavitation technology / O.G. Dubrovskaya, V.A. Kulagin, E.S. Sapozhnikova // Journal of Siberian Federal University. Engineering & Technologies. 2015. Vol. 8. No 2. Pp. 217-223.

[9]. Dubrovskaya O.G. Mathematical modelling of cavitation processes for conditioning of industrial wastewater / O.G. Dubrovskaya, V.A. Kulagin, E.S. Sapozhnikova, Li F., Li J., Zheng Ch. // Journal of Siberian Federal University. Engineering & Technologies. 2015. Vol. 8. No. 3. Pp. 369-376.

[10]. Gimaeva A.R. Sorption of heavy metal ions from water by activated carbon adsorbents / A.R. Gimaeva, E.R. Valinurova, D.K. Igdavletova, F.H. Kudasheva // Sorption and Chromatographic Processes. 2011. Vol. 11. No. 3. Pp. 350-356.

[11]. Zykova I.V. Adsorption of copper ions by ceramic chips from binary and multicomponent solutions / I.V. Zykova, I.V. Lysenko, V.P. Panov // Proceedings of Universities. Chemistry and Chemical Engineering. 2004. Vol. 47. No. 9. Pp. 151-167.

[12]. Tsvetkova A.D. Investigation of the process of adsorption of copper ions on modified silicon dioxide / A.D. Tsvetkova, O.P. Akaev // Bulletin of N.A. Nekrasov Kostroma State University. 2011. No. 2. Pp. 27-30.

[13]. Mukhin V.M. Active coals of Russia / V.M. Mukhin, A.V. Tarasov, V.N. Klushin. Moscow: Metallurgy. 2000. 352 p.

[14]. Timofeev K.L. The kinetics of sorption of indium, iron, and zinc ions by weakly acidic cation exchangers / K.L. Timofeev, A.V. Usoltsev, S.A. Krayukhin, G.I. Maltsev // Sorption and Chromatographic Processes. 2015. Vol. 15. Issue 5. Pp. 720-729.
[15]. Filatova E.G. The integrated technology for extracting nickel and copper ions from industrial wastewater / E.G. Fiatova, O.I. Pomazkina, V.I. Dudarev, N.N. Sheveleva //Water Treatment. 2012. No 12. Pp. 20-25.
[16]. Wang Ji-Zhong Waste from the sewage of heavy metals using natural minerals / Ji-Zhong Wang, Sheng-Rong Li, Bao-Lin Liu, Jing-Gui Tong // Bull Mineral Petrol Geochem. 2005. V. 24. No.2. Pp. 159-164.
[17]. Svilovi'c S. Modeling batch kinetics of copper ions sorption using synthetic zeolite NaX / S. Svilovi'c, D. Ruši'c, R. Stipiši'c // Journal of Hazardous Materials. 2009. V. No 170. Pp. 941-947.
[18]. Demirbas Ayhan. Heavy metal adsorption on agrobased waste materials //
[19]. Journal of Hazardous Materials. 2008. V. 157. № 2-3. Pp. 220-229.
[20]. Lokendra S. Thakur. Adsorption of heavy metal from synthetic waste water by tea waste adsorbent / S. Thakur Lokendra, Parmar Mukesh // International Journal of Chemical and Physical Sciences. 2013. V. 2. No. 6. Pp. 6-19.
[21]. Tumin Najua D. Adsorption of copper from aqueous solution by Elais Guineensis kernel activated carbon / Najua D. Tumin, A. Luqman Chuah, Z. Zawani, S. Abdul Rashid // Journal of Engineering Science and Technology. 2008. V. 3. No. 2. Pp. 180-189.

IOP Conf. Series: Materials Science and Engineering **941** (2020) 012010 doi:10.1088/1757-899X/941/1/012010

Arctic Reindeer Herding and Ecology: When Economy is Impossible without Safeguarding Ethnocultural Distinctiveness

A. Gorbunova [1], M. Zadorin [2], N. Kuprikov [3], I. Saveliev [4]

[1] Association of Indigenous Small-Numbered Peoples of the North, Siberia and Far East of the Russian Federation, 92 Vernadsky Prospekt, sector B, office 260, Moscow, Russia, 119571
[2] Department of International Law and Comparative Legal Studies, Higher School of Economics, Management and Law, Northern (Arctic) Federal University named after M.V. Lomonosov, 58 Lomonosov Prospekt, room 406, Arkhangelsk, Russia, 163000
[3] Scientific and Information Center "Polar Initiative", 1 Volokolamskoye Shosse, Moscow, Russia, 125080, Moscow Aviation Institute (National Research University) , 4-G Volokolamskoye Shosse, Moscow
[4] Department of International Law and Comparative Legal Studies, Higher School of Economics, Management and Law, Northern (Arctic) Federal University named after M.V. Lomonosov, 58 Lomonosov Prospekt, room 406, Arkhangelsk, Russia, 163000

E-mail: m.zadorin@narfu.ru

Abstract. The paper considers environmental issues related to the field of agriculture in the Russian Arctic exemplified by the research into statutory regulation of economic, cultural and ecological aspects of reindeer herding by a small-numbered indigenous people called the Nenets. Using the example of the Nenets Autonomous Area, the authors provide insight into the measures implemented to control the deer-feeding capacity of pastures for the purposes of sustainable use. The article gives an estimate as to the promptness of actions taken to prevent the loss of cattle across the Yamal Peninsula due to the spread of Siberian plague. While giving prominence to the successful regional practices of Yakutia, the authors also emphasize the necessity for the introduction of amendments to the Russian legislation regarding the procedure of ethnological expert examination to the extent that it shall include the opinions given by the officials of organizations representing the interests of small-numbered indigenous peoples inhabiting the territory in question. A legislative gap in the above-mentioned field results in emergency measures replacing preventive measures in the face of potential anthropogenic disasters similar to the one which took place in Norilsk in the summer of 2020. One of the authors was born in the Kanin tundra of the Nenets Autonomous Area and represents an indigenous and nomadic Nenets people.

1. Introduction

In accordance with the Russian Government Decree as of February 13, 2019, No. 207-r "On the Spatial Development of the Russian Federation until 2025", reindeer herding being a branch of agriculture and a specialized economic field in the Nenets, Yamalo-Nenets and Chukotka Autonomous Areas, has been preserved in Russia as part of the unique nomadic culture of the Arctic peoples, their lifestyle and traditional economy.

The Nenets practice reindeer herding in the locations where they originally (traditionally) lived in the Yamalo-Nenets and Nenets Autonomous Areas, Taymyr Dolgano-Nenets Municipal District of the Krasnoyarsk Region, the Komi Republic, and the Arkhangelsk Region. All of these regions belong to the Arctic zone in whole or in part. For instance, around 18 thousand Nenets are leading a traditional nomadic way of life in the Yamalo-Nenets Autonomous Area. According to the data of the Russian Census in 2010, the total number of Nenets is 45 thousand people. The Constitution guarantees them a special legal status in Russia, namely that of a small-numbered indigenous people. Nenets' entire life is related to reindeer, which are not only a source of food: clothes and canvases for traditional portable houses named chums are made of reindeer hides. It is therefore not surprising that half of the reindeer population in world belongs to the Nenets husbandry. Moreover, modern research shows that Arctic reindeer herding can be an effective environment management strategy used for preservation of open tundra landscapes in the conditions of rapid climate change.

Modern understanding of sustainable social development implies growth in three dimensions: society, economy and ecology. As far as the Nenets people are concerned, their social and economic well-being, if not their security as an ethnic group altogether, almost completely depends on ecology. Without pastures serving as reindeer food supply and without clean water bodies, reindeer herding culture defining nomadic lifestyle of the Nenets will disappear together with the people and blend into the globalization map of the contemporary world. Reindeer herding is an ethno-forming industry for the Nenets people [2].

This is why increasing anthropogenic impact on the ecosystem is perceived directly by the Nenets reindeer herders demanding immediate action based upon modern approaches in the fields of state policy and innovative technologies with due regard to the traditional knowledge accumulated in the ethnocultural communities.

2. Economic Appraisal of Current Situation in the Reindeer Herding Industry of the Russian Arctic Regions

Around 800 people, mostly those having nomadic and semi-nomadic lifestyle, are engaged in reindeer herding in the Nenets Autonomous Area. As of January 01, 2018, reindeer population amounts to 177 thousand cattle, including 149.7 thousand owned by agricultural organizations, peasant (farm) households and sole proprietorships.

The population has increased by almost 30 thousand cattle over the last ten years as evidenced by the data obtained from the Socioeconomic Development Strategy of the Nenets Autonomous Area until 2013 (approved by the Decree of the Regional Council of Deputies as of November 07, 2019, No. 256-sd).

As of January 01, 2019, total reindeer population in the Taymyr Dolgano-Nenets Municipal District of the Krasnoyarsk Region amounts to 127,507 cattle. Around 500 families in the region leading a nomadic lifestyle are involved in reindeer herding.

The reindeer population amounts to approximately 800 thousand cattle in the Yamalo-Nenets Autonomous Area. The major feature of reindeer herding in the Yamalo-Nenets Autonomous Area is that 53% of the total reindeer population is owned by the people, and only 47% are owned by agricultural companies. In Tazovsky (81.1%), Yamalsky (43.3%), and Priuralsky (70%) districts reindeer are predominantly owned by the people.

Subject to the Decree of the Komi Republic Government as of December 24, 2010, No. 469, regarding the Regional Reindeer Population Planning Scheme, different types of reindeer husbandries combined comprised more than 89.6 thousand cattle, of which 68.6 thousand were in possession of agricultural organizations, 3.4 thousand belonged to the peasant (farm) households, and 17.6 thousand were owned by private households.

Notwithstanding the above, among the main problems the industry faces in its development are non-availability of a uniform federal standard regulating the field of reindeer herding (i.e. a federal law) [3] and the existing system of disorganized state-owned reindeer farms [4].

Furthermore, the specific nature of food resources (natural forage growing in the Arctic latitudes) suggests that uncontrolled increase in population along with apparent reduction of reindeer pastures might result in degradation of the latter, complication of epidemiological situation as well as other negative consequences. In this regard, regions carry out geobotanical studies of reindeer pastures together with activities aimed at controlling the deer-feeding capacity.

In accordance with the Decree No. 268-p adopted by the Nenets Autonomous Area Government on October 04, 2019, the region shall annually estimate the efficiency of reindeer pastures use based upon the performance of reindeer husbandries during the year.

Estimation of efficiency for the use of reindeer pastures shall apply to each reindeer husbandry by means of examining the following documents:

– materials of geobotanical studies carried out on the territories of reindeer pastures;
– interfarm land management projects for the reindeer pastures;
– acts confirming reindeer count as of January 01 and July 01 of the reporting period;
– acts confirming reindeer fawning count as of July 01 of the reporting period;
– reports on reindeer population movement, both across the husbandry in general (consolidated) and broken down by each reindeer herding brigade (of a peasant (farm) household or a reindeer camp) as of July 01 of the reporting period;
– records of financial and economic performance of agricultural manufacturers for the preceding year.

The estimation in question is conducted in order to comply with the permissible levels of impact on the reindeer pastures for the purposes of their efficient use.

Results obtained from the estimation of efficiency for the use of reindeer pastures form a report comprising the conclusions based upon the estimation results. Reindeer husbandries are provided with recommendations regarding the issues of efficient reindeer pastures use and reindeer population regulation. The report also points out any violations (if found) of the guidelines regulating reindeer feeding, handling, breeding and guarding processes, as well as any non-compliances with recommendations regarding permissible reindeer population size, determined by the interfarm land management project for the reindeer pasture.

3. Agriculture and Climate Change: Global Agenda

While conducting research into the ecological problems in the field of agriculture, it is impossible to ignore such global issues as climate change.

Intergovernmental Panel on Climate Change (IPCC) established by the World Meteorological Organization and the United Nations Environment Programme (UNEP) published its Fifth Assessment Report called *Climate Change 2013: The Physical Science Basis* noting that certain alarming evidence indicate that excess of threshold values leading to irreversible changes in ecosystems and climate system of the planet has already happened. Due to the warming and droughts, the situation in the ecosystems of the Amazonian tropical rainforests and the Arctic tundra has already reached its critical mass.

However, it is of paramount importance that the issues of climate change are addressed through the lens of well-considered decisions of the global community and not from the political standpoint of particular actors. Pursuant thereto, it is impossible to say for certain whether the ecological problems touched upon in the Report (outbreak of Siberian plague in the Yamal Peninsula, diesel oil spill at Norilsk thermal power plant in the summer of 2020) are a direct consequence of the climate change, since they require a separate comprehensive research on the part of expert community.

In July 2016, an outbreak of a zoonotic disease known as Siberian plague started in the Yamalsky district of the Yamalo-Nenets Autonomous Area. In accordance with the data provided by the Federal Service for Veterinary and Phytosanitary Surveillance, the total number of reindeer infected with Siberian plague during the epizootic outbreak (July 16 - August 03) in the Yamal Peninsula amounted

IOP Conf. Series: Materials Science and Engineering **941** (2020) 012010 doi:10.1088/1757-899X/941/1/012010

to 2,650 animals (according to other sources – 2.657 [3]), 2.350 of which died and 300 were subject to forced slaughter.

Siberian plague is an extremely dangerous zoonotic infection, which *Bacillus anthracis* spores can stay alive in the soil for a long time threatening with recurrent outbursts on the exposed territories for many decades to come. The main source of infection for a human being is a sick animal or its carcass, while livestock products, soil and other natural objects serve as transfer factors.

June and July 2016 were marked with anomalous day temperatures reaching 29-34°C, which resulted in deeper seasonal thawing of permafrost and in transfer of anthrax bacteria from the underground layers to the soil surface with intrapermafrost water. Increased number of bloodsucking insects along with the negative impact of heat on the reindeer immune system against the background of ceased vaccination presumably were the cause for mass infection of reindeer with Siberian plague.

Prompt management of necessary works in the Yamalo-Nenets Autonomous Area made it possible to contain the largest focus of Siberian plague in the last 30 years. Eventually, a set of measures was developed with the purpose of preventing Siberian plague outbreaks in the future.

It shall be additionally noted that a recent comprehensive research into the biotic factor in the reindeer population inhabiting several Arctic zone regions has shown a significant increase in the number of opportunistic pathogens, which might be also related to the ration-biomass balance of pastures and epizootic conditions in the herd [6] directly linked to the climate change.

4. Ethnological Expert Examination in the Russian Arctic: Delays are Unacceptable

In order to protect the original living environment, traditional lifestyle, economic activities and folk crafts of Russian small-numbered indigenous peoples, the Federal Law of the Russian Federation No. 82-FZ "On guarantees of the rights of indigenous small-numbered peoples of the Russian Federation" as of April 30, 1999, provides for the right of these peoples and their associations to participate in ecological and ethnological expert examinations of federal and regional state programs for natural resources exploration and environmental protection in the locations where these peoples traditionally dwell and execute their economic activities. However, another Russian Federal Law No. 174-FZ as of November 23, 1995, "On ecological expert examination" does not stipulate for admission of other persons, including representatives of Russian small-numbered indigenous peoples (except experts), to the state ecological expert examination panels.

This is why it is necessary to amend Russian legislation with the following provision: in case business or other activities (subject to ecological expert examination) are intended to be carried out in the locations where small-numbered indigenous peoples traditionally dwell and execute their economic activities, materials of environmental impact assessment shall include an opinion passed by the unions (associations) of small-numbered indigenous peoples or by their representatives' councils in regional and municipal executive bodies.

In case the intended business or other activities are to be developed, negative consequences caused by technogenic impact on the environment along with associated social and other consequences can be reduced by means of conducting ecological expert examinations, which shall be determined as mandatory in the federal legislation. The overriding priority of such tool is to minimize the negative impact of intended business activities on the indigenous living environment and to possibly elaborate alternative ways of attaining the goals of the intended activities in question (e.g. reindeer fawning grounds on the way, etc.). Ethnological expert examination is defined in the above-mentioned federal law on guarantees of the rights of indigenous peoples as follows: Ethnological expert examination is a scientific research into the influence of changes in the original living environment of indigenous small-numbered peoples of the Russian Federation and sociocultural situation on the development of an ethnic group.

Russian President Vladimir Putin instructed the Government to cooperate with the regional executive authorities on the issue of statutorization of the ethnological expert examination procedure and to submit their proposals on the matter (Clause 3 of the Russian President's List of Instructions following the meeting of the Presidential Council for Interethnic Relations as of December 04, 2016, No. Pr-2338).

At the moment, the Federal Agency for Ethnic Affairs, being a responsible executive authority, is developing a project of a statutory instrument regulating the ethnological expert examination procedure.

Nevertheless, a number of Russian regions have already gained successful experience in conducting ethnological expert examination. For instance, the regulation on ethnological expert examination is adopted by the Sakha Republic (Yakutia) Law as of April 14, 2010, 820-Z No.537-IV "On ethnological expert examination in the locations of original dwelling and traditional economic activities of indigenous small-numbered peoples of the Northern Sakha Republic (Yakutia)". According to some researchers, however, the issue of corporate responsibility and proper compensations to the indigenous communities has not been completely resolved yet.

Ethnological expert examination is also important in case of emergencies. Specifically, the situation with the spill of about 20 thousand tons of diesel oil due to a failure of a fuel storage tank at a thermal power plant owned by Norilsk-Taymyr Energy Company on May 29, 2020, was mentioned in the public statement made by the Russian Association of Indigenous Peoples of the North (RAIPON) on June 07, 2020. According to the information available as of the beginning of June 2020, there were not any reindeer husbandries in the Taymyr Peninsula which suffered or reported direct damage. However, due to the delayed environmental impact, particularly that on the conditions of reindeer herding for the local communities as well as on the rivers and water bodies used in traditional industries, it is necessary to conduct an ethnological expert examination in order to estimate the consequences of the spill for the original living environment of indigenous small-numbered peoples, and to determine the compensatory measures aimed at reduction of negative influence on the traditional lifestyle and economic activities of these peoples.

5. Conclusion

Having regard to the above, it seems essential to take certain measures in order to underpin Russia's economic policy in the Arctic with consideration for the distinctive features of the region:

- to adopt a federal law on reindeer herding along with federal subordinate legislation defining common methodology for the estimation of efficiency for the use of reindeer pastures with due regard to the existing experience accumulated in the regions;
- to adopt a federal law on ethnological expert examination, or amendments to the Federal Law No. 82-FZ on guarantees of the rights of indigenous peoples in order to ensure consideration of the opinions given by the communities or associations of indigenous peoples within the set of materials of environmental impact assessment;
- in reliance upon Russia's international experience and its commitment to the global climate agenda, to adopt immediate measures for the development of a nationwide action plan containing guidelines to control the thawing of permafrost and to tackle the threats of man-made disasters in the future.

References

[1] Verma M et al 2020 Can reindeer husbandry management slow down the shrubification of the Arctic? *Journal of Environmental Management* **267** 110636

[2] Yaptik E S 2019 Quality of life of the Nenets people (an inside view of the culture) *Bulletin of Ugric Studies* **9** (1) 153–165 (In Russian)

[3] Ivanov V A 2015 State and trends of reindeer breeding in the arctic region of Russia *Regional Research of Russia* **5** (2) 122–127

[4] Konstantinov Y et al "Uncontrolled sovkhoism": administering reindeer husbandry in the Russian far north (Kola Peninsula) *Acta Borealia* **35** (2) 95–114

[5] Selyaninov Yu O 2018 Phenotypic, biochemical and molecular analysis of Bacillus anthracis strains isolated during the outbreaks of anthrax in the Russian Federation, 2014-2016 (Review) *Sel'skokhozyaistvennaya Biologiya* **53** (2) 404–413 (In Russian)

[6] Popova A Y et al 2016 Outbreak of Anthrax in the Yamalo-Nenets Autonomous District in 2016, Epidemiological Peculiarities *Problems of Particularly Dangerous Infections* **4** 42–46 (In

IOP Conf. Series: Materials Science and Engineering **941** (2020) 012010 doi:10.1088/1757-899X/941/1/012010

Russian)
[7] Ilina L A et al 2018 Comparative analysis of rumen bacterial community of young and adult rangifer tarandus reindeers from Arctic regions of Russia in the summer-autumn period *Sel'skokhozyaistvennaya Biologiya* **53** (2) 355–363 (In Russian)
[8] Gassiy V 2019 Protecting indigenous rights from mining companies the case of ethnological expertise in Yakutia *Sibirica* **18** (3) 92–108

IOP Conf. Series: Materials Science and Engineering **941** (2020) 012011 doi:10.1088/1757-899X/941/1/012011

Environmental Improvement of Saline and Alkali Soils in the Kamyshlovskaya Valley and Revival of Natural Grassland Efficiency

A I Kuzmin[1], Zh A Tusupbekov[1], N L Ryapolova[1] and V S Nadtochiy[1]

[1] Omsk State Agrarian University named after P.A. Stolypin, 1, Institutskaya Ploshchad, Omsk, 644008, Russia

E-mail: gggkiovr@mail.ru

Abstract. Agricultural use of saline soils is possible only through implementing reclamation measures on their desalinization. The removal of salt from soils is a cost-demanding and time-consuming process. Expenses related to the treatment of saline soils are connected with a desalinization method and include costs of water supply to a saline area and its drainage. A reasonable project cost, as well as efficient water consumption depends on the use of natural environment conditions. The reported paper presents the assessment of water recourses in the drainage area of the valley valley, it is focused on the restoration of water bodies in the Kamyshlovsky log, stabilization of their balance, and environmental improvement; the study suggests economic perspectives of saline and alkali soils. The latter is essential to revive the efficiency of natural grasslands in the valley and its use as a disposal of production and household waste water from surrounding rural settlements.

1. Introduction

Geographically, the Kamyshlovsky log stretches over the territory of two states (the Republic of Kazakhstan and Russia); it crosses the north-east of North Kazakhstan Region in Aiyrtau District, Esil District, Akkayin District, and Magzhan Zhumabaev District and passes into Omsk Region, extending over Isilkulsky District, Moskalensky District, Maryanovsky District, and Omsk District (Figure 1). A territory represents dry lowland with numerous lakes in the valley of the Kanyshlovsky log. A total length of the valley from Bolshoy Tarangul Lake is around 380 km, its width ranges 1.5 to 20 km, and the valley cut into the present day topography is up to 15…17 m.

The surface of the valley is a wavy and poorly dissected lowland plain with numerous shallow holes filled with water – lakes [1]. The valley slopes gently to the north-east like the inclination of underlying rocks that used to be a bank of a fresh-water basin in the Tertiary.

Highest areas are located in the south-west of the valley along the boundary with Akmola Region, where the river Kamysakty has its source at elevations of 200 m and higher. Further north-east, absolute heights decrease to 112 m near Bulaevo, North-Kazakhstan and to 100 m near the urban-type settlement of Maryanovka, Omsk Region.

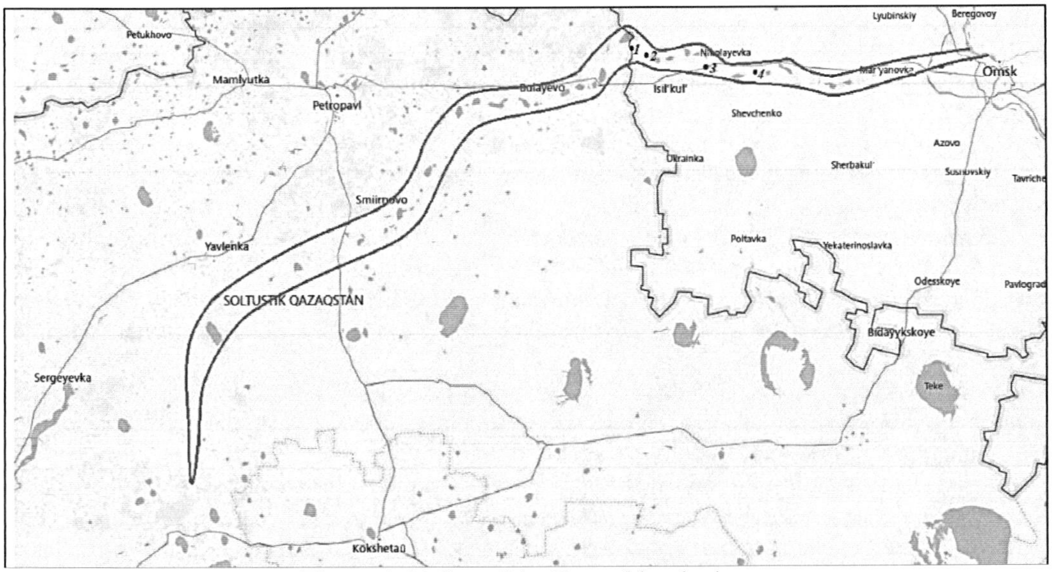

Figure 1. Valley of the Kamyshlovsky log.

To date, the economic use of lands in the Kamyshlovsky log is concentrated on its terraces and slopes and comprises individually used farmland for grain and feed crops or natural pastures grazed by domesticated livestock. The efficiency of agriculturally used lands on these areas is not similar. For instance, arable lands of high terraces differ from zones of the watershed and generally depend on moisture conditions of the areas. Along the Kamishlovsky log there is a strip of saline meadow soils, solod soils, and meadow alkaline soils. The productivity of agricultural fields on terraces creating a gentler slope is lower. It is related to the salinization of soils, which becomes more severe close to the river plain, at a height of 2…3 m from the water edge of lakes they are so poorly productive – even in years with high rainfall amounts a maximum of 3-4 centner/ha (if recalculated to hay) – that they are used for cattle grazing. As for the development of plant communities they have a low value (a content of uneatable species is up to 60%). The reason is salty soils; their percentage is as high as 0.6 to 1.3% in a 0…50 cm layer on sides. Besides it, vast areas on the valley bottom are confined to almost saline and alkaline soils. Typical alkali soils are found around salt lakes.

The efficient land use on this territory is possible provided that desalination actions of saline and alkaline soils are taken; in this case we may increase the productivity of natural grasslands on a total area of above 25 thousand hectares and produce hay of a high quality (15-20 centner/ha). The most optimal measure to reduce a mineralization level of soils is supposed to be the restoration of water bodies aimed at the stabilization of their water-salt balances. Undoubtedly, it will have a positive effect on the efficiency revival of natural meadows in the valley.

2. Hydrographic and hydrologic features of the Kamyshlovsky log

In hydrology, there are mostly bodies with standing water in the valley; a water system represents a chain of lakes with varying sizes and depths, as well as degrees of mineralization. Most lakes in the river basin are without outflow except for the river mouth region with the river Kamyshlovka, which enters the river Irtysh on the left bank at a distance of 1822 km from the river mouth. Water bodies of the rest part have mainly no outlets, only in high-water years some of them may be filled with melt water in spring and turn into running water habitats.

Areas of lake surfaces vary 0.1 to 40 km^2 (Lake Bolshoy Tarangul). A lot of lakes have low and gently sloping shores, and vast areas get overflooded under a high water level rise, some of them get connected with neighboring water bodies at a distance of several kilometers. As a spring flood is over, water levels of lakes fall because of evaporation, many of lakes dry out. Catchment areas of many lakes

IOP Conf. Series: Materials Science and Engineering **941** (2020) 012011 doi:10.1088/1757-899X/941/1/012011

are small – several tens of square kilometers, and the biggest part of their surfaces (50-60%) were ploughed up. Watersheds are not distinct, therefore, poorly recognizable on topographic maps.

Lakes in the valley are mainly salt lakes with a significant area varying 1300 m^2 to 4200 m^2 and a depth of 0.5…1.0 m. A value of total mineralization is in a range 93 to 15 g/l according to the outcomes of survey expeditions carried out by employees of Omsk State Agrarian University in different years. It is related to wide-range water level fluctuations in lakes; in particular, in shallow ones with a large surface area.

Besides salt lakes there are also small but deep (to 2…4 m) fresh-water lakes: Zheltoye, Pokrovka, Presnoye, Pitnoye, Zhelandy, Balykty, etc. A total mineralization level in these lakes varies 0.36 to 3.03 mg/l of chloride and sodium composition with a high concentration of Ca and HCO_3. It relates to the fact that these lakes are deeply cut into the topography and fed by surface streams and subsurface water.

A pattern of fresh-water and salt lakes on the territory is irregular; sometimes they are located close to each other. Fresh-water lakes, in contrast to salt-water lakes, are weeded by marginal aquatic plants, e.g. Phragmites.

For the water balance of a majority of lakes on the territory of interest a lacking outlet is typical, except for infrequent high-water years. The most significant water flowing out was recorded in 1865, when water reached the station of Polydino of the Omsk Railways. Thus, a principal input in the lake balance consists of snowmelt flowing from the catchment area and precipitation on the lake surface.

Rain water with a percentage of 80% in an annual precipitation sum for many years penetrates into soil and evaporates from it without creating a surface runoff because of hot summers and soil dryness. Only lakes with deep basins are fed with ground water.

A main output component of the lake water balance is evaporation on the water surface, which is constant during a year. Within a cold period evaporation on the snow surface is insignificant. Expeditions of State Hydrological Institute carried out in 1956 and 1957 [2] showed an evaporation discharge in the snowmelt time (April) is around 8-10 mm and 550-700 mm within a warm period, respectively.

Lakes are filled when the snowmelt period begins in the first half of April and lasts 10-12 days till the end of April – first decade of Mai. A water level rise in the period of seasonal flood is in a range of 0.3…0.4 m on average and 0.1…0.2 m in low-water years. After filling a high level of water remains in lakes during 2…4 and up to 15…20 days. In the middle of Mai a level of lake water begins to fall gradually and continuously; in summer and autumn it decreases by 0.3…0.5 m on average. The precipitation on the lake surface in a warm year period reduces considerably a water level falling-down rate. The lowest possible level of water in lakes is usually registered in late October. Every year at this time shallow lakes tend to dry out. A level of deeper lakes, which remain filled with water till the end of autumn, hardly changes during winter time. Water bodies with a depth below 1 m freeze solid to the bottom every year.

3. Restoration of Lakes

Lakes in the valley of Kamyshlovka form a complex hydrological system. In the 1920-s they used to be 1.5…2.0 m deep with a water mineralization value of 1.4…6.0 g/l, while nowadays most of them are shallow-water pounds with a depth of 0.4…0.7 m and in certain periods a water mineralization value ranges up to 50 g/l or even 70 g/l. Moreover, many of them dried out completely during drought years (1951…1953, 1974…1975, 1998…2000 and 2010) [3] leaving in the lakebed a 1.0…1.5 cm layer of salts being blown by the wind. Lakes Peschanoye and Severnoye, Maryanovsky District, Lake Kamyshlovoye, Moskalensky District (Elita), etc., tend to dry out every 4-6 years.

A hydrological analysis shows no changes in the total amount of moisture coming to the territory for many years. A water balance of lakes suffers mainly from man-made transformations in the water catchment structure provoked by economic use of lands and road construction of higher categories. Numerous shallow dams in ravines caused the shrinkage of natural water catchment areas, in several cases more than twice. In these conditions a poor natural runoff (7…10 mm) in the residual catchment areas is localized in depressions and fails in reaching lakes. In years with a less changed runoff a degree of water mineralization in lakes remains almost constant 5…15 g/l, in shallow and deep lakes (with a surface up to 20…40 ha) – 1.2…2.0 g/l. Then, a changing hydrological regime led to a dramatically risen water mineralization value in lakes. For instance, a water mineralization level in Lake Solenoye was 11.46 g/l in August 1948 and 14.57 g/l in Lake Krivoye in the same period, whereas in 1953 these values increased to 19.82 g/l and 28.5 g/l, respectively.

The growth of water mineralization is hardly related to a substantial increase of salt resources. A principal factor responsible for mineralization increase in lakes is supposed to be the evaporation on water surfaces, which are too large in changed runoff conditions.

As known, in the Kamyshlovskaya valley there is no runoff into lakes during low-water years with a small amount of rainfall. Naturally, a water level in lakes falls, as well as their volume decreases, at the same time reducing water levels and volumes inevitably provoke the growth (approximately by two times) of water mineralization in lakes and water salinization in the floodplain for ground waters lying near lakes at a depth of 1.0…1.2 m make an important contribution to the evaporation and accumulation of salts in soil. As estimated, a share of salts in the surface of some areas ranges up to 6% and a level of mineralization is 60 g/l.

Therefore, a floodplain part of the valley may be used in economy provided that the desalinization of lands is carried out, and, more importantly, a hydrological regime of lakes is stabilized alongside with their demineralization [4].

Water demineralization in water bodies represents the most complex and challenging issue because it is necessary to drain salt water out of them and supply fresh water from other sources. In this case a reasonable use of fresh water resources from the local runoff makes a problem less difficult and furthers the efficient utilization of the surface runoff.

To assess a possibility to rely on a local runoff in the economic revival of the Kamyshlovka valley we estimated hydrological parameters of the runoff and water balance in sections formed when arranging the valley economy. For this purpose we selected 6 sections within the boundaries of Omsk Region, here, a zero point is coincident with the western boundary of Lake Plovinnoye near the settlement of Pervotarovka, Isilkulsky District. Other sections are determined according to inter-farm dams. The data obtained show insignificant amounts of an annual runoff (Table 1), even under assumption, a natural drainage area is taken into account when calculating a volume of inflow; in fact it is almost twice smaller than values presented (Figure 1).

To evaluate an alternative of stabilizing the water balance of lakes and demineralizing them, a prediction calculation for Lake Piketnoye is conducted (a factual drainage area F_B=56 km^2 and lake surface area f_o=9.25 km^2). As an additional inflow element we considered purified production and household sewage, which also helps demineralize water in the lake.

Table 1. Calculations of annual runoff volumes for different supply (Cs = 2Cv) for selected sections.

№ of sections	Drainage area, km^2	Average long-term runoff, mm	Flow rate Q, m^3/s	Flow rate W, mln.m^3	$Q_{P\%}$, m^3/s			$W_{P\%}$, mln.m^3		
					50	75	80	50	75	80
1	234	7.86	0.058	1.84	0.037	0.014	0.010	1.177	0.443	0.322
2	290	7.86	0.072	2.28	0.046	0.017	0.013	1.459	0.549	0.40
3	1091	7.86	0.272	8.57	0.174	0.066	0.048	5.488	2.066	1.50
4	1202	7.86	0.299	9.44	0.191	0.072	0.052	6.045	2.276	1.653
5	457	10.1	0.146	4.61	0.094	0.035	0.026	2.953	1.112	0.807
6	508	10.1	0.163	5.13	0.104	0.039	0.028	3.282	1.235	0.897

A volume of the lake with a depth of 0.6 m is determined to be W_{O3}=3.7 mln.m^3. For an average mineralization value of water 20 g/l resources of salts in the lake are S_{O3}= 74·10^3 t.

As known, a seriously shrinking real drainage area because of a decreasing runoff has a consequence that significant lake surfaces dry out and a value of their mineralization rises. In such cases it makes sense to reduce a lakescape via deepening. Here, the deepening is planned to 3.0 m and in future the lake can be used for fish-breeding. A surface of the lake will get smaller to 1.23 km^2.

We discussed all possibilities salts can be input into this water body:

- an amount of salts from the drainage area for a layer Y = 7 mm and a value of mineralization q_{CT} = 0.25 kg/m^3 is determined S_{CT}= 95.85 t;
- an annual amount of salts supplied with the precipitation for a layer KX = 320 mm, q_{OC} = 0.05 g/l – S_{OC}= 19.63 t;
- a salt input of Neogenic and Quaternary deposits for a module of subterranean runoff $M_{НЧ}$ = 0.08 l/s per km^2 or 2.52 mm of a runoff layer for a drainage area $F_{НЧ}$ = 54.77 km^2, a value of mineralization $q_{НЧ}$ =1.5 g/l is as much as $S_{НЧ}$ = 207.03 t.
- a salt input of Oligocene waters for a runoff module $M_{ОЛ}$ = 0.01 l/s km^2 or 0.32 mm and a level of mineralization $q_{ОЛ}$ =5 g/l – $S_{ОЛ}$= 87.63 t.

In this case a total amount of salts is calculated:

$$\sum S = S_{\text{ст}} + S_{\text{ос}} + S_{\text{НЧ}} + S_{\text{ол}} = 95.85 + 19.63 + 207.03 + 87.63 = 410.142 \text{ t}.$$

Here, a total volume of water is:

$$\sum W = W_{\text{ст}} + W_{\text{ос}} + W_{\text{НЧ}} + W_{\text{ол}} = (0.3834 + 0.3936 + 0.1382 + 0.0175) \cdot 10^6 = 0.933 \cdot 10^6 \text{ m}^3$$

For the annual evaporation on the surface Z = 700 mm a volume of evaporated water is$W_{\text{ис}} = 0.861 \cdot 10^6 \text{ m}^3$.

Taking into account the evaporation on the lake surface 0.861·10^6 m^3 and a volume of water remained in the lake W_{O3} =3.7·10^6 m^3, a level of mineralization q_{O3} = 20 g/l, an averaged value of mineralization will be q'_{O3}=19.37 g/l in autumn.

Furthermore, in the valley there is a flow of infrabed water in the alluvial apron with an average width 12 m and averaged filtration coefficient 1.5 m/d. In this case a volume of annual ground water input from this part is as high as$W_{\text{ПР}} = 44.6 \text{ m}^3$.

The cut of evaporation losses and involvement of the infrabed flow will support a positive balance of the lake, and a water mineralization value in the restored watercourse will be $q''_{\text{o3}} = 18.644\ g/l$.

In the practice of water use we know that amounts of water taken from water sources should be brought back after certain qualitative or quantitative transformations [5]. Unless water-draining systems are built, there is a problem of sewage recycling. In this case, sewage is stored in special underground tanks – septic system and discharged then. A problem of production and household sewage recycling is to be solved in settlements located along the Kamyshlovsky log, ground waters are frequently overfilled with drained water, what leads to overflooding. In these conditions a use of water bodies located in the valley as an intake structure of purified household sewage can make an additional contribution to the solution of a problem of overflooding. In its turn the discharge of biologically purified water [6] cuts a mineralization value of water bodies. On average, one lake can receive 800 m^3 per day or 292000 m^3 per annum sewage produced by one farm and one farm center; with this volume a value of lake water mineralization in the first year is $q_{1\text{г}} = 17.65$ g/l.

Taking into consideration an average year with all its input and output components of water-salt balance, we can determine a mineralization value of water in the water body. As calculated, a further year reduces a mineralization value of lake water to $q_{2Г}$= 15.36 g/l. Further estimations show that in five years a value of mineralization ranges to $q_{5Г}$=12.23 g/l given that an optimal reorganization plan of water bodies is realized and additional water runoff is taken into account, in 27 years it stabilizes with a value of 4.5…5.5 g/l.

4. Conclusion

The biggest territory of the Kamyshlovsky log is covered with a chain of shallow lakes with no outlets and different values of mineralization; their water levels fall dramatically after seasonal floods and many of them even dry out.

Water bodies with a high level of mineralization are the reason for the expansion of fruitless saline and alkaline soils, their salinity increases near the floodplain.

The environmental improvement and revival of natural grasslands efficiency in the valley are possible if a soil mineralization parameter is reduced through the restoration of water bodies and stabilization of their water balance.

The desalinization of lakes via deepening and reducing their lakescapes, as well as supply of purified sewage from rural settlements is an effective step to improve the condition of soils, although it is a long-term process. A proposed model for Lake Piketnoye is similar to the drainage of 18 lakes located further upstream.

5. Reference

[1] Mokretskaya G P and Kolomin U P 2011 *Interuniversity Bulletin.* **1** 137-41
[2] *Guidelines for the calculation of evaporation from the surface of water bodies* 1969 (Leningrad: Hydrometeoizdat) p 84
[3] *All severe droughts have been observed over the past 40 years* (Omsk hydro meteorological center) http://omsk-meteo.ru/index.php/ru/novosti/118-arkhiv/750-archiv750
[4] Kuzmin A I and Litovchenko A A 1980 Republican Scientific and Technical Conference Scientific *Basics of Land Reclamation in the Creation of Territorial Production Complexes in Siberia* (Krasnoyarsk) pp 369-70
[5] Tusupbekov Zh A, Nadtochiy V S and N L Ryapolova 2019 *Advances in Social Science, Education and Humanities Research, vol. 393. The Fifth Technological Order: Prospects for the Development and Modernization of the Russian Agro-Industrial Sector (TFTS)*
[6] Shlekova I Y and Knysh A I 2018 *Ecology and Industry of Russia* **22(6)** 20-5

Ecological stability of wood species in conditions of root reclamation of salt flats

M Yu Sautkina[1], Yu I Cheverdin[2], A G Akhtyamov[2]

[1]All-Russian Research Institute of Forest Genetics, Breedings and Biotechnology, 105, Lomonosovaave., Voronezh, 394087, Russia
2 Dokuchaev Research Institute of Agriculture in the Central Chernozemic Area, pos. 2 Institutaim. Dokuchaeva, kvartal 5, 81, Talovskii district, Voronezh oblast, 397463, Russia

sautmar@mail.ru; cheverdin62@mail.ru

Abstract. The state of artificial tree planting in 54 years after the meliorative effects on the solonetzic soils (land use, gypsum, manure application) and the transformation of annual soil treatments were studied. There was a general decrease in the forest stand and its quality on saline cells compared to common Chernozem. It is shown that the most salt-resistant tree crops are oak, ash and elm, whose safety is higher compared to other breeds tested in the experiment. Pears, apple trees, and poplars are now almost absent from the forest plantation. In the southern part of the saline section, by the age of 54, an ash-oak plantation with poplar and elm trees was formed (in the aggregate of the first two tiers), and in the Chernozem section, an oak-ash plantation with elm trees was formed. Plantings, according to forest management standards, have a significant difference in the composition of the first tier, in the completeness and totality of crowns and the stock of the stem tree. In the Northern part of the experimental site, the difference in the composition of the first two tiers is more significant, since the proportion of ash in the Chernozem section is 20% lower than in solonetzic soils, where the stand is less dense, full, stock and bonitet than the taxational indicators of the stand in Chernozem.

1. Introduction

In the structure of the soil cover of the steppe part of Russia, a significant place belongs to the solonetzic series of soils [1,2,3,4,5]. The productivity of such lands for agricultural use is at a fairly low level [6,7]. Solonetzic soils in different climatic zones and soil formation conditions differ in terms of forest vegetation. The growth of forest crops depends on many factors, among which the leading ones are the qualitative composition of the soil - absorbing complex of saline soils, salinity features and their water regime.

2. Materials and methods

Forest strip No. 147 (solonetzic cell) is the first experimental land plot, where since 1952 I. A. Yurin, I. F. Porotikov together with scientists of the V. V. Dokuchaev Soil Institute conducted experiments on the reclamation of old solonetzic soils by the forest reclamation method based on the application of Chernozem mass to solonetzic spots in combination with gypsum and manure [1].

In the field, a continuous cross-section of trees with a diameter of 1 cm at chest height was performed on each test area. At the same time, the height of trees was measured, at least 20-25 trees of each breed in three tiers. In some cases, measurements of fallen (windbreak) and dry trees (Tatar maple, ash and poplar) were used.

The diameter of the middle tier tree was calculated using the cross-section areas of each species involved in planting as the weighted average of these indicators. The average height of the tier was determined as a weighted average of the heights of each component of the rock. The average height of the latter was determined based on statistical processing materials and height graphs using the average diameter.

The stock of stem wood was determined by the diameter at a height of 1.3 m and the height of trees for each breed. The total stock is calculated for one hectare of plantings, through which, according to the age of the stand, the average annual increase in the stock was determined, which is a relative indicator of the energy of its growth. The bonitet of the stand was determined by the age and height of trees of the 1st tier.

3. Purpose of work

Based on the analysis of data obtained by researchers for the first time four years of crop growth on different variants of the experiment, to analyze the growth of tree species and the development of stands of plantings on reclaimed salt flats, by carrying out logging operations on variants of former salt flats 54 years after reclamation.

4. Research result

Sections of the studied stands grow on chernozems and solonetzic soils, which were reclaimed by land use with dewatering and gypsum. Based on the data obtained in 1959-1960, the safety of the main tree species (oak, ash and elm) depends on the location of the site in space, and within it on the soil conditions of growth. To date, the plantings belong to the sixth class of age (middle-aged age group) in which the process of forming a stand continues, which affects the preservation of tree species. Thus, in the southern section, the amount of oak in the Chernozem section is 67.2% more than in the solonetzic section, with almost equal density of ash and elm (table 1).

Table 1. Density and safety of wood species by area of experience

Breed	Planted, PCs / ha	Preservation of wood species							
		I plot (C^a)		IIplot (3^b)		IIIplot (C)		IV plot (3)	
		PCs ha	%	PCs ha	%	PCsha	%	PCsha	%
Quercusrobur L.	1560	183	11.7	306	19.6	239	15.3	284	18.2
Fraxinuspennsylvanica	1560	551	35.3	523	33.5	712	45.6	467	29.9
Ulmusparvifolia	1040	55	5.3	50	4.8	61	5.9	145	13.9
Populus balsamifera	1040	6	0.6	-	-	-	-	17	1.6
Acertataricum	3120	111	3.6	150	4.8	95	3.0	56	1.8
Malussylvestris, Pyrus communis subsp. pyraster	520	6	1.2	-	-	-	-	12	2.3
Subtotal	**9520**	**912**	**9.6**	**1029**	**10.8**	**1107**	**11.6**	**981**	**10.3**
Acerplatanoides	-	39	-	139		28		145	-
Cornusalba	-	-	-	-	-	22		-	-
Fraxinus excelsior	-	6	-	-	-	-	-	-	-

Total	**9520**	**957**	**-**	**1168**	**-**	**1157**	**-**	**1126** **-**

Note:
[a]Solonetzic soils not highlighted;
[b]land use in combination with manure and gypsum (Chernozem)

In the Northern part of the solonetzic section, the number of ash trees is 52.1% higher than in Chernozem, where there are more oak and elm trees. The preservation of poplar in the southern section is more related to the black earth section of the forest strip, while in the Northern part of the site it is preserved only in the black earth section. Tatar maple and fruit species were severely thinned during the felling of oak care in 1972. Most of the Tatar maple is dying, and, at present, it is preserved only in the form of dead wood. On both sites, the growth of self-seeding svidina (shrub) and Holly maple, which is part of the second tier of the stand, was noted. Its density in the Chernozem section is 4.4 times higher than the number of trees on solonetzic soils. The common ash we noted in the first section (unmeliorated saltbush) exceeds the growth parameters of downy ash trees, but it is a single tree.

Analysis of the density of preserved tree species shows that in the southern section of the experiment, the number of trees in the Chernozem section is 1.2% higher than in solonetzic soils.

Comparison of the growth parameters of the main tree species growing on the solonetzic soils and chernozems, respectively, sections I and II – III and IV shows that the first-tier oak in the southern half of the experimental site does not have a significant difference in growth, while in the Northern part of the oak on the Chernozem section significantly exceeds the growth rates of trees on the solonetzic soils (table 2).

Table 2. Influence of soil conditions on growth parameterstree species in the first tier of plantings

Region	Number of trees, PCs / ha	The height of the tree (N), m M ± m	Trunk diameter (D), cm M ± m	HCP_{05}on height, m	diameter, cm
			Quercusrobur L.		
I (C*)	122	20.6 ± 0.25	23.9 ± 1.05		
II (3*)	239	21.6 ± 0.22	22.4 ± 0.41	0.7	2.4
III (C)	172	18.0 ± 0.25	19.0 ± 0.66		
IV(3)	239	21.8 ± 0.25	25.1 ± 0.80	0.8	2.2
			Fraxinuspennsylvanica+Fraxinus excelsior		
I (C)	195	19.1 ± 0.37	21.4 ± 0.74		
II (3)	139	18.5 ±0.19	21,3 ± 0.52	0.9	2.2
III (C)	178	16.9 ± 0.26	19,6 ± 0.76		
IV(3)	133	19.5 ± 0.27	20.1 ± 0.90	0.8	2.5
			Ulmusparvifolia		
I (C)	22	17.6 ± 0.28	23.8 ± 0.95		
II (3)	28	20.9 ± 0.26	26.4 ± 0.55	0.8	2.5
III (C)	11	16.5 ± 0.26	24.0 ± 0.72		
IV(3)	39	17.7 ± 0.33	21.1 ±0.39	0.9	1.7

Note:
[a]Solonetzic soils not highlighted;

[b] land use in combination with manure and gypsum (Chernozem)

The growth of downy ash in the southern part of the experimental site does not depend on soil differences, and in the Northern part, trees growing on the Chernozem section (a variant of reclamation) significantly exceed the height of ash on the saline section. And if the number of oak on saline soils without reclamation is 67% less, then ash on these soils has a density 38% higher than on Chernozem sections.

For elm trees in the southern half of the site, the best conditions for growth have been created on land reclamation options, where the parameters of trees significantly exceed those on saline soils. In the Northern part, with a small difference in height, elm on solonetzic soils exceeds the diameter of trees growing on soils with root reclamation. For both sections of the experiment, it is characteristic that the number of trees on solonetzic soils is less than on the sites that include land use with the introduction of manure and gypsum.

Based on the data obtained for the first tier of the stand, which is the main canopy for further development of the plant, we can conclude that the best growth of oak and elm of the first tier on black earth sections and almost equal development of ash on saline differences.

For a more objective judgment about the influence of soil differences on the growth of different tree species, we conducted a comparative analysis of the first and second-tier trees growing on solonetzic and Chernozem sections (with root reclamation).

Summing up the results of research on the growth of tree species on soil differences, we can note a difference in the development of tree species of the first tier, while the indicators of average trees of the first two tiers of the stand are more smoothed.

The development of tree species on different soils had an impact on the taxation indicators of stands of the experimental site. In the southern part, on the solonetzic section by the age of 54, an ash-oak plantation with poplar and elm was formed (in the aggregate of the first two tiers), and on the Chernozem section (reclaimed solonetzic soils) an oak-ash plantation with elm. Plantings, according to forest management standards, have a significant difference in the composition of the first tier, in the completeness and totality of crowns and the stock of stem wood.

In the Northern part of the experimental site, the difference in the composition of the first two tiers is more significant, since the share of ash in the Chernozem section (with reclamation) is 20% lower than insolonetzic soils (without reclamation), where the stand is less dense, full, stock and bonitet than the taxational indicators of the stand on chernozems.

5. Conclusions

In General, according to the aggregate of the main taxational indicators of density, completeness and stock of stem wood, the average formula for solonetzic sections without reclamation (I, III sections) is 5Fraxinus pennsylvanica 4Quercusrobur 1Ulmusparvifolia with a density of 654 trees with a fullness of 16.9 m^2/ha, a wood reserve of 148 m^3/ha and a crown density of 0.65. On the Chernozem sections (solonetzic soils with a limited reclamation), a stand with the composition of 5Quercus robur 2Fraxinus pennsylvanica1Ulmus parvifolia 1Acer platanoideswas formed with the number of trees 782 PCs. / ha, with a fullness of 23.6 m^2/ha, and a wood reserve of 222 m^3 / ha. And the closeness of the crowns is 0.90. On the salt-bearing sections, the average bonit is 7.5 with an annual growth of 2.8 m3/ha of wood, and on the black earth sections, respectively, 1 and 4.1 m^3 / ha. According to the aggregate of the main taxational indicators of plantings, stands on Chernozem soils are 36.9% higher than those on solonetzic soils.

References

[1] N B Khitrov, Y I Cheverdin, I F Porotikov 2009 Solonetzic process in postagrogenic and postameliorative conditions of the Kamennaya Steppe area *Euras. Soil Sci.* **42(11)** 1291-99

[2] N B Khitrov, L V Rogovneva, Y I Cheverdin 2018 Two-dimensional distribution of the properties of verticsolonetz with gilgaimicrotopography in the Kamennaya Steppe *Euras. Soil Sci.* **51(1)**

IOP Conf. Series: Materials Science and Engineering **941** (2020) 012012 doi:10.1088/1757-899X/941/1/012012

1275-87

[3] I N Lyubimova 2018 Postmeliorative transformation of carbonate and gypsum profiles of solonetzic soils in the dry steppe zone *Euras. Soil Sci.* **51(11)** 1288-96

[4] E I Pankova 2015 Salt-affected soils of Russia: solved and unsolved problems *Euras. Soil Sci.* **48(2)** 115-27

[5] N B Khitrov, D I Rukhovivh, N V Kalinina, A F Novikova, E I Pankova, G I Chernousenko 2009 Estimation of the areas of salt-affected soils in the European part of Russia on the basis of a digital map of soil salinization on a scale of 1: 2.5 M *Euras. Soil Sci.* **42(6)** 581-90

[6] N B Khitrov 2004 The choice of diagnostic criteria to judge the development of the solonetzic process in soils *Euras. Soil Sci.* **37(1)** 12-23

[7] I N Lyubimova, A F Novikova 2016 Changes in the properties of solonetzic soil complexes in the dry steppe zone under anthropogenic impacts *Euras. Soil Sci.* **49(5)** 581-90

IOP Conf. Series: Materials Science and Engineering **941** (2020) 012013 doi:10.1088/1757-899X/941/1/012013

Ecological and physiological feature of some microelements and their concentration in vegetable products

K Sidorova[1a], O Dragich[1,2], N Shvets[1b], A Bukin[1], N Ryabova[2], E Klyushnikova[2] and O Kochetova[3]

[1]Northern Trans-Urals State Agrarian University, 7, Respubliki St., Tyumen, 625003, Russia
[2]Tyumen Industrial University, 38, Volodarskogo St., Tyumen, 625000, Russia
[3]Perm Institute of the Federal Penitentiary Service of Russia, 125, Karpinskogo St., Perm, 614012, Russia

E-mail: [a]sidorova.clavdija@yandex.ru, [b]tasha-04-shvets1991@mail.ru

Abstract. In this work, providing human body with essential food components was assessed and the physiological role of chemical elements – iron and manganese – was considered. They are among key nutrients essential for human body. Iron plays an important role in the functioning of immune system; it is necessary for the full functioning of brain and nervous system. Iron deficiency can cause iron deficiency anemia which leads to a slowdown in children's mental and psychomotor development. Manganese has an effect on the functioning of nervous system; it participates in exchange of impulses between nerve cells and regulated reproductive function of the body. This element promotes insulin secretion and metabolism of fats and carbohydrates, it is necessary for blood formation, metabolism, functioning of musculoskeletal system, immunity, hair growth, normal skin, and is an antioxidant. Microelementoses with metabolic disorders of substances such as iron and manganese can be caused by deficiency, excess or redistribution between the reserve and functional funds of microelements under consideration. Based on this, we can summarize that iron and manganese are indispensable microelements for human body; their deficiency results in undesirable pathological metabolic disorders. Considering the relevance of this problem, the analysis of the content of chemical elements in samples of vegetable crops was carried out. These results a high iron content in beets and potatoes what indicates their nutritional value; sufficient consumption of them covers the daily requirements for manganese and iron.

1. Introduction

Mineral substances are important for human body; first of all, they performs various physiological functions. Being nutritional components they are of great concern for metabolism, biochemical processes, formation and building of bone tissue. Macro- and microelements are necessary to maintain acid-base balance of the body, to create a certain concentration of hydrogen ions in tissues and cells, interstitial and intercellular fluids, as well as to give them osmotic properties that provide normal metabolism [1, 2].

The aim of this research was to substantiate the biological significance of iron and manganese for human body and to establish their level in several vegetable crops.

Balanced diet is a pressing issue, since the most part of our country's population lack complete protein, as well as the most important vitamins and minerals that have a direct effect on human health.

Currently, many people are faced with the need for increased amounts of iron and manganese. Iron deficiency is experienced by a large number of rural and urban populations, especially women of childbearing age [3, 4, 5]. It is possible to manage such deficiency with the help of iron-containing products of animal or vegetable origin, or with special additives to the diet [2, 6, 7]. Main sources of manganese for humans are food and water.

Iron and manganese are among the key nutrients vital to the human body. Among a large number of microelements, only nine are indispensable, and their imbalance can lead to the onset of clinical symptoms. They also include copper, magnesium, selenium, chromium, iodine, zinc, cobalt (as a component of vitamin B12) and molybdenum [5, 8].

Manganese has an effect, first of all, on the functioning of nervous system, since it directly influences on the exchange of impulses between nerve cells. It regulates reproductive functions, promotes insulin secretion and metabolism of fats and carbohydrates. Manganese is especially required for blood formation, metabolism, functioning of musculoskeletal system, immunity, hair growth, normal skin, and is an antioxidant. In human body, this microelement is concentrated mainly in bones, brain, liver, pancreas and kidneys [2, 6].

Average consumption of manganese for adults is set at 3–5 mg per day (according to different sources), for children this rate is much lower and should be from 0.5 to 2 mg depending on age [5].

Iron is essential for immune system functioning. An organism with iron in sufficient quantity is able to actively fight infectious diseases [1, 9]. In addition, the rate of wound healing depends on iron [2, 7].

Iron content in human body is not very high – about 5 g [5]. Deficiency of this micronutrient is most often associated with its low content in food and lowered absorption in gastrointestinal tract. Iron takes part daily in many physiological processes, therefore, it is necessary to maintain its quantity and regularly replenish it, first of all, through the consumption of foods high in iron.

Iron deficiency can cause iron deficiency anemia which leads to a slowdown in children's mental and psychomotor development [2]. In addition, iron is necessary for the full functioning of brain and nervous system. Thyroid gland produces its hormones in the required amounts in the presence of this microelement [1].

For a daily balanced diet, one needs to combine products in the right manner. Various elements interacting with each other can both enhance each other's beneficial effect and neutralize it. Thus, iron is most effectively absorbed in intestine in the presence of vitamins B12 and C, as well as folic and other organic acids [8, 10]. For the same reason, meat dishes are best combined with a side dish of boiled cauliflower or potatoes.

It is unacceptable to take iron with calcium which prevents its proper absorption. A similar effect is caused by tannins and oxalic acid. Iron is hardly absorbed in combination with tea or coffee.

Manganese deficiency leads to depression of nervous system, sleep disturbance and anxiety, dizziness. Consequences of the lack of manganese can also include different problems with fetus development, anemia, inability to exercise reproductive function, growth and development retardation [9, 11]. Iron deficiency increases the risk of developing Alzheimer's disease, dementia and other diseases caused by impaired brain activity. As a preventive measure, mineral and vitamin supplements can be added to the main sources of iron [10]. Therefore, in order to provide body with the right amount of manganese for its normal functioning, the constant supply of this microelement is necessary. So, its greatest amount is contained in the products of plant origin: cereals, legumes, greens, horticultural crops, vegetables, nuts (Table 1) [1, 5, 12, 13, 14]. Among products of animal origin, meat, liver and kidneys should be noted.

Table 1. Average content of manganese in food.

Product	Manganese in 100 g, mg	Percent of required daily intake, %
Wheat bran	10–15	300–500
Pine nut	8–10	250–400
Oat bran	5–7	100–230
Oats (grain)	5–6	100–200
'Hercules' oatmeal flakes	3–5	60–100
Rice (grain)	3–6	60–200
Walnut	1–3	20–100
Buckwheat groats (unground)	1–3	20–100
Beans (grain)	1–2	20–65
Basil (greens)	1–1.5	20–50
Spinach (greens)	Under 1	Under 35
Meat (beef, lamb, veal, bacon, poultry, eggs)	0.5–50	10–1670
Fish (salmon, cod, crabs, crayfish)	0.5–2	10–65
Kidney	Under 1	Under 35

2. Equipment and devices used in studies

When assessing the effects of iron and manganese on human body, it is necessary to learn more precisely about the content of this microelement in certain foods. Therefore, we performed studies to determine the quantitative content of manganese in vegetable crops grown in household plots of the city of Tyumen and Tyumen Region.

Laboratory studies were conducted in the conditions of the Agro biotechnological Center of the Northern Trans-Urals State Agrarian University. According to common methods, samples of root crops grown in household plots of the city of Tyumen and Tyumen Region were taken.

All analyzes were performed in accordance with the procedures established by GOST RF (sampling of vegetables and potatoes according to the Standard SEV 4295–83 and ST SEV 4299–83).

Sample preparation for determining the content of toxic elements was performed using mineralization according to GOST RF 26929–94; GOST RF 31218-2003 (ISO 6498: 1998); GOST RF 32343–2013 (ISO 6869:2000).

The content of toxic elements and heavy metals was defined with the help of ContrAA 300 Atomic Absorption Spectrometer and software package.

Processing of obtained data was performed using Microsoft Excel 2007 for Windows.

3. The results of the study and their discussion

As a result of performed studies, the following data were obtained which are shown in Figure 1. In accordance with obtained data, beets and potatoes have the highest iron content among all crops: 6.7 and 6.2 mg $\cdot kg^{-1}$, respectively. Onion contains 48% less iron than beets and 2 times less than potatoes. This iron content in beets and potatoes corresponds to 67% and 61% of physiological iron requirements for adults (10 mg$\cdot$ day^{-1}). It is important to remember that not all iron will be absorbed in the body, so it should be supplemented with iron from foods of animal origin.

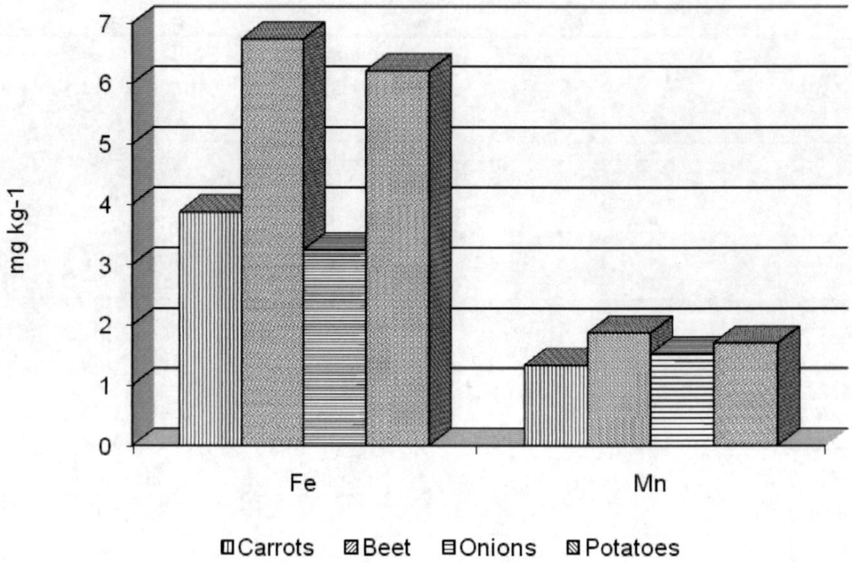

Figure 1. The content of microelements in vegetable crops, mg·kg^{-1}.

In accordance with obtained data, beets and potatoes have the highest iron content among all crops: 6.7 and 6.2 mg· kg^{-1}, respectively. Onion contains 48% less iron than beets and 2 times less than potatoes. This iron content in beets and potatoes corresponds to 67% and 61% of physiological iron requirements for adults (10 mg· day^{-1}). It is important to remember that not all iron will be absorbed in the body, so it should be supplemented with iron from foods of animal origin.

The highest manganese content among studied crops was also found in beets and potatoes and amounted to 1.87 and 1.7 mg· kg^{-1} what indicates the accumulating capacity of these crops. Beets contain 40% higher manganese than carrots and this makes up to 94% of physiological daily intake of manganese (2 mg· day^{-1}).

4. Conclusion

Thus, we can conclude that iron and manganese are indispensable microelements for human body; their insufficient content will result in undesirable pathological metabolic disturbances.

In addition, it should be borne in mind that the Ural Federal District refers to regions with a high content of manganese and iron in soil and water that contributes to their accumulation in plants and animals [5, 10]. Particular attention should be paid to drinking water consumption, since systematic intake of manganese with drinking water in doses exceeding MCL (0.05 mg L^{-1}) [11] increases mortality rate, especially in children, and provokes complications of pregnancy and childbirth [2, 5]

Based on performed studies, high iron content in beets and potatoes was revealed that indicates the nutritional value of these crops; sufficient consumption of them will result in cover the daily requirements for these microelements. Therefore, the studied vegetables should be included in the diet as an additional source of microelements.

References

[1] Baltussen R, Knai C, Sharan M 2004 Iron fortification and iron supplementation are cost-effective interventions to reduce iron deficiency in four sub regions of the world *J Nutr* **134** 2678–84
[2] Dragich Olga A, Sidorova Klavdia A, Sidorova Tatyana A, Naimushina Alla G, Zharkova Marina A 2019 Analysis of Features of Adaptive Reactions of Students in Urban Environment

Available *Proceedings of the International Conference Topical Problems of Philology and Didactics: Interdisciplinary Approach in Humanities and Social Sciences* **312** 108–111

[3] Inácio M, Neves O, Pereira V, Eduardo Ferreira da Silva 2014 Levels of selected potential harmful elements (PHEs) in soils and vegetables used in diet of the population living in the surroundings of the Estarreja Chemical Complex (Portugal) *Applied Geochemistry* **44** 38–44

[4] Kahlcova II, Kahleova II, Levin S, Barnard N 2017 Cardio-Metabolic Ben-efits of Plant-Based Diets *Nutrients* **9(8)** 242–246

[5] Korobeynikova T V 2018 Vegetarianism and micronutrients. Microelements in medicine *Russian society of medical elementology* **19(2)** 34–40

[6] Armini V 2019 Use and improvement of ready-to-use therapeutic food (RUTF) formulas in the management of severe acute malnutrition *Encyclopedia of Food Security and Sustainability* **2** 344–352

[7] Dragich O, Sidorova K, Pashayan S, Glazunova L, Cheremenina N, SidorovaT, Ivakina E, Zobnina S, Domatsky V, Tatarnikova N 2018 Features of Adaptation of Urban and Rural people in the training process in higher educational institutions *Journal of Physical education and Sport* **18** 1108–17

[8] Zhu L, Glahn R P, Chi K Y, Miller D D 2006 Iron uptake by Caco-2 cells from NaFe EDTA and FeSO4: Effects of ascorbic acid, pH, and a Fe(II) chelating agent *Journal of Agricultural and Food Chemistry* **54 (20)** 7924–28

[9] Hurrell R F, Cook J D 1990 Strategies for iron fortification of foods *Trends in Food Science & Technology* **1** 56–61.

[10] Skipin L, Gaevaya E, Zaharova E, Petukhova V, Sidorova K 2016 Biogeochemistry of heavy metals in trophic chain in terms of the south of Tyumen region *Procedia Engineering* **165** 860–868.

[11] Hurrell R F, Reddy M, Cook J D 1999 Inhibition of non-haem iron absorption in man by polyphenolic-containing beverages *British Journal of Nutrition* **81(4)** 289–295

[12] Wenlock R W, Buss D H, Dixon E J 1979 Trace nutrients. Manganese in British food *British Journal of Nutrition* **41** 253–261

[13] Suchánková M, Kapounová Z, Dofková M, Ruprich J, Blahová J, Kouřilová I 2015 Selected fruits and vegetables: comparison of nutritional value and affordability *Czech J. Food Sci.* **33(3)** 242–246

[14] Kostova D, Kanazirska V and Kamburova M 2008 A comparative analysis of different vegetable crops for content of manganese and molybdenum *Agronomy Research* **6(2)** 477–488

IOP Conf. Series: Materials Science and Engineering **941** (2020) 012014 doi:10.1088/1757-899X/941/1/012014

Influence of fish pond culture on the fish fauna of Ladoga Lake

N P Milyanchuk, N V Ilmast, O P Sterligova

Institute of Biology KRC RAS, Petrozavodsk, 185910 Russia

e-mail: ilmast@mail.ru

Abstract. The species composition and biological indices of Ladoga Lake fish near trout farms were studied. Water bodies with commercial rainbow trout production were found to be polluted mainly by nutrients and fish metabolic products that activate eutrophication processes. The functioning of a trout farm was shown to affect the fish population structure of the lake. Fish with spring spawning are preferred. Roach, perch and ruff now predominate in the fish farm area. Fish aggregates in the lake differ considerably in biomass from those in open lake zones. Fish density indices near the fish pond lines were observed to increase almost twofold. Local fish species aggregate near the fish ponds because they eat trout food. The linear and weight growth rates of roach, perch and ruff near the fish ponds were shown to be much higher than those in the remote zones of the lake, because the above fish also eat the leftovers of trout food washed out of the fish ponds. Commercial trout production requires integrated monitoring for assessment of the current condition of the water bodies affected by fish farming.

1. Introduction

Aquatic ecosystems are known to be influenced by both climatic (variations in water regime and temperature) and anthropogenic (commercial fishing, pollution, introduction of alien species, etc.) factors. The influence of economic activities on natural ecosystems is steadily increasing [1, 2, 3, 4, 5, 6]. One type of anthropogenic influence on water bodies is artificial breeding of aquatic organisms.

The need to develop aquaculture, primarily fish farming, is due to a decline in the production of fish and other aquatic organisms in sea water and the unstable condition of the fish stock in freshwater bodies. In the Republic of Karelia, the commercial production of the rainbow trout *Parasalmo mykiss* (Walb.) was launched in the 1980s. By now, its annual production rate has increased to over 20 000 t. Karelia is a national leader in commercial rainbow trout farming (about 70%); annual trout production is expected to increase to 30 000-35 000 t.

Success in fish culture is provided by abundant water resources, a favourable climate, a well-developed transportation network and skilled staff. As fish prices are moderate, high-quality profitable commercial products from fish ponds are made available over a short time (1.5 years). Over 60 trout farms, located mainly in freshwater zones, now function in Karelia.

It is known that the construction of a fish farm is equal in the pollution rate of natural water bodies to the putting in operation of a small factory or plant. Large amounts of food, metabolic products and medicinal preparations from trout farms are discharged into Karelia's lakes. Therefore, a considerable increase in commercial trout production in the northern region may result in their rapid eutrophication. The increasing use of natural water bodies for fish culture disturbs the hydrological, hydrochemical and hydrobiological regimes of aquatic ecosystems [7]. The ongoing functioning of fish farms also affects the aboriginal fish fauna of water bodies. Yet, the scope of scientific research in this field is very small.

The Ladoga is Europe's largest freshwater lake. Natural conditions were conducive for the formation of a lake ecosystem with high-quality water, but in the early 1960s signs of man-induced lake eutrophication were observed.

The Ladoga Lake watershed area is known as a highly developed economic zone with more industrial plants than in other parts of Russia. In the past decade, human economic activities in the watershed area have become more extensive. Therefore, nitrogen and phosphorus concentrations in the water bodies of northern Priladozhye have increased. Eutrophication is most obvious in the most heavily polluted lake bays. There are now several trout farms in the skerries. Active fish farms are local, but they contribute markedly to eutrophication.

The goal of our project is to study the structure and dynamics of Ladoga Lake's fish population affected by large-scale commercial fish farming.

2. Materials and methods

Fish samples were taken from Jakimvaara Bay of Ladoga Lake in 2016–2018. Samples were collected in and away (about 5 km) from fish ponds. Experimental fishing was done using 30 m long and 1.8 m high nets (15-40 mm mesh). The nets were placed at various depths and were exposed for 12 hours. The fish samples were processed by standard methods [8, 9]. The length, body mass, sex and gonad maturity of the samples were estimated. To determine the age of the fish, the scales, gill covers (operculum) and otoliths were sampled and analyzed. Abundance indices, understood as the size of catch per unit of fishing effort, were used for comparing fish abundances in various fishing zones.

Ladoga Lake (61°09′ N, 31°20′ E), located in the Neva River watershed, is part of the Baltic Sea basin. Its water area makes up 40% of Karelia. Depths in excess of 100 m are typical of the northern Ladoga. An important fish farming lake in NW Russia, the Ladoga is fed from the watershed basin covering an area of 258 km^2 [10]. The Ladoga can be described as a mesotrophic type of lake, as indicated by the quantitative characteristics of plankton and benthos. The main morphometric characteristics of Ladoga Lake are shown in Table 1.

Table 1. Main morphometric characteristics of Lake Ladoga

Characteristic	Value
Surface area of the lake, km^2	17700
Islands area, km^2	457
The volume of the lake, km^3	910
Average depth, m	51
Greatest depth, m	230
Catchment area, km^2	258300

3. Results and discussion

The fish fauna of Ladoga Lake consists of 43 species. Whitefish, vendace, smelt, perch, roach, pike-perch, bream, pike, ruff and burbot are of special commercial value.

The current condition of commercial fishing is controlled by both natural and anthropogenic factors. The former are related to a stage at which Ladoga Lake's production potential has decreased. Natural factors are accompanied by the influence of human activities (increased amounts of nutrients and toxic substances discharged into the lake, the poor condition of spawning grounds, the construction of hydropower plants, illegal fishing, etc.). Catches of practically all commercial species are now observed to decline. Some of the species become rare and endangered, while others become less abundant, although they remain commercially valuable.

The condition of the fish fauna in of northern Ladoga Lake (Jakimvaara Bay) in the fish farm area was assessed. Analysis of experimental catches has shown that the fish population of the bay consists

IOP Conf. Series: Materials Science and Engineering **941** (2020) 012014 doi:10.1088/1757-899X/941/1/012014

of 5 species of 4 families: vendace, smelt, roach, perch and ruff. Roach, perch and ruff were most abundant in the fish pond area.

Roach (*Rutilus rutilus*) is an aggregative, abundant, ubiquitous fish. Its abundance is controlled mainly by natural mortality and to a lesser extent by catching. Ladoga roach becomes sexually mature in the fourth year of life, when it has an average length of 13 cm and a mass of 40 g. The age composition of control catches was represented by individuals of 2+ to 10+; 5+ to 10+ fish predominated (over 50%). The fish varied in length from 10.5 to 21.5 cm and in mass from 15 to 180 g (Fig. 1). Comparative analysis of roach growth has shown that its linear-weight indices in the trout pond area are much higher than those of roach from the remote parts of the lake (control).

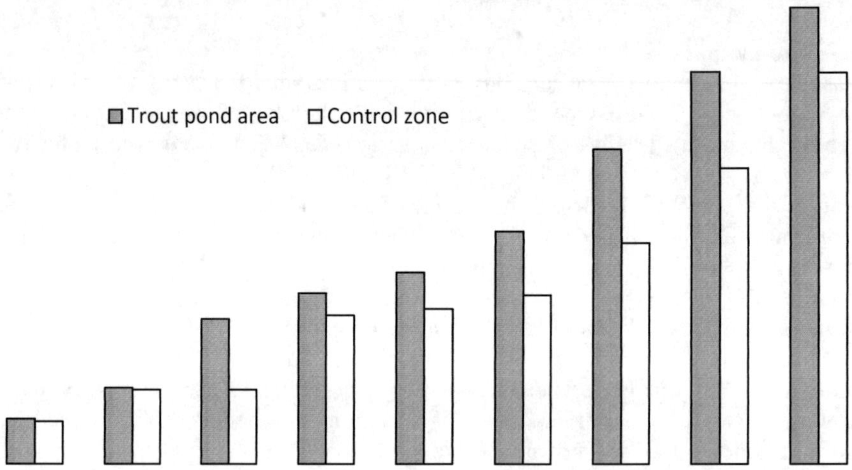

Figure 1. Weight growth of the roach of Lake Ladoga (the y-axis is the mass, g; x-axis – age in years)

Perch (*Perca fluviatilis*) is widespread in Ladoga Lake. It becomes sexually mature in the third year of life, when it has a length of 11-12 cm and a mass of 20-30 g. The age composition of perch catches in the fish pond area consisted of 2+ to 9+individuals; fish in the third and fifth year of life prevailed (about 70%). The samples had a length of 10.5 to 24 cm and a mass of 22 to 230 g (Fig. 2). Comparative analysis of perch growth shows that age 2+ to 6+ fish in the fish pond area display elevated linear-weight indices; fish older than 7 years do not show valid differences in growth.

Ruff (*Gymnocephalus cernuus*) is an aggregative bottom fish which occurs near the shore and in the central zone of the lake. It becomes sexually mature at the age of 2-3 years, when it has a length of 5-8 cm and a mass of 6-7 g. The age composition of control catches consisted of four age groups (1+ - 4+); four-year-old ruff predominated (около 50%). The fish varied from 5 to 11.9 cm in length and from 3 to 26 g in mass (Fig. 3). Analysis of ruff growth has indicated that the samples from the trout pond area display elevated linear-weight indices. For example, age 1+ samples from the fish pond area had a length of 6.6 cm and those from the control group 5.1 cm; age 2+ samples had a length of 8.8 cm and 7.2 cm; aged 3+ samples 9.9 cm and 8.9 cm; and aged 4+ samples – 11.3 cm and 11 cm, respectively.

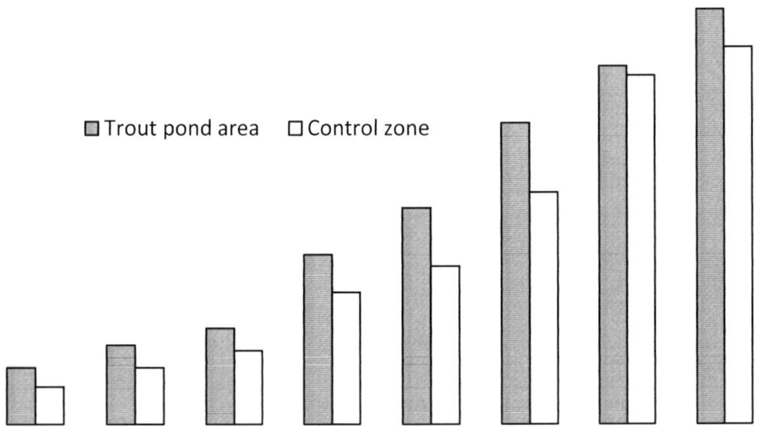

Figure 2. Weight growth of the perch of Lake Ladoga (the y-axis is the mass, g; x-axis – age in years)

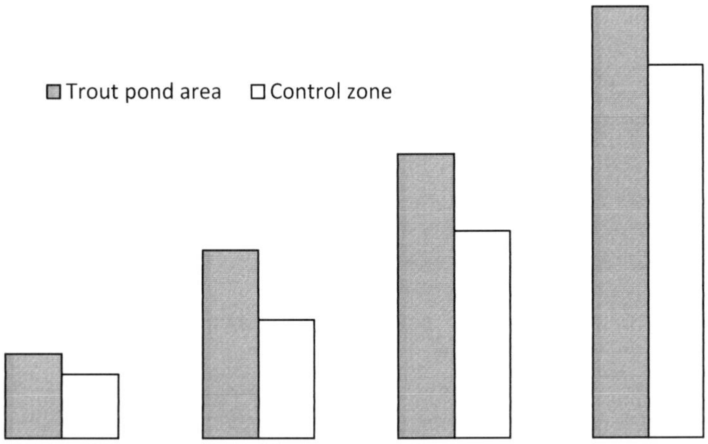

Figure 3. Weight growth of the ruff of Lake Ladoga (the y-axis is the mass, g; x-axis – age in years)

To describe the space-time fish distribution pattern, density indices, understood as catches per unit of fishing effort, were used. A unit of fishing effort is understood as catch (individuals) per unit of time (one hour) per unit area (1000 m^2) of net cloth. Analysis of the fish density index in the trout pond area was 0.867 inds./$m^2 \times h \times 10^3$ and that in the control zone 0.493, respectively.

4. Conclusions

Thus, the analytical results of our studies have shown that the fish population of the trout pond line area in Ladoga Lake is dominated by spring spawning carp and perch species (over 90%). The functioning of the trout farm in the lake is indicated by increased nutrient concentrations; this, in turn, results in more active eutrophication caused mainly by the food and metabolic products of fish. The studies have revealed substantial differences in the biomass of fish aggregates that vary with distance from the farm. Accessible food discharged continuously from the fish farm is the main factor responsible for the formation of fish aggregates near the nursery ponds. The linear and weight growth rate of fish near the nursery ponds was found to be much higher than that of fish from the remote zones of the lake. This is due to extra food, trout food leftovers, washed out from the nursery ponds and made available to the fish. The functioning of the trout farms was shown to affect the hydrochemical and hydrobiological regimes, as well as the fish population of Ladoga Lake. Fish that spawn in spring and eat mainly benthic organisms (roach, perch and ruff) are preferred. Analysis of our results shows that commercial trout farming requires integrated monitoring for assessing the current condition of the water bodies used.

Acknowledgements
The paper is based on research carried out with the financial support of State Program (Projects No 0218-2019-0081, No 0221-2018-0002) and the Russian Foundation for Basic Research (Project No 18-04-00163).

References

[1] Heger T and Trepl L 2003 Predicting Biological Invasions *Biol. Invasions* **5** 313-321
[2] Reshetnikov Yu S, Lukin A A. 2006 Modern state of the diversity of Coregonids from Onega Lake and the problems of their species identification *Journal of Ichthyology* **46** (9) 694-708 doi: 10.1134/S0032945206090025
[3] Sterligova O P and Ilmast N V 2009 Species-colonizers in the water systems of Karelia *Journal of Ichthyology* **49** (4) 331-338 doi: 10.1134/S0032945209040055
[4] Kriksunov E A, Bobyrev A E and Burmenskii VA 2011 Resource availability and its role in development of invasion processes *Biol. Bull. Rev.* **1** (1) 57-70 doi: 10.1134/S207908641101004X
[5] Dgebuadze Yu Yu 2014 Invasions of alien species in Holarctic: some results and perspective of investigations *Russian Journal of Biological Invasions* **5** (2) 61-64 doi: 10.1134/S2075111714020039
[6] Ilmast N V *et al* 2019 Ecological state of aquatic biocenoses in the streams of the Kola basin, Barents Sea *IOP Conf. Ceries: Earth and Environmental Science* **302** 012022 doi: 10.1088/1755-1315/302/1/012022
[7] Kuchko Y A, Ilmast N V and Kuchko T Y 2018 Characteristics of Zooplankton in Syamozero Lake (South Karelia) Based on the Results of Long-Term Monitoring *Biology Bulletin* **45** (5) 497-504 doi: 10.1134/S1062359018050096
[8] Ilmast N V and Sterligova O P 2016 Results of introduction of new fish species into Lake Munozero (southern Karelia) *Russian Journal of Biological Invasions* **7** (4) 333-339 doi: 10.1134/S2075111716040032
[9] Sterligova O P and Ilmast N V 2017 Population dynamics of invasive species of smelt *Osmerus eperlanus* in Lake Syamozero (South Karelia) *Journal of Ichthyology* **57** (5) 730-738 doi: 10.1134/S0032945217050162
[10] *Lakes of Karelia. Reference book* 2013 Petrozavodsk: Karelian Research Centre, Russian Academy of Sciences

IOP Conf. Series: Materials Science and Engineering **941** (2020) 012015 doi:10.1088/1757-899X/941/1/012015

Ecologically safe device technologies of road covering with the using technogenic materials

S O Potapova[1], I I Shepelev[1,3], E N Eskova[1], N N Bochkov[1,2], A M Nemerov[1]
[1] Krasnoyarsk state agrarian university, 660049, Krasnoyarsk, Mira Ave., 90, Russia
[2] LLC «Dolomit», 660049, Krasnoyarsk, Dubrovinsky st., 112, Russia
[3] E-mail: Ekoing@mail.ru

Abstract. The influence of nepheline sludge on the soil of a roadside strip when used in the device of road coverings is studied. Chemical analysis of soil samples taken from the roadside strip of a highway constructed using a base made of nepheline sludge was performed. It is shown that the change in the content of calcium-the main element of the sludge at a distance from 5 to 100 m from the road did not exceed its value in the background sample, and the content of exchangeable aluminum in the soil of the analyzed areas was at a very low level-less than 0.5 mg/kg and practically did not affect its content in the soil at all sampling points from 0 to 150 m from the road. Based on the conducted ecotoxicity studies using single-and dicotyledonous plants, the possibility of using nepheline sludge as a component of road covering in the form of a nepheline-crushed stone mixture has been established and the absence of toxicity in the analyzed soil samples at various distances from the roadside strip has been proved.

Introduction.

In the context of increasing the number of vehicles and the increasing anthropogenic load on the environment, the problem of expanding the raw material base for road construction by maximizing the use of waste from various industries is of particular importance [1,2,3]. In Russian and foreign practice of road construction, local construction materials and soils reinforced with various binders are beginning to be widely used. The bases of road coverings made of materials reinforced with a binder are not only strong, but also economical. Currently, one of the most urgent tasks in the field of ecological safety is research on the creation of new effective composites based on non-toxic technogenic raw materials, characterized by reduced cost and meeting modern requirements for road construction materials [4,5]. The problem of industrial waste disposal is acute throughout the civilized world. Lime, metallurgical slags, waste from mining and volcanic production, and other industrial wastes are widely used in Russia and abroad instead of expensive cement [6-8]. But before using technogenic materials in the device of road coverings it is necessary to study their harmful effect on the roadside soil. According to the results of research conducted at the Research institute of water management in Russia (Yekaterinburg), it was found that the greatest pollution is exposed to a 100-meter strip along roads with high traffic intensity. The main goal of road construction and reconstruction is not only to improve the capacity of vehicles, isolate transit and cargo traffic from residential development, but also to improve the development level of the surrounding area and the possibility of improving the environment. Technical solutions for road construction and reconstruction include other measures that ensure drainage, ecological requirements, and the use of ecologically safe materials for road coverings. Therefore, the problem of ecological pollution of roadside strips is very complex and it is necessary to assess the impact of applied construction materials on the soil objects of the adjacent territory in areas of heavy traffic [11]. In this regard, it should be noted that the impact of materials used in road construction on the environment has not yet been sufficiently assessed. This is especially true for the contamination of the roadside soil and adjacent water basins when using road construction materials of technogenic origin.

One of the promising wastes in Siberia for use in the device of road coverings is nepheline sludge (NS) of JSC "RUSAL Achinsk" [9]. Earlier in the 80-90-ies of the last century, there were attempts to use nepheline sludge in the construction of roads in Siberia, but there was no large-scale use of slime and other wastes of alumina production [10]. In recent years, the volume of nepheline sludge placed on the sludge maps of JSC "RUSAL Achinsk" exceeds 100 million tons and requires immediate solutions to the problem of using this waste.

Results and discussion

The conducted sanitary and epidemiological examination of nepheline sludge for radiation safety indicators showed that the specific activity of natural radionuclides contained in nepheline sludge does not exceed the hygienic standard established for industrial waste used in the manufacture of construction materials of the 1st class and, accordingly, is equal to 66.9 Bq/kg with a standard of 370 Bq/kg [5]. Using x-ray phase analysis, the mineralogical composition of nepheline sludge placed on the sludge map of JSC "RUSAL Achinsk" was studied (table 1).

Table 1. Mineralogical composition of nepheline sludge

Component	$2CaO\cdot.SiO_2$	$Na_2O\cdot Al_2O_3\cdot SiO_2$	$2CaO\cdot Al_2O_3\cdot SiO_2$	$Na_2O\cdot Al_2O_3$	$CaFe_2O_4$, Fe_2O_3	$CaCO_3$
Content, %	80.0-85.0	1.0-3.0	0.5-1.0	0.5-1.0	1.5-4.0	1.0

As can be seen from table 1, the main mineral composition of the nepheline sludge placed on the sludge storage in samples is dicalcium silicate - β-$2CaO\times SiO_2$ (more than 80 %).

It was also found to contain sodium and calcium aluminosilicates ($Na_2O\times Al_2O_3\times SiO_2$, $2CaO\times Al_2O_3\times SiO_2$), iron compounds in the form of ferrites and oxides ($CaFe_2O_4$, Fe_2O_3), calcite ($CaCO_3$) and sodium aluminate ($Na_2O\times Al_2O_3$). Up to 5% of the nepheline sludge contains: calcium magnesium silicates $Cao\times MgO\times SiO_2$, calcium hydrogranates $3CaO\times Al_2O_3\times SiO_2\times 6H_2O$, calcium aluminates $CaO\times Al_2O_3$ [9]. The presence of dicalcium silicate and calcium hydrosilicates in the nepheline sludge confirms the conducted differential thermal analysis.

The study of the granulometric composition of nepheline sludge sample showed a relatively small fraction content of more than 2 mm, which makes it easy to use it in road technologies almost without preliminary preparation. The granulometric composition of nepheline sludge of JSC "RUSAL Achinsk" is shown in table 2.

Table 2. Granulometric composition of nepheline sludge

Fractionsize, mm	Less 0.01	0.01-0.05	0.05-0.1	0.1-0.25	0.25-0.5	0.5-1.0	1.0-2.0	More 2.0
The contents of the fraction, %	1.2	5.7	8.1	20	26	26.6	11.1	1.3

The use of nepheline sludge in road coverings is currently carried out in the form of a nepheline-crushed stone mixture, which is prepared at the production area, then it is laid in the base of the road and rolled with rollers [12]. To simulate the process of forming the base of road coverings, experimental samples were prepared in the form of cubes of a nepheline-crushed stone mixture. When making the control composition of the road mixture, crushed stone from the Masul mine was taken as a large aggregate of the mixture, which meets the regulatory requirements. For forming, the molding process forming with pressing of a nepheline-crushed stone mixture at a pressure of 15 MPa was used. The method provided for holding samples of pressed road mixture for 28 days, after which water extracts from fresh nepheline sludge of current production and its mixture with crushed stone were studied. The analysis showed that in the water extract from fresh nepheline sludge, when crushed stone is added to it, the content of potassium (in 1.6 times) and the content of aluminum (in 3 times) is significantly reduced during the formation of road mixtures. Thus, the model studies have shown that various industrial wastes, such as nepheline sludge and overburden from the Masul limestone mine, can be effectively used as cheap raw materials for obtaining construction material for the construction of road coverings. However, the use of nepheline sludge as a component of road coverings can become effective only in the absence of their negative impact on the soil layer of the roadside strip adjacent, which was considered necessary to study in the course of pilot field studies on the existing road constructed with the use of nepheline sludge.

Analysis of changes in the water extract pH of the roadside soil showed that the pH of the soil selected near the highway in the city of Sosnovoborsk, built using a crushed stone base reinforced with nepheline sludge of JSC "RUSAL Achinsk", has a highly alkaline reaction of the medium (figure 1).

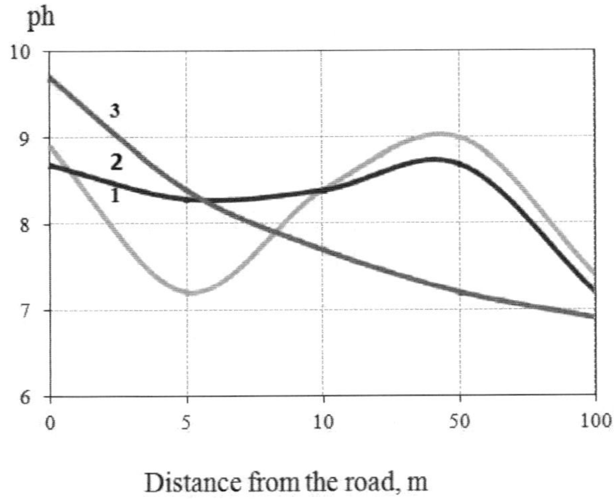

Figure 1. Changing the water extract pHof the roadside soil:
1- road built without the NS (city of Zheleznogorsk);
2 - road with the NS (city of Sosnovoborsk); 3 - road with the NS (village of Zeledeevo)

A characteristic feature of the land from the surveyed territory in the area of Sosnovoborsk is the alkaline reaction of the soil. The water extract pH from various samples taken at a distance of 5, 10, 50, 100 m from the road ranges from 8.2 to 8.8 pH (figure 1), which confirms the increased alkalinity of the soil of this site, even at a great distance from the road. At the same time, the pH value for the background sample taken at a distance of 100 m from the road also had an alkaline reaction of soil water extraction from this area (8.5 pH). The use of nepheline sludge in the reconstruction of the highway at the area near Sosnovoborsk did not affect the pH change, since even in the background sample, the pH value had an alkaline reaction, which is due to soil contamination by nearby industrial facilities.

For comparison, samples were taken on the highway in the village of Zeledeevo, which was also built using a base made of nepheline sludge, but has a longer (15 years) service life and no anthropogenic impact of large industrial facilities. As seen in Fig.1 the water extraction pH of soil samples taken from the roadside of this highway had a smooth character of decreasing to a neutral value almost at a distance of 10-50 m, which confirmed the absence of harmful effects of nepheline sludge on the territory adjacent to the highway.

The nature of changes in the soil pH taken from the roadside strip near the city of Zheleznogorsk, which has a crushed stone base of the road surface made without the use of nepheline sludge showed that within the range of 10 to 50 m, the soil also had a highly alkaline reaction of the medium from 8.3 to 9.4 pH (figure 1). The increased pH values at a distance from the roadside strip are primarily due to the high traffic intensity on this highway, similar to the section near the city of Sosnovoborsk. At the same time, a decrease in the soil pH of the water extraction to neutral (7.4 pH) was noted when analyzing the soil of this area, selected at a distance of only 100 m from the road. The pH value for the background sample at this area was 7.3 units.

Given that the main mineral compound in nepheline sludge is dicalcium silicate (up to 85%), it was considered appropriate to study the calcium content in the roadside soil. Changes in the calcium content in the soil on the analyzed sections of the road with a base constructed with and without nepheline sludge had a similar character at the distance of the road from 5 to 100 m (figure 2).

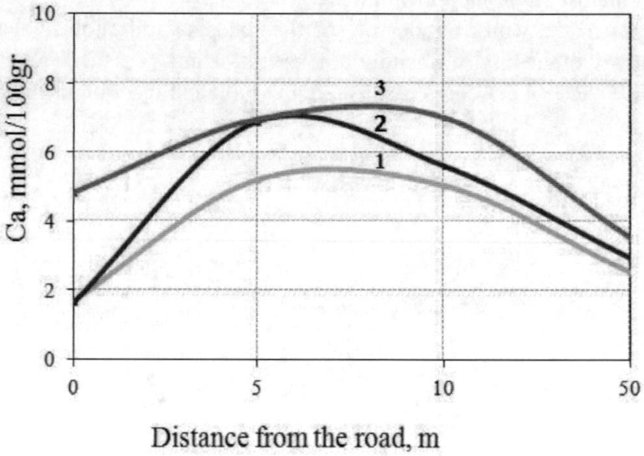

Figure 2. Changing the content of mobile calcium in the roadside soil:
1- road built without the NS (city of Zheleznogorsk); 2 - road with the NS (city of Sosnovoborsk); 3 - road with the NS (village of Zeledeevo)

At the same time, at a distance of 5 and 10 m from the highway, the calcium content was the highest (from 5.12 to 6.79 mmol/100g). At a distance of 50 m, a decrease in the calcium content was observed to a value of 2.9 mmol /100g, which was even lower than the value in the background sample. Tests of soil samples taken at a distance from the roadside proved that the content of exchangeable aluminum in the soil of the analyzed areas was at a very low level (less than 0.5 mg / kg), while the use of nepheline sludge in the device of the base in road coverings practically did not affect its content in the soil at all sampling points from 0 to 100 m from the road.

Additional ecotoxicological studies were conducted to confirm the absence of toxic effects from nepheline sludge. Using the method of phytotesting, the absence of toxic properties from water extracts from a mixture of nepheline sludge with crushed stone (in the ratio of % mass fraction 30:70) and hydrated nepheline sludge was determined. The results of studies obtained on distilled water served as a control. Soft spring wheat (Triticum vulgare L.) sort of Novosibirsk 29 and watercress (Lepidium sativum L.) sort of Donskoy were used as test organisms. Ecotoxicological studies of samples by phytotesting using single-and dicotyledonous plants have established that the nepheline-crushed stone mixture is low-toxic in terms of reducing seed germination and inhibiting root growth of test plants, and the hydrated nepheline sludge stored on the slurry map is practically non – toxic. Using microbial strains, it was found that there was no toxic effect of water extracts from a mixture of nepheline sludge with crushed stone and hydrated nepheline sludge on single-celled green algae Chlorella vulgaris and soil infusoria of the genus Colpoda. It is also noted that fresh nepheline sludge of current production has toxicity, but during its storage in the sludge field, it is hydrated and it does not have a toxic effect on test plants and microorganisms.

Conclusion

The studies performed to assess the impact of nepheline sludge of JSC "RUSAL Achinsk" were confirmed in practice in the process of experimental field work on various sections of roads constructed using this sludge. The change in the content of roadside soil of calcium, as the main element present in the sludge, had a similar character at the distance of the road from 5 to 100 m on all analyzed sections of the road with a base built with nepheline sludge and without sludge. Tests of soil samples taken near the roadside strip proved that the content of exchangeable aluminum in the soil of the analyzed areas was at a very low level (less than 0.5 mg/kg) and did not affect the presence of sludge in the road mixtures. The conducted studies of the soil selected at various distances from the roadside strip proved the absence of toxicity in the analyzed soil samples. Based on ecotoxicity studies using single-and dicotyledonous plants and single-celled algae Chlorella vulgaris and soil infusoria of the genus Colpoda, the possibility of using nepheline sludge as a component of road covering in the form of a nepheline-crushed stone mixture has been established. At the same time, it should be considered that in tight conditions, when rolling the covering base, the nepheline sludge is rapidly hydrated and it passes into a monolithic state similar to concrete [2,5,6].

The widespread use of non-toxic technogenic materials of alumina production in road construction technologies allows us to solve the problem of resource saving of natural building materials and reduce the impact on the environment of industrial waste in places where they are located.

Referenses:

[1] Yudina L V, Turchin V V and Gedeonov P P 1995 *Ecology and resource conservatio* (Izhevsk, IzhSTU) **2** pp 9-24
[2] Tarnarutsky G M, Khromova L M, Botnareva N P and Popolov AS 1979 *Soyuzdor SRI* **24** pp. 63-7
[3] Stolboushkin A Yu and Berdov G I 2011 *University news.Construction* **1** pp 46-53.
[4] Ogorodnikova E N, Nikolaeva S K 2004 *Technogenic soils: Textbook* (Moscow: MSU Publishing house) p 250
[5] Shepelev I I, Kudyakov A I, Bochkov N N and Zhizhaev A M 2017 *Bulletin of TSASU* **1** pp 181-9
[6] Lesovik R V, Kovtun M N and Alfimova N I 2007 *Industrial and civil construction* **8** p 30-2
[7] Roy D M 1999 *Cementand Concrete Research* **29(2)** pp 249-54
[8] Shakarna M H 2013 *Bulletin of BSTU named after V. G. Shukhov* **3** p 38-42
[9] Shepelev I I, Bochkov N N and Sakhachev A Yu 2015 *Higher education. Chemistry and chemical technology* **58(3)** pp 77-81
[10] Beskrovny V M, Dezhina N S 1980 *Highways* **5** pp 23-5
[11] Kruchinina E Yu 1998 Ecological and hygienic assessment of materials for road construction using industrial waste: dis. for the degree of Cand. Biol. Sciences: 14.00.07.Moscow p 186
[12] Shepelev I I, Eskova E N, Potapova S O, Nemerov A M and Bochkov N N *2019 IOP Conference Series: Materials Science and Engineering (MSE)* **537(6)** 062067

IOP Conf. Series: Materials Science and Engineering **941** (2020) 012016 doi:10.1088/1757-899X/941/1/012016

Prospects for the implementation of a separate waste collection system in specially protected areas by the example of the Tunkinsky National Park

N B Lubsanova and L B-Zh Maksanova

Baikal Institute of Nature Management Siberian branch of the Russian Academy of sciences, Russia

E-mail nlub@binm.ru

Abstract. The article discusses the problems of handling specially protected territories by the example of the Tunkinsky National Park. The analysis of approaches to effective waste management in individual national parks is carried out. The rationale for the need and calculations of the effectiveness of using the separate collection system for municipal solid waste using the example of the Tunkinsky National Park.

1. Introduction

There are 12 thousand specially protected territories (SPNA) in the Russian Federation, including 85 in the Republic Buryatia, of which 8 are of federal significance. One of the priority tasks of the development of the system of protected areas of federal significance is determined by environmental education, including the development of ecological tourism [1, 2]. As a result of the implementation of the federal project "Conservation of Biological Diversity and the Development of Ecological Tourism" of the national project "Ecology", it is planned to increase the number of visitors to protected areas from 3.57 million people in 2018 to 7.89 million people in 2024, to 16 million people by 2035 g [3]. It is planned to implement in the territory of at least half of the national parks the economic tourism model, including two in the territory of the Republic Buryatia. [4].

The negative impact of tourism on the environment due to high attendance rates that exceed the capacity of the territory was noted in the works of Gúčik M. and Marciš M. [5], Newsome D. et al. [6], Ragazz I. et al. [7], Ringo J.E. et al. [8], Sasidharan et al. [9]. In particular, one of the most significant negative environmental impacts due to the increase in tourist flow is an increase in the volume of solid waste generated.

2. Problem Formulation

Tunkinsky National Park is one of the largest national parks in Russia, created to preserve the unique ecosystems of the East Sayan and the spurs of the Khamar-Daban on a total area of 1,183,622 hectares. The park is located in the southwestern part of the Republic of Buryatia on the territory of the Tunkinsky district. In the east and north, the park borders on the Irkutsk region, in the west and southwest - with the Republic Mongolia, in the south - with the Zakamensky district of the Republic Buryatia.

IOP Conf. Series: Materials Science and Engineering **941** (2020) 012016 doi:10.1088/1757-899X/941/1/012016

For the Tunkinsky National Park, whose borders coincide with the administrative boundaries of the Tunkinsky district of Republic Buryatia, the problem of solid municipal waste management (MSW) is relevant and is becoming more acute due to the increase in visits to protected areas and the increasing anthropogenic load.

The MSW management system in the Republic Buryatia is based on the burial of the capacity of the generated waste. But on the territory of the protected areas, the construction of facilities for the treatment, recycling, disposal, disposal of waste is prohibited, therefore, the resulting MSW must be removed outside the Tunkinsky district. Only temporary storage for up to 11 months and removal of waste from a temporary storage point (Galbay village) outside the district to the MSW landfill in the village of Nizhny Sayantuy, Tarbagatai district of the Republic Buryatia is allowed. The MSW transportation distance is about 500 km. A long transport shoulder makes garbage collection more expensive, which leads to unauthorized waste disposal by the population and enterprises. According to the Ministry of Natural Resources of the Republic Buryatia, in 2019 there were 10 places of unauthorized placement of MSW in the territory of the district with a total area of 7.15 hectares, on which 2080 m^3 of waste was accumulated.

The prohibition of the construction of facilities for the treatment, utilization, neutralization and burial of waste of the territory of Tunkinsky National Park, the projected increase in tourist flow and, correspondingly, an increase in the volume of waste, necessitate the introduction of a separate collection system for solid waste in the national park.

3. Problem Solution

The main sources of MSW formation on the territory of the Tunkinsky National Park are:

• residents of settlements located in the park;

• tourists.

According to Federal State Statistics Service of Russia, at the beginning of 2019, 20.2 thousand people lived in 14 rural settlements located in the national park (Table 1).

Table 1. Resident population of Tunkinsky district in the context of rural settlements

Name of rural settlements of Tunkinsky district	Population, person
Arshan	2426
Galbai	736
Dalahai	689
Zhemchug	1559
Zun-Morino	1080
Kyrenskoye	5132
Mondy	873
Toltoy	842
Tory	1094
Tunka	2230
Turan	783
Kharbyaty	717
Khoito-Gol	573
Khuzhiry	1445
Total	**20179**

The number of officially registered visitors to the national park in 2017 was 171.8 thousand (Fig.1)

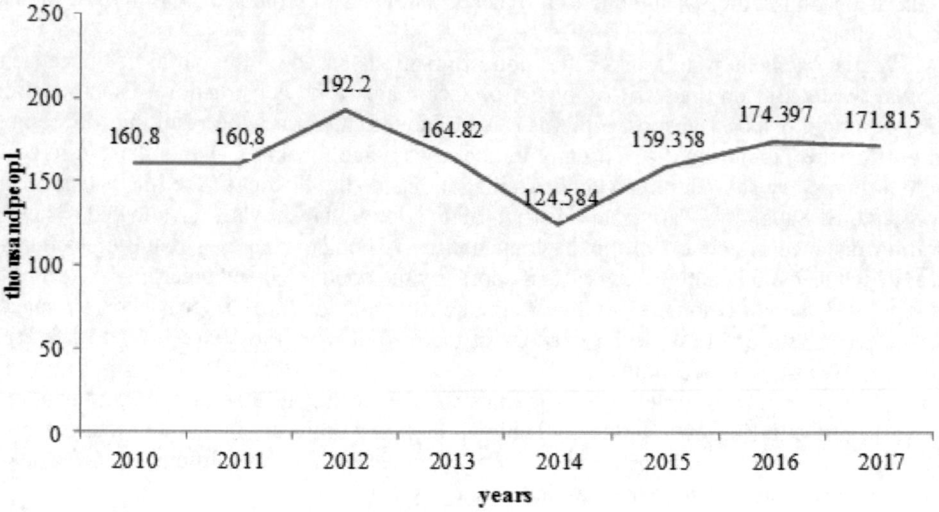

Figure 1. Dynamics of the number of officially registered visitors to the Tunkinsky National Park (according to [10])

According to the Directorate, in 2019, 59.784 thousand m3 of MSW were formed on the territory of the national park [11]. To determine the potential volumes of secondary raw materials and reduce the volume of solid waste collected, the results of the project "National Parks Without Garbage" for the separate collection of solid waste were used. This project implemented since 2019 in the Zaybaybalsky National Park by the Lake Baikal Foundation for Environmental Applications and Research with the support of the Presidential Grants Fund. As part of this project, the morphology of MSW was studied on the territory of the Trans-Baikal National Park in two zones (Ust-Barguzin and Chivyrkuisky Bay). It was determined that "in the average composition of the park's wastes, fractions potentially suitable for recycling account for 42% of all wastes (up to 11% of polymer wastes, 7% of metal, about 24% of glass) produced by tourists and the public" [12]. It is noted that "the predominant fraction is organic waste (food waste), which accounts for 29% of all waste" [12].

Based on the uniform tariffs for the services of the regional operator LLC EcoAlliance for the management of MSW for legal entities and individual entrepreneurs for 2019, which price to 408.6 rubles. per m^3, the potential savings by introducing a separate MSW collection system in the national park were determined to more than 10 million rubles per year, and taking into account the use of composting of organic waste - more than 17 million rubles per year. It should be borne in mind that at the initial stage of introducing a separate waste collection system and ensuring its functioning, additional funding is needed for the construction and equipping of container sites, information support, training of specialists, etc., which will be further compensated by reducing the cost of garbage removal and possible income from the sale of separately collected secondary material resources to utilizing companies.

4. Conclusion

Based on the data obtained, it can be concluded that by organizing separate collection of solid waste and composting food waste in the territory of the Tunkinsky National Park, it is possible to reduce the volume of solid waste collected at landfills by 71% and, accordingly, reduce the cost of their removal. The existing practice of introducing a separate waste collection system in protected areas indicates the prospects in both economic and environmental aspects, as well as in the Stolby State Nature Reserve since 2015, the savings on the export of MSW exceeded 900 thousand rubles. [13]. According to our estimates, the potential savings in the implementation of the separate MSW collection system in the territory of the Tunkinsky National Park will amount to more than 10 million rubles per year, and taking into account the use of composting of organic waste, more than 17 million rubles per year.

Acknowledgments
This work was financially supported by a grant project of the Russian Foundation for Basic Research (RFBR), project 20-010-00665.

References

[1] *The concept of development of the system of protected areas of federal significance for the period until 2020* (approved by the order of the Government of the Russian Federation of 22.12.2011 № 2322-r).
[2] Decree of the President of the Russian Federation of 05.07.2018 № 204 "*On national goals and strategic objectives of the development of the Russian Federation for the period until 2024*".
[3] *Passport of the national project "Ecology"* (approved by the Presidium of the Presidential Council for Strategic Development and National Projects, protocol of 24.12.2018 №. 16).
[4] *The tourism development strategy in the Russian Federation for the period until 2035* (approved by order of the Government of the Russian Federation of 20.09.2019 №2129-r.
[5] Gúčik M and Marciš M 2017 *Ekonomia i Środowisko.* **2 (61)** 76-86
[6] Newsome D, Moore S A and Dowling R K 2012 *Natural Area Tourism: Ecology, Impacts and Management* (Bristol: Channel View Publications) p 457
[7] Ragazz I M, Catellani R, Rada E C, Torretta V and Salazar-Valenzuela X 2014 *Sustainability* **6 (12)** 9080–9095
[8] Ringo J E, Nyahinga T, Mashalla A and Mayengo G 2016 *International Journal of Modern Social Sciences* **5 (2)** 140–152
[9] Sasidharan V, Sirakaya E, and Kerstetter D 2002 *Tourism Management* **23 (2)** 161–174
[10] State report "*On the state of Lake Baikal and measures for its protection in 2017*" 2018 (Irkutsk: ANO "KC Expert") p 340
[11] *Waste management in specially protected natural areas* 2020 (Moskow: Agency for Strategic Initiatives) p 62
[12] *Scientific reports on the first part of the study of the morphology of MSW in the Transbaikal National Park. Part 2* 2019 (Moskow: Lake Baikal Foundation) p 45
[13] *Guidelines for waste management in the protected areas of the Russian Federation* 2019 (Moskow: Lake Baikal Foundation, Agency for Strategic Initiatives) p 174
[14] Khandazhapova L M, Lubsanova N B 2013 *ECO* **3 (465)** 132-137

IOP Conf. Series: Materials Science and Engineering **941** (2020) 012017 doi:10.1088/1757-899X/941/1/012017

Carbonates in the soils of the Aral Sea dried bottom

G V Stulina[1], G V Kharitonova[2], V O Krutikova[2,3], E V Shein[4,5] and N V Berdnikov[3]

[1] Scientific-Information Center Interstate Commission for Water Coordination of Central Asia (SIC ICWC), 11 Karasu-4, Tashkent, 100187, Uzbekistan
[2] Institute of Water and Ecological Problems, Far East Branch of the Russian Academy of Sciences, 56 Dikopoltseva Street, Khabarovsk, 680000, Russia
[3] Institute for Tectonics and Geophysics, Far East Branch of the Russian Academy of Sciences, 65 Kim Yu Chen Street, Khabarovsk, 680000, Russia
[4] Soil Science Faculty Moscow State University named M.V. Lomonosov, Moscow, 1-12 Leninskie Gory, 119991, Russia
[5] Dokuchaev Soil Science Institute, Pyzhevskii per.7, Moscow, 119017 Russia

Email: evgeny.shein@gmail.com

Abstract. The paper considers the first stage of salt deposition, namely, the features of calcium carbonate formation in the soils (solonchak-like formations) of the Aral Sea dried bottom, its south-eastern part. The studied soils (sandy and silty-clay soils) are significantly saline and enriched with calcium carbonate. The content of only mollusk shells can reach 10–20%. The content of mineral CaCO3 in the soils is quite uniform, its content varies by no more than an order of magnitude, which indicates relatively close conditions for its formation in soils with different particle size distribution. However, the types of aggregations of calcium carbonate (with a close size ca. 250 μm) and the mechanisms of their formation in the soils are different. In sandy soils, ooids ($CaCO_3$ cutans on the surface of sand-sized particles) are formed by the template mechanism. In silty-clay soils with a large number of crystallization centers (the predominant particle size is < 50 μm), the precipitation of calcium carbonate leads the formation of $CaCO_3$ microaggregates from silty-clay particles. Separate crystallites and mesocrystals of calcium carbonate (up to 40 μm in size) are found singly.

1. Introduction

Investigation of relatively simple natural systems may contribute significantly to our understanding of more complex phenomena. In soil science, an example of such systems can be soils developed in extreme environments as soils of the Aral Sea (dried bottom). The Aral Sea is a large saline terminal lake situated in the arid and semi-arid Turanian Depression of Central Asia [1, 2]. The area is one of the world's foremost natural laboratories in Holocene hydrology and landscape ecology. Palaeolimnological reconstruction of variability in water levels of the Aral Sea since 11,700 BP indicates a long history of alternating phase of regression and transgression, which have been attributed variously to climate, tectonic and anthropogenic forcing [3–5]. Over the past 60 year the Aral Sea has shrunk drastically, largely due water abstraction from the Amu Darya and Syr Darya rivers for land irrigation. These processes dominate in the shallow south-eastern part of the Aral Sea [6]. All regressions resulted in the deposition of the salts (carbonates and evaporites) by evaporation of the water. Our aim in this article is to study the first stage of salt deposition [7], namely, the features of calcium carbonate formation in the soils of the dried bottom of the Aral Sea.

2. Objects and methods

The study was focused on soils of dried bottom sediments in the south-eastern part of the Aral Sea. The sampling sites are shown on a schematic map (Figure 1). A brief description of selected physical and chemical properties of the soils is given in Table.

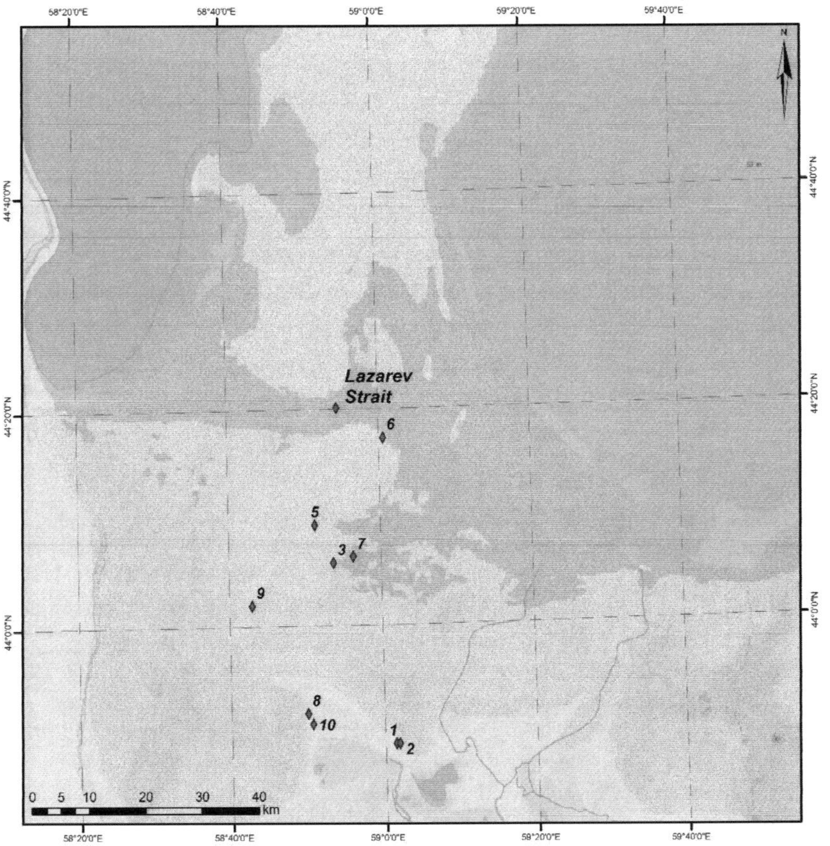

Figure 1. A schematic map of sample site locations.

The major methods were particle size composition, gross analysis and scanning electron microscopy (SEM). The SEM analysis was carried out using VEGA 3 LMH (TESCAN, Czech Republic). For the analysis, the samples, after removing the Mollusca shells, grinding and sieving through a 2-mm sieve, were prepared via pouring, Pt-spraying and magnification of up to 20.000. The X-max 80 energy-dispersive spectrometer (Oxford Instruments, UK) was used to analyze the elemental composition of the most representative regions. The particle size composition (without decomposition of carbonates) determined with the laser diffraction method on a particle size analyzer SALD-2300 (SHIMADZU, Japan). The contents of selected chemical elements were determined via the X-ray fluorescence method (XRF) (Pioneer S4, Bruker AXS, Germany), using the silicate technique. The SEM and XRF analyses were carried out in the Analytical Centre at the Institute for Tectonics and Geophysics, FEB RAS, Khabarovsk.

Table 1. Selected physical and chemical properties of surface soils

Site, depth (cm)		EC ($mScm^{-1}$)	LH[a] (%)	Gross composition (%)					Particle size composition (μm), %		
				SiO_2	CaO	MgO	Na_2O	S	<2	2–50	50–1000
1	0–10	19.6	16.3	44.6	11.6	4.7	6.9	0.7	7.8	76.1	16.1
2	0–10	10.1	27.7	23.5	40.7	2.0	1.9	0.9	3.8	36.0	60.2
3	0	21.1	10.2	51.3	9.8	3.8	10.2	4.7	0	15.5	84.5
	0–10	18.3	17.7	40.3	15.2	4.9	5.1	1.4	7.3	81.1	11.6
5	0–10	2.3	11.3	55.6	14.6	1.7	1.8	0.6	0	7.8	92.2
6	0	23.0	10.4	58.6	8.6	3.5	10.2	3.7	0	1.6	98.4
	0–10	10.9	16.8	43.4	18.9	2.7	4.0	1.4	6.3	30.9	62.8
7	0	1.0	15.0	51.8	20.3	0.7	0.7	0.4	0	0.1	99.9
	0–10	1.7	12.6	54.9	17.3	0.5	0.8	1.6	0	0	100
8	0	0.8	2.4	84.1	2.1	0.8	2.6	0.1	0	0.3	99.7
	*0	0.2	2.2	82.0	2.0	0.7	2.7	0	0	0.6	99.4
	0–10	4.7	10.8	50.4	15.1	1.1	2.2	4.4	0	6.3	93.7
	10–20	1.5	7.5	69.4	9.2	0.5	1.6	0.4	0	0	100
	20–30	0.5	7.2	71.5	9.0	0.5	1.4	0.1	0	0	100
9	0	8.8	9.9	62.8	10.6	2.8	2.8	2.4	0	15.1	84.9
	10–15	0.1	2.9	90.9	3.0	1.0	1.3	0	0	1.2	98.8
10	0–10	4.4	12.6	51.2	12.7	3.5	3.1	0.3	3.8	59.3	36.9
Lazarev Strait		65.9	16.3	27.4	1.1	2.7	48.0	12.0	0	4.6	95.4

[a]loss of heat

3. Results and Discussion

The results of the chemical analysis showed that the soils of the dried bottom are significantly saline and enriched with calcium carbonate. And if salinity (estimation of soluble salts by electrical conductivity) in the studied series of soils varies by more than two orders of magnitude, then the content of calcium carbonate in soils is relatively constant. Its content varies by no more than an order of magnitude from 2 to 40% in terms of CaO, which indicates relatively close conditions of calcium carbonate precipitation.

However, according to the particle size distribution and SEM data, the studied series of samples is clearly subdivided into two subgroups: sand and silty-clay deposits. Figure 2 shows the extreme members of the series both in particle size (figure 2a, c) and in the content of Na_2SO_4 and $CaCO_3$ (figure 2b, d respectively). Significantly smaller compared to the particle size distribution in the case of deposits of the Lazarev Strait is associated with an extremely high content of sodium salts (48% in terms of Na_2O), and the formation of salt microaggregates. Vice versa, the predominance of the sand fraction 50–200 μm in size in the sediment composition of site 2 (peak with a maximum of 140 μm) is caused by aggregation of silty-clay particles with $CaCO_3$ (the content of $CaCO_3$ is more than 40% in terms of CaO, the loss of heat reaches ca. 30%).

In sandy deposits (sites 5-10, Lazarev Strait), $CaCO_3$ precipitation occurs mainly on the surface of quartz grains with the formation of ooids of varying degrees of stability and grain coating (figure 3a–d). Depending on the mineralization conditions (site Lazarev Strait), further successive crystallization of sulfate and sodium chloride may occur (figure 3a). In silty-clay deposits (sites 1–3), $CaCO_3$ precipitation occurs mainly on the surface of clay and silt particles with the formation of microaggregates (figure 3e) up to 250 μm in size. Only at an extremely high concentration of $CaCO_3$ (site 2) was the formation of separate $CaCO_3$ mesocrystals (figure 3f and 3g) observed, which is associated with constrained crystallization conditions [8]. The size of the diagnosed mesocrystals does not exceed 10–15 μm. A rhombohedral calcite crystal measuring ca. 40 μm was diagnosed singly (figure 3h).

Figure 2. SEM images of sandy (a, b) and silty-clay soils (c, d): a – site 8, 20–30 cm (arrows – $CaCO_3$-ooids); b – site in Lazarev Strait, 0–10 cm (arrows – grains of sand size); c – site 3, 0– 10 cm; d – site 2, 0–10 cm (SEM, BSE-detector).

Our data indicate two mechanisms of calcium carbonate formation in the soils as a result of increasing evaporation of water, concentration of salt solution and subsequent drying of the Aral Sea bottom. First, with a small number of crystallization centers (sandy soils), the template mechanism of $CaCO_3$ crystallization is implemented with the formation of cutans of different degrees of stability and coating of sand grains. The formation of predominantly ooids may indicate the orientation of individual $CaCO_3$ crystals perpendicular to the grain surface. In turn, the stability and completeness of the coating with calcium carbonate of rounded quartz grains with a high degree of confidence allows us to assume the formation of aragonite, a polymorphic modification of calcite that is stable at a high salt content [9]. The process of formation of aragonite ooids is noted in the bottom sediments of Grand Salt Lake (USA) with extremely high salinity of water. Needle-like crystals of aragonite are "ideal" for their densest packing on the surface of sand grains and forming the radial structure of ooids. Analysis of Grand Salt Lake ooids showed that a similar mechanism is implemented on biological templates of similar shape and size (*Artemia franciscana* cysts, average size 200–250 µm). It is possible that the ooid shown in Figure 3d is also the result of $CaCO_3$ precipitation on some biological object. The size of *Artemia* cysts, which are widely represented in the waters of the Aral Sea, is suitable. Second, with large number of crystallization centers (silty-clay soils), the precipitation of calcium carbonate leads primarily formation separate clusters (rather not crystals), which bind clay and sand particles into microaggregates.

Figure 3. SEM images of secondary carbonates of sandy (a–d, ooids) and silty-clay soils (e–h): a – site in Lazarev Strait, 0–10 cm; b and c – site 8, surface and 10– 20 cm respectively; d – site 7, 0–10 cm; e – $CaCO_3$-clay microaggregate; f и g – mesocrystals of calcite and vaterite respectively, site 2, 0–10 cm; h – rhombohedral calcite, site 9, 10–15 cm (SEM, BSE-detector).

4. Conclusion

Using the methods of electron microscopy, gross and sedimentation analysis, the composition of the soils of the dried bottom of the Aral Sea, its south-eastern part was studied. It is shown that the soils of the dried bottom – sandy and silty-clay soils – are significantly saline and enriched with calcium carbonate. A distinctive feature of the studied soils is their relative uniformity in the content of the latter. In sandy soils, $CaCO_3$ was diagnosed as part of ooids, in silty-clay soils – as part of microaggregates. Crystallites and mesocrystals of calcium carbonate occur sporadically.

The data obtained indicate two mechanisms of calcium carbonate formation in the soils as a result of increasing evaporation of water, concentration of salt solution and subsequent drying of the Aral Sea bottom. With a small number of crystallization centers (sandy soils), the template mechanism of $CaCO_3$ crystallization is implemented with the formation of cutans of different degrees of stability and coating of sand grains.

With a large number of crystallization centers (silty-clay soils), precipitation of calcium carbonate leads primarily to nucleation (formation of separate clusters of calcium carbonate molecules) and the formation of microaggregates from silty-clay particles.

References

[1] Zavialov P 2005 *Physical Oceanography of the Dying Aral Sea* (Springer-Praxis Publishing, Chichester) p 146
[2] Austin P, Mackay A, Palagushkina O and Leng M 2007 *Quaternary Research* **67(3)** 383–93
[3] Shnitnikov A V 1983 *Paleogeography of the Caspian and Aral Seas in the Cainozoic* part II, ed E G Maev (Moscow, Moscow University Press) pp 106–118
[4] Boomer I, Aladin N, Plotnikov I and Whatley R 2000 *Quaternary Sci.Reviews* **19**(13) 1259–78
[5] Cretaux J-F, Letolle R and Bergé-Nguen M 2013 *Global and Planetary Change* **110** 99–113
[6] Rubanov I V 1991 *History of Lakes Sevan, Issyk-Kull, Balkhash, Zaisan and Aral* ed D V Sevastyanov (Moscow, Russian Academy of Science) pp 231–234
[7] Doner H E, Grossl P R 2002 *Soil Mineralogy with Environmental Applications* vol 7, ed J B Dixon and D G Schulze (Madison, Wisconsin: Soil Sci. Soc. of America) pp 199–228
[8] Cölfen H and Antonietti M 2008 *Mesocrystals and nonclassical crystallization* (New York: John Willey&Sons) p 288
[9] Kharitonova G V et al 2019 *IOP Conf. Series: Earth Environ. Sci.* **368** 012020

Acknowledgment

The work was partly supported by the RFBR, project 19-29-05112_mk, "Digital physics and soil hydrology: the basics of spatial-dynamic analysis, prediction of risks of critical situations and optimal control", projects of basic scientific research, 04 BIOLOGY, 04-180 soil science, Manager of the project Prof. Dr. Shein Evgeny.

IOP Conf. Series: Materials Science and Engineering **941** (2020) 012018 doi:10.1088/1757-899X/941/1/012018

Livestock enterprises waste utilization for heat system supply to rural buildings in Yakutia

V P Druzyanova[1], G E Kokieva[2], I A Savvateeva[1], N S Khiterkheeva[3] and O K Kilchukova[4]

[1] The North-Eastern Federal University named after M.K. Ammosov, st. Belinskogo, h. 58, 677007, Yakutsk, the Republic of Sakha (Yakutia), Russia
[2] The Arctic State Agrotechnological University, sh. Sergelyakhskoe 3 km, h. 3, 677007, Yakutsk, the Republic of Sakha (Yakutia), Russia
[3] The Buryat State University named after D. Banzarov, st. Smolina, h. 24a, 670000, Ulan-Ude, the Republic of Buryatia, Russia
[4] The Kabardino-Balkarian State Agrarian Univarsity named after V.M. Kokov, st. Lenina, h. 1v, 360030, Nalchik, Russia

E-mail: druzvar@mail.ru

Abstract. In recent years, interest in issues of using natural energy sources significantly increased due to limited resources of fossil fuels and state missions of environment protection from pollution. One of the main areas of the researches is heating and hot water supply system development for country house using heat pump systems (HPS) based on non-traditional power sources in moderate climate conditions.

Key words: heat pump systems, power sources, solar power plants, manure processing, ground power, livestock farms, resource-saving technologies, mechanization and automation, methane fermentation, water supply, biogas, ecological system.

1. Introduction

In recent years, interest in processes of biogas production significantly increased. It can be seen in increasing number of planned and under-construction biogas plants, and even in increasing number of farmers' interest, community facilities, businesses, politicians and private households which closely monitor the sector development.

Power industry treats decentralization of production with no caution because of biogas plants construction. Technology of biogas production provides an opportunity for cheap utilization of organic waste and food residues in biogas plants with benefits for agriculture for the food industry, gastronomy, large restaurants, catering establishments and food waste processing enterprises. This technology is gaining great amount supporters among people who has seen its benefits for the environment.

There is a lack of technologies for manure processing in the central Yakutia where nearly 90% of cattle number is concentrated. The manure challenges normal and safe functioning of livestock farms in localities. The lack of technologies for manure processing leads to long-term gathering outside of farms near still water reservoirs that can cause to further heavy pollutions.

Content of nitrates, nitrites, ammonium and phosphates exceeds allowed norms caused by leaked biogenic elements into lakes. Exceedance of nitrates LOK in 110-120 times and phosphates in 70-80

IOP Conf. Series: Materials Science and Engineering **941** (2020) 012018 doi:10.1088/1757-899X/941/1/012018

times was noticed in some lakes of the Zarechnyi districts of the Republic of Sakha (Yakutia). Local people use and drink water from those reservoirs with no purification. There is no doubt that manure is the main factor of infections and invasive diseases spreading among people and animals. It is a source of infections spreading through pasture lands and reservoirs.

Solar power, bioenergy, and ground energy can be used as natural sources of low-grade heat. It is relevant to consider the type of power supply of farming region for further selection the most suitable source. The anaerobic technologies for cattle manure processing require special adaptation for profitable implementation in Yakutia according to several significant technical issues.

One of the major factors for optimal flow of microbiological processes in methane tanks is the microorganisms' residence time in environment (retention time). It is necessary to have enough amount of microorganisms in the substrate; to supply the required residence time in environment for metabolism substrate; to exclude bacteria vanish for complex organic substances decompose to CH4 and CO2.

Therefore, these factors should be decisive when choosing the operation mode of the BEU. Organic substances degradation in methanogenesis conduct as a multi-stage process which requires the participation of at least four groups of microorganisms: hydrolytics, fermenters, acetogens, and methanogens. Methane archaea of the genera Methanosarcina, Methanosaeta (Methanothrix), Methanomicrobium, and others have key roles in organic substances anaerobic degradation to methane.

It is necessary to maintain temperature for normal functioning of anaerobic bio-energetic plant in mesophilic and thermophilic modes. Temperature fluctuations in these modes are immediately reflected in the process indicators. The authors [1-8] note, that as higher the fermentation temperature is, the narrower the CE fluctuations limits are. The admissible variation in a temperature of 38 C is ±2.8 C, and in 53-55 C it is ±0.3 C. Therefore, the modes using will definitely lead to excessive power consumptions required for of optimal temperature maintenance in the bioenergy plant's reactors in conditions of Yakutia.

The humidity of the loaded manure is 90%. Such number of humidity percent leads to good fluidity of manure, reduced storage cost, processing and transportation cost, saving of a clean water and low number of potential environmental pollution possibility percent. Solar plants are widely used in areas in the South of 45° from the North latitude. This zone covers the South of Kazakhstan, republics of Central Asia, the Caucasus and Transcaucasia, regions of Krasnodar and Stavropol and the southern coast of Crimea.

In these regions, the amount of direct solar radiation to the horizontal surface ranges from 1340 to 1850 kW h/m^2 per year. Annual number of sunshine hours is t=2000...3000h. Introduction of low-waste, resource-saving technologies with mechanic and automatic production processes play an important role in protecting the environment from livestock waste in the conditions of their further development.

Nowadays, there are several ways to utilize the waste of agriculture and off-flow. Methane fermentation with the production of organic fertilizers and biogas is the most promising utilization [1-5]. Utilization of animal waste off-flow with use of biogas plants gives a significant socio-economic effect, expressed in improving working conditions, increasing the culture of production and land use.

2. Goal of the study

Natural animal waste has a great value for agricultural production. Using of animal waste in regions with common livestock husbandry is not particular problem. Livestock husbandry is well developed, in this regard, there is a problem of manure utilization in Yakutia [5-10].

It is a big issue according to great animal number in a limited area and imbalance between livestock and land squares. The goal of the research is to study the utilization technology of animal waste by anaerobic treatment for thermal energy withdrawal.

3. Methods of the study

Manure can be as biomass in conditions of agricultural production. It is rational to consider the manure anaerobic digestion process, which produces biogas to assess the potential energy of the biomass (see table 1).

IOP Conf. Series: Materials Science and Engineering **941** (2020) 012018 doi:10.1088/1757-899X/941/1/012018

Table 1 shows that it is possible to save 26.8 million tons of conventional fuel per year due to bioenergy obtained from manure. If we take into account the possibility of obtaining bioenergy from field waste, drain water and household waste, total number increases to 40 million tons. For comparison, we note that the total annual demand of the household sector of the rural population is about 80 million tons of conventional fuel.

Table 1. Analysis.

Farm	Exposure time, day	Organic matter decomposition, %	Energy of produced biogas, mln t of coal equivalent
Cattle breeding	9	25.9	16.9
Pig breeding	8	45.9	2.5
Hennery	43	44.0	7.4
Total	-	-	26.8

Thus, bioenergy would be enough to provide 50% of the heat supply to rural residential buildings. It requires about 15 tons of manure per year for 90% heat supply for one rural house.

4. Main part

The ground can be considered as a huge battery of solar power. Moreover, the power "reservoir" temperature is equal to the environment air average annual temperature. It is characterized by the fact that there is a" pumping " of its thermal energy in summer, and it return to the environment in winter.

To assess the natural heat flow quantity q in summer and winter time, we can assume that at a depth of $L = 2$ m the ground temperature approximately matches with the average annual air temperature, and the maximum difference between the average monthly air temperature and the average annual air temperature is $\Delta t = 15$ °C. When the thermal conductivity γ is equal to 1.4 W/(m*deg), we get

$$Q = \frac{\gamma \Delta t}{L} = 10{,}5 \text{ W/m}^2 \quad (1)$$

The analysis of heat supply systems work on the basis of alternative power sources used in our country in the pilot rural buildings showed, that the most promising systems are those, that meet the following requirements: maximum possible reduction of heat loss through the building envelope (improving thermal insulation); the inclusion of heat pump in the heat supply system; ground low-grade thermal power use as a primary source; the use of cheap solar collectors in moderate climate conditions. Figure 1 shows a schematic diagram of an experimental heat supply system for a rural buildings:

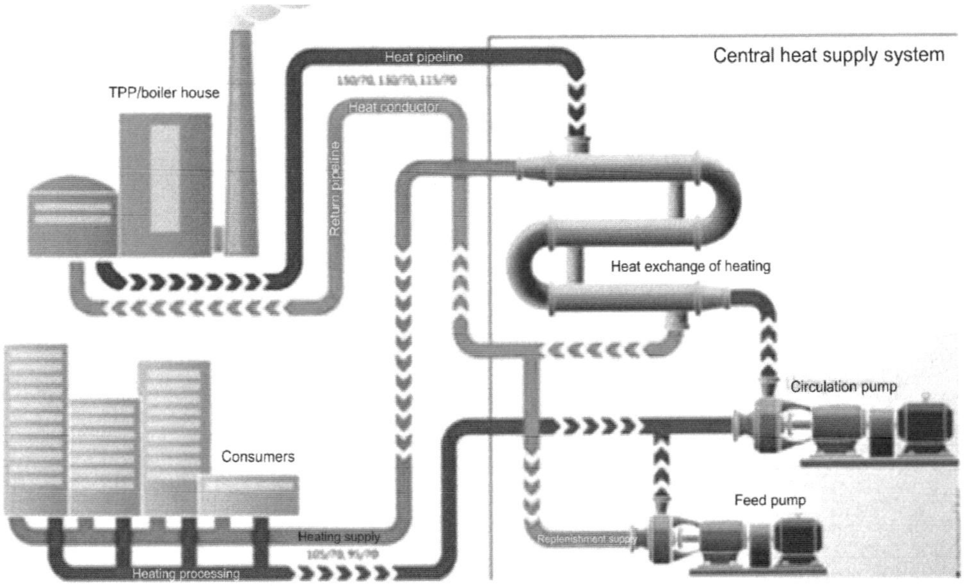

Figure 1. Schematic diagram of an experimental heat supply system for a rural buildings.

It must be noted, there are no clear principles of projecting such schemes in accordance with power releasing total cost developed despite many existing heat supply systems based on non-traditional power sources for individual houses. Many buildings are laboratories for optimal solutions searching in the system's design.

5. Conclusion

To sum up, a rational combination of calculation and experiment allows us to expand the boundaries of research, reduce number of experiments and significantly speed up the work on establishing and enhancement of manure utilization technologies. Thus, the utilization of livestock cattle liquid manure contributes to the establishment of an additional source of fuel as biogas.

6. Literature

[1] Kokieva G E 2013 *Potential of the Baikal region's communications industry development* (Novosibirsk) 39-42
[2] Kokieva G E 2014 Measurement technology of working surfaces under defect apparatuse for the cultivation of microorganisms *Scientific and Technical Volga Region Bulletin* **3** 137-139.
[3] Kokieva G E, Shaposhnikov Yu A and Shagdyrov I B 2016 Ekspluatatsiia fermentatora v sel`skom khozyaistve [Operation of the fermenter in agriculture] (Barnaul: AltGTU Publ.) 103
[4] Kokieva G E, Shagdyrov I B, Shagdyrov B I and Bolokhoev V S 2016 *Method of feed yeast preparation* Pat. of the Russian Federation No. 2580160
[5] Chernoivanov V I 1984 *Sovershenstvovanie tekhniki Dokt. Diss. [Improving technology]* p 54
[6] Semenov V I 2003 *Matematicheskoe modelirovanie plazmy v sisteme kompaktnyi tor. Dokt, Diss. [Mathematical modeling of the plasma in the compact torus. Doct. Diss.]* (Moscow) p 272
[7] Sharifullin V N, Boyarinov A I and Gumerov A.M 1980 *Mass transfer processes and devices of chemical technology* (Kazan) p 17-18
[8] Shebatin V G, Domanskii I V and others 1977 *Airlift transportation of liquids and suspensions/ ZhRKH* **4** p 86-87

IOP Conf. Series: Materials Science and Engineering **941** (2020) 012018 doi:10.1088/1757-899X/941/1/012018

[9] Chernorutskii I G 2004 *Optimization methods in control theory* (Saint-Petersburg: Piter) p 256
[10] Shank H 1972 *Theory of engineering experiment* (Moscow: Mir) p 384
[11] Shaposhnikov Yu A, Chernetskaya N A, Kulagina N A and Kantor S A 2012 Factor effect on interaction dynamics between fertilizers particles and a blade during mixing *Siberian Herald of Agricultural Science* **2 (225)** p 66-71

Production of Biogas using method of agricultural waste fermentation

G E Kokieva[1], G P Protodiakonova, U Z Dondokov[2], D V Shalbuev[3]and SA Voinash[1]

[1] The Arctic State Agrotechnological University, sh. Sergelyakhskoe 3 km, h. 3, 677007, Yakutsk, the Republic of Sakha (Yakutia), Russia
[2] The North-Eastern Federal University named after M.K. Ammosov, st.Belinskogo, h. 58, 677007, Yakutsk, the Republic of Sakha (Yakutia), Russia
[3] East Siberia State University of Technology and Management, st.Klyuchevskaya, h. 40B, 670013, Ulan-Ude, Republic of Buryatia), Russia

E-mail: kokievagalia@mail.ru

Abstract. During using biogas as fuel for diesel engines, it is necessary to clean it from hydrogen sulfide, water and carbon dioxide, as well as to compressing it to fill cylinders more efficiently. Today the production of biogas in agriculture is the disposal of crop waste and manure from livestock farms.This approach is very actual, at the first stage of its implementation was planned to create the necessary material and technical base to widespread using of non-traditional energy sources, including organic biomass energy (fertilizer, crop waste, etc.). In addition, environmental protection, which also requires intensive and rational processing of animal waste, is very closely connected with the problem of waste disposal. This necessitates the search for processing methods that would ensure the comprehensive utilization of its agricultural, energy, feed and other properties. This article presents the technology for processing the utilization of manure using microbial synthesis. When processing manure in anaerobic conditions, the threshold lowered for its smell, germination of weed seeds, helminths die, and nutrients (nitrogen, phosphorus, and potassium) are completely preserved. The resulting biological gas contains up to 70 % methane, and its calorific value is 24 MJ/M3. The introduction of low-waste, resource-saving technologies with mechanization and automation of production processes plays an important role in the environment protection issues from livestock complexes waste in their further development. At present, there are several directions of processing of agricultural and water waste. At the same time, processing of animal waste with the use of biogas plants gives a significant socio-economic effect, expressed in improving working conditions, increasing the production culture and land use. Moreover, the environment protection rask is being solved.

1. Introduction

The first developments of technology and technical means for cleaning, transporting, storing and using manure appeared with the establishment of mechanized farms for milk production, young stock breeding, pigs, and etc. Further research is aimed to establish environmentally friendly systems for cleaning and utilization of manure for all types of farms and complexes, their location zones. Traditional methods of manure storing and using are not acceptable for dilute manure, the share of which in the organic waste structure from livestock farms and complexes is constantly increasing. This makes it necessary to search processing methods that would ensure the comprehensive utilization of its agrotechnical, energy, feed and other properties. Such direction meets the requirements for the environment protection, and it requires much less operations, maintenance personnel, a more even distribution of works volume during the year, that is economically rational to prepare manure for use immediately.

Biogas produces electricity, hot water for heating and other technological benefits. The biogas engines can operate on a mixture of 15 ... 20% liquid fuel and 85 ... 80% of methane. Based on experimental data on biogas production and degree of decomposition of manure, was determined the most effective doses of the average daily loading of the methane tank.

2. The aim of research

Animal waste is of great value for agricultural production. Their use in regions with dispersed livestock production is not particularly difficult. In Yakutia, animal husbandry is well developed, in this regard, there is a problem of manure utilization. This is a big issue because of the high animals concentration in a limited area and the imbalance between its livestock and agriculturally used areas. In this regard the goal of the article is to study the technology of processing animal waste by anaerobic treatment for the thermal energy recovering.

3. Methodology

In methane fermentation process, organic substances are mineralized without air access under the anaerobic microorganisms, which results in a significant increase of mineral nitrogen proportion due to the splitting of its associated forms. Along with preventing of organic nutrients losses of processed mass, unpleasant odors, disinfection and devitalization (weed seeds germination loss) are eliminated. The study of manure physical and microbiological parameters, as well as their dependencies on the processing mode and methods, allowed to analyze the known technological schemes and means for organic waste methane fermentation and obtaining feed additives from it [3 p.2, 7 p. 95].

4. Main part

Scheme of bioenergy plant for 400 animals` farm:

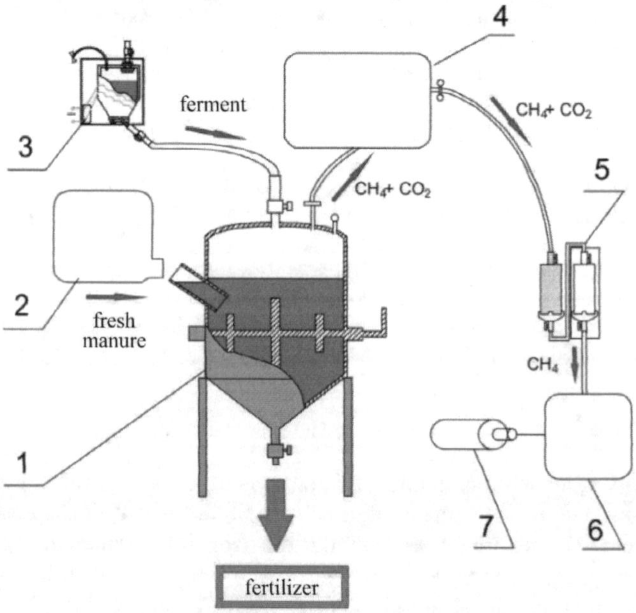

Figure 1. Installation flow-sheet:
1 – methan; 2 – storage tank for fresh manure homogenizing; 3 – adaptable installaion; 4 – has holder; 5 – filter for gas cleaning; 6 – high pressure compressor; 7 – gas bottle.

During fermentation, biogas is released, which enters the gas holder via a gas pipeline or is discharged into the atmosphere through a water seal. From the gas holder, biogas is supplied (as needed) to consumers.

During the testing period, it was processed 215 m^3 of manure with a humidity of 91 ... 93 %, and an ash content of20 %. The total operating time was 4.215 hours, the biogas volume obtained during the tests was 2.360 m^3, and its calorific capacitance was 24 ... 26 MJ/m^3.

As a result of research, biogas output data, its decomposition degree in the mesophilic mode were obtained (Figure 2). The maximum biogas output was 16.2 m^3/day in the first, and 20.3 m^3/day in the second modes.

Analysis of manure physical properties after fermentation showed that the absolutely dry ash-free substance decomposition degree was 22 % with a daily 4.5 % loading dose. The manure humidity and its ash content increase after fermentation, which is explained by the partial expenditure of dry matter for the biogas formation.

Based on experimental data on the biogas output and the manure decomposition degree, it was determined the most effective doses of methane tank average daily loading. It is found that that the maximum value of this indicator should not exceed 4.5 and 9% in mesophilic and thermophilic modes.

When increasing the loading dose, the biogas output decreases and the process is unstable.

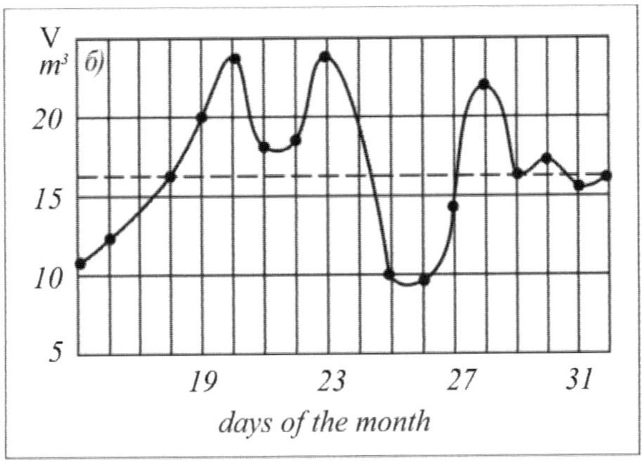

Figure 2. Figure 2. Change in volume of released biogas by days of the month (in the mesophilic mode of operation, 34 ... 35 ° C) with a load of 4.5% (a) and 9% (b).

Figure 2 shows the biogas average yield in the thermophilic mode (20.3 m^3/t) is higher than in the mesophilic mode (16.2 m^3/t). The average daily energy consumption for the process was 59...62 and 1.000...110 kWh. Thus, the energy specific costs for the production of 1 m^3 biogas in the thermophilic mode is 1.5 times higher than in the mesophilic mode.

At the same time, it is known that the manure disinfection and the nonsporeforming microflora destruction are achieved in the thermophilic fermentation mode after 72 hours. Therefore, it is advisable that the equipment set for manure anaerobic digestion function in both modes. The calculated data on the organic substances optimal concentration in the fermentable mass (80...90 g/l) correspond to the manure characteristics used in the experiments (humidity 91...93 %). The calculated fermentation exposure does not significantly coincide with the test results (in the mesophilic mode – 9 days against 22, in the thermophilic mode – 7 days against 11). The calculated degree of organic matter decomposition is slightly higher (25.9 %) than in tests (22%). The dissloution rate was relatively low due to the high ash content in manure (20 % at a rate of 16 %).The calculated biogas output does not correspond to the experimental data (17.5 m^3/t against 16.2 m^3/t). This slight overestimation of the biogas output is explained by the higher ash content in the manure. The process efficiency depends on the climate conditions.

The study of manure anaerobic processing has mainly confirmed the process effectiveness. Based on the research results, we developed a technical proposal for the biogas plant design for a 400 heads farm, and determined its technical and economic efficiency. The initial data were taken as the standards for technological design, according to which on a farm with 430 conditional animals, the excrement putput per day is 23.7 m^3, the sewage total output is 54.8 m^3/day with humidity up to 95 %. The technical requirements provide that anaerobic processing is a flow-through process with cyclical execution of operations.

From the methane tank, the biogas passes to a dry gas holder with 600 m^3 volume, from where it is used for heating water and other needs of the farm. The fermented mass passes through the airlock to the tank farm and to the heat exchanger for heating the initial mass.

5. Research results and discussions

In accordance with the developed methodology, it is possible to determine the process key parameters and process facility.

Manure collector volume

$$V_{mc} = Q\rho t_i r_h = 110\ m^3, \tag{1}$$

whereρ – manure density, $\rho = 1\ t/m^3$; t_i – intake period, t_i = 2days; r_h = 1,5.

Capacity of heating tank:

$$V_0 = L\rho t_0 r_h' = 55\ m^3, \tag{2}$$

where$L$ – daily manure output with 92 % humidity, L = 55,48m^3; t_0 – heating period, day; r_h' = 1.

Methane tank volume:

$$V_{mt} = 100\ L\rho/g = 785\ m^3, \tag{3}$$

where$g$ – daily dose of the methane tank loading, g = 4.5 %.

TakeV_{mt} = 1 000m^3 (close to the standard).

Fermentation length:

$$t_f = 100/q\ =\ 22\ \text{day}, \tag{4}$$

where$q$ – biogas output, per 1 ton of processed manure, q = 20m^3.

Daily biogas output:

$$G_f = Lq\ = 710\ m^3. \tag{5}$$

Gas holder volume:

$$V_g = G_f t_{bs}/24 = 532\ m^3, \tag{6}$$

wheret_{bs} – biogas storage time, h for 1 day.

TakeV_g = 600m^3 (close to the standard).

Total thermal energy of obtained biogas

$$Q_t = G_b C_b = 17\ 000\ MJ, \tag{7}$$

whereC_b – biogas calorific capacitance, C_b = 24MJ/m^3.

The heat consumption for heating the initial manure from $t_1 = 8$ °C to $t_2 = 35$ °C (mesophilic mode temperature)

$$Q_{mm} = L(t_1 - t_2)C_m/\eta = 5\,400\ MJ, \quad (8)$$

whereC_m – manure heat capacitance, $C_m = 4.06$kJ/(kg*°C); η – heating unit efficiency, η = 0.7.

The heat consumption for own use

$$Q_{ou} = Q_{mm} + Q_{hlc}/\eta = 9\,990\ MJ, \quad (9)$$

whereQ_{hls} – The heat consumption for heat loss compensation, $Q_{hlc} = 3\,140$MJ.

Total amount of biogas for own use,

$$G_{bu} = Q_{ou}/C_b = 415\ m^3.$$

Output of marketable biogas

$$G_{mb} = G_b - G_{bu} = 295\ m^3.$$

Discharge ratio of biogasfor own use

$$\eta_b = G_{bu}/G_b = 0{,}58.$$

The processing equipment capacity is selected based on its operating period, which does not exceed the working shift duration($t_p < t_{cm}$), if it can be used for the farm thermal needs.

Minimal heat power

$$W_{min} = G_b q_{ng}/t_{ot} = 2\,150\ MJ/h, \quad (10)$$

whereq_{ng} – calorific capacitanceof natural gas, $q_{ng} = 33.5\ MJ/m^3$.

The closest standard size of a natural gas boiler is KG-1 500 with a capacity of 1 500 kg of steam per 1 hour. When working on biogas, the heat power can be found by the following formula:

$$W_{bl} = 1\,500 q_{ng}/C_6 = 2\,380\ MJ/h \quad (11)$$

Operating time (per day) of the steam boiler for own needs

$$t_{ot} = Q_{ou}/W_{bl} = 4{,}2\ h$$

The technical and economic calculation is based on the benefits determining method, given the profit from the marketable biogas utilization, improving the manure fertilizing properties, and increasing the degree of its disinfection. For a year on a 400 heads farm, it is required to process 12.975 m3 of manure. It is known that 1 ton of well-prepared manure in burts increases the harvest of 100 feed units per crop rotation, and 1 ton of fermented manure is more effective than burt (increasing the harvest) by 10...15 %. Therefore, the harvest annual increase will be 13 * 10^4 feed units. At a cost of 1 food.unit 8. this will give 10 400 roubles of profit.

The effect (E_1) of manure fertilizing properties improving and its disinfection is 52.547 rubles.To determine the overall benefits, a 400-animal farm with an aerobic manure decomposition system is used as the base farm. Manure flows into the pump sump, from where, as fecal pumps accumulate 4NF, it passes to the tanks for aerobic decomposition. The structure and facility costs is 144 thousand rubles. The machinery, facility, and structure costs for the new manure processing technology (E_3) more by 14,126 rubles than the base one.

The total benefit, given all component $E = E_1 + E_2 - E_3 = 52\,547 + 6\,388 - 14\,126 = 44\,809$RUB, where E_2 is the effect of marketable biogas utilization, RUB.

6. Conclusion

Methane digestion of manure in combination with the processing of biogas released into feed protein based on controlled microbial synthesis is a new production process that provides both energy and feed additives. The course of microbial synthesis in devices of different scales in terms of a unit volume of the amount of the obtained product (biomass or metabolic products) will be the same or almost the same in devices of different scales. This technology makes it possible to implement an accelerated cycle of bioconversion of substances at a livestock enterprise parallel to the traditional path of their regeneration in crop production, which provides real opportunities for creating livestock complexes in the form of waste-free production that meets all the requirements of the economy and environmental protection. An analysis of the physical properties of manure after fermentation showed that the decomposition rate of absolutely dry ashless matter was 22% with a daily loading dose of 4.5%. Humidity of manure and its ash content after fermentation increase, which is explained by the partial expenditure of dry matter on the formation of biogas. Using experimental data on biogas yield and degree of decomposition of manure, the most effective doses of the average daily loading of the digester were determined. It was found that the maximum value of this indicator should not exceed 4.5 and 9% in the mesophilic and thermophilic modes, respectively. With an increase in the loading dose, the biogas yield decreases and the process is unstable. In this case, it is possible to ensure the biogas plants effective use on livestock farms and complexes, preserving the block-modular principle of building equipment sets.

References

[1] Druzyanova V P 2010 Selection of insulation material for a biogas plant on processing cattle manure *Materialynaucho-prakticheskoikonferentsiiprepodavatelei, sotrudnikovi aspirantov "Nauchnoeobespechenierealizatsiinatsionalnogoproekta v selskom khozyaistve" [Materials of scientific and practical Conf. of teachers, employees, and postgraduates "Scientific support for the implementation of the national project in agriculture"]* (Ulan-Ude) 58-60 [in Russian]

[2] Druzyanova V P, Osipov D S and Semenov Y S 2010 Processing and utilization of organic agricultural waste *Materialyvserossiiskoinaucnoimolodezhnoishkolnoikonferentsii "Khimiyapoznakom Sigma: issledovaniia, innovatsii, tekhnologii" [Materials of all-Russian scientific youth school conference "Chemistry on the sign of Sigma: research, innovations, technologies"]* (Omsk) 230-231 [in Russian]

[3] Druzyanova V P and Kobyakova E N 2014 Biogas plant for processing waste from private livestock farms in the Republic of Sakha (Yakutia)*Sborniknauchnykhtrudovpo materialammezhdunarodnoinauchno-prakticheskoikonferentsii "Naukaiobrazovanie v XXI veke" [Collection of scientific papers on the materials of the International scientific and practical conference on October 31, 2014: at 17 h. "Science and education in the XXI century"]* (Tambov) 57-61 [in Russian]

[4] Kokieva G E 2014 Study of the cultivation of the microorganism *Scientific and Technical Volga Region Bulletin* **4** 123-125 [in Russian]

[5] Kokieva G E 2006 Feed yeast as a dietary supplement in the feeding of farm animals *Materialy regionalnoinauchno-prakticheskoikonferentsii "Pishchevyetekhnologii, kachestvoi bezopasnostproduktov" [Materials of the regional scientific and practical conference " Food technologies, quality and safety of products"]* (Irkutsk) 42-47

[6] Kokieva GE 2014 Measurement technology of working surfaces under defect apparatuse for the cultivation of microorganisms *Scientific and Technical Volga Region Bulletin* **3** p 137-139

[7] Kokieva G E andShagdyrovI B 2016Issledovaniedrozhevaniyalormovogobelka v oborudovaniipishchevoipromyshlennosti [A study of yeasting of feed protein in the food production equipment] *Bulleton of Buryat State Agricultural Academy named after V.P. Filippov* **2** 95-100 [in Russian]

[8] Sharifullin V N, BoyarinovA I and GumerovAM 1980 Svyazperemeshivaniyai massoperedachinaprimerebarbotazhno-erliftnogoapparata [Connection of mixing and mass transfer on the example of bubbling and airlift apparatus] *Mass transfer processes and devices of chemical technology* (Kazan)p 17-18

IOP Conf. Series: Materials Science and Engineering **941** (2020) 012019 doi:10.1088/1757-899X/941/1/012019

[9] ShebatinV G, Domanskii I V and others 1977 Airlift transportation of liquids and suspensions *ZhRKH* **4** 86-87.

[10] Chernorutskii I G 2004 *Metodyoptimizatsii v teoriiupravleniya [Optimization methods in the control theory / Textbook* (Saint-Petersburg: Piter) p 256

Geoinformation technologies for studying the effects of water level fluctuation of Lake Baikal: the case of Angarsky Sor

T Yu Zengina, S N Kirillov and M V Slipenchuk

Lomonosov Moscow State University, 1 Leninskie Gory, Moscow, 119991, Russia

E-mail: eco-msu@mail.ru

Abstract. Lake Baikal is the world's deepest lake and declared a UNESCO World Heritage Site. The most severe environmental problem facing the lake is the impact of water level regime fluctuation related to natural and human-caused factors. A core region under study is selected a territory of the Angarsky Sor – a shallow bay located in the Northeast of the lake. The research into a changing water content of the territory is based on imaginary provided by the Landsat-8 satellite. The study is focused on photographs taken in the high-water summer 2014 and in the law water level year 2015. The outcome of performed calculations is that an area of marshy and excessively wet regions has decreased more than 1.5 times, whereas there is a slight increase in the area of droughty waterless zones. Principal changes have been detected in the northern and north-eastern part of the Sor territory. Geoinformation technologies and remote sensing methods are considered to be efficient for the monitoring water level fluctuations, changes in the water surface area and vegetation mantle state of land ecosystems in the Sor region, what is essential for the preserving unique wetlands of the Angarsky Sor.

1. Introduction

Lake Baikal is declared a UNIESCO World Heritage Site. To date, however, a unique ecosystem of the lake has been affected by a number of natural and man-induced risks but their mapping helps keep the situation under control [1, 2]. Level fluctuations in the water body may lead to dramatic consequences.

Most factors of natural origin, which cause a water level change in Lake Baikal, have a cyclic character and are related to the shifting of water balance. For instance, recent climate changes have considerably altered the relationship between input and output water balance constituents of Lake Baikal, what may result in long-term gradual rising and falling of water surface, as well as cause the difference in water volume of the lake. Seasonal water level variations in Lake Baikal depend on the flow fluctuation of rivers entering the lake. It is reported also on wind-driven up and down surges induced by short-term (30-40 hours) ebb-and-flood with a level difference up to 10 cm.

Water level fluctuations in Lake Baikal observed over recent low-water years have impressive consequences for the environment since they pose a real threat to the unique ecosystem of the lake and may provoke a number of ecological and economic problems in the region. For instance, a water shortage brings about the lowering of a ground water level, causes difficulties in water supply and leads to drying up of wells; in addition, it is supposed to be the reason for a high forest fire danger and intensive exogenous geodynamic processes on the lake shoreland – increasing abrasion, shoreline erosion, flooding and bogging up of low-lying shore areas [3].

Several studies have highlighted the necessity of monitoring and forecasting consequences of such situations because of their importance for an economic land use [4]. However, it is a quire challenging issue as the extended Sor area and lacking connection with many of its regions restrict the use of surface techniques to some territories only. Geoinformation technologies and remote sensing methods may be quite effective in seeking for solutions to this problem and provide valuable information for studying spatiotemporal changes of many factors characterizing the state of ecosystems in the Sor.

Water level fluctuations and their consequences in Lake Baikal have been investigated on the example of the Angarsky Sor. As a principal method of analysis we used geoinformation technologies [5] and remote sensing methods. In addition, we collected all necessary data in field surveys, file materials and literature reports.

2. Materials and Methods

The Angarsky or Verkhneangarsky Sor is a shallow bay with an area of 377.3 km^2 located in the North of Lake Baikal. It includes deltas of the Verkhnyaa Angara and the Kichera; the sand Yarki Island stretching over 11 km and numerous small islands separate it from Lake Baikal, in addition, it is quite unattractive for tourists [6]. In terms of hydrology the Sor is a strongly related to water regime and complex system comprising rivers, creeks, lagoons, esturian lakes, and bogs alternating with dry land areas. A maximal depth in the Sor is about 4 m, ranging to 5.5 m in river beds.

In the early 1960-s prior to the building of Irkutsk Hydropower Plant an overgrowth area of the Sor water zone with aquatic plants was estimated to be around 75%, and rare high and low lands were used for hay production. When constructing the hydropower plant, a water level rose, as a consequence, 120 km^2 low lands in the Sor were flooded, a total overgrowth area decreased considerably, having affected a natural and economic potential of the territory.

As the territory of interest borders the Baikal-Amur Mainline, it has been an arena of dynamic economic activities [7]. Within its boundaries there are several inhabited settlements with a population of around 7 000 people, the biggest one is an urban-type settlement of Nizhneangarsk. The population of the region has been traditionally employed in fishing; hay production and cattle farming used to be main economic activities in the past. A water level rise in Lake Baikal has increased a water surface area of the Angarsky Sor [8, 9], provoking the flooding and dramatic shrinkage of hay and pasture fields, the expansion of poorly-accessible areas in the Sor and damaging of sand beaches, and accelerated the degradation of the Yarki Island, etc. From the historical literature we know that 100 years ago hay production was a principal economic activity on the entire lowland territory from the Angarakan creek to the Sor water area. Nowadays, these lands are impassable or barely-passable marshes. Fishing conditions have also been changed enormously. These circumstances have given rise to transformations in the structure of ecosystem exploitation and influenced an economic potential of the territory [9].

On the contrary, even a slightly fluctuating water level in recent years may be a reason for serious problems. A striking example is the level lowering in the late autumn 2014 related to the significantly falling water content in the Verkhnaya Angara – main feeding river of the Sor. A water shortage was observed till the autumn 2015. As a consequence, in summer 2015 a shoreline became shallow, serious changes were detected in the watering degree of Sor ecosystems, as well as in the vegetation cover, among other consequences there is the overgrowth of the basin with aquatic plants, fish kill, and prohibition on the navigation of capture and small size vessels. To illustrate, in 2014 fishing vessels used to enter the Angarsky Sor through channels in the Yarky Island, but in August 2015 the entrance into the Sor, as well as into the Verkhnaya Angara and Kichera was rather difficult due to low depths (somewhere less than 70 cm).

To assess the potential of remote sensing materials [10, 11] for analyzing a watering degree fluctuation on the territory we used imaginary data provided by the Landsat-8 satellite (camera OLI), a spatial resolution of zonal images is 30 meters. We studied pictures taken on July, 17 2014 (before a sharp drop of the water level) and those of the low-water summer in the year 2015 (July, 20).

At the preliminary stage of satellite data processing we carried out radiometric calibration in the software package ArcGis 10.3 using tool Raster Calculator in section Map Algebra (menu ToolBox). Then, fragments of a needed territory were generated and color images were synthesized (Figure 1).

Figure 1. The Angarsky Sor in the synthesized Landsat-8 image (July, 20 2015; a combination of channels 6, 5, and 2). The blue line indicates a boundary of the zone under study.

To analyze how the watering degree behaves in the Sor ecosystems we needed to select an area maximally sensitive to water degree fluctuations and ecosystems related to it. For this purpose horizontals were drawn and a map of relative altitudes was developed for the region of interest in the software package Global Mapper using data provided by Digital Terrain Model SRTM X-band (a cell size 30x30 m) and contrasted with DTM ASTER GDEM (a cell size 15x15 m). A horizontal countering the level and matching with a high flood plain was accepted a basic one, but a boundary zone was slightly expanded and first terraces of an alluvial-lacustrine plain of the the Angarsky Sor southeastern part were included because the region under study represents an important commercial fishing and environment-oriented object (Figure 1).

A review of literature focused on methods of exploring wetlands and determining a water surface area with the help of remote sensing techniques has revealed a number of approaches and methods of image processing can be adapted for this purpose: identification of multichannel spectral indexes (so-called water indexes), subject-based classification combined with learning, linear division, single-channel classification related to the difference threshold, etc. Interpretation methods involving multichannel water indexes are used most widely.

The most popular method currently is NDWI index – Normalized Difference Water Index (Water Level Index) introduced for the first time in 1996 by S.K. McFeeters as an index capable for the separating water surface and land [12, 13, 14]. The index is frequently applied to evaluating the water content in a green biomass, detecting wet, excessively and insufficiently wet zones. NDWI indexing uses green and near-infrared wavelengths of images Landsat-8 (GREEN and NIR).

3. Results and Discussion

The construction of Irkutsk Hydropower Plant in 1959-1962 is thought to be the most serious man-induced interference; as a result, a water level in Baikal rose by 1.2 m. Since a Resolution of the Government of the Russian Federation was adopted in 2001 Baikal has been an artificially controlled lake having a water level difference of one meter in a range of 456-457 m as determined in the Pacific Scale (Figure 2).

Nevertheless, to date, one of the serious problems in the region of Lake Baikal is reported to be a registration of water level falling in the lake. For instance, in April from 2010 to 2014 a water level was critically low, 456.04 m. The water level dropped abnormally in the autumn 2014, as a consequence, a low minimum ever of 455.86 m (i.e. 14 am less than a maximally permitted value) was registered in the late April 2015; a water level in summer was also very low.

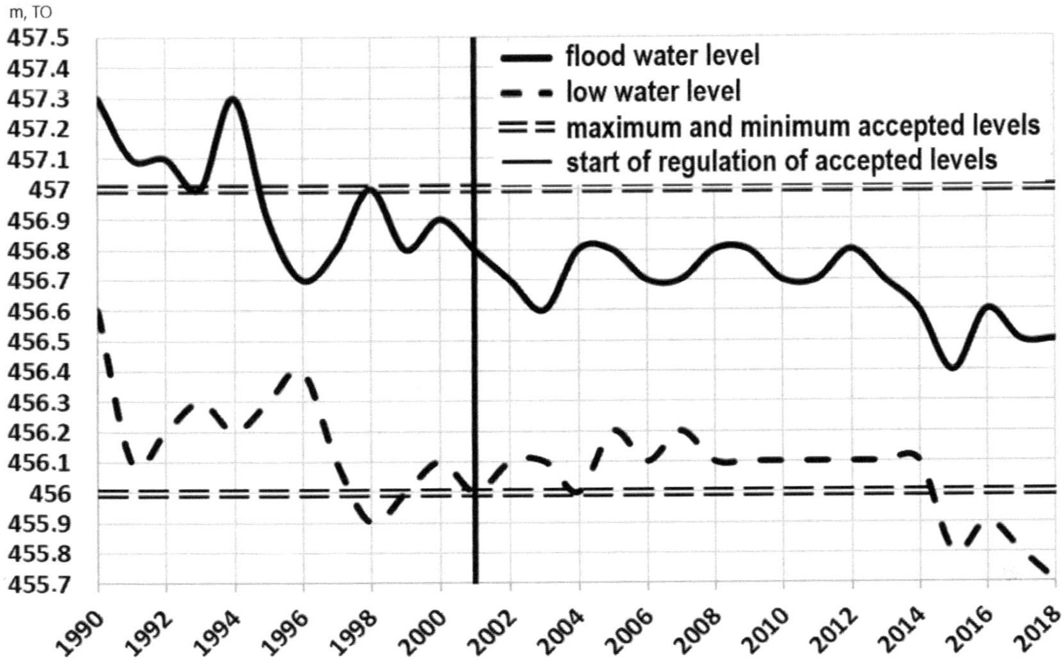

Figure 2. Water level in Lake Baikal before and after introducing in 2001 a range of level regulation from 456 to 457 m of the Pacific Scale (data provided by Irkutsk Territorial Administration for Hydrometeorological and Environmental Monitoring Rosgidromet and State Reports "On the state of Lake Baikal and measures of its protection") [8].

For a preliminary assessment of the changing open water surface over two years under study we addressed to the asynchronous synthesis procedure. For this purpose we used zonal images in near infra-red wavelengths for 2014 and 2015, respectively. From the data given in Figure 3 it is apparent in which zones on the territory of interest a water level dropped maximally and a shoreline dried up. For instance, maximally extending dried up zones shown red in the picture are located in the delta of the Verkhnaya Angara, where numerous small zones between river arms and a shoreland of the Yarky Island turned to be dry.

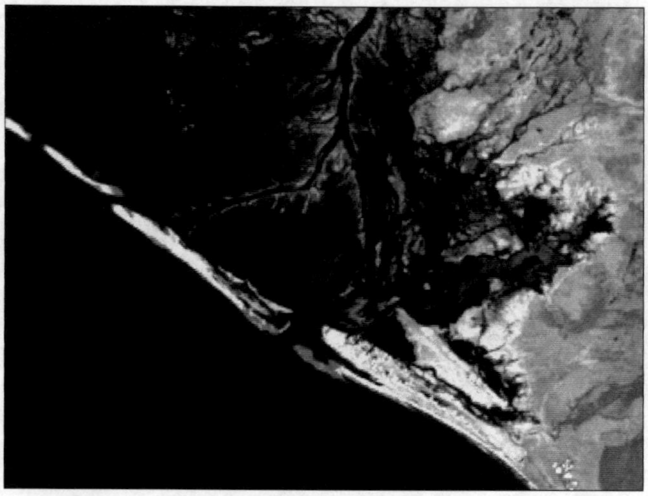

Figure 3. Asynchronous synthesis of Landsat-8 imaginary over 2014 and 2015 taken above the territory of Angarsky Sor carried out in the software MultiSpec – synthesis of fifth channels (zones dried up in 2015 as opposed to 2014 are shown in red).

Further, we used NDWI images to evaluate and compare a watering degree for two investigated dates. Calculated images of the index are presented in Figure 4, and estimated areas of zones with various watering degrees are summarized in Table 1. To find a ratio between watering degree and numerical value we adapted formulae provided on the website of Harris Geospatial Solutions (http://www.harrisgeospatial.com).

Table 1. Compared watering degrees of the territory

	NDWI index value	2014		2015	
		km^2	%	km^2	%
Droughty, non-aqueous surfaces	-1 to -0.3	225.81	46.84	240.08	49.8
Moderate droughty, non-aqueous surfaces	-0.3 to 0	35.94	7.46	25.38	5.27
Flooding, humidity	0 to +0.2	19.66	4.08	12.5	2.59
Water surface	+0.2 to +1	200.65	41.63	204.1	42.34

In addition we calculated a NDVI image (Figure 5), which was used in the analytical assessment alongside with the data on zones with various watering degrees. NDVI index (Normalized Difference Vegetation Index) is the most frequently used criterion for an amount of photosynthetically active biomass. It is calculated by formula:

$$\mathbf{NDVI = (NIR - RED) / (NIR + RED)}$$

where NIR – reflection in near infra-red wavelength; RED – reflection in the red region of spectrum.

Figure 4. NDWI-based watering degrees calculate for July, 17 2014 and July, 20 2015

NDVI index is a tool to assess the green cover productivity and state, the efficiency of forestry and agriculture, biological recourses in the region, etc. It can provide data for assessing the dislocation and suppression of vegetation communities. A combination of NDVI index and water index makes possible a more detailed spatial interpretation of results when analyzing a territory-related structure of the vegetation cover in wetlands and helps reveal plant communities in harmful moisture conditions. Furthermore, an application of the index allows a distinct separation of plants from other nature objects. Water bodies shown with the help of this index have negative values in images (below 0), whereas other nature complexes – positive values (0 to 1).

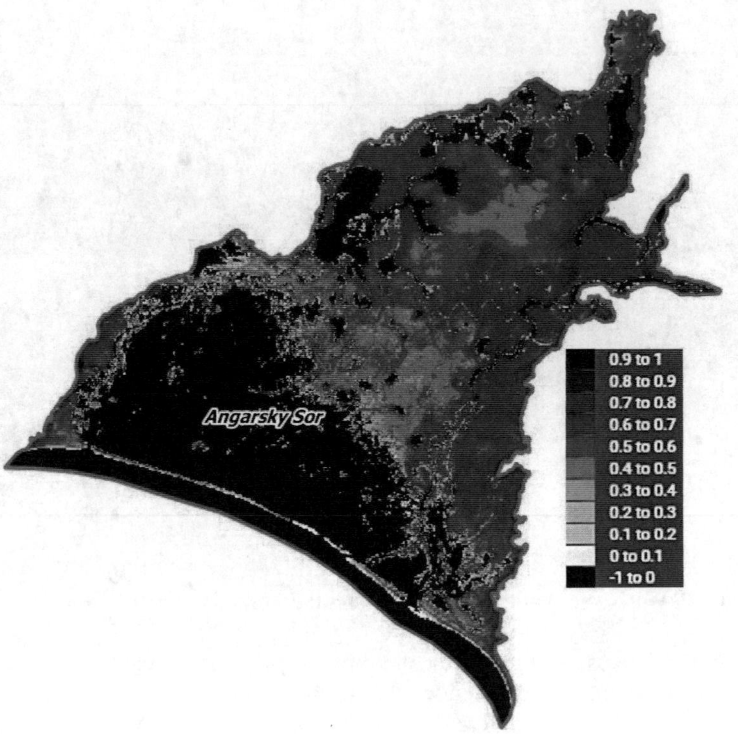

Figure 5. An example of calculating a NDVI index on July, 20 2015

From the data obtained it is seen that a water level fall in 2015 provoked certain changes within the boundaries of the Angarsky Sor. To be more precise, an area of marshy and excessively wet zones shrank by more than 1.5 times, areas of moderate droughty and poorly watered zone decreased slightly. A 3% increase of droughty, non-aqueous surfaces is reported. A maximal expansion of the water surface in 2015 is supposed to be related to a suppression state of aquatic and wetland vegetation caused by a falling water level. As a consequence, water surfaces covered in the past with floating plants (Nymphoides peltata, Nuphar pumila, Nymphaea tetragona, Sagittaria natans, Persicaria amphibian, etc.) were not categorized as water according to spectral characteristics, turned out to be opened in 2015, increasing therefore an area of water surface in a low-water year. Dramatic changes in the watering were recorded in the eastern and northeastern part of the territory under study in the region of the Verkhnaya Angara, as well as near Lakes Sikili, Poliguevskoye, Bludnoye. The watering change on a territory makes a difference for the state, productivity, and species composition of vegetation communities. A deeper investigation of this issue needs an application of high spatial resolution multi-zone images and further field surveys.

4. Conclusion

The study has revealed that Sor regions – unique shallow bays in the delta of the Selnga [15] and the Verkhneangarsky Sor, territories with outstanding biological productivity and biodiversity are most sensitive to potential dangers related to a falling water level in Lake Baikal. These are areas for the spawning of Baikal endemic species, as well as countries for a great number of amphibian and mammal species, aquatic and semi-aquatic birds. The drying up of wetlands in these regions may distort a water regime of ground waters and existing self-cleaning mechanisms of Baikal waters. A further drop of the Baikal level may cause a gradual extinction of wetlands.

Water level fluctuations in Lake Baikal and a watering degree in wetlands near it are closely related to natural factors [16]. An unstable hydrological state leads to hazardous situations for ecosystems in wetlands and poses a threat to animal and bird species living there, makes difficult an economic use of

IOP Conf. Series: Materials Science and Engineering **941** (2020) 012020 doi:10.1088/1757-899X/941/1/012020

the territory. Therefore, the constant monitoring of the situation is needed, as well as appropriate and timely protective measures. Since a majority of wetlands are extended and poorly accessible territories, remote sensing methods may be quite effective for the in-process monitoring of their state and its dynamics. Data of multi-zone satellite acquisition provide valuable and continuously updated information about the entire territory. Geoinformation and remote sensing techniques can make an important contribution to the preserving wetland ecosystems of the Angarsky Sor, because with their help it is possible to monitor water level fluctuations, the water surface behavior, as well as the dynamics of the vegetation cover in land ecosystems.

Acknowledgements
This research has been funded by the Russian Foundation for Basic Research (RFBR), project No 17-29-05078 "Impact of socio-economic processes on ecosystems, ecological and economic value of the Baikal natural territory".

References
[1] Kuznetsova T I and Lopatkin D A 2019 IOP *Conference Series: Earth and Environmental Science* **381** 012053
[2] Batuev A R, Beshentsev A N, Bogdanov V N, Dorjgotov D, Korytny L M and Plyusnin V M 2015 *Geography and Natural Resources* **36(1)** 1-12
[3] Tulokhonov A, Garmaev Y and Tsydypov B 2013 *Geography, Environment, Sustainability* **6(2)** 20-7
[4] Kirillov S, Slipenchuk M and Vorobyevskaya E 2020 *Advances in Engineering Research* **191** 103-7
[5] Gachenko A S, Minaev V, Mikhailov A A, Khmelnov A E and Fereferov E S 2016 *Geography and Natural Resources* **6** 174-8
[6] Slipenchuk M, Kirillov S, Vorobyova T and Zengina T 2015 *International Multidisciplinary Scientific GeoConference Surveying Geology and Mining Ecology Management SGEM* **2(2)** pp 979-986
[7] Vorobyevskaya E, Kirillov S, Slipenchuk M, Tulskaya N, Ustyantsev A and Tsymbal M 2018 *Ecology and Industry of Russia* **22(12)** 68-73
[8] Piotrovsky A A and Zengina T Yu 2018 *Problems of regional ecology* **2** 102-9
[9] Zengina T Yu and Piotrovsky A A 2019 *Natural and technical sciences* **8** 111-8
[10] Hwang C, Yu-Chi K and Tangdamrongsub N 2011 *TAO: Terrestrial, Atmospheric and Oceanic Sciences* **22(2)** 9
[11] Ivanov A Y 2012 *International journal of remote sensing* **33(23)** 7579-98
[12] McFeeters S K 1996 *International Journal of Remote Sensing* **17** 1425-32
[13] Xu H 2006 *International Journal of Remote Sensing* **27** 3025-33
[14] Ji L, Zhang L and Wylie B 2009 *Photogrammetric Engineering and Remote Sensing* **75** 1307-17
[15] Chalov S, Thorslund J, Kasimov N, Aybullatov D, Ilyicheva E, Karthe D and Pietron J 2017 *Regional environmental change* **17(7)** 2039-53
[16] Beshentsev A N and Tsibudeeva D T 2018 *Water Resources* **45(1)** 127-32

IOP Conf. Series: Materials Science and Engineering **941** (2020) 012021 doi:10.1088/1757-899X/941/1/012021

Innovations in the technology for producing asphalt concrete based on neutralized catalyst as a condition for improving the quality of road surfaces

G I Egorova[1], S P Semukhin[1], A N Egorov[1] and N A Kachalov[2]

[1]Tyumen Industrial University, 38 Volodarskogo Street, Tyumen, 625100, Russia
[2] National Research Tomsk Polytechnic University, 30 Lenin Ave., Tomsk, 634050, Russia

E-mail: egorovagi@list.ru

Abstract. The conditions of safety and environmental friendliness of the operation of roads as an international key problem of our time are disclosed. Objective reasons are shown that reduce the quality indicators of the construction and operation of roads. A call to the technology for producing asphalt concrete based on spent alumina-chromium catalysts is a timely need, due to the increasing requirements of international and Russian standards for the quality of roads and their operating conditions. The main scientific idea is to obtain asphalt concrete based on reduced aluminum-chromium catalyst to improve the quality of road surfaces.
A safe method is proposed for converting toxic waste into safe raw materials, taking into account the conversion of chromium (VI) compounds to chromium (III), which makes it possible to use it as part of asphalt concrete coatings. The optimal amount of alumina-chromium catalyst in the composition of asphalt concrete (6.4 ÷ 12.5 wt. %) is shown. It is proved that the use of reduced aluminum-chromium catalyst in the composition of asphalt concrete increases the proportion of waste in the total raw material balance, expands the range of building materials used in atmospheric conditions.
Samples of asphalt concrete mixtures were prepared and tested according to ISST 9128-97 with the use of spent catalyst additives.
The analysis of the current state of road surface coating technology is carried out. The optimal formulation for the preparation of asphalt mixes based on the spent catalyst having indicators of physical and mechanical properties is better than a standard sample of asphalt concrete. An effective method is proposed for removing toxic hexavalent chromium from spent catalyst as waste by transferring it to non-toxic trivalent chromium. One of the long-term problems of improving the quality of road surfaces with the inclusion of reconditioned spent aluminum-chromium catalysts, as a solution to the industrial and environmental safety of an industrial enterprise and improving the quality of operation and durability of road surfaces has been solved.

1. Introduction

Reconstruction and hopelessness - these are two brief statements for Russian roads, which are still relevant. Statistics show that 82% of modern roads require reconstructive modifications, which proves the need for the introduction of innovative achievements in both coating technology and material technology [1].

Today, well-known technologies have gained distribution: the use of asphalt concrete and cement concrete. At the same time, road surfaces are whole or assembled of their individual slabs. If we talk about the stages of construction operations, several obvious steps are obvious: preparatory work on the foundation of the road (compaction of the soil, distribution of the pillow - a layer of sand, then crushed stone).

To improve the quality and strength of coatings, a number of materials already created by scientists are used [2]. The most relevant is geotextile as a special protective material from erosion. It is possible to use other materials, for example, a geogrid, which is built between sand and gravel, which affects the durability of the use of roads. Next, the preparation of the base of the road cover is poured with a certain composition based on the bitumen mixture, and then it is required to apply asphalt concrete coating or make cement concrete coating according to the technology.

Cement concrete is the most advantageous coating for European countries and America. When conducting a literature review, we were interested in analyzing what discrepancies materials have for covering roads in America and Russia. The total percentage of cement concrete roads in America was 80%, which is very different from Russian data. In Russia, the total percentage of concrete pavement is only 3.2%.

2. Results and discussion

Our literature and patent searches have shown options for modern technological processes from classic to innovation, which are used in the construction of road surfaces [3].

Technological features of the production of classic coatings are modified: saturation of asphalt with special fibers (conductive), the inclusion of iron micro particles to eliminate cracks. The first technology is related to the fact that the fibers that conduct the current heat the bitumen, and it, in turn, glues the formed crack.

The second technology is associated with the action of an alternating magnetic field on iron particles.

If we take into account natural and climatic phenomena, then it is appropriate to note road options that are efficient in operation with the following components:

- water drainage (ground) to reduce rainfall;
- the use of special voids for the penetration of water and its quick discharge through a special trench;
- inclusion of communication with hot water circulation in the structure of the road surface;
- the use of special panels (Solar Roadways) with heating elements and LEDs operating at constant current (48 V);
- manufacture of special plastic waste (Plastic Road - environmentally friendly), which will be completely replaced with asphalt pavement (Holland);
- the inclusion of special polymer compounds for roads (Excusal Asphalt Protection) to reduce road risks in operation, eliminate icing (Holland).

With the advent of electric vehicles, the question arises of their timely charging. The construction of special rails included in the road surface for the transmission of electricity is no longer a new technology on the roads of Europe, America, and Russia.

Improving the quality of road marking materials is also an urgent task. Thus, the safety and environmental friendliness of the operation of roads is an international key problem of our time [4].

The technologies considered on the one hand will ensure safety and durability in moderate temperatures, in areas with a mild climate, which is typical for European countries. On the other hand, these are not economically viable technologies, since they are expensive processes.

There are various climatic zones on a Russian scale, which must be taken into account both within the framework of the technology for creating materials and as part of the operation of roads.

One can point to a number of existing objective reasons that reduce the quality indicators of construction and operation of roads: a) rationing, operating rules require processing; b) losses in the procurement of materials; c) low-quality bitumen, asphalt concrete, rapidly aging during operation; g) violations when covering asphalt concrete [5].

The existing reasons determined the relevance of the study - obtaining asphalt concrete from their petrochemical and chemical waste, to obtain better pavement, this is on the one hand, and on the other hand, solving the issue of rational disposal of spent catalyst as production waste will ensure the high environmental effect and the level of industrial safety at the enterprise.

These effects will ensure the quality of the environment through systemic, organized waste management; reduction of toxic components through the neutralization of toxic chromium; reduction of waste diversity (including hexavalent chromium); reduction in the number of unauthorized landfills; elimination of the source of environmental degradation of the enterprise, places of population. Today, official documents have been adopted on the strategy and future tasks for the development and safety of roads, with an implementation period of up to 2024 [6].

In addition to improving the quality of the environment and industrial safety of the enterprise, we note the fact that the waste itself - the spent catalyst contains valuable components that we use as secondary resources. In addition, this type of waste does not require additional construction of landfills, disposal tanks, which provides a high resource and economic effect to the modern enterprise. In addition, we note that the saved land, as practice shows, does not fall out of economic circulation for at least half a century.

However, the landfills used for solid industrial waste are special facilities designed to isolate and render waste harmless and do not guarantee the sanitary and epidemiological safety of the population around the industrial enterprise.

The main scientific idea is to obtain asphalt concrete based on reduced alumina-chromium catalyst to improve the quality of road surfaces [7].

In the processes of dehydrogenation of paraffin, isoparaffins in a fluidized bed of an aluminum-chromium catalyst, a significant amount of toxic waste is formed - a dust-like aluminum-chromium catalyst from 100 to 200 tons per month. This waste contains 15-20% of chromium oxide Cr_3 + (Cr_2O_3), alumina Al_2O3, silicon oxide SiO_2, potassium oxide K_2O. The water-soluble part is about 5% (compounds of chromium, potassium, iron).

2.1 Methodology of work. Spent aluminum-chromium hydrocarbon dehydrogenation catalysts contain highly toxic chromium compounds Cr_6 + in an amount of from 0.1 to 1.5 wt. %, which limits their use as secondary raw materials.

An indispensable condition for environmental safety is the absence of toxic components in their composition. In this regard, the first stage in the preparation of the spent alumina-chromium catalyst is the stage of removal of toxic hexavalent chromium compounds from it.

We have developed an original method for the conversion of toxic hexavalent chromium compounds in the composition of the spent catalyst into nontoxic trivalent chromium compounds. For this, a chemical reduction method using an active reducing agent was chosen.

The aim of the research was to maximize the reduction of Cr_6 + content in the spent catalyst up to the absence. The goal was achieved by converting toxic Cr_6 + compounds into non-toxic Cr_3 + compounds by treating the spent catalyst with aqueous hydrazine solutions at temperatures up to 100 °C, or rather, even at room temperature. Neutralization was carried out on two samples of catalysts - granular and dust-like.

The analysis of the content of Cr_6 + in the catalyst was carried out according to TU 6-68-170 "Determination of the mass fraction of chromium oxide (VI)".

The analysis of the content of Cr_6 + in the filtrate was carried out according to the measurement procedure: "Photometric determination of chromic anhydride and chromic acid salts in the air." The processing of spent chromium-containing catalysts by the proposed method allows achieving the degree of removal of toxic Cr_6 + compounds up to 100%.

The spent alumina-chromium catalyst was used by us as part of the component of asphalt concrete mixtures.

The method for preparing the asphalt mixture was to mix heated mineral materials, introduce viscous bitumen and mix it with mineral materials in a heated state.

The aluminum-chromium catalyst, crushed in a ball mill, had a grain composition, wt. %: smaller than 1.025 mm - 99.9; other indicators of the properties of the mineral powder corresponded to ISST R 52129-2003 (porosity,%, not more than 35, swelling of samples from a mixture of powder with bitumen,% - not more than 2.5, water-resistance of samples from a mixture of powder with bitumen,% - not more than 0.7, bitumen consumption index, g - not more than 80, humidity,% by mass - not more than 1.0). Natural sand corresponded to ISST 8736-93; as a binder used bitumen road class BND, ISST 22245-90. Samples of asphalt concrete mixtures were prepared and tested according to ISST 9128-97 with the use of spent catalyst additives.

The results showed high-quality asphalt mixtures.

The optimal formulation of the asphalt mix, % mass: bitumen - 10%, spent catalyst 20%, sand up to 70%.

The manufacture and testing of samples (1-5) was carried out according to ISST 12801-98. Based on the results obtained, it can be concluded that samples of asphalt concrete mixtures, which include reduced alumina-chromium catalyst, have better physical and mechanical properties than standard samples (Table 1).

In the formulations of the mixture, the optimal amount of chromium-chromium catalyst should be in the range of 6.4 ÷ 12.5 wt. % (Samples 2-4). A decrease or increase in the catalyst content in the mixture leads to a decrease in physical and mechanical parameters: water saturation and water resistance coefficient (samples 1-5) (Table 1).

Table 1. Test results of samples of asphalt concrete mixtures based on reduced toxic chromium-containing waste.

Indicators	Standard	1	2	3	4	5
Density, g / cm^3	2.34	2.32	2.33	2.34	2.33	2.38
Water saturation about, %	0.73	2.58	2.57	0.8	1.99	6.13
Swelling, vol. %	0.0	0.8	0.0	0.0	0.1	1.3
Tensile strength in compression, MPa, at temperature: 20°C	5.1	4.1	5.1	5.3	5.3	3.6
50°C	2.7	2.1	2.6	2.8	2.6	1.6
Water resistance coefficient	1.02	0.8	1.03	1.12	1.01	0.8

The maximum value of the compressive strength at a temperature of 50 °C is achieved with the following range of values: the percentage of crushed stone (44 ÷ 51.1); water resistance coefficient (0.825 ÷ 0.864). The technical result obtained is the use of spent catalyst in the composition of asphalt concrete is reflected in patent 2403217 RU C1 C04B026 / 26.

Prototypes of asphalt concrete were tested during the operation of pavements in the southern regions of the Tyumen region. The features of asphalt concrete of such compositions were laid hot, after which the coating was leveled and compacted by equipment. According to ISST 9128-2009, the temperature of the asphalt mix during installation was at least (103 °C). Lying was built in layers (from 5 to 7 cm), bitumen was used for the adhesion process.

3. Conclusions

As part of our research, we conducted a patent and literature review to identify proprietary technologies for improving the quality of asphalt concrete for road surfaces.

The analysis of the current state of road covering technologies is carried out; their advantages and disadvantages are revealed.

An effective method for removing toxic hexavalent chromium from the spent catalyst as a waste by converting it to non-toxic trivalent chromium is proposed. Optimal recipes for preparing asphalt concrete mixture based on spent catalyst have been developed, which have indicators of physical and mechanical properties better than the standard sample of asphalt concrete.

One of the long-term problems of improving the quality of road surfaces has been solved, taking into account the inclusion of recovered spent aluminum-chromium catalysts, as a solution to industrial and environmental safety of an industrial enterprise and improving the quality of operation and durability of road surfaces.

References

[1] Arkhipov A 2010 *Terra economics* **3** pp 36-46
[2] Gladkov In 2009 Standards and quality **5** pp 94–97.
[3] Silyanov In 2008 *Transport and operational qualities of roads: a textbook for higher education institutions* (Moscow: Academy) R 352
[4] Egorov A 2016 *Waste from petrochemicals enterprises. Rational disposal methods.* Conference materials Energy saving and innovative technologies in the fuel and energy complex (Tyumen: Tyumen State Petroleum University) pp 265-268
[5] Ushakov V and Yarmolinsky In 2019 *Guidelines for the installation of asphalt concrete pavements under adverse weather conditions* editor Zakharova M (Moscow: Rosavtodor) p 46
[6] National project passport 2018 *Safe and high-quality roads* (Moscow: Rosavtodor) p 66
[7] Egorov A 2013 *Oil refining and petrochemicals* **3** pp 91 - 99.

State of Humus in Soils under Rice Cultivation

A Kh Sheudzhen[1,2] and O A Gutorova[1,2]

[1]Kuban State Agrarian University named after I.T. Trubilin, Kalinina, 13, Krasnodar, 350044, Russian Federation
[2]Federal scientific rice centre, Belozerny, 3, Krasnodar, 350921, Russian Federation

E-mail: ashad.sheudzhen@mail.ru

Abstract. The study investigates an anthropogenous influence on the humus content in meadow and meadow-bog soils under rice cultivation in Adygea (Gleyic Phaeozems and Calcic Chernic Gleysol). A field experiment focused on the use of mineral fertilizers $N_0P_0K_0$ (control); $N_{90}P_{60}K_{45}$; $N_{120}P_{80}K_{60}$; $N_{150}P_{100}K_{75}$ and $N_{180}P_{120}K_{90}$ is carried out on rice paddies. Soil samples are taken from the plow layer before seeding and after harvesting of rice; a total humus content and fraction-group composition are studied. An increase in fulvic acids is detected in soils as a consequence of dominant reduction and oxidative degradation of humic acids. Heightened amounts of fertilizers make humus more fulvic. A labile fraction in the total humus content is determined to be 7.97-11.02 % in meadow soil and 5.38-8.33 % in meadow-bog soil. In labile humus of meadow soils fulvic acids prevail over humic acids (HA-1: FA (1a+1) = 0.71-0.89). A share of fulvuc compounds is higher in meadow-bog soils (HA-1: FA (1a+1) = 0.38-0.55). Mineral fertilizers taken in ratios of $N_{150}P_{100}K_{75}$ and $N_{180}P_{120}K_{90}$ raise a percentage of newly formed humus substances.

1. Introduction

The soil use for rice crops is associated with an additional anthropogenous factor – long-term soil flooding. This fact necessitates an annual 4-5 month recovery of paddy soils. This dramatically changing environment leads inevitably to transformations in essential elementary soil processes and affects the soil balance. Humus-related impoverishment or enrichment of paddy soils depends on the amount and biochemical composition of a coming organic matter and its decay rate. The intensification of crop residues mineralization when irrigated is supposed to be a natural process because there is a specific dynamic balance in humus of soils not irrigated for a long time. Its content and composition are in line with certain bio-hydrothermal conditions, transforming, they make old humus irresistible to the new environment, facilitating its degradation. Over time, properties of humus substances tend to change and they are not in conflict with new biochemical and hydrothermal conditions [1-3].

The humus lability develops in periodically flooded paddy soils. Processes of reduction to evolve in excessively moistened soils decrease a share of humus for water soluble organic matters are taken away, and humification shifts increasingly to the formation of fulvic acids [4-7]. Both a total quantaty of humic acids and their calcium-related content decline, i.e. the most active humus fraction important to soil buffering is removed. At the same time, groups of tightly bound humic acids and humin, i.e. low-active humus rise [8].

The study aims to explore an influence of anthropogenous factors on the humus content of paddies.

IOP Conf. Series: Materials Science and Engineering **941** (2020) 012022 doi:10.1088/1757-899X/941/1/012022

2. Objects and methods

The research was carried out on meadow and meadow-bog soils of rice paddies in Takhtamukaysky District of the Republic of Adygea (Gleyic Phaeozems and Calcic Chernic Gleysol in WRB [9]). In geomorphology the northern part of the territory belongs to the Kuban delta and floodplain, which includes low highlands and quite wide and extended depressions. The southern region is a part of the Pre-Kuban sloping plain. The hydrographic network covers the river of Kuban with its tributaries - mountain rivers: Psekups, Afips, Tsitsa, Apgas, Shunduk, etc.

Soil-geographical regionalization attributes the district to steppe lands of the chernozyom region in the Azov-North Caucasian steppe facies of the soil area in the lower reach of the Kuban River. Basically, the soil mantle consists of meadow soils, i.a. meadow gleic and meadow-bog types. Soils are referred to as classified in [10].

Geographically, meadow soils are found in the floodplain of the Kuban and its tributaries, meadow gleic soils – in lower zones, and meadow-bog ones in closed low floodplains and ravine bottoms.

A field experiment on fertilization was carried out in a rice paddy. Fertilizers were applied before rice seeding in ratios: $N_0P_0K_0$ (control); $N_{90}P_{60}K_{45}$; $N_{120}P_{80}K_{60}$; $N_{150}P_{100}K_{75}$; $N_{180}P_{120}K_{90}$.

In soil samples taken before seeding and after harvesting a content of total humus was estimated by the Tyurin method, and to characterize the humus composition we used the Tyurin method with modifications by Ponomaryeva and Plotnikova. A state of labile and inert humus was analyzed according to fractions of humic and fulvic acids.

The Tyurin method includes sulfur-chromium oxidizing of an organic matter. A fraction of organic carbon in the soil was corrected to humus content using a coefficient of 1.724 [11, 12].

When assessing the humus composition as in the modification by Ponomaryeva and Plotnikova, an alkaline extract was taken from a soil weighted sample and various fractions of organic matters were sequentially extracted from other weighted samples of the same soil type. Within this method it is possible to divide soil humus into three fractions of humic acids – HA (fraction HA-1 free and bound with non-silica sesquioxides; HA-2 calcium-related; HA-3 – with stable forms of sesquioxides and clay minerals), and four fractions of fulvic acids – FA (fraction FA-1a free and bound with non-silica sesquioxides or "aggressive"; FA-1 – bound with fraction 1 of humic acids; FA-2 – bound with fraction 2 of humic acids; FA-3 – bound with fraction 3 of humic acids) and humin (insoluble part) [11].

3. Results and discussion

Before rice seeding humus content in the plow layer of meadow and meadow-bog soils was as high as 3.70 and 3.90 %, respectively. The use of mineral fertilizers had a positive effect on the humus state of soils – at the end of a rice growing season a humus content increased by 0.08-0.14 % and by 0.10-0.17 %, respectively, as compared with the control data (Table 1-2).

An important characteristic of soil humus is its group and fraction composition. Studies with the focus on quality composition of soil humus in rice paddies are of theoretical and practical significance, contributing to the control of pedogenesis processes, which are relevant for the humus state.

A correlation between humic and fulvic acids in humus composition provides information on a humus-formation type, a resultant of natural factors and anthropogenous influence on soils. A content of carbon in humic acids is higher than that in fulvic acids in the group composition of meadow and meadow-bog soil humus. When a rice vegetation season was completed and irrigation water removed from paddy fields, a content of fulvic acids increased from 18.9 and 20.0 to 19.6-21.1 and 20.8-22.9 % vs. total carbon, respectively. The fertilization didn't make a difference for a content of humic acids but a fraction of fulvic acids increased in the composition of humus. In addition, an increased use of fertilizers makes humus more fulvic, especially in meadow-bog soils, where humic type of humus alters to the fulvic and humic one. Meadow soil humus is humic (C_{ha}: C_{fa} = 2.17-2.49), and that of meadow-bog soil – humic (C_{ha}: C_{fa} = 2.10-2.28) or fulvic and humic (C_{ha}: C_{fa} = 1.94-2.00). The latter correlation C_{ha}: C_{fa} is narrower, indicating higher mineralization resistance of humus.

Humin characterizes the strength humus matters bind with a mineral soil element, i.e. its silt fraction; its hydrolysis and mineralization are hard. In soils of rice paddies under study a low amount of non-hydrolyzed residue was detected. They display no significant differences. Fertilizers reduced an insoluble residue from 34.1-34.5 (control) to 32.4-33.8 % vs. C_t on fertilized areas.

Transformations in humus and its growth are well characterized using labile forms of humus substances. These are "young" forms of humus with a high nitrogen concentration. They tend to quick mineralization, supplying nitrogen to plants. A ratio between labile and inert humus fractions in soils makes possible the estimation of humus content dynamics and its resistance to negative natural and anthropogenous factors, and their dehumification rates [13].

In contrast to meadow soils humus of meadow-bog soils is characterized with low lability, an inert part in its composition amounts to 91.7-94.6 % vs. total humus. This percentage in meadow soil is less significant, being 89.0-92.0 % of total humus (Table 1-2).

Table 1. Total, labile, and inert humus in the meadow soil

Variant	Determination period	Total humus (%)	Labile humus (%)	Inert humus (%)	Labile vs. total (%)
$N_0P_0K_0$	prior rice seeding	3.70	0.34	3.36	9.19
	after harvesting	3.64	0.29	3.35	7.97
$N_{90}P_{60}K_{45}$	after harvesting	3.66	0.38	3.28	10.38
$N_{120}P_{80}K_{60}$	after harvesting	3.78	0.38	3.40	10.05
$N_{150}P_{100}K_{75}$	after harvesting	3.78	0.40	3.38	10.58
$N_{180}P_{120}K_{90}$	after harvesting	3.72	0.41	3.31	11.02

In labile humus of meadow soil fulvic acids prevail over humic acids in a ratio HA-1: FA (1a+1), equal to 0.71-0.89. Importantly, a content of humic and fulvic acids was equal before flooding: HA-1: FA (1a+1) = 0.10.

Labile humus of meadow-bog soil contains more fulvic compounds with a ratio HA-1: FA (1a+1) = 0.38-0.55, what indicates the considerable predominance of fulvic acids. Before flooding a concentration of labile humus was low with dominating fulvic acids: HA-1: FA (1a+1) =0.50.

In inert humus of meadow soil calcium-related fractions are most important (34.7-37.8 %) in a ratio HA-2: FA-2 = 7.67-8.75, as well as humin in a percentage of 32.8-34.6 % of total humus. Fractions of clay minerals and stable forms R_2O_3 are least important with a share of 18.6-21.0 % in total humus. Fulvic acids prevail there with a ratio HAK-3: FA-3, equal to 0.82-0.96.

Inert humus in meadow-bog soil is similar to that of meadow soil. Calcium-related fractions are in a percentage of 35.1-39.0 % (HA-2: FA-2 = 4.73-5.83), non-hydrolyzed residue amounts to 32.3-34.4 %; fractions of clay minerals and stable forms R_2O_3 – 21.3-24.0 % of total humus (HA-3: FA-3 = 1.00-1.12).

Table 2. Total, labile, and inert humus in the meadow-bog soil

Variant	Determination period	Total humus (%)	Labile humus (%)	Inert humus (%)	Labile vs. total (%)
$N_0P_0K_0$	prior rice seeding	3.90	0.21	3.69	5.38
	after harvesting	3.82	0.21	3.61	5.50
$N_{90}P_{60}K_{45}$	after harvesting	3.92	0.26	3.66	6.63
$N_{120}P_{80}K_{60}$	after harvesting	3.99	0.29	3.70	7.27
$N_{150}P_{100}K_{75}$	after harvesting	3.96	0.33	3.63	8.33
$N_{180}P_{120}K_{90}$	after harvesting	3.93	0.32	3.61	8.14

The use of mineral fertilizers raised a share of labile humus in soils. Before rice seeding and flooding of paddies its amount was 0.34 in meadow and 0.21 % in meadow-bog soils, on vegetation completion – 0.29 and 0.21 % (control) and 0.38-0.41 and 0.26-0.33 % on fertilized areas, respectively. The increased usage of fertilizers $N_{150}P_{100}K_{75}$ and $N_{180}P_{120}K_{90}$ contributed to the enrichment of soils with labile humus. At the end of rice cultivation period its content in total soil humus was as high as 10.58 and 11.02 % in meadow and 8.33 and 8.14 % in meadow-bog soil, respectively (Table 1-2).

To summarize, mineral fertilizers raise an amount of newly formed humus substances, which are immediate reserves of fertilization elements for plants and microorganisms. Therefore, the total humus content and its labile part represent most important indicators of the humus state in rice paddy soils.

4. Conclusion

When growing rice, fulvic acids increased in meadow and meadow-bog soils owing to dominant reduction processes and oxidative degradation of humic acids. The use of mineral fertilizers contributed to humus preservation in soils. In humus of soils under consideration compounds of humic acids or substances related to fulvic and humic acids dominate. An increase of fertilizers makes humus more fulvic. A labile fraction in total humus amounts to 7.97-11.02 % in meadow and 5.38-8.33 % in meadow-bog soil, respectively. In labile humus of meadow soil fulvic acids prevail over humic acids (HA-1: FA (1a+1) = 0.71-0.89). Humus of meadow-bog soil is more fulvic (HA-1: FA (1a+1) = 0.38-0.55). Fertilizers contribute to the development of "young" labile humus substances. The increased usage of fertilizers $N_{150}P_{100}K_{75}$ and $N_{180}P_{120}K_{90}$ intensify processes of humus formation.

References

[1] Bahmanyar M A 2007 The Influence of Continuous Rice Cultivation and Different Waterlogging Periods on Morphology, Clay Mineralogy, Eh, pH and K in Paddy Soils *Pakistan Journal of Biological Sciences* **10** pp 2844-9

[2] Huang L M, Thompson A, Zhang G L, Chen L M , Han G Z and Gong Z T 2015 The use of chronosequences in studies of paddy soil evolution: a review *Geoderma* **237-238** pp 199-210

[3] Winkler P, Kaiser K, Kölbl A, Kühn T, Schad P, Urbanski L, Fiedler S, Lehndorff E, Kalbitz K et al 2015 Response of Vertisols, Andosols, and Alisols to paddy management *Geoderma* **261** pp 23-35

[4] Sheudzhen A and Gutorova O 2020 Changes in the Magnetic Susceptibility of Kuban Soils Depending on Soil Formation Conditions *Advances in Engineering Research* **191** pp 264-8

[5] Sheudzhen A Kh, Gutorova O A, Shein E V and Romanenkov V A 2019 Agrogenic soil evolution of rice agrolandscapes IOP *Conf. Series: Earth and Environmental Science* **368** 012044

[6] Maie N, Watanabe A and Kimura M 1997 Origin and properties of humus in the subsoil of irrigated rice paddies. I. Leaching of organic matter from plow layer soil and accumulation in subsoil *Soil Science and Plant Nutrition* **43 (4)** pp 901-10

[7] Mitsuchi M 1974 Characters of humus formed under rice cultivation *Soil Science and Plant Nutrition* **20** pp 249-59

[8] Nikolayeva S A 1996 Stability of delta ecosystems under intensive irrigation for rice cultivation *Eurasian Soil Science* **28 (11)** pp 320-8

[9] *World Reference Base for Soil Resources 2014. Update 2015. International soil classification system for naming soils and creating legends for soil maps* (World Soil Resources Reports No. 106 FAO, Rome) p 192

[10] Egorov V V, Fridland V M, Ivanova E N, Rozov N N, Nosin V A and Friev T A 1977 *Classification and soil diagnostics of the USSR* (Moscow: Kolos) p 221 (in Russian)

[11] Orlova N E, Bakina L G and Orlova E E 2007 *Study methods of humus composition and content* (Saint-Petersbur: Publishing of Saint-Petersburg University) p 145 (in Russian)

[12] Blair G J, Lefroy R DB and Lise L 1995 Soil Carbon Fractions Based on Their Degree of Oxidation, and the Development of a Carbon Management Index for Agricultural Systems *Australian Journal of Agricultural Research* **46** pp 1459-66

[13] Akhtyrtsev B P and Akhtyrtsev A B 2002 Changes in the humus state of forest-steppe and steppe chernozems resulting from their long-term plowing and sealing under burial mounds *Eurasian Soil Science* **35(2)** pp 123-131

Innovation Clusters in the Arctic Zone of Russian Federation

A A Pankratov[1], R A Musaev[1] and S V Badina[2,3,4]
[1] Faculty of Economics, Lomonosov Moscow State University (MSU), 1 Leninskie Gory, Moscow, 119991, Russia
[2] Plekhanov Russian University of Economics, 36 Stremyanny Street, Moscow, 117997, Russia
[3] Peoples' Friendship University of Russia, 6 Miklukho-Maklaya Street, Moscow, 117198, Russia
[4] Institute of Economic Forecasting, Russian Academy of Sciences, 47 Nakhimovsky Prospekt, Moscow, 117418, Russia

E-mail: pankratov_aleksey_ml@mail.ru

Abstract. The article discusses the problems, directions and opportunities for the development of federal level innovation clusters in the Arctic zone of Russian Federation (AZRF), taking into account the Russian state long-term plans for Arctic territories development and implementing major investment projects. The article analyzes the current territorial structure of supported clusters in the AZRF, explores their industry and production specifics, and identifies the main distinctive features and characteristics of them. Based on the analysis of Russian Arctic regions socio-economic development indicators, the regions, municipalities and cities for the prospective cluster formations development are identified; directions, key specializations and cluster development factors are distinguished. The article justifies the relevance of innovative clusters formation in the context of task innovative, territorial, technological, information and communication, socio-economic development by AZRF.

1. Introduction

The advanced development of the AZRF is one of the most important national priorities for the Russian Federation socio-economic and territorial development in the 21st century [1, 2]. The list of territories within the AZRF is approved by the Russian President Decree – "On the Land Territories of the Arctic Zone of the Russian Federation". According to it, the AZRF includes the territory of the Murmansk Region, the Nenets, Yamalo-Nenets, Chukotka Autonomous Districts, as well as certain municipalities of the Karelia, Komi and Yakutia Republics, the Krasnoyarsk Krai and Arkhangelsk Region.

2. Federal level strategic planning documents in the field of Arctic zone of Russian Federation development

The special status of the AZRF is established by federal level strategic planning documents. Thus, according to the Spatial Development Strategy of the Russian Federation until 2025, the AZRF, as well as the Far East, Crimean Peninsula, North Caucasus and Kaliningrad Region are recognized as geostrategic territories of the Russian Federation. Geostrategic territories are defined as territories within the borders of one or few regions that have significant importance for ensuring sustainable socio-economic development, territorial integrity and security of the Russian Federation, characterized by specific living conditions and economic activities.

The special importance of the Russian Arctic in terms of solving the strategic tasks of economic development and completing the basic transport, energy, information, and military infrastructures formation is noted in the Russian Federation National Security Strategy. The Russian President Decree – "On the Fundamentals of Russian Federation State Policy in the Arctic until 2035" defines the long-term directions and priorities of the AZRF socio-economic development until 2035. The federal budget provides a separate state program for "Socio-economic development of the Arctic zone of Russian Federation until 2025", which determines budget financing for the implementation of large investment projects in the AZRF: the total budget for this program is about 7 trillion rubles until 2025. The current

IOP Conf. Series: Materials Science and Engineering **941** (2020) 012023 doi:10.1088/1757-899X/941/1/012023

Development strategy for the AZRF and national security until 2020 will be completed in 2020. However, as of June 10, 2020, a new strategic planning document on the federal level has been presented a draft Strategy for the development of the AZRF until 2035 was approved by the Russian Government and is being finalized at the Russian Security Council.

3. Directions of the AZRF socio-economic and spatial development

The main national interests in the AZRF is socio-economic and spatial development for the period until 2035: the Russian Arctic development as a strategic resource base and its sustainable use; transport infrastructure development, including an increase in Northern Sea Routes cargo turnover up to 80 million tons per the year by 2024, the use of the AZRF transit potential; state support of entrepreneurship [3]; environmental protection and conservation, environmental safety and sustainable use of natural resources; adaptation of the population and economy to global climate change and its consequences [4, 5].

The largest and most capital-intensive projects in the Russian Arctic are projects related to the main transport infrastructure development, included in the separate federal project "Development of the Northern Sea Route" as part of the Comprehensive Plan for the Modernization and Expansion of the Main Infrastructure until 2024:

Northern Sea Route development – around 906 billion rubles;
Northern latitudinal railway construction – 233 billion rubles;
Sabetha seaport facilities construction (Yamal Peninsula) – 144 billion rubles;
Murmansk transport hub development – 74 billion rubles;
Marine transhipment complex for liquefied natural gas in the Murmansk region – 60 billion rubles;
Chaika coal terminal in the Dikson seaport (Krasnoyarsk Krai) – 19 billion rubles, others.

Due to government measures, the AZRF contribution to the total socio-economic development in the Russian Federation is growing. An evidence therefore is the increasing share of gross regional product (GRP) produced in the AZRF, of the overall Russian GRP, as well as some other important socio-economic indicators [6].

However, according to the innovative and technological development target indicators for the AZRF, in recent years there was a negative trend, which is primarily determined by the strengthening of the raw material economy specialization (table 1):

Table 1. Socio-economic development indicators of the AZRF in 2015 and 2019[a].

Indicator / unit of measurement	2015	2019	Dynamics
The share of Arctic GRP in the total Russian regions GRP /%	5	5.8	16% increase
Average life expectancy / years old	71.02	72.39	2% increase
The share of the population using the information and telecommunication network "Internet" in the total AZRF population / %	83.4	91.2	9% increase
The value added share of high-tech and knowledge-based economy sectors in the AZRF GRP / %	7.5	6.1	19% reduction
Percentage of science-based innovative goods and services of organizations in the total goods and services volume / %	0.17	0.05	3.4 times reduction

[a]Source: Rosstat, official statistical information on the AZRF socio-economic development, compiled by the authors

4. Results and Discussion: innovation clusters in the AZRF: current state and development prospects

Within the framework discussed in section 2, 3 the federal level strategic documents for AZRF development, covers the tasks of Russian Arctic industrial potential accelerating development, increasing the manufacturing industries share (including high-tech products with high added value).

Despite these tasks set, in current conditions the raw material sector of the Arctic economy is developing much more intensively than manufacturing sector, but this remark does not remove the need to solve the problem of accelerating the development of the manufacturing sector in the AZRF territory.
Innovation clusters can be considered as one of the instruments ensuring industrial, innovative and technological development in the Arctic territory. The main problem of the cluster development in the Russian Arctic lies in the fact that the clusters formation and support, as a rule, carried out in large cities and developed territories specializing in manufacturing and innovative economic activities, while the creation of clusters in hard-to-reach territories with a raw material economy specialization is widespread to a lesser degree and represents a difficult management task [7, 8].
As of June 2020, according to the Russian Cluster Observatory, 118 federal level clusters are supported in the Russian Federation, and only four of them are located in the AZRF: 2 each in the Murmansk and Arkhangelsk regions (table 2):

Table 2. Federal level clusters supported in the territory of the AZRF, 2020[a].

Cluster	Key specialization	Number of residents	Employed, people	Year of creation
Shipbuilding cluster of the Arkhangelsk region	Shipbuilding	23	50427	2012
Timber cluster of the Arkhangelsk region "PomorInnovaLes"	Forestry and woodworking; pulp and paper industry	31	20110	2014
Tourist and recreational cluster of the Murmansk region	Tourism (entertainment and leisure industry, art, sports)	12	59	2015
Northern Design Cluster of the Murmansk Region	Business services	12	90	2018

[a]Source: Russian Cluster Observatory

The main centers of cluster development in the Russian Arctic are Murmansk and Arkhangelsk regions. At the same time, the potential for cluster development in other parts of the AZRF remains highly unrealized.
Taking into account the state's plans for the Arctic territories development, innovative clusters can be formed and supported in the following industries and regions: transport sector – maritime safety, marine navigation systems and ship repair (Murmansk and Arkhangelsk regions, Karelia Republic); ecology and environmental protection, ecological nature management (Yamal-Nenets Autonomous Okrug); timber industry sector, deep wood processing, furniture production (Arkhangelsk region, Karelia Republic); biotechnology (Arkhangelsk region); integrated processing of coal and industrial wastes, coal chemistry (Nenets Autonomous Okrug, Komi Republic); information and communication technologies (Yamalo-Nenets Autonomous Okrug); military-industrial sector (Murmansk and Arkhangelsk regions); nuclear and radiation technologies (Murmansk region); electric power industry (Murmansk region); new materials (Murmansk and Arkhangelsk regions, Krasnoyarsk region); natural hazard management and protection (Murmansk and Arkhangelsk region, Republic of Karelia) [9, 10].
Perspective centers of cluster development in the AZRF, primarily, can be considered large settlements – cities and urban-type settlements with year-round transport accessibility. As an important cluster development factors should also consider the city population size, as well as the presence in it of large educational institutions and research organizations.
In total, 32 cities with a 1.84 million people population size, which is 78% of the total Russian Arctic population, located in the AZRF. Of the 32 Russian Arctic cities, only eight have a population size over 50 thousand inhabitants: Arkhangelsk, Murmansk, Severodvinsk, Norilsk, Novy Urengoy, Noyabrsk, Apatity, Salekhard. These cities are the most promising centers of cluster development in the AZRF (table 3):

Table 3. Perspective centers of cluster development in the AZRF[a].

City	Region	Population, 2020, thousand people	Number of Universities
Arkhangelsk	Arhangelsk region	359	3
Murmansk	Murmansk region	298	4
Severodvinsk	Arhangelsk region	185	1
Norilsk	Krasnoyarsk region	179	1
New Urengoy	Yamal-Nenets Autonomous Okrug	115	-
Noyabrsk	Yamal-Nenets Autonomous Okrug	107	1
Apatity	Murmansk region	55	2
Salekhard	Yamal-Nenets Autonomous Okrug	51	1

[a]Source: Rosstat, Ministry of Education and Science of Russia, compiled by the authors

5. Conclusions

Currently, a small amount of federal-level innovation clusters is supported in the AZRF. The specific features and characteristics of this instrument, its connectedness to the developed territories and large cities, explains the low demand for cluster formations in the AZRF. This study substantiates the relevance of innovation clusters creating in the most important in terms of socio-economic development sectors and major urban centers in the Russian Arctic. Cluster entities as objects of innovative infrastructure will contribute to solving the problem of developing the Russian Arctic industrial sector and increasing the high-tech industries share in the total AZRF economy.

Acknowledgements

The reported study was funded by RFBR, project number 19-310-90081.

References

[1] Zamyatina N and Pilyasov A 2018 *Regional Research of Russia* **3** 215–224
[2] Zamyatina N and Goncharov R 2018 *Area Development and Policy* **3** 293–308
[3] Pilyasov A 2020 *GEOGRAPHY, ENVIRONMENT, SUSTAINABILITY* **1** 46–56
[4] Baburin V, Tikunov V, Badina S and Chereshnia O 2018 *Regional Science Inquiry* **2** 37–44
[5] Baburin V, Badina S, Goryachko M, Zemtsov S and Koltermann K 2016 *Vestnik Moskovskogo Universiteta, Seriya 5: Geografiya* **6** 71–77
[6] Zamyatina N and Pilyasov A 2016 *Foresight and STI Governance* **3** 53–64
[7] Zemtsov S, Barinova V, Pankratov A and Kutsenko E 2016 *Foresight* **3** 34–52
[8] Shiklomanov N, Streletskiy D, Suter L et al 2020 *Land Use Policy* **93** 2–11
[9] Romanovsky V, Smith S, Isaksen K et al 2019 *Bulletin of the American Meteorological Society* **9** 153–156
[10] Shiklomanov N, Streletskiy D, Suter L et al 2017 *Polar Geography* 1–18

Dynamics of Agrarian Lands of the Republic of Tyva (Russia) in 1890–2019

A. Sambuu, D. Dabiev

Tuvinian Institute for the exploration of natural resources of the Siberian branch of the Russian Academy of Sciences, Kyzyl, Russia

E-mail: Sambuuanna@sambuuanna.yaconnect.com

Abstract. The article is devoted to the analysis of changes of croplands dynamics of the Republic of Tyva during traditional, Soviet, crisis and post-crisis periods. It is shown that despite the fact that the leading branch of agriculture of the Tyva People's Republic (TPR) was animal husbandry, crop production developed no less dynamically, and crop area in 1944 increased by 4.3 times over 17 years. During the period of the Tuva Autonomous region (TAR), crop area increased significantly, so in comparison with 1945, by 1960, the area of grain crops increased more than three times, amounting to 279.8 thousand hectares, including wheat – 193.6 thousand ha, barley – 21.7 thousand ha, oats – 34.7 thousand has, millet – 27.1 thousand ha. The crop area of the feeding grounds has increased by an order of magnitude. The acreage of vegetable and melon crops, including potatoes, has significantly increased. But the unjustified plowing of virgin and fallow lands in the 50s and 60s of the twentieth century led to soil erosion and destruction, and the loss of more than a million hectares of agricultural land. During the period of the Tuva ASSR, there is a further increase in crop area. The largest area of grain crops was observed in 1980, when there were 370 thousand hectares in the Republic. At the same time, the Central zone accounts for about 2/3 of the crop area of grain crops, the Western zone – 20%, the Southern zone – 5%, and the Eastern zone – only 1%. With the beginning of market reforms, there is a decrease in the crop area of grain crops. Lack of state support, violation of economic relations, price disparity, lack of tax incentives and preferential loans, and other market and non-market reasons led to the ruin of large farms in the Agro-Industrial Complex (AIC), which accounted for more than 99% of agricultural products. The crop area of grain crops in 2000 compared to 1980 decreased by 8.4 times, wheat by 6.7 times, and fodder crops by 11 times. At present, there is again a slight increase in croplands.

1. Introduction

The steppe region of Central Asia has supported human culture for thousands of years and during this period has been subject to major climatic and social changes. In the past, this area was sparsely populated and the land fully provided for the needs of the local population, leading a nomadic lifestyle. Folk customs strictly regulated the nature, timing, cycles and duration of land use. Also, due to repeated rest periods, the condition of the land was good [1].

The steppes of the Asian part of Russia are a significant agricultural region, but little attention is paid to it in the literature. Therefore, research on the problems of rational use of agricultural land in Tuva, analysis of changes in the structure of arable land and agricultural land in the arid zone of risk agriculture is particularly relevant.

The Republic of Tyva is the youngest subject of the Russian Federation, which became part of the country in 1944. It is part of the Siberian Federal district, in the South and South-East it borders with the Mongolia, in the North-East – Irkutsk region, in the North-West – the Republic of Khakassia, in the East – the Republic of Buryatia, in the West – the Republic of Altai, in the North – Krasnoyarsk region. Total area of the region is 168.6 thousand km^2. The Ubsunur basin in the southern part of Tuva belongs to the Basin of Large lakes of Mongolia.

For thousands of years, land use in Tuva has been conditioned by the unique natural and climatic conditions and the development of pasture-based animal husbandry. At present, animal husbandry is the main traditional sector of the Republic's economy. Agriculture, along with animal husbandry, has been a vital branch of Tuva's agriculture since time immemorial. It is very surprising that in ancient times the size of the acreage on the territory of the Republic was not inferior to modern ones. In the last century, political and economic changes in Russia, in the republics of Tuva, Khakassia and Buryatia caused changes in the population density of the regions, in the ways of organizing the economy and the state of the land. Unfortunately, today the size of Tuva's croplands is even lower than the level reached by the Tyva People's Republic (TPR). In our study, we will try to consider the dynamics and structure of the acreage of agricultural land in the Republic in different historical periods, and the socio-economic reasons that influenced these changes [2].

The purpose of the study is to identify the ecological and economic reasons for changes in the croplands of agricultural crops on the territory of Tuva for the period 1890 to 2019.

2. Materials and methods

The objects of research are agricultural lands in the intermountain basins of Tuva.

Tuva is located in the Central part of Asia between 50–56° N. l. and 89–99° E. l. From an orographic point of view, the Tuva landscape region of the Altai-Sayan mountain country has clearly defined borders that separate the mountainous part from the low mountains and foothill depressions [3]. A characteristic feature of the terrain is the presence of high ridges and highlands located on its outskirts and intermountain basins [4]. In General, mountain systems cover more than 80% of the entire territory of Tuva, and about 20% comes from intermountain basins: dry-steppe Central Tuva, consisting of the Kyzyl, Ulug-Khem and Khemchik depressions; steppe Turan-Uyuk; taiga-forest Todzha and Tere-Khol; semi-desert Ubsunur. The complex terrain of the Republic has a strong influence on the regime of meteorological elements and, especially, on the surface temperature.

According to the climate essay by N.A. Efimtsev [5], Tuva reflects the influence of neighboring territories: from the North and North-East – taiga Eastern Siberia, from the South and South-East – desert-steppe regions of Mongolia, from the West – mountain-taiga Altai. Due to this geographical location and the sharp division of the terrain, the natural features of Tuva, including the climate, differ in significant contrast. According to E.M. Murzaev [6] the Northern border of the deserts of Central Asia passes here.

Low-snow winters, hot and dry summers, low precipitation and high amplification of absolute and daily temperatures are characteristic features of the Tuva climate. Thus, the average temperature of January under the influence of the Siberian anticyclone in the basins is minus 28–35°C, July – 15–20°C, the average annual temperature is from -3.3 to -6.1°C, the height of snow cover is 10–20 cm. The amplitude of extreme values of air temperature for a year is 82–90°, the amplitude of extreme values of temperature on the soil surface is 118°. The sum of temperatures above +10°C is 1812–2086°. On average, the frost-free period lasts about 185 days (from April 15 to October 18). The period of active vegetation with daily temperatures not lower than 10°C lasts on average 134 days (from may 8 to September 19). During this period, the sum of active temperatures with 80% security is 1864°. The cold period lasts 205–225 days, 50–80% of these days are windless. Moisture in the basins is poor (Hydrothermal coefficient – 0.4), there is an average of 200–300 mm of precipitation per year, in the mountains – 700 mm [7, 8].

Agricultural land is occupied by intermountain steppe basins with heights of 550–1200 m above sea level, the lower parts of mountain slopes, and high terraces of river valleys. Large tracts of the steppes are characteristic of the Central Tuva and Ubsunur basins.

To analyze the structure and dynamics of croplands and farmlands in the administrative areas of the Republic of Tyva (RT) was involved literary, stock and statistical materials, gathered in the Federal service of state statistics [9], Federal service for state registration, cadaster and cartography for the Krasnoyarsk region, the Republic of Khakassia and the Republic of Tyva [10], the Ministry of natural resources and ecology of the Republic of Tyva [11–13].

3. Results and discussion

At the beginning of the bronze age, the tribes of southern Siberia made the first steps in the development of cattle breeding. Tuva was dominated by a settled pastoral and agricultural forms of economy. In the middle of the first Millennium BC, cattle breeding became the main occupation of the population of the steppe regions of Tuva, and caused the transition from a settled pastoral-agricultural life to a semi-nomadic one. A large number of livestock could be kept on the significantly expanded pasture areas. Increasing pasture load required seasonal migrations [14].

Typical features and features of the Tuvans' farming in the *pre-Soviet period* were traditional nomadic cattle breeding, which was maximally adapted to the local natural conditions. Along with animal husbandry, Tuvans also engaged in agriculture, which was a subsidiary occupation. Thus, since 1890, small areas (about 3 thousand hectares) of irrigated arable land were created in the river valleys of the Yenisei and Khemchik. No more than 25% of peasant (in Tuvan «Arat») farms located mainly in the Ulugh-Khem and Khemchik depressions were engaged in agriculture. The cultivation of crops was very primitive and at the beginning of the XX century only 3.5 thousand ha of land were sown (Fig. 1). In the spring, the Arats watered, loosened the land with a wooden plough, scattered the seeds and then closed them up with scrapers of the shrub Caragana [15].

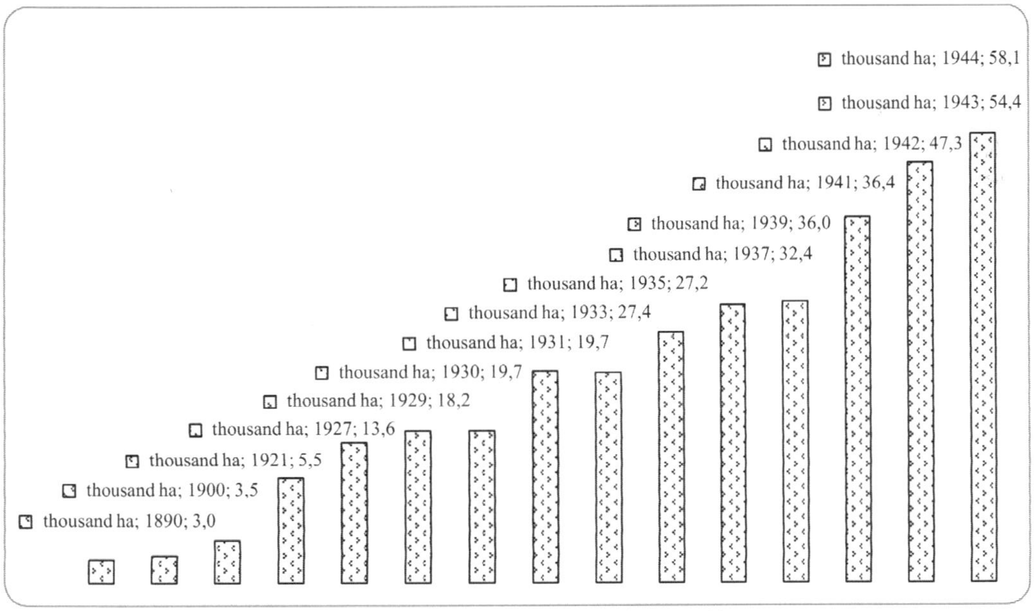

Figure 1. Dynamics of acreage areas of TPR in 1890–1944, thousand ha.

The period of the Tyva People's Republic (1921–1944). The Leading branch of agriculture was also animal husbandry of the TPR. The number of livestock in the period of the TPR increased significantly and according to the data of 1945 was 759 thousand heads, including 542 thousand heads of sheep and goats, 129.8 thousand heads of cattle, 61.7 thousand heads of horses, 25 thousand heads of other animals (yaks, camels, deer, pigs) [16].

However, crop production developed no less dynamically, and acreage areas also increased. In comparison with 1927, acreage areas increased by 4.3 times and amounted to 58.1 thousand ha.

Unfortunately, the history of TPR agriculture has not yet been sufficiently studied, and specific data on acreage areas for the structure of agricultural crops of this period are not yet available.

But according to the Tuva agricultural and demographic census conducted by Soviet scientists, crop area of grain in 1931 was 19.4 thousand ha, including wheat 7.5 thousand ha, barley 3 thousand ha, oats – 4.5 thousand ha, millet – 3.7 thousand ha. The area of vegetable and melon crops was insignificant and amounted to only 0.2 thousand ha, including potatoes 0.05 thousand ha, forage crops – 0.02 thousand ha [17, 18].

The period of the Tuva Autonomous region (1944–1961). After the TPR became part of the Soviet Union in 1944, agriculture received a new impetus for its development. On the one hand, the nomadic population of the Republic was transferred to a settled life, which was carried out simultaneously with the collectivization of agriculture – personal farms were replaced by collective and state farms. These moments had both positive and negative sides. On the other hand, new cities and towns were built, and new branches of the national economy were developed.

Significant investments were made in both industry and agriculture. Widespread use of mechanization in agriculture, with the creation of machine and tractor stations, has significantly increased the plowing of area under crops. Grass ecosystems are undergoing a profound transformation, when the number of livestock increases simultaneously with plowing of virgin and pasture lands.

So, in the period of the Tuva Autonomous region, acreage areas increased significantly: compared to 1945, by 1960, the area of grain increased more than three times, amounting to 279.8 thousand ha, including wheat – 193.6 thousand ha, barley – 21.7 thousand ha, oats – 34.7 thousand ha, millet – 27.1 thousand ha (Fig. 2). The acreage areas of forage lands has increased by an order of magnitude. The croplands of vegetable and melon crops, including potatoes, which by 1960 amounted to 6.2 thousand ha, and vegetables – 0.6 thousand ha, increased significantly (Fig. 3).

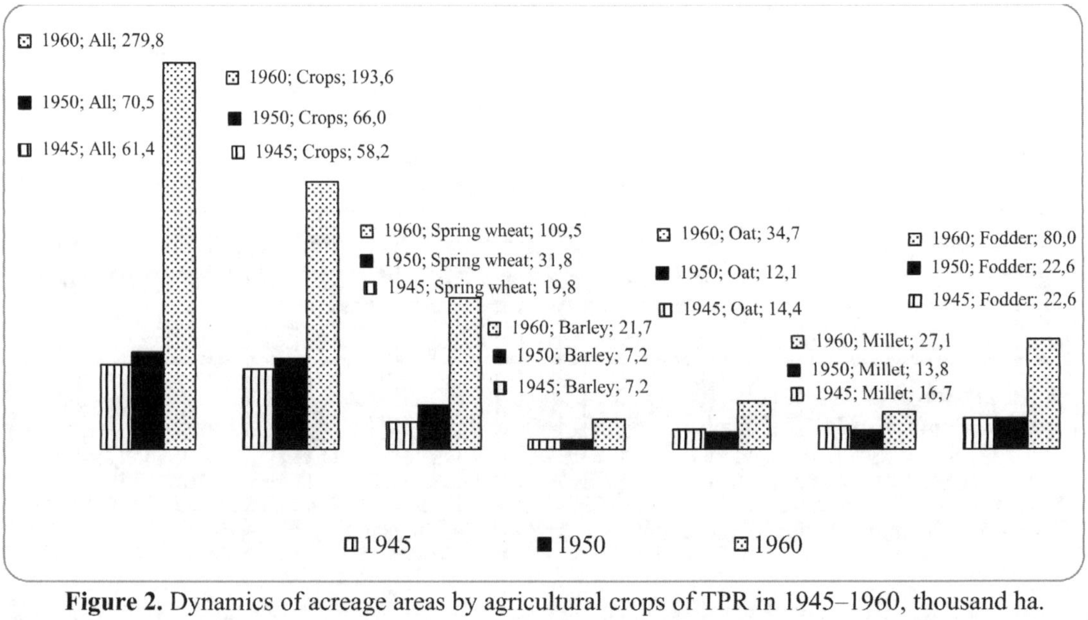

Figure 2. Dynamics of acreage areas by agricultural crops of TPR in 1945–1960, thousand ha.

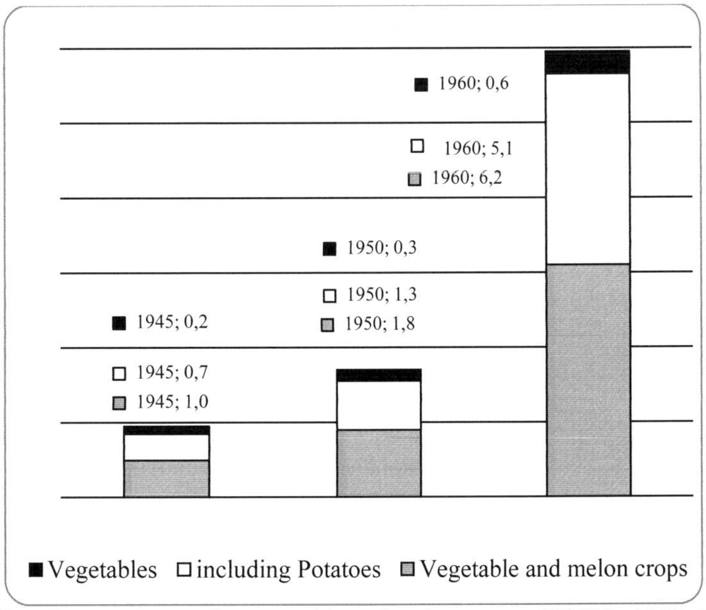

Figure 3. Dynamics of acreage areas for vegetable and melon crops of TPR in 1945–1960, thousand ha.

In 1950-60, the development of virgin land in the country began, which did not pass by Tuva. Of course, the development of virgin land was an erroneous course in agriculture, imposed by party ideology, recognized as such a little later [19]. Nevertheless, it should be noted that there are erroneous works concerning the development of agriculture in the Republic in the 60s of the XX century. Some authors reported that plowing of virgin and fallow lands should be considered significant achievements of agriculture in Tuva [20], while the volume of agricultural land was increased by 1.4 times, plowed in turnover by 4.2 times, and 4.7 times. In addition, the authors pointed to the need to use maize as the main forage crop [21, 22].

More than a million hectares of agricultural lands were lost during the virgin land period (1953 to the 1960s), decreased from 5622 thousand ha to 4489 thousand ha (Fig. 4). In later studies, it was shown that the main factors of wind erosion in Tuva were the unjustified expansion of crops and forage crops, especially maize, the predominance of spring plowing and the use of seed varieties that should not be used in arid climatic conditions. Moreover, it was found that all types of soils are subject to wind erosion in areas of insufficient moisture, "...in sandy and sandy loam soils, erosion begins 2–3 years after the development of virgin land, in light-loam and medium-loam – 4–5 years, and in clay and heavy-loam after 7–10 years" [23]. Thus, the campaign for the development of virgin lands in the country between 1954 and 1963 had a particularly strong impact on vegetation and soil, a significant anthropogenic impact in the steppes will be traced for several centuries. Almost simultaneously with the beginning of large-scale agricultural use of virgin and fallow lands, soil degradation and climate change occur [24].

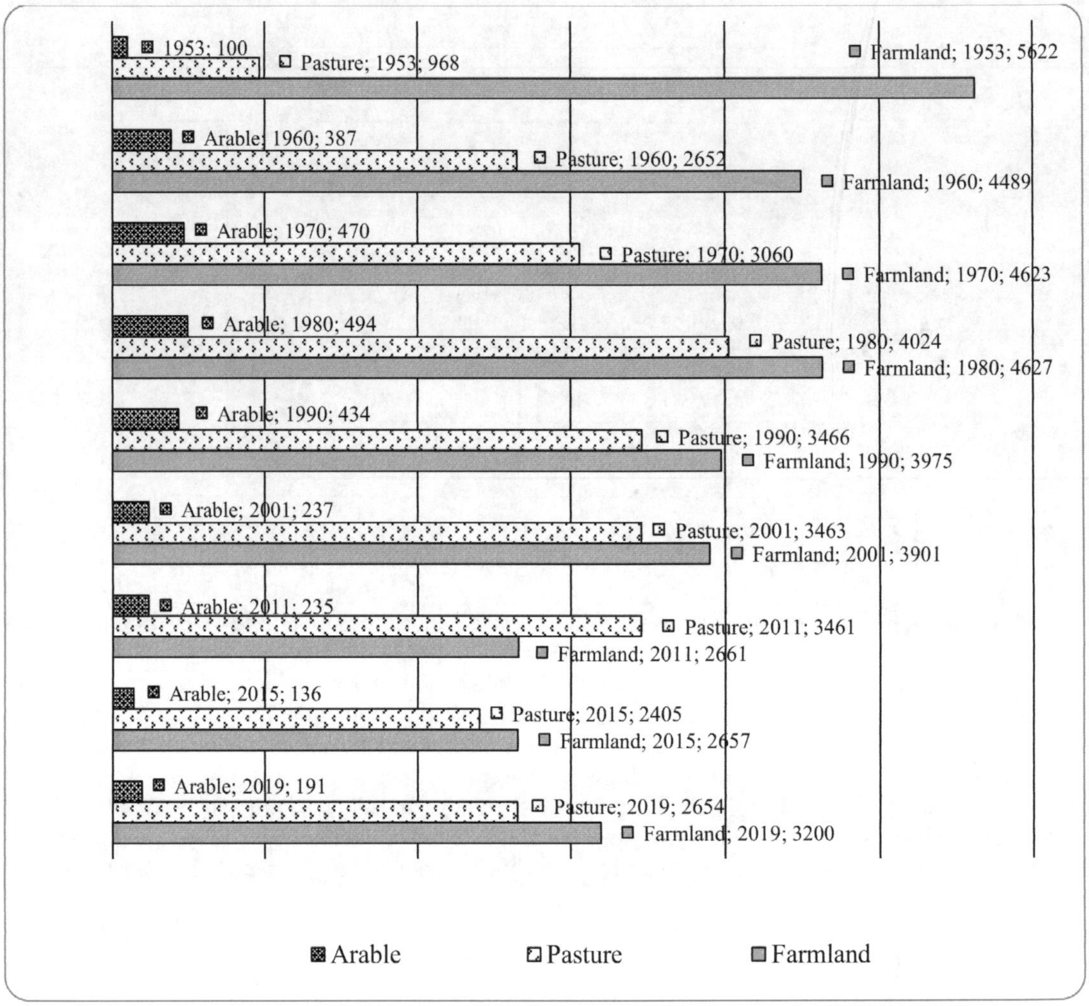

Figure 4. Dynamics of agricultural lands, pastures and arables in Tuva in the period 1953–2014, thousand ha.

As early as in 1967, the large scale of eroded land in Tuva was estimated at 175 thousand ha, including 79 thousand ha of arable land, 65.6 thousand ha of pastures and 30.4 thousand ha of fallow lands [25] (Table 1). It is shown that most of it is caused by wind erosion, which accounted for more than 161 thousand ha or 92% of the area of eroded land. At the same time, more than 2/3 of the land is subject to severe and average wind erosion, including 65.6 thousand ha of pastures and 45.5 thousand ha of arables.

Table 1. Structure and area of eroded lands of the Tuva ASSR according to data from 1967, thousand ha

Croplands	Area of eroded land	Including those subject to							
		Wind erosion				Water erosion			
		Total	of them by degrees			Total	of them by degrees		
			weak	middle	strong		weak	middle	strong
Arables	79.0	75.5	30.0	32.5	13.0	3.5	1.0	2.5	-
fallow lands	30.4	20.3	-	-	20.3	10.1	-	6.0	4.1
Pastures	65.6	65.6	-	29.0	36.6	-	-	-	-

Total croplands	175.0	161.4	30.0	61.5	69.9	13.6	1.0	8.5	4.1

Geographically, zones of erosion are common in the Ubsunur and Central-Tuva basins. The most affected areas are the Erzin, Tes-Khem, Tandy, Kaa-Khem, Ulugh-Khem and Dzun-Khemchik districts, which account for more than 70% of the area of eroded lands. Approximate losses of natural nutrients and fertilizers during soil erosion, which negatively affect productivity, are given: during wind or water erosion of soils on an area of 1 ha, when 2.5 cm of chernozem is blown out and washed away, phosphorus is lost – 196 kg, nitrogen – 981 kg, and organic substances – about 15 tons [26]. In general, erosion processes reduce the yield of agricultural crops by 40–70%, and the efficiency of fertilizer application by 20–30%. At the same time, the yield of areas that are not provided with irrigation systems, depending on weather conditions (wet or dry year), can vary from 1.5–2.5 centner/ha to 12–16 centner/ha.

The period of the Tuva ASSR (1961–1991). By the beginning of 1960, Tuva's agricultural enterprises were being enlarged and grouped into agricultural zones depending on natural and economic conditions: the Central zone included such districts as the Kyzyl, Ulug-Khem, Tandy, Piy-Khem and Kaa-Khem districts, the southern – Erzin, Tes-Khem and Ovyur districts, and the Western – Dzun-Khemchik, Barun-Khemchik, Sut-Khol and Mongun-Taiga districts, to the East – Todzha district. The reasons are financial difficulties and the growing debt of collective farms to the state from year to year. It became possible to increase the acreage [27]. In all categories of farms they reached about 278.3 thousand ha or increased almost 4 times to the level of 1950 (Fig. 5).

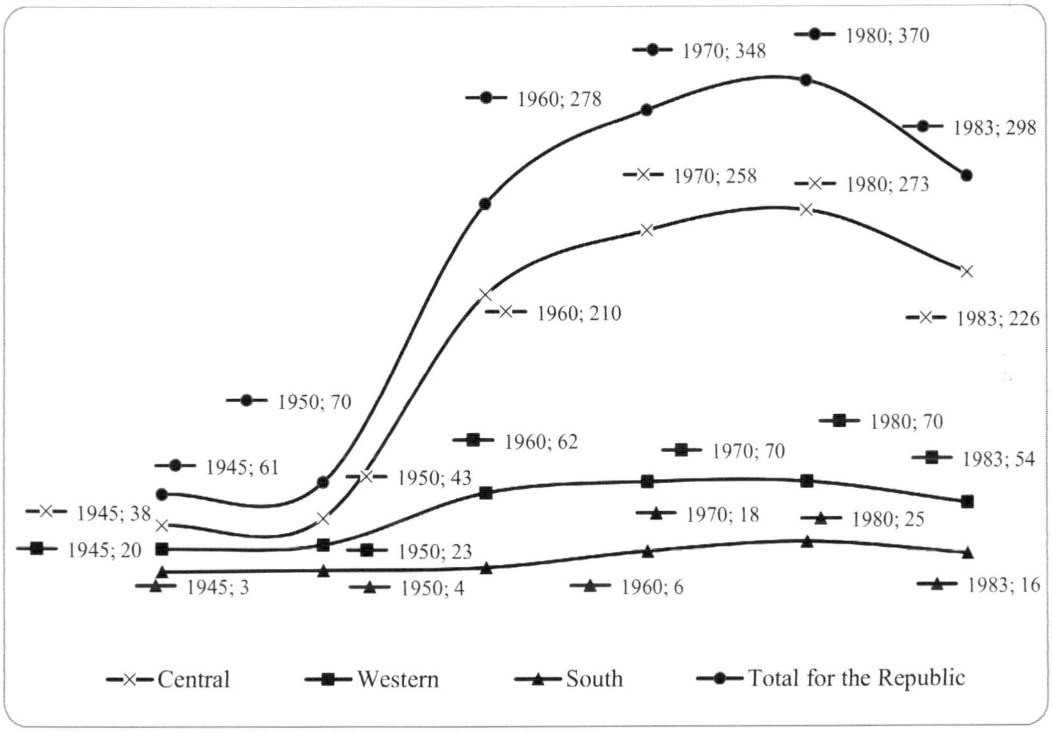

Figure 5. Dynamics of acreage areas of grain crops Tuva's agricultural zones (1945–1983), thousand ha.

The study of the structure and dynamics of acreage areas of grain crops in the agricultural zones of Tuva shows that in the 50-60s of the XX century, there is a significant increase in sown areas [19]. Thus, in 1966–1970, in the structure of crops in Tuva, cereals accounted for 230.8 thousand hectares. In

General, the 60–70 years are characterized as the years of formation of large agricultural enterprises in the Republic with a strong production base and labor resources. The development of virgin lands began during the years of a rather acute food crisis in the country. According to the conclusion of leading scientists, the virgin epic as an extensive model for solving the problem has not justified itself either economically or environmentally [28–31].

By 1980, the acreage in the Republic reached 370.7 thousand has, mainly due to an increase in the acreage of cereals and forage crops. During this decade, 1972 was the most productive year – the gross harvest of grain crops amounted to 298.1 thousand tons. Thus, the 70–80 years of the last century in the Republic's agriculture were years of intensification, specialization and concentration of production, wide introduction of economic calculation and other methods of production. However, from 1976 to 1980, there was a downward trend in the rate of agricultural development in the Republic. The main reason for the relatively slow development of the industry was a decrease in the yield of the main agricultural crops: grain, potatoes and vegetables. The reasons were both socio-economic and adverse weather and climate conditions.

The largest area of grain crops was observed in 1980, including 273 thousand ha in the Central region, 70 thousand ha in the West, and 25 thousand ha in the South. At the same time, the Central zone accounts for about 2/3 of the sown area of grain crops, the Western zone – 20%, the southern zone – 5%, and the Eastern zone – only 1%. The same structure of acreage areas of grain crops remains in the future.

According to the structure and dynamics of potato acreage in Tuva, the picture is somewhat different. Interestingly, the largest areas of sown areas were observed in the 1960s – during the virgin land period, when they were more than 5 thousand hectares. since 1965, the sown area of potatoes has not decreased much, amounting to 4.2 thousand hectares in 1970 (Fig. 6). In the structure of potato acreage, the largest share belongs to the Central agricultural zone (79%), followed by the Western zone (18%) and the southern zone (3%). In the Eastern zone, the acreage of cereals and potatoes is minimal.

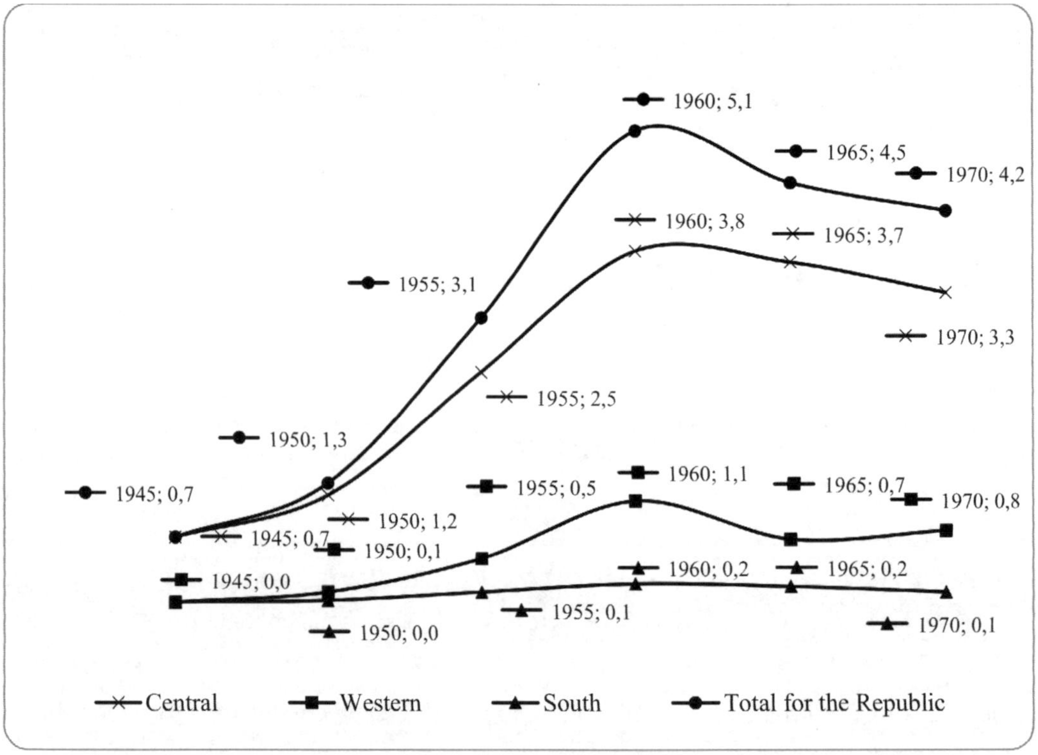

Figure 6. Dynamics of potato acreage in Tuva's agricultural zones (1945–1983), thousand ha.

It should be noted that before 1950, the largest share of sown areas of grain crops was in Ulug-Khem district (12.5 thousand ha), followed by Tandy (12.3 thousand ha), Piy-Khem (9.4 thousand ha), Kaa-Khem (7.2 thousand ha), Barun-Khemchik (7.5 thousand ha) and other districts. Since the 60s of the last century, the situation has changed somewhat – the Tandy district is increasing its grain acreage at a faster pace and by 1983 it accounts for almost a quarter of all the Republic's grain acreage (Table 2). A significant share of acreage also falls on the Kaa-Khem, Piy-Khem, and Ulug-Khem districts, which traditionally belong to the favorable agriculture of the Central Tuva basin.

Table 2. Acreage areas of grain crops in the districts of Tuva (1945–1983), thousand ha

Areas	1945	1950	1960	1970	1980	1983
Mongun-Taiga	0.2	0.2	-	-	0.3	0.3
Bai-Taiga	3.6	4.1	9.1	8.3	9.8	7.8
Barun-Khemchik	6.8	7.5	18.3	18.9	14.0	13.8
Ovyur	1.2	1.3	1.1	1.7	2.8	1.2
Sut-Khol	3.0	4.2	6.7	9.6	15.0	9.1
Dzun-Khemchik	5.9	6.9	28.2	33.3	30.9	23.2
Ulug-Khem	12.2	12.5	48.4	43.7	47.5	32.0
Tandy	10.5	12.3	68.0	77.7	82.8	72.9
Piy-Khem	9.1	9.4	34.6	36.3	37.9	37.5
Todzha	-	0.1	0.3	2.1	2.5	2.2
Kaa-Khem	6.4	7.2	40.2	54.3	55.9	44.8
Tes-Khem	1.4	1.9	3.4	8.1	10.8	6.7
Erzin	0.7	0.8	1.3	8.5	11.1	7.7
Kyzyl	-	1.7	18.7	45.7	48.4	38.9
Total for the republic	**61.0**	**70.1**	**278.3**	**348.2**	**369.7**	**298.1**

The relatively low starting conditions of the Tuva economy in the pre-reform period, management errors in the planning period, disruption of production and technological links of most enterprises, and many other factors led to a catastrophic reduction in acreage, including grain, between 1980 and 1990. The reduction of grain acreage in 1990 by 1980 was about 40%, including grain by more than two times (Fig. 7).

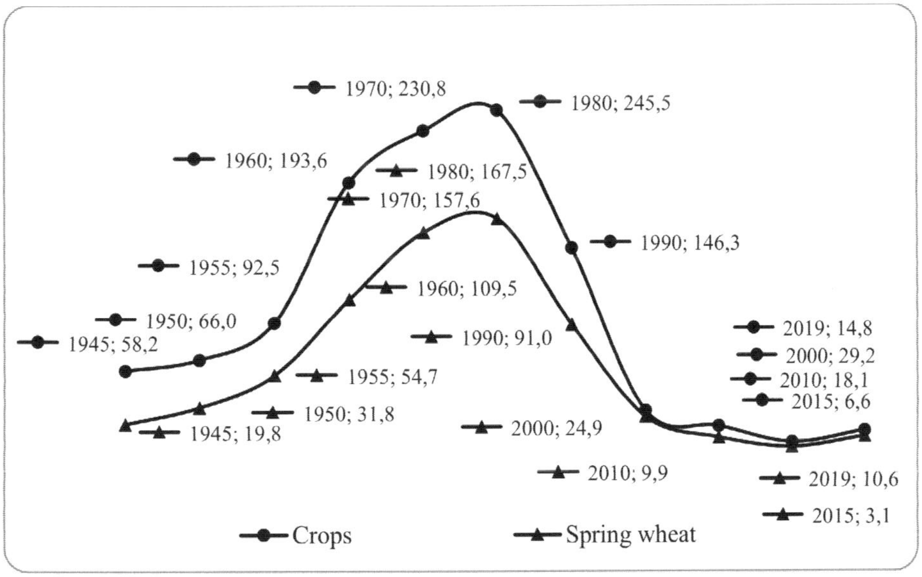

Figure 7. Dynamics of acreage areas of grain crops, including wheat in Tuva for the period 1945–2000, thousand ha.

In General, the years 1960–1990 were the period of formation and development of collective and state farm production in the Republic's agriculture. Years of gradual development of the planned economy. The industry had a good material and technical base. Over the past 30 years, the Republic has achieved a fairly dynamic development of agriculture. At the same time, it is noteworthy that more than 30 thousand people were employed in the agricultural enterprises of the Republic on an average annual basis and had a permanent basis for motivated work.

As a result of the systemic crisis that engulfed Russia in the late 1980s and early 1990s, huge tracts of croplands were withdrawn from circulation [32]. During the period of reforms in the 1990s, negative trends in the country's economy had a more acute impact on the socio-economic development of the Republic of Tyva, given the complex of unresolved social problems, as well as the relative transport isolation of the Republic, which is the only region in Russia that does not have a railway connection with the rest of the country. Such factors as violation of economic relations between agricultural enterprises, processing and sales enterprises, lack of government support, disbandment machine-tractor parks as well as quick wear, disparity of prices for agricultural equipment compared with prices of agricultural products, the lack of tax incentives and soft loans led to the ruin of large farms of agro-industrial complex, which accounted for more than 99% of agricultural products [33] In the new market conditions in Tuva, there is an even greater decrease in the acreage of grain crops. Thus, in 2000, in comparison with 1980, the sown areas of grain decreased by 8.4 times, wheat by 6.7 times, and fodder crops – 11 times. "...Thousands of arable land turned out to be unclaimed and, gradually overgrown with weeds and other vegetation, began to turn into a Deposit" [34]. According to the materials of D.I. Lyuri and others [29], the Republic of Tuva experienced an unprecedented reduction in the area of arable land – 2.034 thousand ha, which corresponds to 56.9% of the former fund in Tuva before 1990.

Currently, Tuva lags behind most regions of the Siberian Federal district and Russia in many socio-economic parameters. However, there is a slight increase in acreage, but not reaching the acreage even in 2000.

Conclusions

Thus, it is shown that despite the fact that the leading branch of agriculture in the TPR was animal husbandry, crop production developed no less dynamically. During the TAR period, the acreage of cereals, forage land, and vegetable crops, including potatoes, increased significantly. But the unjustified plowing of virgin and fallow lands in the 50s and 60s of the twentieth century led to soil erosion and destruction, and the loss of more than a million hectares of agricultural land. During the period of the Tuva ASSR, there is a further increase in acreage. With the beginning of market reforms, there is a significant decrease in the sown area of grain crops. At present, there is again a slight increase in acreage.

Acknowledgments

The reported study was funded by RFBR according to the research project № 19-29-05208\19 mk

References

[1] Steppes of Central Asia. 2002. Gadzhiev, I.M., Korolyuk, A.Yu., Titlyanova, A. A., Andrievsky, V.S., Bayartogtoh, B., Grishina, L.G., Kosykh, N.P., Kyrgyz, Ch.O., Mironycheva-Tokareva, N.P., Romanova, I.P., Sambuu, A.D., Smelyansky, I.E.: monograph. Novosibirsk: Publishing house of the SB RAS, 299 pp. (in Russian).

[2] Natural resources of the Republic of Tuva. 2018. Sambuu, A.D., Krasnoborov, I.M., Sevastyanov, V.V., Sevastyanova M.G., Sukhova, N.G., Andreychik, M.F., Kudryavtsev V.I., Kudryavtseva, A.I., Zabelin, V.I., Archimaeva, T.P., Kurbatskaya, S.S., Gurkova, E.A., Prudnikova, T.N., Zaika, V.V., Arakchaa, L.K., Kuksin, A.N., kuksina, D.K. Vol. 1.: monograph. Novosibirsk: Garamond, 488 pp. (in Russian).

[3] Zyatkova, L.K. 1969. Tuva. Altai-Sayan mountain region, Moscow: Science, 333–362. (in Russian).
[4] Kushev, S.L. 1957. Relief. Natural conditions of the Tuva Autonomous Region. Proceedings of the complex expedition. Issue 3. Moscow: AS USSR, 11–14. (in Russian).
[5] Efimtsev, N.A. 1957. Climate essay. Natural conditions of the Tuva Autonomous region. Moscow: Science, 46–65. (in Russian).
[6] Murzaev, E.M. 1948. The basin of Large lakes in Western Mongolia and the occurrence of its landscapes. Proceedings of the 2nd AllSoviet geograph. Congress. Vol. 1. M., 367–378. (in Russian).
[7] Agroclimatic handbook of the Krasnoyarsk territory and Tuva ASSR. 1974. L., Hydrometeoizdat, 215 pp. (in Russian).
[8] Agroclimatic resources of the Krasnoyarsk territory and Tuva ASSR. 1974. L.: Gidrometeoizdat, 211 pp. (in Russian).
[9] Statistical yearbook of the Republic of Tuva. 2019. Stat. compl. Krasnoyarsk Stat., 425 pp. (in Russian).
[10] Agriculture, hunting and forestry. 2020. The Federal service of state statistics. M. https://www.gks.ru/enterprise_economy?print=1
[11] State reports "About the state of the environment of the Republic of Tuva". 1999–2019. Ministry of natural resources and ecology of the Republic of Tyva, Kyzyl. (in Russian).
[12] Report on the state and use of land in the Republic of Tuva. 2005–2016. Territorial authority of the Federal state statistics service for the Republic of Tuva. Kyzyl. (in Russian).
[13] Report on the state and use of land in the Republic of Tuva. 2017–2020. Department of the Federal service for state registration, cadaster and cartography for the Krasnoyarsk territory, the Republic of Khakassia and the Republic of Tuva. Krasnoyarsk. (in Russian).
[14] History of Tuva. 2001. Vol. 1. 2 pub. under the general editorship of S.I. Weinstein, M.Kh. Mannai-ool: monograph. Novosibirsk: Science, 367 pp. (in Russian).
[15] Sedegov, V I., Krasovsky, S.A. 1946. Tuva agriculture to the 25th anniversary of the national liberation revolution. Twenty-five years of the Tuva national liberation revolution. Compilation of scientific articles. Kyzyl: Tuva regional publishing house, 46–53. (in Russian).
[16] Sambuu, A.D., Dapyldai, A.B., Kuular, A.N., Khomushku, N.G. 2012. Desertification problem of the Republic of Tuva. *Arid ecosystems* 18. № 4 (53), 35–44. (in Russian).
[17] Tuva Autonomous region in numbers. 1946. Twenty-five years of the Tuva national liberation revolution. Compilation of scientific articles. Kyzyl: Tuva regional publishing house, 99–101. (in Russian).
[18] Tuva agricultural and demographic census of 1931. 1933. Moscow: Publication of the research Association for the study of national and colonial problems, 169 pp. (in Russian).
[19] Jubilee statistical compilation for the 100th anniversary of the unity of Russia and Tuva. 2014. Kyzyl: TyvaStat, 208 pp. (in Russian).
[20] Kongar, N.M. 1967. State farms of the Tuva Autonomous region. *Scientific notes* XII. Kyzyl, 221–235. (in Russian).
[21] Dostak-ool, M.B. 1963. Main issues of agriculture in Tuva. *Scientific notes* X. Kyzyl: Tuva book publishing house, 116–124. (in Russian).
[22] Bakulin, A.K. 1964. Tuva agriculture is on the way to intensification. Proceedings of the XI. Kyzyl: Tuva book publishing house, 137–151. (in Russian).
[23] Kuznetsova, E G. 1963. Maize – a wide road in Tuva. *Scientific notes* X. Kyzyl: Tuva book publishing house, 124–130. (in Russian).
[24] Falcone, G., Stillitano, T., Montemurro, F., De Luca, A.I., Gulisano, G., Strano, A. 2019. Environmental and economic assessment of sustainability in Mediterranean wheat production. *Agronomy Research* 17(1), 60–76.
[25] Nazyn-ool, V.D. 1967. Fight against wind erosion of soils. Kyzyl, 24 pp. (in Russian).

[26] Nazyn-ool, V.D. 1971. Eroded soils and ways to use them. *Scientific notes* XV. Kyzyl: Tuva book publishing house, 209–218. (in Russian).
[27] Soldatov, V.P. 1967. Placement and specialization of agriculture in the Tuva ASSR: author's abstract diss. ... Ph. D., 20. (in Russian).
[28] Agroecological assessment of land, design of adaptive landscape systems of agriculture and agricultural technologies. 2005. Ed. Kiryushina, V.I. M.: Rosinformagrotech, 794. ISBN: 5-7367-0525-7
[29] Lyuri, D.I., Goryachkin, S.V., Karavaeva, N.A., Denisenko, E A., Nefedova, T.G. 2010. Dynamics of Agricultural lands of Russia in XX century and Postagrogenic Restoration of vegetation and soils: monograph. M: GEOS, 416 pp. (in Russian).
[30] Kiryushin, V.I. 2013. Agronomic soil science. SPb: KVADRO, 678.
[31] Ecological and economic assessment of land degradation. 2016. Edited by A.S. Yakovlev, O.A. Makarov, S.V. Kiselev, E.N. Molchanov: monograph. Moscow: MAKS Press, 252. ISBN 978-5-31705437-3
[32] Lyuri, D.I., Nekrich, A.S. 2019. Changes of the dynamics Agrarian lands of Russia in 1990-2014. *Bulletin of the Russian Academy of Sciences. Geographical series* 3, 64–77. (in Russian).
[33] Dabiev, D.F., Dabieva, U.M. 2015. Problems and prospects of development of rural economy of Tuva. *Problems of modern economy* 3, 280–282. (in Russian).
[34] Zhulanova, V.N., Chuprova V.V. 2009. Current state of land resources and agricultural soil fertility in Tuva. *Bulletin of Krasnoyarsk state agrarian University* 2, 57–63. (in Russian).

Formalized model of agricultural insurance development strategy as an element of industry management digitalization

K Zhichkin[1], **V Nosov**[2], **L Zhichkina**[3], **A Tarakanov**[4], **V Zhenzhebir**[2], **F Sterlikov**[2]

[1] Department of Economic Theory and Economics of AIC, *Samara State Agrarian University*, 2 Uchebnaja Street, 446552, Kinel, Russia
[2] Department of Economics and Management, *K.G. Razumovsky Moscow State University of technologies and management*, 73 Zemlyanoy val, 109004, Moscow, Russia
[3] Department of Land Management, Soil Science and Agrochemistry, *Samara State Agrarian University*, 2 Uchebnaja Street, 446552, Kinel, Russia
[4] Department of State Law Disciplines, *Institute of Public Administration*, 80 Leninsky Prospekt, 119261, Moscow, Russia

E-mail: zskirill@mail.ru

Abstract. The article discusses the possibilities of agricultural production management digitalization with formalized models, using the example of state support for agricultural insurance. The aim of the study is to adapt the BCG model for use as a tool for designing a development strategy for agricultural insurance with state support. Tasks: - determine the possibilities of using the BCG model to design an agricultural insurance development strategy; - offer a model parameters system; - define a set of strategies in the framework of the goal. Putting the model into practice will make the existing state support system more transparent and predictable, depending mainly on the objectives of state policy and the actual indicators achieved by the industry. At the costs high level offered by these measures, the system involves the creation of a sustainable market for insurance services, regulated by both economic and administrative mechanisms. The presence of such decisions totality will automate the decision-making process at the industry management level; ensure the agricultural policy continuity in the long term.

1. Introduction

Digitalization of agricultural production is the most important task for the agro industry development. However, it should include not only automation of production processes themselves (parallel driving systems, processing of crops with plant protection products and automatic fertilizing, etc.), but also managerial decision-making automation, both at the level of individual enterprises and industry as a whole [1-8]. The introduction of MRP II / EPR, APS standards and their further development make it possible to use methods for optimizing the enterprises production activities [9, 10]. However, all these approaches are mainly used at the operational and tactical planning level and do not affect the strategies development level [11]. This is due to difficulties in obtaining quantitative information currently used in management software products.

IOP Conf. Series: Materials Science and Engineering **941** (2020) 012025 doi:10.1088/1757-899X/941/1/012025

2. Methods and Materials

Formalized strategic analysis models, such as BCG, GE / McKinsey models can be used as the basis of a strategic planning system for an industry, production that allows formalizing the strategy development, adapting them to the characteristics of a particular industry or a specific task. An example of this approach is the state support systems for agricultural insurance.

The aim of the study is to adapt the BCG model for use as a tool for designing a development strategy for agricultural insurance with state support.

Tasks: - determine the possibilities of using the BCG model to design an agricultural insurance development strategy; - offer a model parameters system; - define a set of strategies in the framework of the goal.

The object of the study is the economic aspects associated with determining the development strategy of the industry in crop insurance with state support. The research methodology consists in analyzing the economic mechanism for the development of agricultural insurance with state support. During the study, the abstract logical method, situational and system analysis, economic and statistical methods were used.

3. Results and Discussion

According to the World Bank, most of the agricultural insurance market in the world falls on crop insurance. More than 90% of the collected premium is generated through insurance in the crop industry [12]. In the Russian Federation, the crop sector accounts for more than 95% of the damage from emergencies in agricultural sectors [13-16]. All this testifies to the importance of agricultural insurance for the agro-industrial complex of Russia. Moreover, in the Samara region conditions crop insurance with state support remains minimally widespread. Since 2014, the number of concluded agreements on crop insurance with state support has been reduced (from 114 in 2014 to 0 in 2017) [17]. The region is located in the risky farming zone. The environmental and climate factors have a major impact on crop production in agricultural organizations. At the same time, the loss of agricultural insurance varied from 8.2% in 2014 to 33.0% in 2013. That is, for the period under review by 1 ruble insurance premiums agricultural producers received 0.287 rubles compensation (table. 1).

Table 1. Data on agricultural insurance in the Samara region

Index	2012	2013	2014	2015	2016	2017	2018
Insurance premium							
Agricultural insurance total (million rubles)	185.3	227.9	367.0	305.3	444.9	(19.6)	104.0
Crop insurance with state support (million rubles)	144.7	198.6	355.3	298.9	441.7	(27.1)	87.1
including subsidies received (million rubles)	56.9	90.0	143.8	121.5	96.6	0.0	46.3
Number of Contracts							
Total agricultural insurance (pcs)	1378	1104	858	592	384	251	241
Crop insurance with state support (pcs)	77	50	114	76	30	0.0	24
Insurance payments							
Insurance payments under agricultural insurance contracts, total (million rubles)	102.4	104.1	93.5	158.0	77.3	109.6	0.11
Insurance payments (crop insurance with state support) (million rubles)	34.8	65.6	29.2	88.9	76.9	109.4	0.0
Sum insured under crop insurance contracts (million rubles).	1998.4	2742.4	4905.4	4127.3	6098.5	0.0	1724.6
Loss of agricultural insurance (%)	55.3	45.7	25.5	51.7	17.4	-	0.1

Loss of crop insurance with state support (%)	24.0	33.0	8.2	29.7	17.4	-	0.0

The model serves to formulate the state strategy directions in the field of supporting the agricultural insurance system at the regional level (for countries with a significant variety of climatic conditions and, accordingly, with the agricultural production risk level) or the country as a whole (for small countries) [18-19].

As a basis, it is proposed to divide the regions into 4 groups in accordance with two indicators - the agricultural insurance spread, the risk degree (the possibility of losing the yield part). The first indicator characterizes the agricultural producers' protection degree in relation to external influences (first of all, to the climatic conditions that are forming in this territory). The widespread insurance use allows enterprises, even with significant losses, to continue normal economic activity, making it unnecessary to develop emergency assistance programs in the adverse circumstances case (for example, as a result of the 2010 drought, the state was forced to help in several areas - eliminating the drought effects, purchasing feed grain, compensation for the purchasing seeds cost, etc.) As a rule, these measures are more dependent on the needs of the region or country leadership in supporting a specific population group than on the real need.

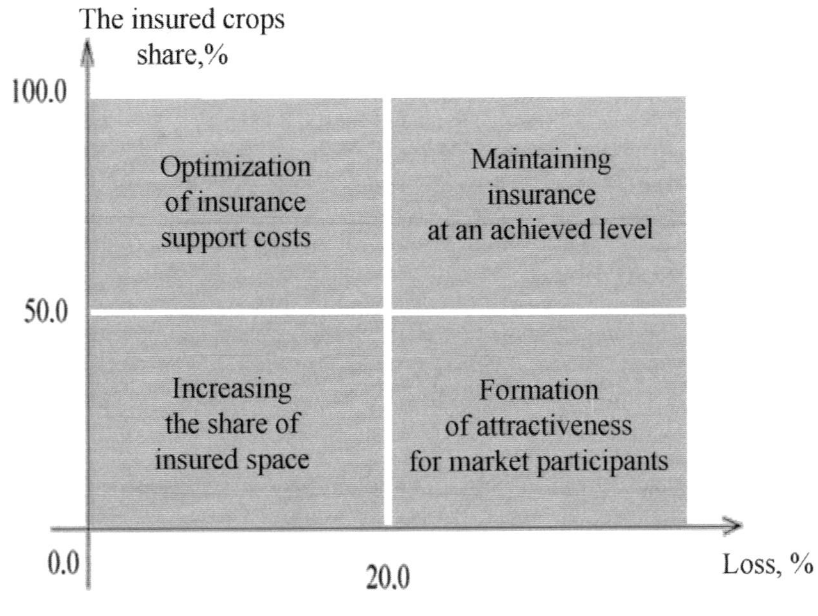

Figure 1. A formalized model of the state support strategy for agricultural insurance.

The second indicator - the risk degree - is proposed to be assessed in terms of the products share that are annually lost as a result of natural or man-made emergencies. The value of this indicator (in accordance with the existing methodology) is calculated as the arithmetic mean of the total loss over the past five years. To reduce the impact of inflationary processes and the existing pricing system, it is necessary to calculate this indicator in physical terms for each culture separately, taking into account the occupied space. We propose a boundary value of 20%.

In view of the foregoing, the model should have the form (Figure 1).

Strategy "Increasing the share of insured space". With a losses low level and a small share of insured sown area. It is enough for the government to compensate for a part of the insurance premium and organize a monitoring service for the activities of insurance companies that would detect violations by insurers and policyholders. The compensation amount should be determined by the predicted (preferred) rate of the insurance market development. The result of this strategy should be a shift to the "Cost optimization" position.

Strategy "Optimization of insurance support costs". Having reached the desired state of the market (the share of insured areas should be at least 70%), the region (country) administration can proceed to fix the situation achieved, optimizing costs. The condition for this should be the achievement of the market equilibrium state, in which a possible reduction is compensated by the new participants appearance. With a possible reduction in the insurance premium, the state should offer new insurance products that would be attractive to agricultural producers without state support (for example, crop insurance as collateral for lending).

Strategy "Formation of attractiveness for market participants". Unlike the two previous strategies, the proposed situation is much more complicated. Significant amounts of annual damage make the situation unattractive for insurance companies. Therefore, the priority for the state is to create the basis for the effective and mutually beneficial activities of insurers and policyholders. To reduce potential damage, it is proposed to introduce a basic state insurance program similar to the US CAT program [20-23]. A commodity producer pays only administrative expenses in the amount of 1.000 rubles / culture and can claim a minimum compensation for damage. In addition, the program is the basis for participation in other types of state support. Such an insurance product will reduce the amount of possible damage by 30%, which makes insurance activities more attractive [24-26].

Strategy "Maintaining insurance at an achieved level". The state, having created an insurance system corresponding to this level, ensures the stable operation of the agro-industrial complex in the event of any emergency. In addition to the measures proposed in the previous strategy, it is necessary to solve additional problems - ensuring payments in the event of large-scale climatic disasters, ensuring the interest of insurance companies in providing insurance services with significant compensation for damage, etc. [27-29] Therefore, this position proposes the implementation of the following measures: control over the insurance fund formation; the compensation of administrative costs and part of the damage exceeding the capabilities of the insurance fund and the reinsurer, who took part in the risks under the contract; the principle implementation to "insurance self-compensation", linking the possibility of providing highly profitable insurance services (for life insurance, medical insurance, property insurance and civil liability) with more risky (and less profitable) types (agricultural insurance).

4. Conclusion

Putting the model into practice will make the existing system of state support more transparent and predictable, depending mainly on the objectives of state policy and the actual indicators achieved by the industry. At the costs high level offered by these measures, the system involves the creation of a sustainable market for insurance services, regulated by both economic and administrative mechanisms. The presence of such decisions totality will automate the decision-making process at the industry management level; ensure the agricultural policy continuity in the long term.

References

[1] Petera P and Šoljaková L 2020 *Economic Research-Ekonomska Istrazivanja* **33 (1)** 46-67 doi: 10.1080/1331677X.2019.1697719

[2] Arunruangsirilert T and Chonglerttham S 2017 *Asian Review of Accounting* **25 (1)** 85-105 doi: 10.1108/ARA-11-2015-0107

[3] Bhimani A and Langfield-Smith K 2007 *Management Accounting Research* **18 (1)** 3-31 doi: 10.1016/j.mar.2006.06.005

[4] Bromwich M 1990 *Accounting, Organizations and Society* **15 (1-2)** 27-46 doi: 10.1016/0361-3682(90)90011-I

[5] Cescon F, Costantini A and Grassetti L 2019 *Journal of Management and Governance* **23 (3)** 605-636 doi: 10.1007/s10997-018-9431-y

[6] Cinquini L and Tenucci A 2010 *Journal of Accounting & Organizational Change* **6 (2)** 228-259 doi: 10.1108/18325911011048772
[7] Nixon B and Burns J 2012 *Management Accounting Research* **23 (4)** 229-244 doi: 10.1016/j.mar.2012.09.004
[8] Zhichkin K, Nosov V, Zhichkina L, Tkachev S and Voloshchuk L 2020 *E3S Web of Conferences* **161** 01060 doi:10.1051/ conf/202016101060 e3s
[9] Zhichkin K, Nosov V, Zhichkina L, Grigoryeva O, Kondak V and Lysova T 2020 *IOP Conference Series: Earth and Environmental Science* **433** 012004 doi:10.1088/1755-1315/433/1/012004
[10] Siska L 2018 *Engineering Economics* **29 (5)** 591-600 doi: 10.5755/j01.ee.29.5.17463
[11] Verezubova T A, Zhichkin K A, Mukhitbekova A M, Penkin A A and Zhichkina L N 2020 *BIO Web of Conferences* **17** 00003 doi: 10.1051/bioconf/20201700003
[12] Belkina T D, Shcherbakova E M and Protokalistova L V 2019 *Studies on Russian Economic Development* **30 (6)** 672-681 doi: 10.1134/S1075700719060029
[13] Lobell D B, Schlenker W and Costa-Roberts J 2011 *Science* **333 (6042)** 616-620
[14] Porfiryev B N and Bobylev S N 2018 *Studies on Russian Economic Development* **29 (2)** 116-123 http://www.springer.com/sgw/cda/frontpage/0,11855,1-40117-70-63638732-0,00.html?changeHeader=true%22%22=%22%22%20target=doi: 10.1134/S1075700718020119
[15] Ray D K, Gerber J S, Macdonald G K and West P C 2015 *Nature Communications* **6** 6989
[16] Kovshov V, Lukyanova M, Galin Z, Faizov N and Frolova O 2019 *Montenegrin Journal of Economics 15 (3)* 179-188 doi: 10.14254/1800-5845/2019.15-3.13
[17] Roberts M J, Schlenker W and Eyer J 2013 *American Journal of Agricultural Economics* **95 (2)** 236-243
[18] Zhichkin K, Nosov V and Zhichkina L 2019 *IOP Conference Series: Earth and Environmental Science* **403** 012073 doi:10.1088/1755-1315/403/1/012073
[19] Bogachev D V 2015 *Regional Research of Russia* **5 (4)** 392-401 doi: 10.1134/S2079970515040073
[20] Coppola A, Ianuario S, Chinnici G, Di Vita G, Pappalardo G and D'Amico M 2018 *Agris On-line Papers in Economics and Informatics* **10 (2)** 33-47 doi: 10.7160/aol.2018.100204
[21] Rosenzweig C, Elliott J, Deryng D, Ruane A C, Müller C, Arneth A, Boote K J, (...), and Jones W 2014 *Proceedings of the National Academy of Sciences of the United States of America* **111 (9)** 3268-3273
[22] Medvedeva T N, Artamonova I A, Baturina I N, Farvazova E A, Roznina N V and Mukhina E G 2019 *IOP Conference Series: Earth and Environmental Science* **341** 012010 doi:10.1088/1755-1315/341/1/012010
[23] Hammouda M, Wery J, Darbin T and Belhouchette H 2018 *Computers and Electronics in Agriculture* **155** 167-179 doi: 10.1016/j.compag.2018.09.036
[24] Liu Y, Zhang Z and Wang J 2018 *Dili Xuebao/Acta Geographica Sinica* **73 (2)** 203-218 doi: 10.11821/dlxb201802001
[25] Luthra S and Mangla S K 2018 *Resources, Conservation and Recycling* **138** 194-206 doi: 10.1016/j.resconrec.2018.07.005
[26] Tack J B and Ubilava D 2015 *Agricultural Economics (United Kingdom)* **46 (2)** 245-257
[27] Reiff M, Surmanová K, Balcerzak A P and Pietrzak M B 2016 *Journal of International Studies* **9 (3)** 62-74 doi: 10.14254/2071-8330.2016/9-3/5
[28] Huffman W E, Jin Y and Xu Z 2018 *Agricultural Economics (United Kingdom)* **49 (4)** 463-479
[29] Rejesus R M, Coble K H, Miller M F, Boyles R, Goodwin B K and Knight T O 2015 *Journal of Agricultural and Resource Economics* **40 (2)** 306-324

IOP Conf. Series: Materials Science and Engineering **941** (2020) 012026 doi:10.1088/1757-899X/941/1/012026

Investment attractiveness of the territory, greening and self-recreation of citizens: Northern regions

Yu O Vladykina [1], N J Donets [2] and N V Rozumnaya [3]

[1,2] St. Petersburg State Institute of Agricultural, Tsarskoye Selo, Petersburg Highway, 2, 196601 Saint-Petersburg, Russia

[3] St. Petersburg State Institute of Technology, 24-26/49 lit.B Moskovsky prospect, 191023 Saint-Petersburg, Russia

E-mail: j_vladikina@ngs.ru

Abctract:

Interest in involving countryside territories in economic turnover occurs the various ways. Housing construction, development of agro-industrial complexes, introduction of innovative technologies, redistribution of the land Fund, etc. The main problems: lack of coordination of actions, unidirectionality of agroeconomic, working in areas of risky farming, seasonal employment. At the same time, socio-economic and infrastructure issues remain out of attention: priority and procedure for financing social facilities, strategic planning of village territories, industry interaction and total environmental impact on the territory. The lack of interaction between stakeholders allows us to solve only operational tasks. Increasing the investment attractiveness of the territory requires a comprehensive approach that provides a synergistic effect of industries. Systematic redistribution and involvement of free population flows in rural labor allows not only to satisfy the growing interest of citizens in active forms of recreation, but also to ensure the introduction of organic forms of agriculture in agricultural activities. The study of recreational motivation in combination with branding of rural areas allows us to predict not only investment activity, but also the reorientation of citizens towards environmentalization and self-employment [1].

1. Introduction.

To determine the investment attractiveness of rural territories of certain regions of Russia, the article examines and analyzes the performance indicators of twelve subjects of the Siberian Federal district and the North-Western Federal district. The choice of territories is determined by the similarity of climatic, territorial, logistical, and socio-economic factors. The analysis of investment activity in the regions for 2016-2018 was conducted in order to identify investment leaders and track the strategic validity of investments in conjunction with the indicators of investment attractiveness of the hotel business as an indicator of business activity in the region. Reorienting the flow of recreants within regions can not only help the development of rural areas, but also solve modern social and environmental problems [2].

Analysis of investments in fixed capital in the hotel business (table 1) showed a decrease in the overall indicator in the period from 2015 to 2017 [3]. We can assume that this is due to sanctions and a reduction in business activity in the country as a whole. However, the figure of 2018 significantly exceeded the values of all previous years, which is largely due to the activities for the provision of food. Investments in the provision of temporary accommodation also showed growth, but did not exceed the value of 2015. Compared with 2014, in the prices of the current year, investments were a third less. It can be concluded that investments in fixed capital in Russia are more dependent on the economic situation in the country, the planned events (Olympic games, Sochi-2014, world Cup-2018), the state monetary policy, the stability of the national currency and the overall investment climate. At the same time, capital investments in the activities of travel agencies and other organizations providing services in the tourism sector showed a slight decrease only in 2017 [3].

Table 1. Investments in fixed capital in the Russian Federation by type of economic activity, bill. rub.

	2014	2015	2016	2017	2018
activities of hotels and catering establishments	105.5	92.1	93.9	86.7	112.0
from it:					
activities to provide temporary accommodation	95.8	67.5	64.5	57.0	66.1
food and beverage activities	9.7	24.6	29.4	29.7	45.9
activities of travel agencies and other organizations providing services in the field of tourism, billion rubles.	0.5	0.5	1.9	1.7	3.8

According to Rosstat [3]

Let us consider the development of the tourism market over the past 5 years in the period from 2013 to 2017 inclusive, using official data of Rosstat [3].

Table 2 Real and nominal increase of internal tourist market of Russia.

Indices	Calculated formula	2014/2013	2015/2014	2016/2015	2017/2016
Real increase of market	$Iq = \frac{\sum q_1 p_0}{\sum q_0 p_0}$	1.024	1.34	1.148	1.31
Nominal increase of market	$I_{pq} = \frac{\sum q_1 p_1}{\sum q_0 p_0}$	1.065	1.98	0.97	1.46
Paashe indices (price influence)	$Ip = \frac{\sum q_1 p_1}{\sum q_1 p_0}$	1.04	1.68	0.85	1.12
Quantity factor	$A = \frac{\sum q_1}{\sum q_0}$	1.023	1.34	1.149	1.31

If we calculate the indices, we can conclude that the real growth of the domestic tourist market due to the increase in the number of sold packages, was especially high in 2015 and increased by 34% and in 2017 – by 31% .

We will analyze the change in prices for nominal growth of the market (on average in all directions): in 2014, inflation in this area amounted to 4%, in 2015, prices for domestic tourism products increased by 68%, and in 2016 there was a decrease in prices by 15%. In the last reporting year 2017, the growth of the nominal market due to the increase in prices amounted to 12%, with an average inflation rate of 2.5% in the country. Such significant changes in prices in 2015 are caused by a sharp depreciation of the national currency.

Comparing tables 1 and 2, it can be concluded that the Russian tourist market shows real growth, as well as the indicator of investment in the activities of travel agencies that sell vouchers to health resorts in the Russian regions and offer their own tourist routes in Russia, including accommodation in local hotel enterprises. Thus, tour operators operating in the domestic market are also indicators of the profitability of the regional hotel sector, despite the decline in capital investment in the provision of temporary accommodation [4].

Let us consider some indicators of investment attractiveness of the hotel business in the regions, using the data of the previous period. This approach allows us to draw a conclusion about the relationship of current investments with the trends of profitability and stability of the hotel business in the regions.

Investment attractiveness of the hotel business in the international market is estimated in the following indicators: room revenue, occupation, average daily room rate (ADR), revenue per available

IOP Conf. Series: Materials Science and Engineering **941** (2020) 012026 doi:10.1088/1757-899X/941/1/012026

room per day (RevPar), revenue per available customer (Rev Pac), double occupation. In addition, the international network business is considered to be indicators Gross operating profit (GRP) Gross profit margin (GOP), Net operation profit (NOP) and others. Using the open data of the Rosstat database, you can calculate the indicators in General. Some of these indicators, in addition to the investment attractiveness of the object also indicates the quality of strategic management of the hotel or hotel chain.

Consider the rate of profitability of a hotel room in several Russian regions. For the analysis and study were taken separate regions of the Siberian and North-Western Federal district, comparable in natural and environmental, transport and logistics and socio-economic indicators. They were investigated the following data:

- the ratio of total income and expenditure, in relative terms. In international practice, this indicator is considered to be gross margin profit (GPM);
- hotel room profitability indicators in the region (RevPar);
- - income per guest, by region (Re Pac);
- status level of hotels (prevalence of hotels of a certain star rating in the region);
- the average length of the guest's stay in the hotel and, as a consequence, in the region, etc.

For all indicators, the dynamics of growth is considered and the spread of stability of growth or decline is taken into account, which is taken into account in the average deviation (table.3).

Table 3. Analysis of profitability indicators of hotel business in the regions of Siberian and North-Western Federal district.

Regions	**Gross margin**		**Regions**	**Gross margin**	
	Relative average growth, %	**Average deviation, %**		**Relative average growth, %**	**Average deviation, %**
Altai territory	8%	179%	Arkhangelsk region	7%	147%
Kemerovo region	16%	69%	Vologda region	19%	82%
Novosibirsk region	9%	127%	Leningrad region	16%	198%
Omsk region	22%	57%	Murmansk region	9%	166%
Tomsk region	13%	80%	Novgorod region	17%	48%
Altai republic	17%	138%	Pskov region	17%	64%
Irkutsk region	15%	56%			
Krasnoyarsk region	5%	223%			
The Republic Of Khakassia	7%	225%			
Republic of Tuva	-14%	-191%			
Siberian federal district	11%	64%	**North-Western Federal district**	**14%**	**26%**

2.Results and Discussion.

Analysis of the activity of hotels in the Siberian and northwestern Federal districts, with the aim of identify the most attractive for investment in the region, showed that the best performance of gross margin of the Siberian Federal district for the period 2000-2015 was observed in the Omsk, Kemerovo, Irkutsk, the Republic of Altai: 22,16,15%,17%, respectively; in the North-West Federal district more uniform dynamics of indicators of the Vologda, Novgorod, Pskov, Leningrad region (19,17,17,16%) and are leaders in growth of profitability of the hotel enterprises in the region (table 3).

Omsk, Kemerovo, Irkutsk regions have the most stable situation in the dynamics of marginal profitability (57,69,56%, respectively), while Altai Krai, which periodically demonstrates explosive growth dynamics, is very unstable and risky in terms of gpm formation (138%), which means that it is the most risky region for investing in the hotel sector. In relation to the North-Western Federal district, stable dynamics of profitability growth is observed in Novgorod (48%), Pskov and Vologda regions (64 and 82%), while the Leningrad region, with 16% of the average relative growth of GPM, has the highest level of instability in the region of 198%. The situation in the Leningrad region may indicate both the volatility of investment activity in the region and the development of long-term investment projects related to the activities of the Federal center of St. Petersburg (table 3).

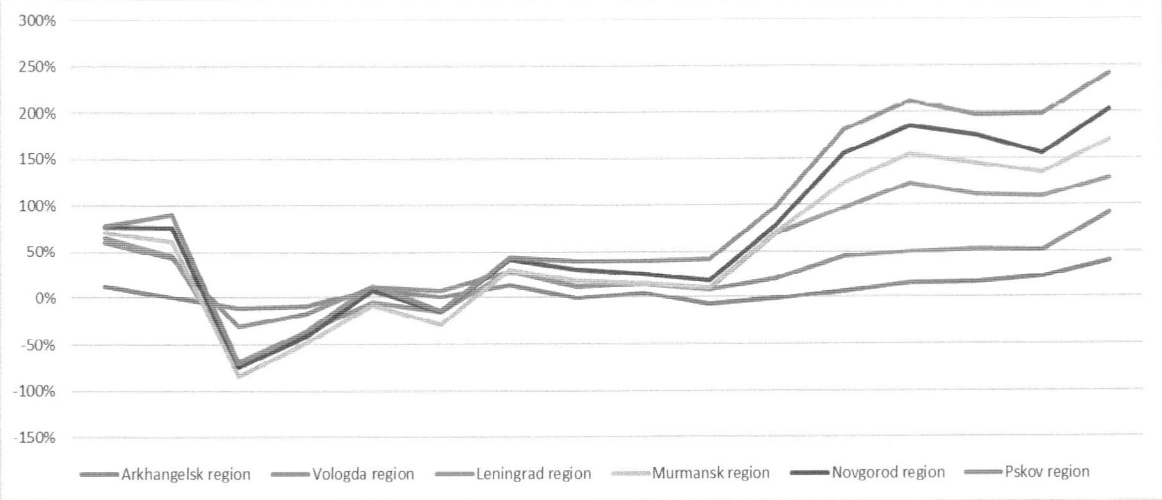

Figure 1. Dynamics of gross profit margin in the hotel enterprises of the North-West Federal district [3]

Considering the two regions as a whole, it can be concluded that the gross margin profit in the period under review developed more steadily in all the territories of the North-Western Federal district (Figure 2). The dynamics of the indicator in the Siberian Federal district looks less stable (Figure 1), which is also reflected in the relative average growth and average deviation by region. The abrupt development of the industry, reflected in the form of GPM, most of all refers to changes in the political and economic situation. As can be seen from the diagram (Figure 1), all regions of the Siberian Federal district are more or less synchronous in the growth and decline of profitability of the hotel business during the study period. However, the Tomsk region, the Republic of Tuva and Altai, which demonstrate a higher level of GPM in short periods, are obviously more dependent on external factors than Omsk, Kemerovo and Novosibirsk, where the latter does not "pass" the criterion of average deviation (127%) (table 3).

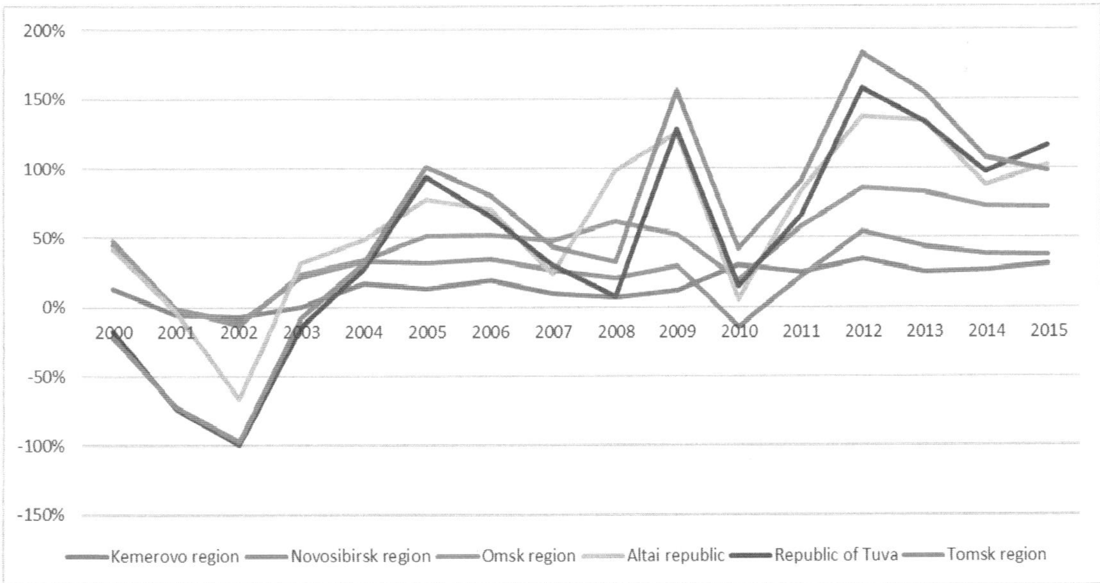

Figure 2. Dynamics of gross profit margin in the hotel enterprises of the Siberian Federal district

IOP Conf. Series: Materials Science and Engineering **941** (2020) 012026 doi:10.1088/1757-899X/941/1/012026

This situation is a "risk zone" for investment, but it can also demonstrate a periodic increase in business income caused by the opening of international network chains in terms of their interaction with the local budget. The issue of profitability of international network business for the Russian territories is the most important in assessing the prospects of investment. All income indicators of the receiving party are relevant only in terms of accumulation of funds in the territory, in terms of tax payments. Net profit of chain hotels or Net profit margin (NPM) is the income of chain corporations and has no relation to the Russian territory.

The mass opening of hotel chains, timed to coincide with the 2018 FIFA world Cup, is also an indicator of the investment attractiveness of the hotel business in the regions. But it is necessary to take into account the fact that the new network enterprises of international business do not take into account the total traffic of guests in the region, relying most of all on their image and the existing system of loyal customers. And if we take into account the factor of the currency imbalance of the ruble against foreign currencies, the growing regional flow of tourists from year to year is quite a natural reality. The most important thing in this situation is the need to determine the place and role of domestic Russian business and Russian chain hotels in servicing the growing tourist traffic [5].

Then the indicators of profitability of the room as a direct indicator of the investment attractiveness of hotel enterprises were considered. The combination of the relative average growth and the average deviation allowed to distinguish the leaders: Novosibirsk, Omsk, Tomsk, Irkutsk, Kemerovo regions in the SFO and Novgorod, Pskov, Murmansk regions in the northwestern Federal district. It should be noted that the intensive growth of the indicator also gives instability in the investment relation. Business risks in this case should be insured by additional financing and strategic decisions.

The ratio of the number of rooms to the number of beds, demonstrating the status and stardom of hotels, was also analyzed in dynamics. The star rating of hotels is about 1.7-2.3 *, in all regions of the Siberian Federal District, except for Tomsk Oblast, the Altai Republics and Khakassia, has remained stable over the past ten years or more. In the Tomsk region one can see a sharp decline in the status of hotels, as can be seen from the increasing number of multi-bed rooms. In Tyva, on the contrary, a sharp increase in the number of stars and the prevalence of single-bed and suite rooms. At the same time, hotels are small, with an average of 40-60 rooms, by region of the district. The situation is stable for ten years in the Northwestern Federal District, where the average star level of hotels in the region is in the range of 1.5-2.5, with the exception of the Vologda and Leningrad regions, where the average upper level approached the range of 3.1-3.5. *. The average hotel capacity is slightly higher than in the Siberian Federal District, from 55-90 rooms, which also indicates the scale of the business.

Similarly, the indicator RevPac, the dynamics of the growth of income and expenses in the hotel enterprises of the Northwestern and Siberian federal districts was analyzed. The results were compared with the dynamics and volumes of investments in fixed assets over the period 2013-2018 in order to determine the strategic feasibility of active operations in the hotel business of the studied territories. The detailed results of this study will be reviewed separately.

3.Conclusion.

After analyzing the interdependence, investment attractiveness of the territory, investment in fixed assets, as well as business and recreational activity of the region, the authors came to the following conclusions:

- investment activities and investments in fixed assets are related to the business and political activities of the territory;
- the most promising areas for the formation of recreational traffic are the centers of business activity in the regions;
- redistribution of free movement of citizens is optimally within the local micro-region, in the current situation of self-isolation;
- the profitability of a single arrival can vary from 100 to 1000 dollars, depending on the length of stay and the level of service in suburban areas;

- organizing reorientation processes for active recreationists, including self-employment, is the responsibility of local governments and rural communities.

It should be noted that investment in fixed capital and the degree of their influence on the development of rural areas require a separate study. The variety of factors that affect the investment attractiveness of regions requires a comprehensive approach based on the use of multi-factor econometric models [1,4].

References

[1] Y Wei, B Davidson, D Chen, R White, 2009, Agriculture, ecosystems & environment 131 (3-4), 263-273

[2] C Su, B Fu, Y Wei, Y Lü, G Liu, D Wang, K Mao, X Feng, 2012, Sustainability Science 7 (1), 17-32

[3] Faddeenkov A, Vladykina J, 2015, Applied methods of statistical analysis. Nonparametric approach: proc. of the intern. workshop, Novosibirsk, NSTU, P. 338-343.

[4] Faddeenkov A, Vladykina J, 2018, 14th International Scientific – Technical Conference APEIE, 2018, NSTU. [Access mode: http://apeie.conf.nstu.ru/apeie2018/NSTU], P.73-76

[5] Vladykina J, Rozumnaya N, 2018, Siberian Development 2018. IOP Publishing IOP Conf. Series: Earth and Environ-mental Science Vol. 204. [Access mode: https://iopscience.iop.org/issue/1755-1315/204/1], P.48.

IOP Conf. Series: Materials Science and Engineering **941** (2020) 012027 doi:10.1088/1757-899X/941/1/012027

Thermophysical properties of lowland Histosols of the Upper Volga woodland (on the example of the Yakhroma river valley)

N V Sorokina[1], E V Shein[1,2], A D Pozdnyakova[2], A P Shvarov[1]
1 Lomonosov Moscow State University, GSP 1, Leninskie Gory, Moscow 119991, Russia
[2] Dokuchaev Soil Science Institute, Pyzhevskii per.7, Moscow, 119017 Russia

E-mail: rebelde1502@mail.ru

Annotation. The dependence of thermal diffusivity on volumetric humidity for Histosols developed on wood peat, underlain by sediment peat and mineral peat, was studied using Histosols from the Yakhroma Valley (Moscow Region). The determination was carried out at the stages of dreinage and humidification by the laboratory method of Kondratiev. For the stage of draining, the dependence has a more pronounced dome-shaped form than for the stage of saturation. The highest values of thermal diffusivity were achieved with a volumetric soil water content of 50-55% for Histosols on wood peat and 40-45% for agromineral peat. The approximation of the data was carried out according to the equation of Chang and Horton. As a result of comparing the approximation parameters, it was revealed that they do not significantly differ from each other, which indicates the absence of hysteresis of the thermal diffusivity from humidity for the thermal diffusivity curve, the unambiguous values of which can be used to simulate the temperature regime.

1. Introduction

Climate change affects the course of intra-soil processes, soil functions in the biosphere. The impact of global climate change and local anthropogenic impacts has not been studied enough, both in Russian and in world science. In this regard, first of all, problems arise of the influence of temperature and humidity on soil productivity, their stability and carbon balance of peat soils, as the main intrazonal soils with a high content of organic matter. Given that Histosols are currently actively used in agriculture and determine the carbon balance in the surface layer of the biosphere, the temperature factor for biological activity is often limiting. There is little information to regulate and predict changes in the hydrothermal regime of Histosols, which are based on the dependence of soil thermal diffusivity on moisture. Histosol is distinguished by its specific physical properties: low density, hydrophobicity in the dry state, sharp swelling, so laboratory research is difficult [1-3]. This article presents the influence of the processes of desiccation and moistening on the dependence of the thermal diffusivity on moisture with an assessment of the possibility of a hysteresis of this dependence in these soils.

Under hysteresis in soil physics, we understand the change in the shape of the curve, depending on the direction or history of the process [4-5]. Most modern studies are aimed at identifying hysteresis for the water retention curve of mineral soils, while data on the presence or absence of hysteresis for the thermal diffusivity curve are currently practically absent. As known, one of the most likely causes of

IOP Conf. Series: Materials Science and Engineering **941** (2020) 012027 doi:10.1088/1757-899X/941/1/012027

hysteresis is a complex pore space. The structure of the pore space of peat is significantly different from mineral soils. Histosols and peat are a macroporous system. When filling its pores with water, significant swelling occurs, while at the same time, drying out can lead to a state of hydrophobicity. In this case, a "re-arrangement" of particles occurs, and the structure of the pore space changes. To regulate the temperature regime, it is necessary to know whether this affects the position of the curve of the dependence of thermal conductivity on volumetric humidity and whether this should be taken into account in the calculations.

The aim of the work is to study the dependence of thermal diffusivity on humidity at the stages of moistening (humidification) and drainage (desiccation), for the quantitative calculation and prediction of the hydrothermal regime of peat soils. Tasks: conduct laboratory studies of curves during wetting and drying for Histosols samples of the Yakhroma floodplain; approximate the data in the program Statistica; compare the obtained approximation parameters and identify if there are any significant differences between them.

2. Objects and methods

The object of the study is the lowland peat of the Dmitrov branch of the Federal Scientific Center “Soil Science Institute named after VV Dokuchaev”, located in the Yakhroma River Valley. Histosols developed on wood peat, underlain by a deposit of grass and sedge peat, located in the 8th quarter of Dalniy hospital and agromineral Histosols mosses from the 1st quarter of the Middle hospital were selected. Both hospitals are now in intensive agricultural use, “Middle” since 1914, “Far” from 1961-1965.

The selected wood peat is located in the central part of the floodplain, is well drained and less moisturized, consists of the remains of woody vegetation, and the remains of bog vegetation are also found [6,7]. These Histosols have a high degree of decomposition and high ash content (Table 1). Ash content is also affected by the continuous application of fertilizers. Groundwater is enriched in iron.

Table 1. Some physical properties of peat soils.

Name of Histosols and underlying peat	Depth, cm	Decomposition degree, %	Ash content, %
Histosols developed on wood peat, underlain by grass sedge peat	5 - 15	81.6	24.8
	25 - 40	69.5	20.1
Agromineral Histosol	5 - 15	79.8	23.2

As for agromineral peat, they are located in the near-river part of the valley and were formed during the floods of the river. Due to intensive agricultural activities, such as plowing slopes, peat formation stopped there and a layer of mineral sediment was deposited on top [6]. The layers of buried deluvium and alluvium have a great influence on the presence of minerals. They contain a higher content of potassium, phosphorus and magnesium compared to other peat layers [7].

In connection with the active use of these peat soils in agriculture, it is necessary to know their features of the thermal regime. This will allow you to choose more suitable crops, and the correct prediction of the temperature regime will increase yield. All this can be possible only in the case of a

IOP Conf. Series: Materials Science and Engineering **941** (2020) 012027 doi:10.1088/1757-899X/941/1/012027

detailed study of the dependence of the thermal diffusivity of peat on moisture and the preparation of forecast models [8].

The thermal diffusivity was determined by the laboratory method of Kondratiev [9, 10]. For the same sample, measurements were carried out at the stage of drying and moistening. The first curve was obtained by drying a pre-saturated peat column; upon wetting, the goal was to return to the original humidity. On the basis of the obtained data, two curves of the dependence of the thermal diffusivity on humidity were obtained, which, according to statistical processing, have no significant differences.

3. Results and Discussion

Figure 1 (a, b) shows the data of the dependence of thermal diffusivity on the moisture content of peat, developed on wood peat, underlain by a deposit of grass, sedge peat for layers 0-20 cm and 20-40 cm, obtained at the stage of drying and the stage of moistening. The graphs have the form of a dome-shaped dependence characteristic of mineral soils [10-12]. Naturally, with an increase in the volumetric moisture content, thermal diffusivity increases, reaching the highest value at a humidity of 50-55% in both cases. The highest values of thermal diffusivity in both cases are close to 4.16 cm^2 / h and 3.35 cm^2 / h for layers 0-20 and 20-40 cm, respectively. Reaching the maximum point, the values of thermal diffusivity begin to decrease. For graphs at the humidification stage, this is less pronounced. This type of dependence can be explained by an increase in the amount of water between the particles, and, accordingly, a decrease in the number of solid-phase contacts.

Figure 1 (c) shows similar curves obtained for agromineral peat. The graphs also have a domed character. The maximum value of thermal diffusivity is achieved at values of volumetric humidity of 40-45%, and after reaching the maximum thermal diffusivity begins to decrease. It should be noted that the highest values of thermal diffusivity are 4.9 cm^2 / h and 4.63 cm^2 / h at the stage of drying and moistening, respectively.

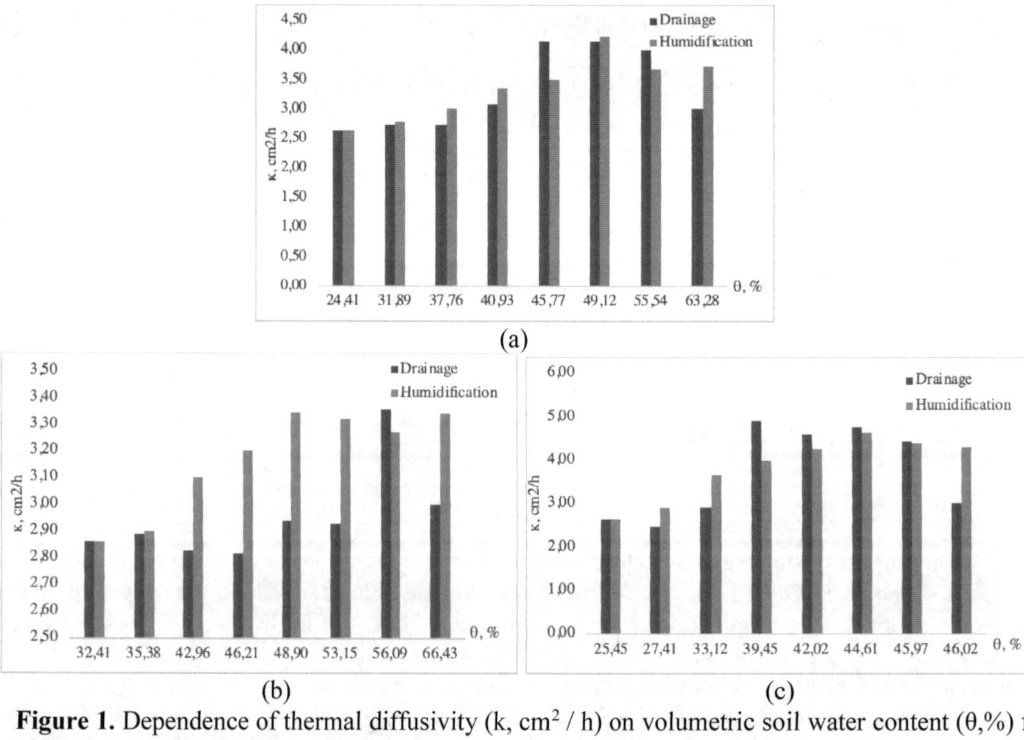

Figure 1. Dependence of thermal diffusivity (k, cm^2 / h) on volumetric soil water content (θ,%) for layers 0-20 cm (a) and 20-40 cm (b) of Histosols developed on wood peat, underlain by grass sedge peat and for a layer of 0-20 cm (c) agromineral Histosol.

Table 2 presents the main parameters of the approximation of the dependence of thermal diffusivity on humidity, obtained by nonlinear approximation. For the approximation, we used the polynomial model of Chang and Horton $\kappa(\theta) = b_1 + b_2 + b_3 \times \theta^{0,5}$, where b_1, b_2 and b_3 are the approximation parameters that are used in mathematical modeling of heat transfer in soils [13-14].

It is important to note that the obtained parameters at a significance level of 0.05 are not significant. Probably, it is necessary to increase the number of repetitions and refine the methodology for determining thermal diffusivity by laboratory methods.

Table 2. Statistics of approximation parameters (b_1, b_2 and b_3) of thermal diffusivity (κ, cm^2/h) versus volumetric humidity (θ, %) according to the equation $\kappa(\theta) = b_1 + b_2 + b_3 \times \theta^{0,5}$ [15].

Name of peat	Depth, cm	Process	Parameter	Value	Standard error
Histosols developed on wood peat, underlain by grass sedge peat	0 - 20	Drainage	b_1	-9.957	9.591
			b_2	-0.262	0.232
			b_3	3.774	3.005
		Humidification	b_1	0.197	10.907
			b_2	0.0005	0.292
			b_3	0.49	3.626
Histosols developed on wood peat, underlain by grass sedge peat	20 -40	Drainage	b_1	2.648	4.859
			b_2	0.01	0.104
			b_3	0.026	1.424
		Humidification	b_1	-3.923	2.308
			b_2	0.051	0.051
			b_3	0.691	0.691
Agromineral Histosol	0-20	Drainage	b_1	-36.522	37.251
			b_2	-0.967	1.066
			b_3	12.557	12.677
		Humidification	b_1	-10.345	12.854
			b_2	-0.234	0.364
			b_3	3.759	4.356

Knowing the approximation parameters, we can statistically compare the dependences of the temperature conductivity on humidity for the processes of drying and moisturizing, and we can conclude that they differ. We did this using the t-test, which is calculated by the formula [13]:

$$t = \frac{|b_n^{'} - b_n^{''}|}{\sqrt{(S_{b_n^{'}})^2 + (S_{b_n^{''}})^2}},$$

where $S_{b_n^{'}}$ and $S_{b_n^{''}}$ – are the standard deviations of the parameters $b_n^{'}$ and $b_n^{''}$. Accordingly, if the t-criterion turns out to be more tabular for a given degree of freedom and significance level (traditionally 0.05), then the parameters of the two samples significantly differ from each other. In this case, it can be argued about the reliability of the differences in the corresponding characteristics of the process. In our case, the calculations showed that the curves do not differ significantly. This suggests that there is no hysteresis for the thermal diffusivity of peat soils and we may not take it into account in the calculations.

4. Conclusion

Histosols in their physical properties are very different from mineral ones. At the same time, they are actively used at the moment in agriculture, so the study of their properties is especially important. Temperature is often a limiting factor, and due to lack of data it is difficult to regulate. Due to the strong swelling and hydrophobicity of peat during desiccation, it is important to know that the dependence of thermal diffusivity on volumetric moisture, Histosols are characterized by an unambiguous dependence of thermal diffusivity on humidity, does not have hysteresis, and this dependence can be used to predict the thermal regime of soils both when they are moistened and when dried.

Acknowledgments

The reported study was funded by RFBR according to the research project No. 19-29-05112 mk "Digital Physics and Hydrology of Soils: Basics of Spatial-Dynamic Analysis, Prediction of Risk of Critical Situations and Optimal Management". Project Manager Professor E.V. Shein.

References

[1] Johnson M J, Lee K Y and Scow K M 2003 *Geoderma* **114** 279–303
[2] Rabinovich G Yu, Pozdnyakov L A and Dubrova M S 2016 *Materials of the international scientific conference* (Moscow: VNIIA) pp 241–5
[3] Pozdnyakov L A 2008 *Eurasian Soil Science* **10** 1217-23
[4] Shein E V, Skvortsova E B , Dembovetskii A V , Abrosimov K N, Il'in L I and Shnyrev N A 2016 *Eurasian Soil Science* **3** 344-54
[5] Shein E V 2005 *Course Of Soil Physics* (Moscow: Moscow State University Press) p 432
[6] *Environmental Engineering in Polesye* Vol 2 Edited by edited by Yu Mazhayskiy, A Rokochynskiy et al 2019 (Ryazan : Meshchersk office of VNIIGiM of A N Kostiakov) p 503
[7] Pozdnyakov A I , Kovalev N G et al 2014 *Peat and eutrophic torfozems in long-term agricultural use* (Tver: TVTSU) p 356 p
[8] Shein E V , Pozdnyakova A D , Sorokina N V , Dembovetsky A V , Shvarov A P , Il'in L I *Eurasian Soil Science* **11** 1363–8
[9] *Theories and Methods of Soil Physic*s Edited by Shein E V, Karpachevskiy L O l 2007 (Moscow: Grif and K) p 616
[10] Arkhangelskaya T A, Lukyashchenko K I , Tikhonravova P I 2015 *Eurasian Soil Science* **2** 177–82
[11] Ghader A 2014 *Int J Adv Appl Sci* **6** 31–6
[12] Usowicz B, Lipiec J, Łukowski M, Marczewski W, Usowicz J 2016 *Soil Till Res* **164** 45–51
[13] Shein E V and Ryzhova I M 2016 Mathematical Modeling in soil science (Moscow: IP Marakushev AB) p 377
[14] Šimůnek J, van Genuchten M T and Šejna M 2008 *Vadose Zone J* **7** 587
[15] Mahdavi S M, Neyshabouri M R and Fujimaki H 2016 *Eurasian Soil Science* **49** 915–25

IOP Conf. Series: Materials Science and Engineering **941** (2020) 012028 doi:10.1088/1757-899X/941/1/012028

The effectiveness of vermigumates under the conditions of vegetation experience

E V Kallas[1], A V Rodikva[1], S P Kulizhskiy[1]

[1]National Research Tomsk State University, 36 Lenin Ave., Tomsk, 634050, Russia

E-mail: rodikovaav@mail.ru

Abstract. The effectiveness of biofertilizer, which was obtained by extracting humic acids from vermicompost 0.02 n NaOH, is shown. The increase in the productivity of the test culture in the growing experiment against the background of watering with the vermigumate (dilution 1:50) was 73% in relation to the control.

1. Introduction and Background

At the present stage of development of human society, the problem of restoring and improving the fertility of cultivated soils is more acute than ever. One of the ways to solve this pressing problem is vermicomposting. This would allow to get environmentally friendly highly effective organic fertilizer based on almost any organic substrates, including food waste, in large quantities which would be generated in the course of human life. Furthermore, the latter allows us to solve another global problem of society. It is the disposal of organic waste. Vermicomposting is widely used in the USA, Western Europe, Japan. However, in our country there are no more than 5 large vermifarms that work.

The scientific literature actively discusses issues which are related to the production technologies and the effectiveness of vermicomposts, their influence on the yield of cultivated plants and the chemical, physico–chemical and physical properties of soils. The properties of vermicomposts, which were obtained in the processing of various types of manure [1, 2] and leaf litter [3] are relatively well studied. However, the agrochemical properties of vermicomposts that are based on food waste have practically not been studied. The issues related to the technology of using liquid agents, including vermigumate that was extracted from vermicomposts, have not been sufficiently studied. It is humic acids that play a crucial role in increasing plant productivity, which is associated with such mechanisms of their influence on living organisms as the activation of protein synthesis in cells [4], the effect on the rate of electron-donor-acceptor reactions, interaction with cell membranes and effects on biochemical processes [5], penetration of humic acids into cells and interaction with intracellular components [6, 7].

The purpose of this study is to obtain an organic fertilizer that is based on vermicomposting food waste, to study its properties and to identify the effect of different doses of vermigumates on the productivity of the green mass of the test culture (arugula) under the conditions of a growing experiment.

2. Materials and Methods

The objects of study were: vermicompost which was obtained by the method of processing (within 8 months) food waste by the California worm Eisenia andrei Bouche; vermigumates that were represented by humic compounds that were extracted with 0.02 n NaOH from vermicompost; cultivated horizon of agro–gray soil (horizon material was used as a substrate in the growing experiment); test culture for the growing experience of arugula Indau Poker.

For the preparation of vermicompost, food waste of plant origin was used: bananas, carrots, beets, fruits, zucchini, pumpkins, as well as tea and coffee. Vermigumates were prepared as follows: 200 g of vermicompost was poured with 2 liters of 2n NaOH, shaken and filtered after 24 hours. In the resulting extract, the organic carbon content was 2.5%.

To characterize the properties of agro–gray soil and vermicompost, the main agrochemical indicators (content of C, N, P_2O_5, absorbed bases, qualitative composition of humus, pH, particle size distribution) were determined by methods and techniques generally accepted in soil science.

The vegetation experiment was carried out in plastic ("seedling") vessels with a capacity of 250 ml, filled with soil material that were removed from the cultivated horizon by agro–gray soil. Irrigations with solutions of vermigumates with different concentrations were carried out from above until the soil was completely saturated with a frequency of 5 days, combining them with watering. Excess irrigation water flowed through the openings at the bottom of the vessels. The experiment was carried out in 4–fold repetition according to the following scheme:

Control (watering).

Variant 1 (watering: 2 ml of vermigumates + 200 ml of water, dilution 1: 100).

Variant 2 (watering: 4 ml of vermigumates + 200 ml of water, 1:50 dilution).

Variant 3 (watering: 6 ml of vermigumates + 200 ml of water, dilution 1:33).

The arugula of the early-ripening Indau Poker variety (20–25 days), which has unpretentiousness and rapid growth, was chosen as a test culture. After emergence on the 7th day, 4 plants were left in each cup, the rest were removed. The calendar terms of the experiment are from June 20 to July 17, 2019 (duration 27 days).

During the biometric observations, the following morphological indicators were recorded (at the time the experiment was completed): the number of true leaves; the maximum plant height; parameters of lamina (maximum length and width).

Productivity was estimated by the weight of the aboveground (green) part of the plants.

Mathematical processing of the results was carried out in Excel.

3. Experimental Section

The agrarian soil used in the growing experiment was characterized by medium loamy fine-arenic coarse–grained granulometric composition with a content of current–dispersed (<0.001 mm) particles of 14%, a high content of adsorbed cations (41 mEq / 100 g), close to neutral pH (6, 54). The content of organic carbon is 3.2%, total nitrogen is 0.25%, total phosphorus is 0.28%. In the composition of humus, the proportion of humic acids (HA) is about 26% of the total organic carbon (C_{total}), fulvic acids (FA) is 23%, the ratio of carbon HA to carbon FC (Cg : Cfc) is 1,11 (type of humus is fulvate-humate). The characteristics of vermicompost indicate a high content of organic carbon (23%), total forms of nitrogen (2.13%) and phosphorus (0.63%), enrichment with easily hydrolyzed forms of nutrients (nitrogen content 345 mg / 100 g, phosphorus 500 mg / 100 g), strongly alkaline reaction of the medium (pH more than 9 units). The high fertilizer value of vermicompost is indicated by the narrow ratio C: N (10,9).

The fractional group composition of the system of humic substances of vermicompost differs from that in agro–gray soil with a high proportion of HA, 2 times the proportion of FA, which makes it possible to classify humus as a humate type (Cg : Cfc = 2.29). The absolute content of humic acids is high, about 7.8% of vermicompost, which is more than 33% of total organic carbon. HAs are represented mainly by the most valuable fraction, the calcium humates (black HAs), their share reaches 28% of the total.

The results of the growing experiment showed that irrigation with a vermigumates in all of the applied concentrations led to an increase in the biometric parameters of the test culture and its productivity compared to the control one. At the time of completion of the experiment, in all of the variants of the experiments, the number of real leaves in plants that were irrigated with biofertilizer was 4–5 pcs., whereas in all of the replications of the experiment in the control there were no more than 3–4 pcs. The maximum height of the plants was also greater (by an average of 2.8 cm), as affected by vermigumates, although differences in the experimental variants were observed. The greatest effect on this indicator was noted in variant 1 with a dilution of 1:100 – 18.1 cm (in the control 14.6 cm), and the smallest (16.3 cm) was in variant 3 with a dilution of 1:33. The dimensions of the arugula leaf lamina in the variants with vermigumate irrigation were also significantly larger than in the control: length was on average 1.3 cm, width was 0.5 cm. The greatest effect was observed in variant 2 with a dilution of 1:50. There the increase of the width of the leaf lamina was 0.7 cm. In the same embodiment, the highest productivity of the test culture was revealed. The weight of the aerial part of 4 plants on average of 4 replicates was 8.78 g, while in the control it was 4.71 g. In the variants with a lower and higher concentration of the agent, the productivity was also higher than in the control, but the increase was slightly less (figure 1).

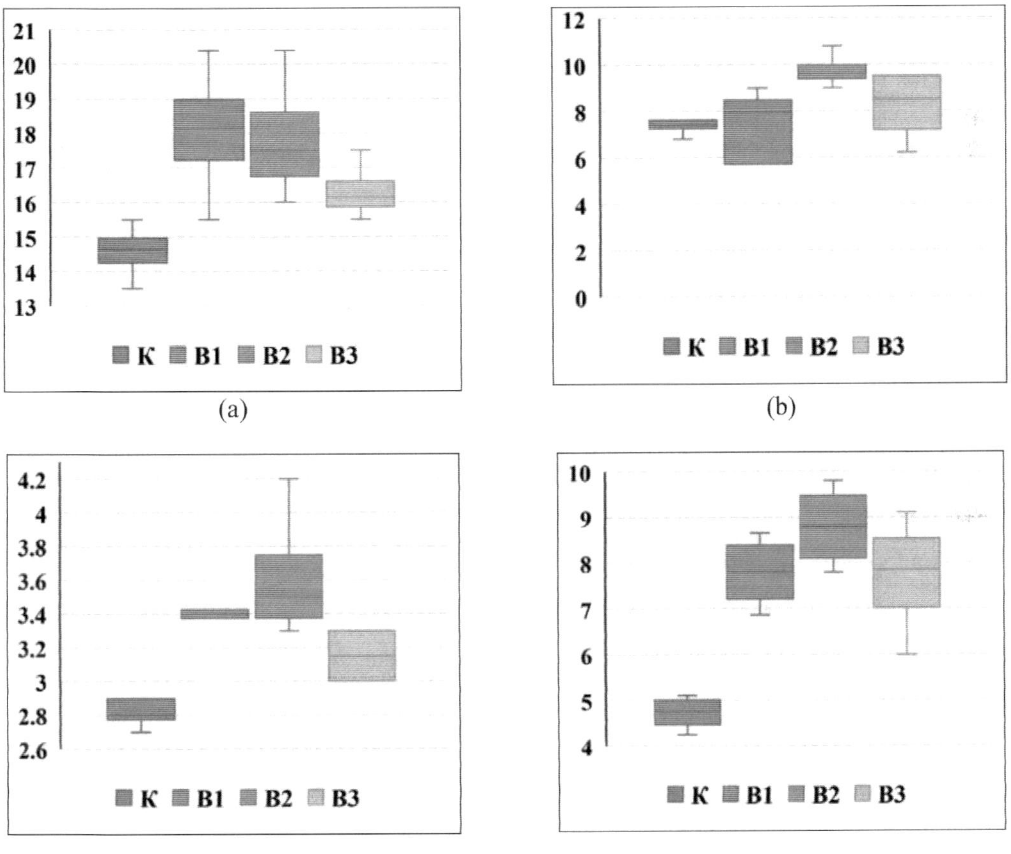

(a) (b)

(c) (d)

a – maximum plant height (cm); b – lamina length (cm); c – lamina width (cm); d – aboveground (green) biomass (g).

Figure 1. Biometric characteristics and arugula productivity in experiment (K – control, B1-B3 – experiment variants).

Thus, it was concluded that the use of vermigumates is effective and has a positive effect on the growth and development of Indau Poker arugula. The best indicators of vegetative growth and development of the test culture were noted in variant 2 (4 ml of the drug + 200 ml of water, 1:50 dilution), where the productivity of arugula, compared with the control, was 73% higher. When the vermigumates concentration was 1.5 times higher (variant 3), a slight decrease was observed in relation to variant 2 of the main parameters characterizing the development of culture.

4. Result and Discussion

Stimulation of plant growth and development by vermicomposts and vermigumates is associated with improved water-physical properties and optimization of the soil nutritional regime, as shown by numerous studies [8], as well as with a positive effect on plant physiological processes [9]. Another reason for increasing the productivity of crops against the background of the use of vermicomposts and vermigumates, as N.N. Tereshchenko et al. [8] emphasize, lies in the biological characteristics of vermicompost products. These authors consider the calling card of earthworms, which is the presence in the substrates of their habitat of such groups of microorganisms as nitrogen bacteria and nitrifying bacteria. Along with the function of regulating the nitrogen nutrition of plants, the azotobacter also produces biologically active substances: plant growth and development stimulants and biofungicides.

The positive effect of irrigation with a vermigumates is most likely due not only to the effect on the physiological processes occurring in plants, but also to the optimization of conditions for the life of the soil biota and the development of the root system, which has contributed to an increase in the amount of root excretions and metabolites in the soil. The fact of enhancing the biological (including enzymatic) activity of the soil under the influence of root secretions of plants receiving humic substances has been shown by many researchers [9, 10, 11].

The Growth-stimulating agents, which include vermigumates, can likely, at high doses, reduce the effectiveness of the action and even have an inhibitory effect on plants. The data available in the literature show that low doses of humic preparations (0.001–0.010% solutions) are more effective, while increased doses are generally less economical and may be toxic [9, 10, 11], which requires a cautious approach to the application.

5. Conclusion

The studies expand the understanding of the properties of vermicomposts which were obtained on the basis of food waste, especially in the field of fractional group composition of the humic substances system. HAs are the dominant group, and their composition is dominated by the fraction of calcium humates, which plays a leading role in the processes of structuring soil fine earth, creating favorable water-air, physicochemical and chemical properties of the soil. Vermigumates extracted with 0.02 n NaOH from vermicomposts have growth-promoting properties and are highly effective. An analysis of the yield structure of the test culture of the Indau Poker aragula variety showed that the use of vermigumates (under the conditions of the growing season) positively affected the height of the plants, the length and width of the leaf blades, which contributed to an increase in productivity by 73%.

References

[1] Chan P L S and Griffiths D A 1988 *Biological Wastes* **24** pp 57–69
[2] Atiyeh R M, Edwards C A, Subler S and Metzger J D 2001 *Bio-resource Technology* **78(1)** pp 11–20
[3] Petrochenko K A, Kurovsky A V, Babenko A S and Yakimov Yu E 2015 *Tomsk State University Journal* **2(30)** pp 20–34
[4] Trevisan S, Francioso O, Quaggiotti S and Nardi S 2010 *Plant. Signal. Behav.* **5(6)** pp 635–43
[5] Vaccaro S, Ertani A, Nebbioso A, Muscolo A, Quaggiotti S, Piccolo A and Nardi S 2015 *Chem. Biol. Technol. Agric.* **2(5)** pp 1–12
[6] Schiavon M, Pizzeghello D, Muscolo A, Vaccaro S, Francioso O and Nardi S 2010 *JOEC* **36** pp 662–9

[7] Voronina L P, Yakimenko O S and Terekhova V A 2012 *Agrochemistry* **6** pp 50–7
[8] Tereshchenko N N, Bubina A B and Pisarenko S V 2008 *Tomsk State University Journal of Biology* **2(3)** pp 47–60
[9] Stepanov A A, Yakimenko O S, Gosse D D and Smirnova M E 2018 *Agrochemistry* **6** pp 36-43
[10] Pukalchik M A, Terekhova V A, Yakimenko O S, Kydralieva K A and Akulova M I 2015 *Soil Science* **6** pp 1–10
[11] Pukalchik M A, Panova M I, Terekhova V A, Yakimenko O S and Fedoseyeva E V 2017 *Agrochemistry* **8** pp 84–91

IOP Conf. Series: Materials Science and Engineering **941** (2020) 012029 doi:10.1088/1757-899X/941/1/012029

Development of clonal micropropagation technique for *Gloxinia hybrida* hort. "Strawberry Ice-Cream"

A A Zaripova, A A Mukhametvafina, K A Urazbakhtina, A I Shigapova, A Sh Akhmetova, Z Kh Shigapov

South-Ural Botanical Garden-Institute, Ufa Federal Research Centre, RAS, Mendeleev str., 195/3, Ufa, 450080 (Russia)

E-mail: zaripova.al@mail.ru

Abstract. The reported paper presents outcomes of clonal micropropagation of ornamental variety *Gloxinia hybrida* hort. "Strawberry Ice-Cream". In the study conditions supporting *in vitro* implanting leaf blade fragments are developed with an outcome of 97 % viable explants. An optimal composition of the Murashige and Skoog growth medium is suggested; it contains BAP (0.5 mg/l) and IAA (0.2 mg/l) and brings about efficient adventive sprout regeneration. Implantation takes place in a growth medium; no root stimulating hormones are applied. The survival of regenerative plants is reported to be high after transplanting *ex vitro*.

1. Introduction

To date, plant tissue and organ culture is widely used all over the world in commercial flower growing. This technique has a number of advantages over traditional plant propagation, such as output of genetically homogenous planting stock; production of virus-free plants owing to meristem culture; a high propagation coefficient; a possibility of year-round works and efficient use of working areas for young plants production; and easy transportation of tube stock plants to any destination [1].

Clonal micropropagation makes it possible to reduce a growth period of plants till they are ready for selling, e.g. 1 – 1.5 months shorter for foliage begonias and ornamental flowering saintpaulias, gloxinias, and 3 – 4 months less for chrysanthemums, lilies, carnations, and orchids. Propagating a new variety, several millions of plants can be output annually; in 2 – 3 years of growing a planting stock of high quality is produced. It might take over 10 years given that conventional methods are used [2].

On the world scale a majority of ornamental flowering plants are grown with the help of clonal micropropagation techniques. The Netherlands is a West-European leader in the field of clonal micropropagation (about 70 laboratories are involved into micropropagation production). This fact is related to cultivation of ornamental plants – a global market niche traditionally dominated by the Netherlands. Approximately a half of cut flowers and ornamental flowers exported worldwide are output in this country [3]. Among most important groups propagated *in vitro* in the Netherlands there are pot plants, orchids, and bulbous plants.

Recently, Russia has an extensive experience in the micro-cloning most demanded cultures. Biotechnology laboratories are opened almost in all scientific and research institutes, selection centers with the purpose to improve a valuable planting material and promising varieties as well as propagate them via micro-cloning [3, 4].

Our study aimed at developing a cloning micropropagation technique for *Gloxinia hybrida* hort. "Strawberry Ice-Cream" in culture *in vitro.* To achieve the purpose our study sought for a production method of a sterile culture, an optimal composition of the growth medium supporting adventive shoot regeneration and implantation *in vitro*, conditions of transplanting regenerative samples *ex vitro*.

IOP Conf. Series: Materials Science and Engineering **941** (2020) 012029 doi:10.1088/1757-899X/941/1/012029

2. Materials and Methods

The technique of plant tissue and organ culture is adapted to the clonal micro-propagation [5].

Our study is focused on a double-flowered variety of *Gloxinia hybrida* hort. "Strawberry Ice-Cream" selected in the Netherlands; its fascinating big double flowers have the colour of strawberry juice and a white petal brim. The plant has a strong medium length flower-bearing stem and a compact shrub. The variety tends to flowering in bouquets and grows in the collection of South-Ural Botanical Garden-Institute, Ufa.

For experiments we selected undamaged leaves of generative plants, washed them in a mild soap solution for 15 minutes and in flow tap water for 30 minutes. Then the surface of a planting material was treated with sterilizing 0.1 % diacid and 70 % ethanol solutions in the aseptic laminar flow cabinet. The exposure time to sterilizing agents was set 0.5 to 20 minutes. Finally, the planting material was washed in sterile distilled water for 15 minutes. As explants we took 2-4 cm^2 fragments of leaf blades with a midrib.

Explants were placed into the growth medium in test tubes or retorts. Growth media were prepared and sterilized as recommended [5, 6]. The Murashige and Skoog growth medium (MS) was used for cultivation [7], its composition and concentration were modified with growth stimulators 6-benzylaminopurine (BAP), kinetin, indole-3-butyric acid (IBA), and 3-indoleacetic acid (IAA). Plants were reproduced in such controlled conditions as a 16-hour photoperiod, a temperature of 24°C, and air humidity of minimum 70 %.

Regenerative plants were transplanted into containers with a soil substrate and kept in the vegetation chamber.

3. Results and Discussions

A state of an explant is one of key factors when implanting into culture *in vitro*. Here, particular attention is to be paid to the physiological age and size of an explant.

Our observations showed mature developed and grown leaves of gloxinia were more capable to morphogenesis than young and growing old leaves.

Another important factor of successful micro-propagation is an explant size. Bigger explants of gloxinia were viable longer, a leaf tissue did not shrink as a consequence, and more meristematic zones developed.

As known, micro-propagation begins from selecting and disinfecting an explant to obtain an aseptic culture. Explants of gloxinia leaves had gentle and soft tissues, a planting material was disinfected in aseptic conditions as given in Table 1; accurately, trying to avoid any damages.

Table 1. Disinfection solutions vs. sterility and living ability of explants *Gloxinia hybrida* hort. "Strawberry Ice-Cream"

Disinfection solution	Exposition, min.	Number of sterile explants, %	Number of viable explants, %
70% ethanol 0.1% diacid	0.5 8	80	80
70% ethanol 0.1% diacid	0.5 10	97	97
70% ethanol 0.1% diacid	0.5 20	100	30

From the data of experiments it is apparent the exposition time in the first diacid-containing disinfection solution is short because a percentage of sterile explants are low. Using the third disinfection solution, we obtained less viable explants in spite of a high percentage of sterility. An optimal disinfection variant appears to be the exposure of a planting material in 70% ethanol solution for 0.5 min. and 0.1% diacid solution for 10 min. A maximal number of viable and sterile explants are a product of this disinfection variant.

A lot of species representing Gesneriaceae family: Saintpaulia, Cape primrose, Episcia, Gloxinia, etc. regenerate plants from leaf blades, sprigs, stems. In traditional vegetative propagation one *Gloxinia* leaf gives one plant. Using the tissue culture technique, mass propagation of a particular Gloxinia variety is possible through inducing the adventive bud formation *de novo* from small leaf segments, i.e. origination of organized structures immediately from specialized epidermal and sub-epidermal tissues, leaf mesophyll, etc.

Under appropriate disinfection conditions all explants of *Gloxinia* hort. "Strawberry Ice-Cream" are capable for regeneration after aseptic treatment. Adventive buds form from plant tissue due to supplementing growth regulators in different amounts to the nutritive medium. In most cases, this process takes place in culture media containing either cytokinin or its combination with auxin. As cytokinin BAP or kinetin is used, IAA is frequently a source of auxin [8].

Sterile Gloxinia explants were introduced into five growth medium variants MS (Table 2). First morphological transformations were observed during the first cultivation week. Explants became thicker, especially on the cut zone. In two weeks first adventive sprouts developed. In a month of cultivation without a passage sprouts formed 2 to 4 leaves (Fig. 1).

Table 2. The effect of growth regulators on the morphogenesis of *Gloxinia hybrida* hort. "Strawberry Ice-Cream" in conditions *in vitro*

Growth regulators, mg/l				Morphogenesis type
BAP	Kinetin	IAA	IBA	
-	-	-	-	Growth of a single sprout with a lot of roots (hemmorisogenesis)
0.5	-	0.2	-	Continuous regeneration of sprouts (hemmogenesis)
-	-	0.5	0.5	Growth of roots (rhizogenesis)
1.0	1.0	1.0	-	Continuous regeneration of sprouts (hemmogenesis), into weeks after their formation we observe vitrification
0.5	-	0.5	-	Continuous regeneration of sprouts (hemmogenesis)

Optimal growth media supporting the formation of sprouts are variants containing BAP (0.5 mg/l) and IAA (0.2 and 0.5 mg/l) (Table 2). In these nutritive media the highest regeneration coefficient is obtained. Approximately 50 sprouts form per 1 cm^2 leaf explant during one month after visual regeneration begins. Further, we observed continuous budding, explant tissue did not die. Adventive sprouts tend to form even more intensively in the growth medium MS containing an equal concentration of BAP, kinetin, and IAA (1.0 mg/l). However, after 6 weeks of cultivating explants in this medium we observed partial vitrification of forming sprouts. Moreover, stems got thicker; leaves elongated and tended to deformations because of morphological transformations in tissues and cell expansion related to water stress. One of the reasons inducing such transformations is supposed to be low transpiration caused by abnormal evolution of mouth-like openings and exodermis structure. To solve the problem of vitrification we reduced a BAP concentration in the medium to 0.5 mg/l or transplanted them into a hormone-free nutritive medium Murashige and Skoog (MSO), where sprouts developed a normal morphology over two weeks of cultivation. 4 weeks later bigger sprouts formed a root system.

In aseptic conditions of laminar flow cabinet a conglomerate of sprouts was taken out of culture tubes, sprouts with formed roots were separated; sprouts without roots were transplanted for rhizogenesis into the medium MS containing IAA (0.5 mg/l) and IBA (0.5 mg/l). During 3 – 4 weeks micro-sprouts developed roots, and a habit of regenerative plants expanded (Fig. 2). There was no significant difference in the development of plants in growth media with and without hormones. A growth medium containing auxins at the stage of rooting brought forward the rhizogenesis by 1 – 2 weeks compared with a hormone-free nutritive medium.

IOP Conf. Series: Materials Science and Engineering **941** (2020) 012029 doi:10.1088/1757-899X/941/1/012029

Figure 1. Adventive sprout regeneration of *Gloxinia hybrida* hort. "Strawberry Ice-Cream" in a growth medium MS, 0.5 mg/l BAP and 0.2 mg/l IAA over 1 month.

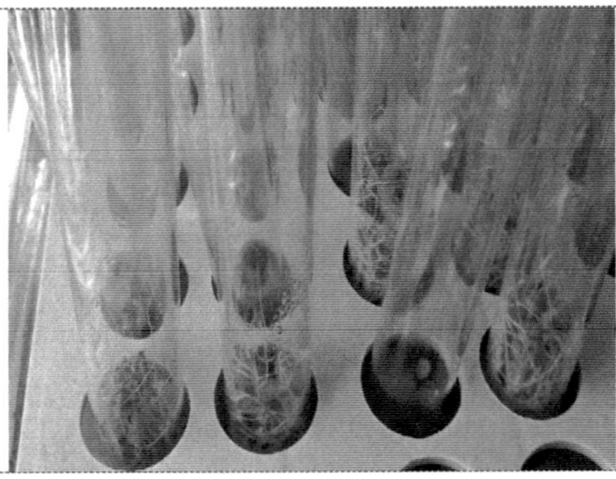

Figure 2. Adventive sprout regeneration of *Gloxinia hybrida* hort. "Strawberry Ice-Cream" in a growth medium MS, 0.5 mg/l IAA and 0.5 mg/l IBA over 3 weeks.

In the process of further growing regenerative plants nodules formed and developed. All *Gloxinia* plants grown in the hormone-free medium continued forming branch sprouts (Fig. 3), although less numerous than in the propagation medium. This fact indicates a high morphogenesis potential of Gloxinia.

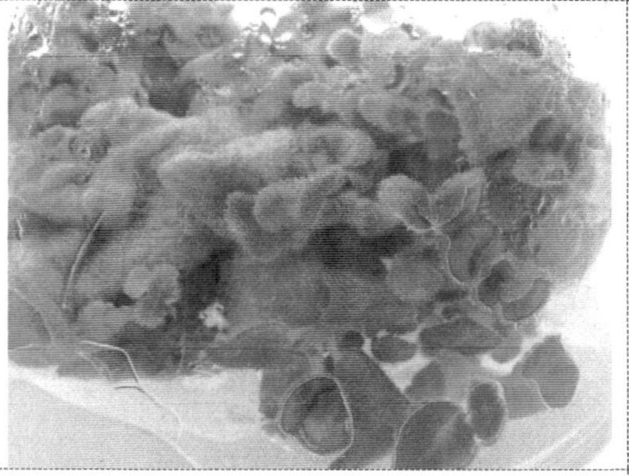

Figure 3. Evolution of regenerative plants *Gloxinia hybrida* hort. "Strawberry Ice-Cream" in a hormone-free growth medium MS.

Figure 4. Cultivation of plants *Gloxinia hybrida* hort. "Strawberry Ice-Cream" in a soil substrate.

In the next propagation phase plants were transplanted into the soil substrate and kept in the vegetation chamber. *Gloxinia* plants with a well-developed root system were transplanted into containers with a soil blend (Fig. 4) made up of garden mold, organic matter, and bank sand taken in equal amounts. First two weeks plants *ex vitro* were covered with a horticulture fleece "Agrotex". The fleece "Agrotex" is breathable, maintains humidity and protects from temperature fluctuations, furthermore, it is not heavy to cause plants bending. The survival ability of plants was as high as 90 – 95%. Over 3 weeks regenerative plants were transplanted into individual vegetative pots (Fig. 5). Transplanted *Gloxinia* samples developed forming flower buds and started flowering 1 – 1.5 months later (Fig. 6).

IOP Conf. Series: Materials Science and Engineering **941** (2020) 012029 doi:10.1088/1757-899X/941/1/012029

Figure 5. Transplanted into individual pots regenerative plants of *Gloxinia hybrida* hort. "Strawberry Ice-Cream".

Figure 6. Generative regenerative plants of *Gloxinia hybrida* hort. "Strawberry Ice-Cream".

Conclusion

A clonal micropropagation technique of *Gloxinia hybrida* hort. "Strawberry Ice-Cream" is developed. Double-flowered *Gloxinia* varieties are slowly reproducible. If reproduced through leaf cuttings, one leaf gives one sprout only. A plant starts flowering in 1.5 years. In culture *in vitro* one medium-sized leaf provides on average 10 – 12 fragments (2 cm^2). Further, more than 100 sprouts are regenerated on each explant placed into a growth medium. Owing to up to date biotechnology methods, clonal micropropagation, in particular, one leaf gives above one thousand identical plants per 6 months, making a propagation time one year shorter.

References

[1] 2017 *Fundamentals of Plant Biotechnology* / edit. R G Farkhutdinov et al. (Ufa: RIC Bashkir State University) 244 p (*in Russian*)
[2] Kataeva N V and Butenko R G 1983 *Clonal Micropropagation of Plants* (Moscow: Nauka) 96 p (*in Russian*)
[3] Milekhin A V and Rubtsov S L 2015 *Young scientist* **24** 335-8 (*in Russian*)
[4] Butenko R G 1999 *Cell Biology of Higher Plants in Vitro and Biotechnologies on their Basis* (Moscow: FBK-PRESS) 160 p (*in Russian*)
[5] Butenko R G 1964 *Isolated Tissue Culture and Physiology of Plant Morphogenesis* (Moscow: Nauka) 272 p (*in Russian*)
[6] Kalinin V F, Sarnatskaya V V and Polishchuk V E 1980 *Tissue Culture Methods in Physiology and Biochemistry of Plants* (Kiev: Naukova Dumka) 488 p (*in Russian*)
[7] Murashige T and Skoog F 1962 *Physiol. Plant.* V. 15 **13** 473-97
[8] Akhmetova A Sh, Mironova L N and Zaripova A A 2015 *Agro XXI* **4-6** 37-9 (*in Russian*)

IOP Conf. Series: Materials Science and Engineering **941** (2020) 012030 doi:10.1088/1757-899X/941/1/012030

Development of biotechnology for cultivating *Potentilla* L. plant material with antivirus and antibacterial activity

L I Tikhomirova[1] and AA Zaripova[2]
1Altai State University, Lenina str., 61, 656049 Barnaul (Russia)
2South-Ural Botanical Garden-Institute, Ufa Federal Research Centre, RAS, Mendeleev str., 195/3, 450080, Ufa (Russia)

E-Mail: L-tichomirova@yandex.ru

Abstract. The reported paper presents a biotechnology developed to produce phytomass of *Potentilla* L. species (*P. alba, P. fragarioides, P. fruticosa, P. chrisantha, P. longifolia).* Regenerative *Potentilla* plants are good at extracting bio-friendly elements (Pb, Cu, and Zn) from the Murashige and Skoog growth medium and accumulating them.
Water extracts of plant material *P. alba, P. fragarioides, P. fruticosa, P. chrisantha, P. longifolia* have shown a good biological activity against herpes simplex virus type II. Extracts of bio-technologically produced *Potentilla alba* plants have demonstrated an antibacterial activity against *E. coli, K. pneumoniae, S. typhimurium, St. aureus* taken in a minimal dose of 37.5 μg/ml.

1. Introduction
The genus Potentilla (*Potentilla* L.) is one of the biggest and polymorphous species in the Rosaceae family representing the flora of West Siberia and the Altay Republic, in particular. Several plant species have been used in ethnoscience and academic medicine so far, although there are little in-depth studies of their phytochemical composition. To date, a few species representing the genus *Potentilla* L., such as tormentilla (*Potentilla erecta* L.) and silvery cinquefoil (*Potentilla argentea* L.) are officially allowed herbs [1 – 4].
Pentaphylloides fruticosa (L.) – dumetosous cinquefoil, marsh cinquefoil or shrubby cinquefoil is a heavily-branched bush, 10 to 150 cm high, its branches have a red-brown or grey-brownish exfoliating bark, and young branches are silky-pilose. Dumetosous cinquefoil is allowed in food industry. It is a component in tea-like native or fermented beverages under a trademark "Sibirskiy chai". Production processes have been developed to utilize plants of dumetosous cinquefoil instead of traditional tea [5].
Potentilla chrisantha Trev. is a perennial plant with a height of 15 – 40 cm. Thin, ascending stems are with a few leaves, dychotomically branched, and covered with short fur and longer squarrose hairs [6].
Potentilla longifolia Willd. is a perennial plant. A.I. Shreter (1975) claimed rhizomes were used in the Russian ethnoscience as a pain killer, blood-stopping and stomachic medicine to cure scurvy, diarrhea, and dysentery [7].
Another promising species seen as a phytomass source for a new raw material is *Potentilla fragarioides* L. Its stems, 5 to 25 cm high, are weak, with a few leaves; similarly to leaf stalks they are covered with long squarrose hairs, frequently found on rumples. Bottom leaves are pinnate with 2 – 3 pairs of toothed leaves. Leaves of the upper pair, including the top one, are 1 – 6 cm long, and 0.6 – 3 cm wide. Fibrous root system. Plants contain tannins (6.2 %), and flavonoids [8].
A considerable amount of literature has been published on *Potentilla alba* L. – a perennial herb, 8 – 25 cm high, with a thick and poorly-branched long black-brownish rhizome. These studies focus on

chemistry and pharmacology of the plant. Medicaments based on *P. alba* influence the thyroid, regulate its function, neutralize diffuse changes, and decrease toxicity in the body. Furthermore, *P. alba* is good for prevention and cure of liver deceases, cardiovascular system, and gastro-intestinal tract, e.g. stomach ulcers, as well as an antiseptic and wound healing medicine [1].

This study aims to develop a production technology for *Potentilla* L. species phytomass and assess an antivirus activity of obtained extracts.

2. Materials and Methods

Tissue culture. In the research we used regenerative plants, selected and grown at Biotechnology Department of Altai State University. Our studies were focused on rhizomes of *Potentilla alba* intact plants applied for manufacturing at ZAO Evalar (Biysk, Russia). Intact plants were collected near Novoaltaysk, Altay Krai in 2015 – 16.

Methods of research. Data on the content of chemical elements were obtained using inverse voltammetry with an analyzing device TA–Lab (OOO "NPP "Tomanalit""). Zn, Cd, Pb and Cu mass concentrations were measured with a voltammetry analyzing device in preliminary prepared homogenized samples of medical herbs. We prepared probes via wet mineralization of a weighted portion of 1.0 g (heated with HNO_3 and 30 % solution of hydrogen peroxide) and subsequent full digestion in the furnace chamber PDP-Lab at a temperature of 450 ^{0}C. Before analysis a residual matter was dissolved in 1.0 ml concentrated formic acid, a solution was mixed up with a glass stick, kept resting 2 – 3 min., 9.0 ml bi-distillated water was added and mixed [9].

When assessing the intensity of accumulating chemical substances in organs of regenerative plants we determined accumulation factors (Fa) – ratios between element concentrations in organs and in a medium. To classify elements according to this coefficient we used groups suggested by A.I Perelman: 1) intensive accumulation (100> Fa ≥10); 2) strong accumulation (10> Fa ≥1); 3) weak accumulation and moderate capture (1> Fa ≥0.1); 4) weak capture (0.1> Fa ≥0.01); 5) significantly weak capture (0.01> Fa ≥0.001) [10].

Assessment of antivirus activity of Potentilla extracts against herpes virus. An analysis was carried out in State Research Center of Virology Biotechnology VEKTOR via measuring the cell uptake of an intravital stain – neutral red (NR) [11]. Ethanol (96%) and water ejected extracts were used to estimate the antivirus activity.

3. Results and Discussion

The micro-cloning of *P. fragarioides, P. fruticosa, P. chrisantha, P. longifolia.*

As explants we took ripen seeds from the collection of South-Ural Botanical Garden-Institute *(P. fruticosa*) and South-Siberian Botanical Garden (*P. fragarioides, P. chrisantha,* and *P. longifolia).* Before being sterilized seeds were washed in flow water for 15 – 25 minutes. The 10-minute sterilization was conducted in a laminar flow cabinet using a 1% sulphochlorontine solution. Then they were washed three times with sterile distillated water. Owing to this sterilization procedure 70% explants were sterile and capable of living. Growth media for the introduction into tissue culture were prepared without phytohormones as stated in the Murashige and Skoog protocol (MS) [12].

Over 10 – 15 days *Potentilla* young sprouts were transplanted into propagation media MS enriched with 1.0 – 0.5 μM 6-furilmethylaminopurine, 0.25 μM IBA (indole-3-butyric acid) and 0.05 μM GA (Gibberellic Acid). The growth medium for *P. chrisantha* was prepared on the base of MS with 1.0 μM BAP (6-benzylaminopurine) 0.5 μM IBA and 0.05 μM GA.

Conglomerates of micro-sprouts were easily divided into single ones. They were transplanted into new growth media every 30 days. Long-termed cultivation of an active proliferous crop requires a rotation of media containing low and high cytokinin concentration every second passage. As a result, a sterile crop with a stable propagation coefficient was grown. A number of sprouts per one explant was as high as 2 to 15 each passage. Sprouts were implanted into Murashige and Skoog media with 1.0 μM IBA.

The intensity of elements accumulation from the growth medium MS (Table 1) was determined using a Fa of bio-friendly elements (Cu, Zn) and its V (variation coefficient). Importantly, these elements are in the group of intensive accumulation (100> *Fa* ≥10) for all analyzed *Potentilla* varieties. It appears

IOP Conf. Series: Materials Science and Engineering **941** (2020) 012030 doi:10.1088/1757-899X/941/1/012030

to be the consequence that elements in a growth medium are easier consumable by plant tissues than in the soil. However, Cu and Zn variation coefficients of (Fa) are quite high because the accumulation of elements by different plant varieties is not similar in the same growth medium. Our data have proven the importance of a genetic factor to the chemical composition developing in plants (Table 1).

Table 1. Accumulation factors of chemical elements in regenerative plant organs

Variety	Cu		Zn	
	Content, mg/kg	Fa	Content, mg/kg	Fa
P. fragarioides	1.1±0.4	110	136±53	71.8
P. chrysantha	1.1±0.4	110	124±48	65.3
P. alba	2.0±0.8	200	118±46	62.1
P. fruticosa	1.1±0.4	110	39±15	20.7
Potentilla longifolia	1.1±0.4	110	76±30	40.2
V, %		29.41		36.55

In our study we compared the element composition in regenerative plants with intact plants (*P. fragarioides, P. chrysantha, P. fruticosa*). We revealed a specificity of bio-friendly microelements Pb, Cu, and Zn (Table 2).

There is 1.1 – 1.3 times less Pb in regenerative plants *P. chrysantha, P. fruticosa* than in intact plants. The concentration of this element in *P. fragarioides* was not related to the growing method (Table 2). In all varieties the concentration of Cd did not exceed 0.0015 mg/kg.

Intact plants contain twice more copper than regenerative samples (Table 2). Naturally, the concentration of copper is in a range from 3 to 40 mg/kg in air-dry mass. An appropriate concentration of copper is relevant both for plants themselves and their use in food of humans and animals.

Zn tends to variation within the species *Potentilla* L. Regenerative plants *P. fragarioides* and *P. chrysantha* contained 2.5 times more zinc than intact ones, while regenerative *P. fruticosa* – 1.7 times less (Table 2).

Table 2. Chemical elements in organs of regenerative plants vs. intact samples

Variety	Chemical element content, mg/kg					
	Pb		Cu		Zn	
	Regenerative	Intact	Regenerative	Intact	Regenerative	Intact
P. fragarioides	0.24±0.09	0.22±0.09	1.1±0.4	3.6±1.4	136±53	55±21
P. chrysantha	0.21±0.08	0.36±0.14	1.1±0.4	3.4±1.3	124±48	56±22
P. fruticosa	0.11±0.04	0.17±0.07	1.1±0.4	2.0±0.8	40±16	66±26
Note: Cd content is below 0.0015 mg/kg						

Medicaments on the base of *P. alba* have a radioprotective effect. Researchers suggest it is a radioprotective activity of *Potentilla alba* roots that helped inhabitants of Belarus Polesie consuming a restorative drink with *Potentilla alba* instead of tea avoid so many endemic goiter diseases after the Chernobyl NPP disaster because far less cases were registered in this area than in other zones near that region.

Our production method of medical herbs is a combination of micro-cloning and hydroponic growing [13]. This method was applied to the all year-round cultivation of *Potentilla alba* in a hydroponic line "Minivit" and a wet biomass output of the medical herb was as high as 10.05 kg/m^2 (Figure 1).

a) b) c)

Figure 1. a, b) A phase of micro-cloning *P. fragarioides;* c) regenerative plants *P. alba* when growing in the line "Minivit" for 45 days. Dry weight of a plant 10.51 g, weight of roots 3.85 g.

The toxicity of *Potentilla alba* extracts and their activity against herpes simplex virus type II were studied in a green monkey kidney continuous cell line VERO. As seen in Table 3, water extracts of *P. alba* showed a significant antivirus activity. With a low toxicity value, both regenerative and intact plants had a high selectivity index: 85 and 64, respectively. Alcohol extracts of these herbs demonstrated a weak antivirus activity (ED_{50} equals to dilution 1:160), and a low selectivity index (>4) in spite of a weak toxicity value (CD_{50} >1:40). It is important that data on intact and regenerative plants are consistent; being approximately only two times different for all characteristics, it is within the error of the method (Table 3).

Table 3. Toxicity and antivirus activity of *P. alba* extracts ejected from plants grown differently

Raw materials for extraction	Solvent	Toxicity, CD_{50} (dilution of extract)	Antivirus activity, ED_{50} (extract dissolution)	Selectivity index, IS
Intact plants	Water	1:120	1:10240	85
Regenerative plants	Water	1: 80	1:5120	64
Intact plants	96% ethanol	>1:40	1:160	>4
Regenerative plants	96% ethanol	>1:40	1:160	>4

Besides an activity against herpes simplex virus type II extracts of bio-technologically produced *Potentilla alba* plants demonstrated a good antibacterial activity against *E. coli, K. pneumoniae, S. typhimurium, St. aureus* taken in a minimal dose of 37.5 μg/ml. Our study was focused on the biological activity of water extracts *P. fruticosa, P. chrisantha, P. longifolia*. Interestingly, a good antivirus activity against herpes simplex virus type II was revealed only in extracts of biotechnologically produced *P. fruticosa and P. chrisantha* (Table 4).

Table 4. Cytotoxicity and activity against herpes simplex virus type II *in vitro* of water extracts ejected from *Potentilla* L. species

Material for extraction, (plants)		Toxicity, CD_{50} (μg/ml)	Efficiency, ED_{50} (μg/ml)	Selectivity index, IS
P. longifolia	Regenerative plants	927.58	86.96	10.6

	Intact plants	1318.12	164.76	8
P. chrisantha	Regenerative plants	2638.75	329.84	8
	Intact plants	Antivirus activity not detected		
P. fruticosa	Regenerative plats	660.0	61.81	10.6
	Intact plants	Antivirus activity not detected		

Conclusion

To sum up, a biotechnology is developed to produce phytomass of *Potentilla* L. species (*P. alba P. fragarioides, P. fruticosa, P. chrisantha, P. longifolia).* Regenerative *Potentilla* plants are good at extracting bio-friendly elements (Pb, Cu, and Zn) from the Murashige and Skoog growth medium and accumulating them.

The concentration of Pb in regenerative plants *P. chrysantha, P. fruticosa* is 1.1 – 1.3 times lower than in intact samples. Its concentration in *P. fragarioides* is not related to the growing method. Intact plants contain twice more copper than regenerative samples. Zn varies within the species *Potentilla* L. Regenerative plants *P. fragarioides* and *P. chrysantha* contain approximately 2.5 times more zinc than intact ones, while regenerative *P. fruticosa* – 1.7 times less.

Water extracts of plant material *P. alba, P. fragarioides, P. fruticosa, P. chrisantha, P. longifolia* show a good biological activity against herpes simplex virus type II. Extracts of bio-technologically produced *Potentilla alba* plants demonstrate also an antibacterial activity against *E. coli, K. pneumoniae, S. typhimurium, St. aureus* taken in a minimal dose of 37.5 μg/ml.

Acknowledgements

The authors greatly appreciate assistance provided by T.N. Il'icheva, Doctor of Biology, Professor, Head of Laboratory, State Research Center of Virology Biotechnology Vector when analyzing an activity of *Potentilla* extracts against herpes simplex virus type II.

The authors are also grateful for assistance and support of L.V. Zatonskaya, PhD in Chemistry, Department of Analytical Chemistry, Federal State Budgetary Educational Institution Altay State University provided when studying a content of chemical elements in regenerative plants.

References

[1] Tihomirova L.I., Bazarnova N.G., Sysoeva A.V. Phytochemical analysis of biotechnologically produced Potentilla L. plants // Chemistry of plant raw material. – 2018. – № 1. – pp. 145-154. – (*in Russian*) – Tihomirova L.I., Bazarnova N.G., Sysoeva A.V. Fitohimicheskij analiz biotehnologicheskogo syr'ja predstavitelej roda Potentilla L. // Khimija rastitel'nogo syr'ja. – 2018. – № 1. – pp. 145 – 154.

[2] Tomczyk M., Latté K. P. Potentilla – a review of its phytochemical and pharmacological profile // J. Ethnopharmacol. 2009. Vol. 122. P. 184 – 204. doi: 10.1016/j.jep.2008.12.022.

[3] Smyk G. K. // New cultures in economy and medicine. 2 parts. Kiev, 1976 – (*in Russian*) – Smyk G. K. // Novye kul'tury v narodnom hozjajstve i medicine: V 2 ch. Kiev, 1976.

[4] Tomczyk M. Pleszczynska M., Wiater A. Variation in total polyphenolics contents of aerial parts of Potentilla species and their anticariogenic activity // Molecules. 2010. Vol. 15 (7). pp. 4639–4651.

[5] Miftahova S.A., Skrockaja O.V., Zajnullina K.S. Biology of a rare species – shrubby cinquefoil чая (Pentaphylloides fruticosa) – in the Northern culture // Bulletin of Komi Research Center UrB RAS. 2017. № 2(30). pp. 30-36. – (*in Russian*) – Miftahova S.A., Skrockaja O.V., Zajnullina K.S. Biologija redkogo vida – kuril'skogo chaja (Pentaphylloides fruticosa) – v kul'ture na severe // Izvestija Komi nauchnogo centra UrO RAN. 2017. № 2(30). pp.30-36.

[6] Tikhomirova L.I., Bazarnova N.G., Sysoeva A.V., Shcherbakova L.V. PHYTOCHEMICAL ANALYSIS OF BIOTECHNOLOGICAL RAW MATERIALS OF REPRESENTATIVES OF THE GENUS POTENTILLA L // Russian Journal of Bioorganic Chemistry. 2019. V. 45. № 7. pp. 942-949.
[7] Shreter A. I. Medical flora of the Soviet Far East. Moscow: Medicina. 1975. 323 p. – (*in Russian*) – Shreter A. I. Lekarstvennaja flora sovetskogo Dal'nego vostoka. Moskva: Medicina. 1975. 323 p.
[8] Flora of Siberia. Rosaseae / edited. A.V. Polozhij, L.I. Malysheva. Novosibirsk: Nauka, 1988. V. 8. 199 p. – (*in Russian*) – Flora Sibiri. Rosaseae / A.V. Polozhij, L.I. Malysheva. Novosibirsk: Nauka, 1988. T. 8. 199 p.
[9] FR.1.31.2004.00.986 Methods of measuring a mass concentration of zinc, cadmium, lead, copper in food products, food raw materials, foodstuffs and products of their processing, dietary food supplements, biological objects by the method of voltammetry with analyzing devices TA–Lab (certificate of attestation № 31-04/04 issued 26.12.2003). – (*in Russian*) – FR.1.31.2004.00.986 Metodika vypolnenija izmerenij massovoj koncentracii cinka, kadmija, svinca i medi v pishhevyh produktah, prodovol'stvennom syr'e, kormov i produktov ih pererabotki, biologicheski aktivnyh dobavkah k pishhi, biologicheskih ob#ektah metodom inversionnoj vol'tamperometrii na analizatorah tipa TA–Lab (svidetel'stvo ob attestacii № 31-04/04 ot 26.12.2003 g).
[10] Perel'man A.I. Geochemistry. M.: Vysshaja shkola, 1989. 528 p. – (*in Russian*) – Perel'man A.I. Geohimija. M.: Vysshaja shkola, 1989. 528 p.
[11] Finter N. B. Dye uptake methods for assessing viral cytopathogenicity and their application to interferon assays // J. Gen. Virol. – 1969. – № 5. – P. 419-427.
[12] Murashige T., Skoog F. A Revised Medium for Rapid Growth and Bio Assays with Tobacco Tissue Cultures // Physiol. Plant. 2006. Vol. 15, № 4. P. 473-497. DOI: 10.1111/j.1399–3054.1962.tb08052.x.
[13] Tihomirova L.I., Il'ichjova T.N., Bazarnova N.G., Sysoeva A.V. Production process of medical plant material *Potentilla alba* L. in hydroponics conditions // Chemistry of plant materials. – 2016. – № 3. – pp. 59-66. – (in Russian) – Tihomirova L.I., Il'ichjova T.N., Bazarnova N.G., Sysoeva A.V. Sposob poluchenija lekarstvennogo rastitel'nogo syr'ja lapchatki beloj (Potentilla alba L.) v uslovijah gidroponiki // Himija rastitel'nogo syr'ja. – 2016. – № 3. – p. 59-66.

IOP Conf. Series: Materials Science and Engineering **941** (2020) 012031 doi:10.1088/1757-899X/941/1/012031

Active substrate as a tool to improve the efficiency of green technology for seeds production

V.V Popov1, V A Afonkina1 and E I Krivosheeva1.

[1]Federal State Budgetary Educational Institution of Higher Education «South Ural State Agrarian University», Troitsk, Russia

E-mail: Ntc-es@mail.ru

Abstract. The most common and important method of agricultural products processing is drying. The quality of finished product, energy intensity and material intensity of the entire processing depend on the drying process. Drying the seeds of different crops is influenced by many factors: the structure, size, properties and condition. We studied the process of drying oil-bearing small-seeded crops taking rapeseeds as samples. Rapeseed is covered with a hard shell that prevents the release of moisture; in this case the seeds should be dried at low temperatures. High temperature causes rapeseeds to have low quality indicators. Analysis of the existing equipment has shown that nowadays there are no special systems for drying small-seeded crops, and the drying process is carried out on grain dryers which are characterized by large seeds losses, considerable energy costs and high temperatures. The main advantages of the developed infrared dryer of conveyor cascade type are the lack of hot zones, the uniform heating and the minimum percentage of seeds losses because of their blowing out and removal. In addition, the heat carriers used in the dryer of conveyor cascade type help to perform the drying process at the required wavelength, that can affect only the moisture in the seeds.

1. Introduction

At present in the Russian Federation there is an acute problem of increasing the efficiency of agricultural production, which should be carried out under the condition of rational use of material and energy costs. Therefore, the issue of resource- and energy saving is of great interest now, since its solution helps to find the most effective ways to improve energy efficiency.

Post-harvest processing of seed stock is an integral part of agricultural production. Consequently, the preservation of seed material and the problem of reducing the energy intensity of the process related to drying small-seeded crops are considered to be urgent tasks. It is rather difficult to ensure effective drying of seeds that have hard shells and at the same time to minimize energy consumption [6, 15, 18, 20]. The process of drying small-seeded crops by infrared radiation (IR) depends on many parameters:

- heat treatment mode [5];
- state and properties of small-seeded crops [1, 9];
- design parameters of drying equipment [7, 16, 17].

The developed infrared dryer of conveyor cascade type [14] using the electric film heaters EFH [10] completely meets all the requirements for the technology related to the drying process of small-seeded oil-bearing crops.

2. Theoretical studies

The desire to minimize energy costs and to make profit and super-profit from production often leads to a negligent attitude to the quality of the finished product; therefore, it is recommendable not to forget about the post-harvest preparation of small-seeded crops for storage, sale or hydration.

Relatively easy destruction of organic matter is a characteristic of small-seeded crops; hence, the moisture removal by IR drying depends on the correct choice of the limiting temperature, which depends on the energy irradiation and the absorptive capacity of the irradiated material surface (with the growth of these values the temperature also increases; Tiller and Garber's formula) [3].

$$T_{max} = \frac{AE_e}{aS} + T_0, \tag{1}$$

where T_{max} is the limiting temperature; T_0 is the ambient temperature; A is the absorption capacity of the irradiated surface; E_e is the energy irradiation of the surface; a is the convection heat transfer coefficient; S is the surface area involved in heat exchange between the material and air.

It is also necessary to take into account the irradiation field uniformity depended on the density of the incident radiation flux which has a significant impact on the quality indicators of small-seeded crops [4, 13]. During the research it has been found that the flux density and the irradiation uniformity depend not only on the distance between the electric heater and rapeseeds, but also the substrate under the seeds significantly affects these indicators [12, 19].

The design parameters of a dryer are considered as the important factors related to high-quality drying of small-seeded crops. In addition to the length, width, height of the emitter suspension, the speed of the conveyor belt and other basic quantities [11], the following factors should also be included:

- the distance between the substrate and the crop being dried;
- the substrate and the material from which it is made;
- the material of a conveyor belt.

The effect of the distance between the substrate used in the design of the dryer and a small-seeded crop was studied earlier [15].

The description of the heat generation scheme shows that the temperature distribution in the product irradiated by infrared rays will depend on its thickness, reflectivity and absorption, as well as on the optical properties of the substrate on which it is placed (Fig. 1, *a-d*).

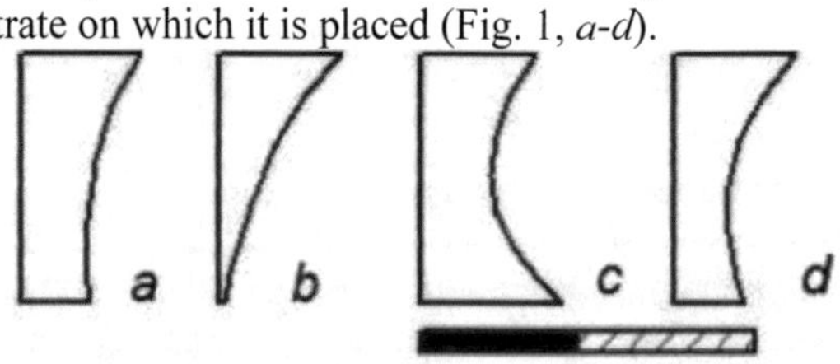

Figure 1. Temperature distribution in the layer of irradiated rapeseeds (infrared radiation is directed from above).

In a thin layer of irradiated side of the product, which weakly absorbs radiation, the temperature is slightly higher than on the opposite side, and besides it is lower in the cross section of the layer than at the boundaries (Figure 1, *a*). A completely different temperature distribution is observed on one side of highly absorbent rapeseeds irradiation. The outer irradiated side heats up quickly, whereas on the inner side the temperature does not rise (Figure 1, *b*). The substrate can significantly affect the temperature distribution in the layer of rapeseeds (Figure 1, *c, d*).

The optical properties of the substrate material can significantly affect the process of removing moisture, so using a blackened substrate gives better results than using a light substrate [9]. When analyzing the curves of mass loss we observed that higher final results were obtained using a blackened substrate, and the use of a light substrate was noticeably inferior in results. Thus, we use an electric film heater (EFH) as a substrate for the crops being dried, in which nichrome is applied as a resistive heating and radiating element [8]; that gives EFH blackness degree ε=0.96 and brings it closer to the blackness degree of an absolutely black body.

The use of an electric film heater helps to achieve the effect of an active substrate, as it not only reflects infrared radiation, but also acts as a radiator while remaining within the temperature limits recommended when processing rapeseeds. Darkened substrate-reflector accelerates the drying process since it removes the shielding of dried rapeseeds, provides a high reflectance coefficient of radiant energy and makes the temperature distribution in the monolayer of seeds more uniform. The substrate-reflector doubles this effect, and as a result the difference in heating temperatures of rapeseeds in a conveyor cascade dryer decreases during the drying process; this indicates a more uniform distribution of irradiance.

The uniform temperature distribution helps not only to reduce the time of rapeseeds drying with infrared radiation in a conveyor-cascade dryer [2], but also to increase the thickness of the filled seed layer on the conveyor belt of the dryer, that is to increase the productivity.

3. Research methodology

The research was carried out on an experimental conveyor-cascade dryer made for small-seeded crops drying. It involved checking the curves of infrared drying of seeds with and without a substrate under continuous irradiation; the electric film heater (EFH) served as a substrate material. The initial moisture content of the material was 12% (Figure 2). Rapeseeds were transferred to the dryer in a dense layer.

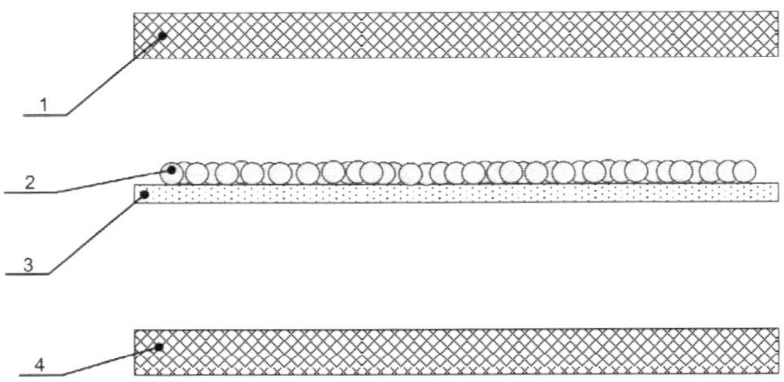

Figure 2. System with the active substrate
1 - EFH, 2 - rapeseeds, 3 - conveyor belt, 4 - active substrate

Figure 3 illustrates the experimental curves for drying rapeseeds with and without a substrate. Analysis of the curves obtained reveals that at the beginning of the process, when the moisture content of the material decreases along a convex curve, there is a short-term stage of heating the material. Then, under all drying conditions studied, the process proceeds in the second period (the drying speed drops). The presence of a foil substrate speeds up the drying process, since it removes the shielding of dried seeds and provides a high reflection coefficient of radiant energy.

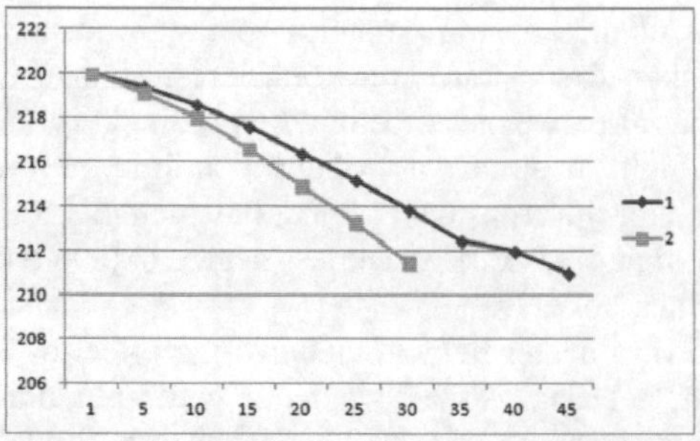

Figure 3. Curves of rapeseeds drying with the substrate and without it.

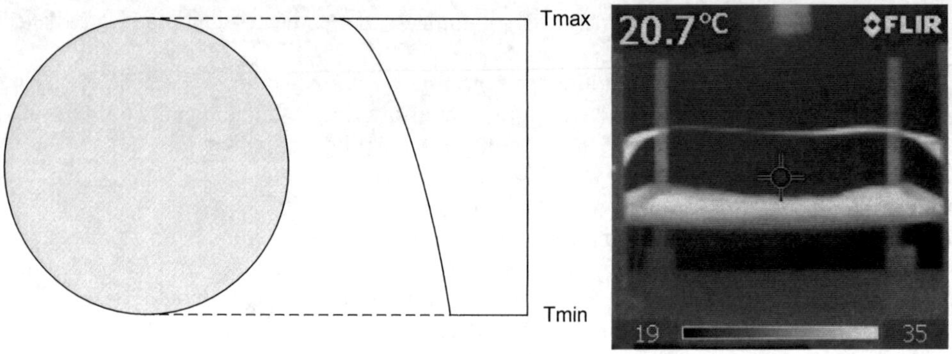

Figure 4. Turning on the radiator only (without substrate).

Distributing the temperature in a thin monolayer of rapeseeds provided only the EFH emitter is turned on, we can observe the following picture presented in Figure 4. The outer irradiated side of rapeseeds quickly heats up while on the inner side the temperature is slightly lower.

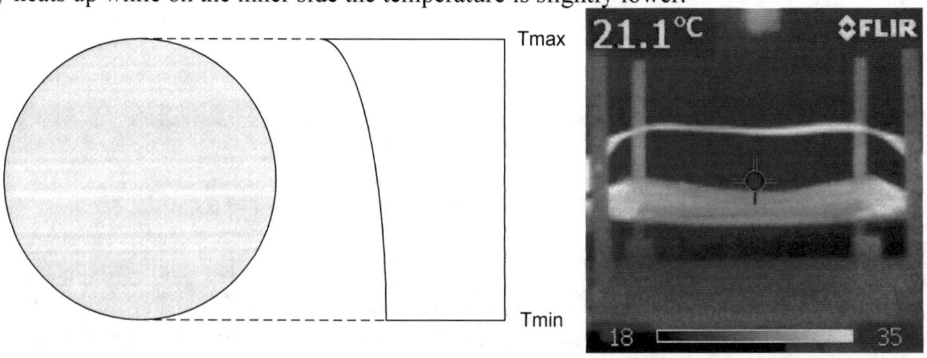

Figure 5. Turning on the radiator only (with substrate).

A different picture is observed when there is a substrate. When the radiation reaches the substrate, a greater amount of heat is emitted at the boundary of the material than in the substance layer, which leads to a uniform temperature distribution throughout the layer of material. The change in the temperature of rapeseeds can be observed in Figure 5.

However, this option in the system also does not dry the product effectively due to the temperature difference between the upper and lower layers of the material.

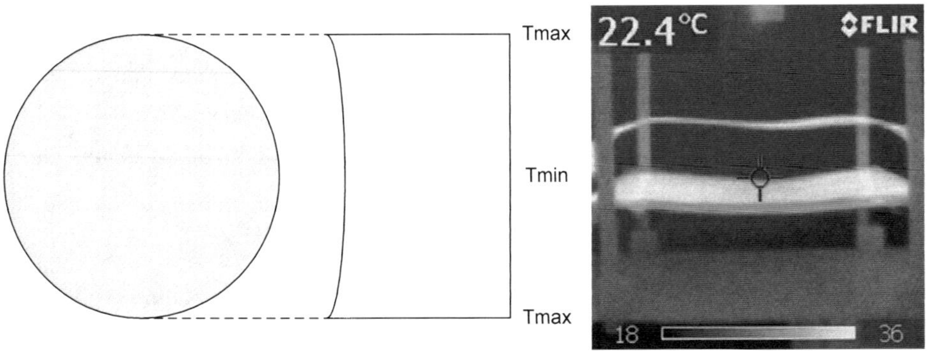

Figure 6. Turning on the emitter and the substrate

Changing the substrate into the active mode, i.e. turning on the substrate as another source of infrared radiation makes it possible to equalize the temperature between the outer layers of the material, thereby intensifying the drying process without increasing the temperature (Figure 6).

It should be noted that when using foil as a substrate for rapeseeds, no changes in colour or appearance of the seeds were observed which makes this drying option acceptable for practice.

4. Conclusions

1. Technical means that work on the principle of convective heat exchange, using heat obtained from burning liquid, solid, or gaseous fuels are widely used for drying various crops. The mode of operation is characterized by high temperatures, high energy intensity and is considered environmentally unsafe.

2. Technological regulations of infrared drying in a conveyor cascade driers using an active substrate can be shorter in time; this results in energy and cost savings, reduces working time without loss of the final product quality.

3. Among all IR heaters the electric film heaters (EFH) with sputtering having body blackness of 0.96 are best suited for drying small-seeded oil-bearing crops according to the optical characteristics and the magnitude of the radiation flux.

4. The experiments have proved that low-temperature electric film heaters can effectively dry crops with a layer thickness of 1 or 0.8 cm.

5. The results of the experiment carried out on the installation of conveyor cascade type show that by changing the thickness of seed layer and technological parameters of drying, it is possible to change the temperature distribution in rapeseeds (since the initial seed moisture is different).

6. The conducted studies suggest that during the continuous mode of drying rapeseeds in a dense layer, the use of the substrate will lead to a significant intensification of the process.

References

[1] Altukhov I V and Ochirov V D 2009 *Irkutsk, Journal of IrSAA* **37** 43-9
[2] Afanasyev, V A 2002 *Theory and practice of special processing of grain components and compound feed technology*. (Voronezh: Voronezh State University) p 296
[3] Borkhert, R and Yubits 1963 *V Technique of infrared heating* (Moscow: Gosenergoizdat) p 278
[4] Volonchuk S K, Sapozhnikov A N and Shornikova L P 2011 *Novosibirsk - Barnaul, Polzunovskiy Journal* **2/1** 167-171
[5] Ginzburg A S 1973 *Fundamentals of the theory and technology of drying food* (Moscow: Food industry) p 528
[6] Malin N A 2004 *Energy-saving drying* (Moscow: KoloSS) p 240

IOP Conf. Series: Materials Science and Engineering **941** (2020) 012031 doi:10.1088/1757-899X/941/1/012031

[7] Machkashi A and Banhidi L 1985 *Radiant heating* (Moscow: Stroiizdat) p 464
[8] Polevoy B G and Popov V M 2010 *Patent 100353*
[9] Popov V M, Afonkina V A and Shukshina E I 2012 *Science and education: experience, problems, development prospects* **2** 144-6
[10] Popov V M, Afonkina V A and Shukshina E I 2012 *News of St. Petersburg State Agrarian University* **26** 387-91
[11] Popov V M, Afonkina V A and Shukshina E I 2014 *LIII International scientific and technical conference Achievements of science to agricultural production* **III** 357-63
[12] Popov V M, Afonkina V A, Shukshina E I and Levinskiy V N 2014 *LIII International scientific and technical conference Achievements of science to agricultural production* **III** 363-7
[13] Popov V. M, Afonkina V A, Shukshina E I and Mayorov V I 2017 *Patent 2638690*
[14] Popov V. M, Afonkina V A, Shukshina E I and Mayorov V I 2016 *In: International scientific-practical conference in the framework of the XXVI international specialized exhibition "Agrocomplex-2016"* **III** 45-9
[15] Popov V. M, Afonkina V A, Shukshina E I, Mayorov V I and Levinskiy V N 2015 *International research journal* **9 (40)** 109-12
[16] Popov V. M, Afonkina V A, Shukshina E I and Khatmullin Sh A 2014 *X International scientific and practical conference "Vȩda a vznik". Prague: Publishing House "Education and Science"* **32** 30-3
[17] Romanenko G A 1987 *Oil-bearing crops* **4** 2-5
[18] Sekanov Yu P 1985 *Moisture measurement of agricultural materials. All-Union Academy of Agricultural Sciences named after V. I. Lenin* (Moscow: Agropromizdat) p 160
[19] Tsuglenok N V and Manasyan S K 2003 *Resource-saving mechanization technology. Krasnoyarsk. Appendix to the "Bulletin of KrasGAU"* **1** pp 122-5
[20] Skvortsov A A, Pshonkin D E, Luk'yanov M N and Rybakova M R 2017 Solid *State Phenomena.* **269** 31-6
[21] Goroshko D L, Shevlyagin A V, Chusovitin E A, Galkin K N, Chernev I M and Galkin N G 2016 *Solid State Phenomena* **247** 61-5

Physical and microbiological aspects of the soil seed bed in the early formation of the barley root system (*Hordeum vulgare* L.): tomographic studies

A V Suzdaleva[1], E V Shein[1,2], N V Verkhovtseva[1] and K N Abrosimov[2]
1 Lomonosov Moscow State University, GSP 1, Leninskie Gory, Moscow 119991, Russian Federation
2 Dokuchaev Soil Science Institute, per. Pyzhevskii 7, Moscow, 119017 Russia

E-mail: avsuzdaleva@gmail.com

Abstract. Barley seeds (Mikhailovsky sort) in a model physical experiment with a two-layer density soil bed (the range of soil density from 0.7 to 1.2 g/cm^3) of silty loam arable soil (Albic Glossic Retisols (Lomic, Cutanic), Moscow Region) were placed for germination at the boundary of different-density layers at optimum moisture content. During the period from planting to 7-10 days. the dynamics of the root system development was studied using an x-ray microtomograph "Bruker SkyScan 1172G" (Belgium) while simultaneously studying the composition of soil biota, which was reconstructed from microbial markers (fatty acids and their derivatives).The roots of the germinating seed mastered the entire pore space of the soil, regardless of the density zones studied in experiments. The number of bacteria increases by the fifth day with a noticeable dominance of actinobacteria (aerobacteria *Actinobacteria* spp., *Rhodococcus equi)* and Firmicutes (anaerobes *Ruminococcus* sp.) due to the destruction of the seed coat with its subsequent decrease on the seventh day.

1. Introduction

The complex soil-root-micro-organism consortium differs from the entire soil biome due to specific trophic chains, moisture potential distribution, and redox reactions. Activity and functioning, localization within the root system of rhizosphere biota are diverse and dynamic. It is known that groups of microorganisms can differentially populate the root system [1-2] and are distributed differently around the root. How micro-organisms soil biota are changed as the plant grows and the root system forms , is an extremely complex task, especially if it is solved in a regular dynamic aspect. The root system performs the mechanical function of fixing, absorbing moisture, nutrients, and interacting with the environment, which is determined by the morphology of the root system. Root morphology also affects yield, stress tolerance to factors such as drought, soil compaction, and nutrient deficiencies. [3] Plant roots have a great influence on the physical and biochemical properties of the surrounding soil. They stimulate the growth and activity of microorganisms and, as a result, affect the regulation of decomposition of soil organic matter [4]. The spatial and temporal dynamics of processes occurring underground in real time is complicated by both the opacity of the soil and the sensitivity of roots to extraction from the soil, damage to small lateral roots. All these properties have made the rhizosphere the subject of many studies involving non-invasive methods of volumetric visualization [5-7]. It is believed that the architecture of the root system is the result of the interaction of the genotype and growth

conditions. For example, it was found that rice has a growth-specific response to changes in soil structure [8]. A non-invasive computed tomography method was used in combination with microdialysis to determine the absorption characteristics of nitrogen forms in the rhizosphere.
This combination of complementary methods provides a unique ability to measure the nutrients available to roots without destruction and almost in real time [9].

Studies using computed tomography still only inspire hope for conducting quantitative experiments and obtaining stable relationships, in particular, the use of non-destructive methods of x-ray tomography [10-13]. At the moment, there are practically no non-destructive methods for simultaneous research of the formation of root systems, rhizosphere and microbiota composition in the zone of the emerging root system of plants.

The aim of this study: to study the root system of barley seedlings in the early stages of development, while using x-ray computed tomography methods and studying groups of microorganisms in the basal biota.

2. Objects and methods

Soil aggregates 3-5 mm of silty loam arable soil (Albic Glossy Retisols (Lomic, Cutanic), Moscow Region, Eldigino village were used in the experiment. 56°08 ' 023" North latitude, 37°048'116" East longitude) obtained by dry sifting. Basic properties Albic Glossic Retisols is shown in table 1.

Table 1. Some physical properties of arable horizon Albic Glossic Retisols (Lomic, Cutanic) (Moscow region)

Depth, cm	Texture of soil, % mass			Field capacity,%	Saturation conductivity, cm/day	C_{org}
	<0.002	0.002–0.05	>0.05			
0–10	10.50	84.78	4.72	30.1	21.6	1.23

The model seedbed was created in roentgen-transparent plastic cylinder with a diameter of 1.3 cm, height 4 cm There was a filter paper drain at the bottom. The bottom layer is compacted soil from sifted aggregates up to a density of 1.2 g / cm^3. The top layer consisted of separate aggregates of 3-5 mm, the bulk density of this layer is 0.7 g / cm^3. Between the layers was placed barley seed (*Hordeum vulgare* L.). The average values for the studied object for germination energy were 92%, and for germination-61%. The seedlings were grown for 7 days starting from the day of planting, in complete darkness, at a constant air temperature of 26℃.

Watering was performed once at the beginning of the experiment. Water saturated the soil and excess water flowed freely through the drainage. At the beginning of the experiment, the soil moisture content was close to the field capacity (30%). but at the end, on the 7th day, the humidity was near 14%.

The granulometric composition of the studied objects was determined by laser diffraction using a laser particle size analyzer "Analysette 22 comfort" (FRITCH, Germany). Determination of the carbon content in soil samples was carried out by dry combustion in the air stream using the Express analyzer "AN-7529 "(manufacturer "Meter", Gomel) [14]. Soil density, the soil water field capacity, and filtration coefficient were determined by traditional classical methods described in the literature [15].

2.1 Tomographic survey

Tomographic studies were performed using the equipment of the center for collective use of scientific equipment "Functions and properties of soils and soil cover" V. V. Dokuchaev Soil Science Institute. An x-ray microtomograph "Bruker SkyScan 1172G" (Belgium) was used for tomography. Tomographic survey was performed on the 1 ,3, 4, 5, 6, 7th days at a resolution of 16.1 µm.

The study parameters, such as sample size and shooting settings, were repeatedly worked out in other studies [16-21]. In this experiment, the priority was the speed of shooting, because due to the rapid growth of roots, there is a risk of getting a distorted image. Tomography of one segment (out of five in the sample) at optimal settings took about 40-60 minutes, but in our case we managed to reduce the time to 10 minutes. As a result of computer tomography data processing, a volume structure with clearly distinguishable x-ray contrast phases (soil, pore space, grain with sprouts and roots) was obtained, but complicated by high image noise. Automatic phase segmentation (Otsu method) [22] is partially functional in such conditions, and it reliably allows segmenting only the mineral phase (soil). In preparation for the analysis, manual threshold and Median filter, Despeckle were used, which made it possible to separate the pore space and the organic matter of the germinating seed with a high degree of confidence.

The calculation of seed and root volumes was performed using Bruker CT analyzer (CTan) software during all study periods. A tool in CTan – Structure separation was used to calculate the pore size. This tool is a method for calculating the local thickness of an object by filling it with virtual spheres from the inside [23].

2.2 *Microbiological research.*

The composition of the microbial community was reconstructed from microbial markers (fatty acids and their derivatives-fatty _ydroxyl acids and aldehydes), which were determined after acid methanolysis of soil samples by the molecular method of gas chromatography – mass spectrometry (GC-MS). The analysis was performed on the GC-MS system "HP-5973 Agilent Technologies" (USA). The analysis method is described in detail in these sources [24-25]. Samples were prepared for analysis: control-the initial soil in which no barley was grown; soil from samples with sprouting barley seed on the 5th and 7th days. Measurements were performed in three-fold repetition.

3. Results and Discussion

Figure 1 shows tomographic horizontal sections of soil in the box for underground and aboveground layers. It is clearly morphologically seen that subseminal layer had fine pores mainly with a diameter near 0.02 mm, while for above seminal layer the individual units and interaggregate pores size with the diameter up to 0.6-0.8 mm are characteristic, facilitating good aeration above seed layer.

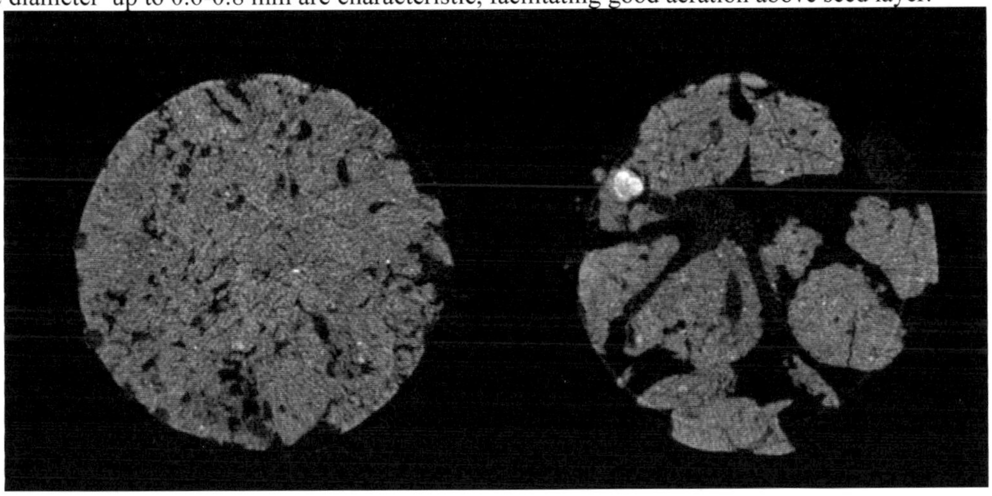

Figure 1. Tomographic horizontal sections of soil in the subseminal (left) and supraseminal (right) layers of soil (solid phase gray, pores – black). 3 day of the experiment . d=1.3 cm.

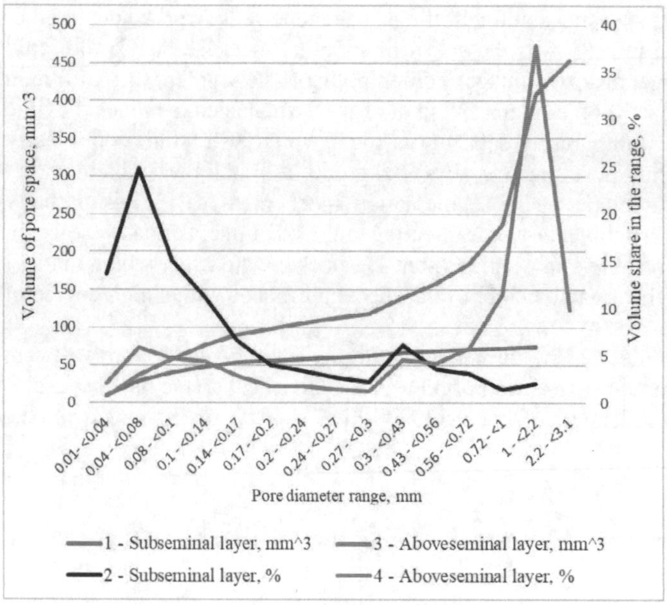

Figure 2. Pores for subseminal layer in volumes of pores (mm^3- 1) and in % of the the total porosity (2), for aboveseminal layer (3 and 4, respectively).

Figure 2 shows the tomographic data on pore distribution in layers allowed us to identify the main pore ranges. In the subseminal layer compacted to 1.2 g/cm^3, the pore diameter did not exceed 2 mm, while the largest volume of pore space (71%) comes to the pores up to 0.17 mm, which may be due to the destruction of aggregates. In the bulk layer of 3-5 mm aggregates, 52% of the pore space is occupied by pores up to 1 mm, 38% is occupied by pores of 1-2, 2 m in size.

Using tomographic imaging, we observed a decrease in the volume of seed for the period from 3 to 7 days from 50% to 36% of the total volume of the plant located in the survey zone. At the same time, the change in the share of the volume occupied by the root mass was from 22% to 61% over the same period.

Volume calculations were performed based on the volume structure corresponding to the sprouted seed. The calculation tool in CTan – Structure separation is used to calculate the pore size and thickness of seedlings.

The composition of the microbial community reconstructed from the fatty acid analysis of bacterial membrane structures showed the presence of 38 species of microorganisms belonging to 5 phylums, and a species of mycorrhizal fungi Gigaspora was also isolated. The experiment was repeated three times. Table 2 shows the result.

Among the microorganisms, 3 phylum *Actinobacteria*, *Proteobacteria* and *Firmicutes* dominated, the other two – Bacteroidetes and Cyanobacteria were represented in a relatively small number and small variety of species.

Table 2. Total abundance of soil microbial species of the model seedbed at different stages of barley germination, control-without planting

Composition of soil biota, cells·× 10^6/g		Control	5 day average	7 day average
Actinobacteria	***Actinobacteria* spp.**	43.8	382.6	63.7
	***Arthrobacter* sp.**	12.6	10.6	12.3
	***Cellulomonas* sp.**	11.3	19.1	7.9
	***Corynebacterium* sp.**	3.2	2.0	3.0
	***Propionibacterium* sp.**	8.7	4.8	9.0
	***Mycobacterium* sp.**	3.5	4.9	5.6
	Rhodococcus equi	31.3	29.6	27.3
	Rhodococcus terrae	0.6	0.9	0.0
	***Pseudonocardia* sp.**	3.0	4.4	2.5
	Streptomyces-Nocardiopsis	14.2	9.2	13.2
	Nocardia carnea	2.8	2.7	3.5
	Actinomadura roseola	0.6	0.3	1.2

Bacteroidetes	***Sphingobacterium spiritovorum***	1.7	1.5	1.8
	Bacteroides hypermegas	0.2	0.1	0.2
	Bacteroides ruminicola	1.3	1.1	1.5
	***Cytophaga* sp.**	2.4	2.1	2.2
	***Riemirella* sp.**	1.5	1.3	1.2
Firmicutes	***Bacillus subtilis***	5.1	5.8	4.6
	Bacillus sp.	0.0	0.9	0.0
	C.pasteurianum	19.9	49.6	11.0
	C.perfringens	0.1	0.2	0.1
	Clostridium* OPA	7.3	17.7	3.9
	Ruminococcus* sp.+*	81.9	107.2	43.5
Proteobacteria	***Acetobacter sp.***	9.0	6.8	7.4
	Agrobacterium radiobacter	2.5	0.4	0.9
	Pseudomonas fluorescens	3.8	4.2	7.6
	P.putida	0.9	0.6	2.0
	P. vesicularis	1.4	1.2	1.6
	Sphingomonas adgesiva	1.0	1.5	2.0
	Sphingomonas capsulata	2.0	2.1	3.0
	***Xanthomonas* sp.**	4.7	3.6	4.9
	Enterobacteriaceae	5.7	2.0	13.5
	Aeromonas hydrophila	7.8	9.8	6.5
	WARB**	3.0	2.7	2.5
	***Desulfovibrio* sp.**	6.1	5.9	5.3
	***Nitrobacter* sp.**	17.4	25.5	13.3
	***Caulobacter* sp.**	5.0	4.0	6.1
Cyanobacteria	***Anabaena (cyanobacteria)***	6.5	6.0	9.0
	Total	333.8	734.9	305.1

**Clostridium* OPA - *C. omelianskii* , *C. pasteurianum, C. acetobutyricum*

***Ruminococcus – Acetobacterium,* AMF *Glomus,Scutellospora* ,железоредукторы от Лавли (по общему маркеру 16:1ω5). 16:1 – 16 - the number of carbon atoms,: 1 – the number of double bonds, ω5-the position of the double bond of an unsaturated fatty acid, counting from the carboxyl group

*** *Wolinella-Acholeplasma-Roseomonas-Burkholderia*

The phylum *Actinobacteria* dominated throughout the experiment, the peak number was reached on day 5, and then the number of microorganisms approached the original, but still remained higher. This trend was observed in aerobic species *Actinobacteria* spp. The peak number on day 5 and its decrease by day 7 was observed in *Cellulomonas* sp, *Pseudonocardia* sp., while the number of *Rhodococcus equi* decreased gradually, which is probably due to the destruction of the seed shell. The decrease in seed volume was observed for the period from 3 to 7 days from 50% to 36% of the total volume of the plant located in the survey area, as can be seen from the results of tomography. These species have an active complex of cellulolytic enzymes that decompose complex organic compounds (the seed coat) under aerobic conditions. Products of cellulose hydrolysis can be carbon dioxide, water. The number of *Streptomyces-Nocardiopsis* decreased by 7 days. This may indicate that only certain types of actinobacteria in the soil near the germinating seed participate in the hydrolysis of the seed coat.

The General trend within the phylum of *Proteobacteria* is expressed by the peak number on the 5th day, with a gradual decrease in the number, which by the 7th day became lower compared to the control sample. This applies to the anaerobic hydrolytic *Ruminococcus* sp and the aerobic nitrifier *Nitrobacter* sp.. The increase in these species indicated the initial destruction of the cellulose shell of the grain and the processes of fixation and conversion of nitrogen in the microbiota of the germinating seed, necessary for the formation of a certain C/N ratio. Abundance of anaerobic *Ruminococcus* sp. both types of iron reducers can be explained by the presence of $Fe(OH)_3$ in the soil absorbing complex, which is restored by anaerobic iron reducers. As with other anaerobes, their numbers decrease by the 7th day. These species in the process of their anaerobic metabolism form such gaseous components of soil air as hydrogen, carbon dioxide, oxygen.

A gradual increase in numbers by the 7th day was observed in aerobic species such as *Pseudomonas fluorescens*, *Pseudomonas putida*, *Pseudomonas vesicularis*, and *Sphingomonas adgesiva*. This also indirectly indicates a change in the gas composition towards aerobic conditions. As the gas regime changed, the number of aerobic and anaerobic microorganisms changed, respectively.

The total number of phylum *Firmicutes* species decreases by 7 days, passing through the peak number on the 5th day. This is the case with the number of *Clostridium pasteurianum* (which is presumably related to the desiccation of the soil and the change of conditions towards aerobic) and facultative anaerobes *Bacillus subtilis*, which participate in the processes of nitrogen transformation (denitrification and ammonification) and stimulate the activity of rhizospheric nitrogen fixators [26]

Among the fungal microbiota, the mycorrhizal species *Gigaspora* stood out mainly, the content of Which in the initial period (during seed planting) reached an average of 5.3 mg per gram of soil.

Then it increased to 8.3 mg/g of soil, and then it decreased to 3.4 mg/g of soil by the 7th day. Apparently, the decrease in the mass of fungi was primarily affected by a decrease in soil moisture, which at the end of the experiment was 14.4% (about 50% of the lowest moisture content). In addition, mycorrhizal fungus to the seventh day passes, apparently, from the rhizosphere state to the rhizoplane of the developing root system of the seed.

In the chosen method, along with the obvious and undeniable advantage of being able to observe changes in the architecture and growth of the root system in the entire volume of the sample without destruction, there are also disadvantages. For example, the complexity and subjectivity of the process of manual segmentation of tomograms, the restriction on the size of the sample. The small volume of the studied model of the seedbed allows us to study the earliest processes of formation of underground organs and better use the advantages of the tomographic method. There is a belief that the morphological features and" behavioral " differences of roots found in seedlings will be inherent in the adult plant [27], while attention is drawn to the fact that it is often the physical, biological and chemical environment of the soil that largely has a dominant influence on the growth and development of young roots [27].

Conclusion

Thus, the number and composition of soil biota undergoes dynamic changes at the initial stage of seed germination. These processes are physiologically consistent with the growth of the root system and changes in the seed. So, it is noticeable that the desiccation of the soil leads to an increase in the number of aerobic species by the 7th day and a decrease in anaerobic species. The decrease in the volume of the seed shown on the tomographic survey for the period from 3 to 7 days from 50% to 36% of the total volume of the plant occurs when the biota is constantly dominated by actinobacteria (aerobacteria *Actinobacteria* spp., *Rhodococcus equi*) - "active hydrolytics" that decompose complex organic compounds (the seed shell) in aerobic conditions. At the same time, the total number of bacteria increases by the 5th day. In the next 2 days, it decreases (the number of actinobacteria decreases), which is probably due to the destruction of the seed coat.

The interrelatedness of the methods of visual research results of computed tomography and quantitative research method molecular complex of microorganisms around germinating seed overlap in the visual field determine the dynamics of its hydration and the formation of pores filled with water and air when changing water, air and gas regimes and the accumulation of metabolic products of aerobic and anaerobic species of microorganisms.

Acknowledgment

The work was partly supported by the RFFI, project 19-29-05112A, "Transformation of properties and structure of intra-aggregate pores of loamy soils under variable humidity and external load", projects of basic scientific research, 04 BIOLOGY, 04-180 soil science; RNF, project 19-74-10070 "Universal method of segmentation of tomographic images of soil pore space based on machine learning: from modern local algorithms to synthetic tomograms", as well as with the involvement of equipment Of the center for collective use of the Soil Institute named V. V. Dokuchaev "Functions and properties of soils and soil cover".

References

[1] Marilley L and Aragno M 1999 *Applied soil ecology* **13** 127–36
[2] Yang C H and Crowley D E 2000 *Applied and environmental microbiology* **66** 345–51
[3] Metzner, R, van Dusschoten D, Bühler J, Schurr U and Jahnke S *Front Plant Sci* **5** 469 https://doi org/10 3389/fpls 2014 00469
[4] Gregory P 2006 *European Journal of Soil Science* **57** 2-12 https://doi org/10 1111/j 1365-2389 2005 00778 x
[5] Lafond J, Han L, Dutilleul P 2015 *Front Plant Sci* **6** 1111 doi:10 3389/fpls 2015 01111
[6] Zappala S, Mairhofer S, Tracy S et al 2013 *Plant Soil* **370** 35–45 https://doi org/10 1007/s11104-013-1596-1
[7] Mooney S, Pridmore T, Helliwell J et al 2012 *Plant Soil* **352** 1–22 https://doi org/10 1007/s11104-011-1039-9
[8] Rogers E, Monaenkova D, Mijar M, Nori A, Goldman D and Philip N 2016 *Benfey Plant Physiol* 2028–40 doi: 10 1104/pp 16 00397
[9] Brackin R, Atkinson BS, Sturrock CJ and Rasmussen A 2017 *Plant Cell Environ* **40** 3135–42 https://doi org/10 1111/pce 13072]
[10] Mawodza T, Burca G, Casson S and Menon M 2020 *Geoderma* **359** 113988 doi: 10 1016/j geoderma 2019 113988
[11] Gerke K, Skvortsova E and Korost D 2012 *Eurasian Soil Sci* **45** 700–9 doi:10 1134/S1064229312070034
[12] Ivanov A, Shein E and Skvortsova E *Eurasian Soil Sci 2019* **52** 50–7 doi: 10 1134/S106422931901006X
[13] Jiang Z, van Dijke M I J . Geiger S, Ma J, Couples G D and Li X 2017 *Advances in Water Resourses* **107** 280–9 10 1016/j advwatres 2017 06 025
[14] Shein E V, Milanovskii E Ju, Khaydapova D D, Pozdnjakov A I, Tjugai Z N, Pochatkova T N, Demboveckii A V 2017 *Workshop on solid phase physics of soils: tutorial* (Moscow: Buki-Vedi) p 119
[15] *Theories and Methods of Soil Physics*. Edited by Shein E V, Karpachevskiy L.O.l 2007 (Moscow: Grif and K) p 616 (in Russian)
[16] Skvortsova E, Shein E, Abrosimov K, Romanenko K, Yudina A, Klyueva V, Khaidapova D and Rogov 2018 *Eurasian Soil Science* **51** 190–9 doi: 10 1134/S1064229318020102
[17] Ivanov A, Shein E and Skvortsova E 2019 *Eurasian Soil Science* **52** 50–7 doi: 10 1134/S106422931901006X
[18] Müller K, Katuwal S, Young I, McLeod M, Moldrup P, de Jonge L W and Clothier B 2018 *Geoderma* **313** 82–91 doi:10 1016/j geoderma 2017 10 020
[19] Helliwell J R, Sturrock C J, Grayling K M, Tracy S R, Flavel R J, Young I M, WhalleyW R and Mooney S J 2013 *European Journal of Soil Science* **64** 279–97 doi:/10 1111/ ejss 12028
[20] Borges J A, Pires L F, Cássaro F A, Roque W L, Heck R J, Rosa J A and Wolf F 2018 *Soil Tillage Research* **182** 112–2 doi: 10 1016/j still 2018 05 004
[21] Wildenschild D and Sheppard A P 2013 *Advances in Water Resources* **51** 217–46 doi: 10 1016/j advwatres 2012 07 018
[22] Otsu N 1979 *IEEE Trans Sys . Man . Cyber : journal* **9** 62—6 doi: 10 1109/TSMC 1979 4310076
[23] Yudina A . Romanenko, K 2019 *Mechanistic Understanding of Soil Hierarchical Structure.* EGU General Assembly 2019, Conference poster DOI:10 13140/RG 2 2 26167 16809

[24] Verkhovtseva N V and Osipov G A 2002 *Microbiology of composting, Springer-Verlag Berlin Heidelberg* 99–108 doi: 10 1007/978-3-662-08724-4_8 (http://www scirp org/journal/abb/)
[25] Shekhovtsova N V, Marakaev O A, Pervushina K A, Osipov G A Advances 2013 *Biosci Biotechnol* **4** 35–42
[26] Pishchik V N, Vorobyev N I, Moiseev K G, Sviridova O V and Surin V G 2015 *Eurasian Soil Science* **48** 77-84 https://doi org/10 1134/S1064229315010135
[27] Liao H and Yan XL *Acta Bot Sin* **43** 1161–6
[28] Hargreaves C E, Gregory P J and Bengough A G 2009 *Plant Soil* **316** 285–297 https://doi org/10 1007/s11104-008-9780-4

IOP Conf. Series: Materials Science and Engineering **941** (2020) 012033 doi:10.1088/1757-899X/941/1/012033

Ultrastructure Aspects Of Acquired Resistance In Plants Under Viral Pathogenesis

S N Lega[1], I N Tikhonova[2], V N Orobinskaya[3],

[1]Medical and Pharmaceutical Institute (Affiliate), Volgograd State Medical University, Pyatigorsk;
[2] Stavropol State Pedagogical Institute, Essentuki;
[3]Institute of Service, Tourism and Design (Affiliate), North-Caucasus Federal University, Pyatigorsk.

e-mail: stasjairustix@mail.ru

Abstract: The reported paper analyzes mechanisms of plant resistance to biotic factors with a focus on virus-induced lesions. Ultrastructure transformations occurring in leaf cells of virus-infected plants are studied. The work provides data on morphological features of cellular organelles in systemic and local viral lesions of virus-hypersensitive plants.

1. Introduction

To date, natural response reactions (immunity) to hazardous biotical and abiogenous environmental factors developing in a plant have become an important issue. However, a systematic understanding of this problem is still lacking. At the same time, a solution to this issue may make a certain contribution to the progress of selection, increase an agricultural yield, further the preservation of biodiversity, as well as facilitate research with a focus on plant protection. The phenomenon of interest has received critical attention because of ever-increasing environmental pollution, global climate changes, and inability of living organisms to adapt to these factors.

To explore the immunity evolvement to biotic factors in plants it is reasonable to consider an ultrastructure aspect of this question.

2. Main part

In co-evolution of plants and pests a variety of natural protective mechanisms have developed to form the adaptive immunity. From our point of view, a plant-virus interaction is a future-oriented research domain for there has been little discussion of virus-induced diseases so far.

The adaptive immunity develops if an organism is in immediate contact with a pathogen and may be either systemic or local.

The systemic acquired resistance of plants shows whether they are re-infected by the same virus (re-infection). It is related to the development of defensive reactions, limiting propagation and accumulation of a virus [1,2,3,4]. However, for the evolution of systemic resistance, developing defensive mechanisms are far to be sufficient to localize an infection in the virus entry area and prevent it from spreading through an entire plant, whereas local acquired resistance may isolate a virus in its entry spot. The adoptive immunity is apparent at different levels of a species: organism, tissue, ultrastructure, and molecule.

At a level of organism an immune response of plants to a virus entry affects their morphological characteristics on the whole, e.g. a variety of diagnostic signs caused by diverse diseases appear on plant leaves. For a variety of plants and pathogens these signs can be in a form of mosaic light yellow or green spots, leaf tissue can tend to curling up or crooking, knobs can develop on leaf ribs, so called yellow sickness – yellowish or light green leaves – chlorosis may be seen.

At a molecular level the resistance of a plant may show as the synthesis of certain compounds. For instance, PR-proteins are produced in plants to respond to bacteria, fungi, and viruses. These compounds were detected in leaves when systemic or local types of virus-induced resistance developed there. They were discovered for the first time when exploring hypersensitive tobacco plants infected with tobacco mosaic virus [5]. It was established these compounds to connect with processes furthering defensive reactions of plants to a pathogen entry. For now, PR-proteins represent molecular markers used to reveal the systemic induced resistance [6].

There are also other chemical compounds, which influence directly or indirectly the effect of acquired virus-induced resistance. For instance, it is reported on a so-called "anti-virus factor" (AVF) [7] targeting the inhibition of a virus replication process [8]; the research has highlighted activated hydrolases (RNase and protease) in the apoplast for a case of virus-induced resistance [9]. In addition, chemical resistance inductors were detected in plants; e.g. active oxygen forms, salicylates, jasmonates, and ethylene. Jasmonic acid is thought to be a key molecule in the warning system activating defensive mechanisms in plants stressed by damaging factors.

In her research A.M. Polyakova [10] demonstrated that plants may respond to a virus entry in a different way: from morphological to ultrastructure transformations in infected leaves. The researchers reported on a combination of paling ribs and morphological changes in cellular organelles in cut Samsun tobacco leaves totally infected with tobacco mosaic virus. In local and systemic virus infections such responses of plants to pathogen entries were related to the promoting protein-synthetic activity of cells and processes of catabolism (lysosomal activity) [10]. For instance, a high degree protein-synthetic system activity manifested itself in the occurrence of large nuclei with expanded nucleoli in infected cells, as well as numerous ribosomes in cytoplasm, and a well-developed system of mitochondria and chloroplasts. At the same time, a number of researchers [10, 11, 12, 13] pointed at active lytic processes in infected cells (activated Golgi apparatus, a significant number of small vacuoles, residues of lysed cell bulk, etc.).

The evolving local acquired resistance is proposed to be one of the most important criteria indicating the developing acquired virus resistance.

The local acquired resistance results in the necrosis on infected leaves and a hypersensitivity reaction, which localizes a pathogen in its entry area, blocking its propagation from a necrosis through the plant [14]. Therefore, no particles of virus are detected in tissues between necroses on infected leaves [15]. However, a great number of virus particles were found in a local necrosis, they are in a form of small or large blocks with a varying number of intact virions [13].

The re-infection (re-inoculation) of leaves with hypersensitivity to a pathogen caused by a homologous virus initiated less extended secondary necroses between primary necrosis zones with a space above 1-2 mm from their edges. On the contrary, neither secondary necroses nor virus particles are detected 1-2 mm from primary local necroses. Investigating local necrosis areas infected with tobacco mosaic virus in 1961, A. Ross suggested the resistance to a virus infection to develop in these tissues. He offered to refer to these zones as ad-necrosis ones. K. Yarwood [16] and Ross [17] demonstrated the so-called acquired resistance to evolve in ad-necrosis areas infected with tobacco mosaic virus.

At the same time, the resistance is induced between ad-necrosis zones in tobacco leaves infected with tobacco mosaic virus; a number and diameter of secondary necroses decrease in re-inoculation. This type of resistance was offered to refer as a systematic one.

So far, there has been no detailed investigation on the realization of systemic and local virus-induced resistance. Recent studies have established that under the local acquired resistance ultrastructure transformations of cells in and out of local necroses and ad-necrosis areas are quite different, indicating a degree and target of their metabolic activity. The important finding to emerge from the study is that these transformations depend upon the position of cells relative to a local necrosis edge [11]. Thus, the most significant destructive changes in the ultrastructure of cell organelles were detected in the necrosis center, where most of cells were subject to necrotization. In a varying degree necrotic cells were found close to a necrosis edge, although ultimately intact cells with considerable destructive changes of organelles formed there.

At a distance from a necrosis edge (in a space between necroses) the morphology of cell organelles was quite normal, demonstrating no signs of necrotization [11]. For instance, nucleoli in some cells close to the necrosis edge suffered from necrotization, but in a varying degree. At the same time, necrotization of nuclei was not registered in more distant cells in an ad-necrosis zone (1-2 mm), although they displayed considerable morphological deviations. These transformations represented characteristics of organelles typical for higher metabolic activity of a cell, i.d. a twisted nucleolemma, coarse nucleoli, and heterochromatin.

Endoplasmic reticulum in Datura Stramonium L. leaf tissue infected with tobacco mosaic virus had a variety of destructive changes. For instance, cisternas of agranular endoplasmic reticulum in a local necrosis and close to it were swollen. Here, swollen cisternas in the middle of a necrosis were somehow brazed-in into the dark collapsed cytoplasm. Swollen cisternas of endoplasmic reticulum may evidence deteriorating defensive properties in their membranes and suggest entering of hydrolases into cytoplasm [18]. It is a most probable reason for a cytoplasm collapse.

Furthermore, researches pointed at destructive transformations of vacuoles close to a necrosis edge for their tonoplast was unusually electron-dense. Destructive changes were also detected in chloroplasts. Their edges looked amorphous. Mitochondria also deviated from the norm since their crista was expanded or narrowed, other changes were found as well.

No necrotization signs were reported at a further distance from the necrosis edge (1-2 mm). The crista in mitochondria was well-developed and embedded in a relatively dense matrix, that fact may be considered as an evidence of its activity. There were more membranes of rough endoplasmic reticulum detected in palisade cells there than in non-infected cells. Moreover, the well-developed granular endoplasmic reticulum was found in cells of ad-necrosis areas. It was reported on an increased amount of Golgi bodies and activation of its apparatus; hence, a lot of vesicles were formed. The study highlighted a heightened number of peroxisomes, multivesicular bodies, and coated vesicles. In addition, a morphometric analysis disclosed a high volume of cytoplasm and volume density of mitochondria, micro-bodies, membranes of granular endoplasmic reticulum in cells of ad-necrosis areas, except for chloroplasts.

To sum up, despite the fact cell organelles remained morphologically intact in ad-necrosis areas, where the local acquired resistance developed, their ultrastructure differed somehow from not-infected tissues. These changes, as apparent from morphological parameters, are typical for catabolism processes, evolving immediately at a necrosis edge, and anabolism processes occurring at a distance from a local necrosis area.

Such ultrastructure transformations in the morphology of cell structures revealed in various parts of Datura Stramonium L. leaves infected with tobacco mosaic virus might indicate the realization of acquired resistance mechanism with a focus on virus localization and damaging.

Hence, the research into mechanisms of responses evolving in cells of virus-infected plant leaves at the ultrastructure level may provide data to determine a degree of plant resistance to pathogen due to a direct correlation between plant lesions and developing lytic processes [11].

To date, there are published data suggesting that under the local acquired resistance a virus is localized thanks to necrotization of tissues, as a consequence, a local necrosis forms, and other processes taking place in tissues around it. This assumption was confirmed as virus particles were detected beyond a local necrosis, close to its edge, in cells remaining morphologically intact [18].Therefore, mechanisms furthering tissue necrotization restrict a propagating virus but fail to localize it totally [19]. There is still no consensus concerning mechanisms, which initiate the development of a local necrosis area. Some data [20,21] suggest necrotization is caused by products of phenols oxidation – quinones accumulating in tissues. However, the further research pointed out this is not a main reason for the development of necroses for they originate before the activation of polyphenol oxidase [22] and peroxidase [23] ferments is detected. H.W.J. Ragetli's experiments [24] showed hydrolytic ferments, etc. contribute to the development of hypersensitivity reaction. It was suggested local necroses may be caused by mechanical factors, such as blocked plasmodesm with callose [25, 26] However, this phenomenon is also observed if a systemic infection develops [26] In some works it was reported on the formation of lignin under evolving necrosis [27].

To conclude, there has been no comprehensive investigation of mechanisms initiating the development of necroses, biological importance of this process, as well as mechanism isolating viruses so far; therefore, they should be closely examined. Nevertheless, there are published data suggesting the potential biological importance of necrotization is not limited by virus isolation in its entry zone but it promotes the synthesis of some signal molecules under cell death. They might initiate reactions which totally localize infections. This assumption is indirectly proved by findings of ultrastructure studies on local necroses developing on Datura Stramonium L. leaves infected with tobacco mosaic virus; they demonstrated both collapsed cells and intact cells, which were not subjected to the hypersensitivity reaction [11, 13]. The ultrastructure remained in these cells in spite of being in the center of a local necrosis; in addition, there were virus particles in them.

Several researchers suggested that completely collapsed and intact cells may co-exist in a local necrosis as fully necrotized cells produce enough signal molecules, which initiate reactions of total pathogen isolation [11, 13], so destruction of other cells is unnecessary. In some studies oligosaccharines are suggested to be such signal molecules [28, 29].

When isolating a virus, so-called antivirus proteins may be very important [14, 29, 8]. In their work Moser et al. [31] assumed that the isolation of a virus may be associated with the destruction of a transport protein. A search of literature revealed findings of other researchers making their contribution to the understanding of this process. However, a systematic understanding of the virus-induced local resistance is still lacking. Therefore, an ultrastructure aspect of this phenomenon may provide new insights into mechanisms of acquired systemic and local resistance development.

3. Conclusions

To sum up, a study with a focus on mechanisms of acquired resistance in plants to harmful biotical environmental factors has revealed ultrastructure transformations in cells of plants are in line with principles and orientation of metabolic processes. Furthermore, morphological characteristics of transformations attributed to the development of local acquired resistance in Datura Stramonium L. leaves infected with tobacco mosaic virus are different due to prevailing anabolism (ad-necrosis cells) or catabolism (necrosis cells) processes. This fact might be used for assessment of plant resistance to viruses.

The target of both systemic and local acquired resistance is to isolate a virus in a plant and damage it.

References

[1] Zhuravlev Ju N 1979 *Phyto-viruses in an entire plant and in model systems*. M.: Nauka 247 p.

[2] Matthews R F 1981 *Plant virology. Second edition*. New York, London, Toronto, Sydney, San Francisco: Acad. Ppress. 897 p.

[3] Fulton R W 1982 The protective effects of systemic virus infection *Active defense mechanisms in plants*. New York, London: Plenum press. P. 231245.

[4] Reunov A V 1999 *Viral pathogenesis and plant defense mechanisms*. Vladivostok: Dalnauka 175 p.

[5] Malinovsky V I 2010 *Resistance mechanisms of plants to viruses* Vladivostok: Dalnauka 324 p.

[6] Vasjukova N I, Ozereckovskaja O L 2007 Individual resistance of plants and salicylic acid *Applied bio-chemistry and microbiology*.. V 43 **4** 405–411.

[7] Sela I, Applebaum S W 1962 Occurrence of antiviral factor in virus-infected plants *Virology*.. **V.17. 5** 543-548.

[8] Sela I 1981 Plant-virus interactions related to resistance and localization of viral infections *Adv. Virus Res* **V.26** 201-237.

[9] Malinovsky V I 1998 *Mechanisms of resistance of hypersensitive tobacco plants to tobacco mosaic*

virus: Author's abstract. dis. ... doct. biol. sciences. Vladivostok.45 p.

[10] Poljakova A M 2000 *Ultrastructure analysis of plant cells infected with viruses (tobacco mosaic virus, potato virus X, soya bean mosaic virus) and its application to assessment of plant liability to diseases*: Author's abstract of thesis … candidate biological sciences Vladivostok 22 p.

[11] Reunov A V 1989 *Cytopathology of a plant cell infected with tobacco mosaic virus, potato virus X and plant resistance*: Author's abstract of thesis … doctor biological sciences. Kiev 36 p.

[12] Nagorskaja V P 2000 *Ultrastructure aspects of viral pathogenesis and plant resistance induced with 1.3; 1.6- β-d-glucan*: Author's abstract of thesis … candidate biological sciences. Vladivostok 24 p.

[13] Lega S N 2002 *Reactions of virus-infected plant cells and their role in protection of plants and pathogenesis*. Thesis … candidate biological sciences. Vladivostok. 150 p.

[14] Loebenstein G 1972 Localization and induced resistance in virus-infected plants *Ann. Rev. Phytopathol.* **10** 177-206.

[15] Israel H W, Ross A F 1967 The fine structure of local lesions induced by tobacco mosaic virus in tobacco. *Virology* **33** 272-286.

[16] Yarwood.C E 1960 Localized acquired resistance to tobacco mosaic virus *Phytopathology* **V 50 10** 741-744.

[17] Ross A F 1961 Localized acquired resistance to plant virus infection in hypersensitive hosts *Virology* **V 1. 3** 329-339.

[18] Reunov A V, Lega S N 1984 Electron-microscopy research into cells of mesophyll-resistant zones to tobacco mosaic virus around local necroses induced with tobacco mosaic virus on Datura Stramonium *L. leaves. Cytology and Genetics* **4** Kiev.

[19] Loebenstein G., Spiegel S, Gera A 1982 Localized resistance and barrier substances *Active defense mechanisms in plants*. New York, London: Plenum Press 211-230.

[20] Farkas G L, Kiraly Z, Solymosy F 1960 Role of oxidative metabolism in the localization of plant viruses *Virology* V.12. **3** 408-421.

[21] Rubin B A, Arcihovskaja E V 1968 *Bio-chemistry and physiology of plant immunity* M.: Higher School 416 p.

[22] Wagih E E, Coutts R H A 1982 Peroxidase, polyphenol oxidase and ribo-nuclease in tobacco necrosis virus infected or mannitol osmotically-stressed cowpea and cucumber tissue. I. Quantitative alterations *Phytopathol. Z.* **104. H l** 1-12.

[23] Conti G G, Bocci A M, Sprocati M T 1982 Peroxidase and polyphenol oxidase isoenzymes, hypersensitive reaction and systemic induced resistance in *Nicotian aglutinosa* L. infected with tobacco mosaic virus *Rivista Patol. Veget.* Ser. **18** 83-102.

[24] Ragetli H W J 1967 *Virus-host interactions, with emphasis on certain cytoplasmic phenomena* Can. J. Bot.. **V. 45. 8** 1221-1234.

[25] Wu J H, Dimitman J E 1970 Leaf structure and callose formation as determinants of TMV movement in bean leaves as revealed by UV irradiation studies *Virology* **V. 40 4**. 820-827.

[26] Shimomura T, Dijkstra J 1976 Effect of eosin Y on the formation of local lesions and on the accumulation of callose in "Samsun NN" tobacco and French bean leaves inoculated with tobacco mosaic virus *Neth. J. Plant Pathol* **V 82 3** 109-118.

[27] Kimmins W C, Wuddah D 1977 Hypersensitive resistance: determination of lignin in leaves with a localized virus infection *Phytopathology* V. 67 **8** 1012-1016.

[28] Albersheim P, Darvill A G, Augur Ch., Cheong J J, Eberhard S, Hahn M G, Marfa V, Mohnen D, O'Neill M A, Spiro M D, York W S 1992 Oligosaccharins: oligosaccharide regulatory molecules *Accounts Chem. Res* **25** 77-83.

[29] Ozereckovskaja O L, Romenskaja I G 1996 Oligosaccharines as regulatory molecules in plants *Plant Physiology* **V 43 5** 743-752.

[30] Ladygina M E, Tajmla Je A 1979 Role of proteins in plant defense against phyto-viruses *Success of contemporary biology..* **V. 88. Issue 3(6)** 387-400.

[31] Moser O, Gagey M J, Godefroy-Colburn T, Stussi-Garaud C, Ellward-Tschurtz M, Nitschko H,

Mundry K W 1988 The Fate of the transport protein of tobacco mosaic virus in systemic and hypersensitive tobacco hosts *J.Gen. Virol* **V. 69 6** 1367-1373.

IOP Conf. Series: Materials Science and Engineering **941** (2020) 012034 doi:10.1088/1757-899X/941/1/012034

Reproduction methods of *Iris Spuria* varieties in the forest steppe of the Altai Territory

Z V Dolganova[1] and L I Tikhomirova[2]
[1]Federal Altai Scientific Centre of Agro-Bio Technologies, 656910, Altai Territory, Barnaul, Nauchnyy gorodok, 35
[2]Altai State University, Lenina str., 61, 656049 Barnaul (Russia)
Email: zdolganova@inbox.ru

Abstract. The subgenus of beardless *Iris Spuria* (SPU) includes 19 species. The entire spectrum of flower colors of *Iris Spuria* species selected in Russia and abroad is used for gardening in the forest steppe of the Altai Territory. A low vegetative reproduction coefficient of *Iris Spuria* is reported: 1–4 for three-year-old plants, and 1–6 for four-year-old plants. A suggested production method of *Iris Spuria* planting material involving the clonal micro-propagation and adaptation in aeroponics conditions is commercially applicable, furthers the growing high quality young plants and simplifies technological stages. Using this cultivation technique, one mature plant may supply 2048 young plants in a year given that a reproduction coefficient is 2.0. The tissue culture reduces labor and financial costs and production areas.

1. Introduction

The genus *Iris* L. comprises 8 subgenera, 5 of them are beardless Iris species, i.a. the subgenus of *Iris Spuria* (SPU). The subgenus *Spuria* is divided into 19 species [1]. *Iris spuria* L. (Limniris) is a herbaceous perennial plant, reaching a height of about 30-90 cm. The pedicel is slightly aplanate, thick, roundish, kneed, and subramous, bearing three to eight flowers. 30-40 cm long and 1-2 cm wide leaves are upright-expanded, widely linear to linear-ensate, lanceolate, leathery and malodorous when ground up. Primordial leaves are sharply-keeled and leather-like. Flowers reach a length of 4-5 cm and 6-8 cm in diameter, lilac or violet-blue with dark veins, scentless, stalkless or almost stalkless. Flowering is in May to June. Fruiting is in July to August. Rhizomes are 1-1.5 cm thick. The plant is hygrophilous but drought-tolerant with a habitat in Central and Eastern Europe and Asia [2].

Since the latter half of the twentieth century researchers in the USA, Canada and Australia have been actively engaged in the selection of *Iris Spuria* [3, 4, 5]. Generally, they use species representing the flora of Armenia, Azerbaijan, Iran, Turkey, and Greece [4]. In Russia, 17 species, varieties and sub-varieties of *Iris Spuria* have been explored (E. Regel and Ya. Kesselring). Varieties selected from the Russian species *Iris halophila* Pall., and the French *Iris spuria subsp. maritima* (Dykes) P. Fourn appeared to be resistant. In his study G.I. Rodionenko has found out *Iris Spuria* varieties selected abroad fail a long-termed cultivation at the latitude of Saint-Petersburg [5].

In the Altai Territory, in M.A. Lisavenko Scientific-Research Institute of Horticulture of Siberia SPU varieties selected abroad were introduced in the 21st century. These are most awarded varieties of outstanding plant-growers: B. Blyth, B. Hager, Mc Gown, D. Niswonger grown in collections of A.Yu. Shikuts, A.I. Shokhin, etc [6]. White and blue early flowering varieties grown by G.I. Rodinenko (Lenkoran', Frigia, Moldova) were included into the gardening assortment for the forest steppe in the Altai Territory. The productivity of abroad varieties appeared to be worse than that of

Russian ones; however they demonstrated more diverse flower colors. 10 *Spuria* varieties selected abroad: Belissinado, Bay of Silk, Wyoming Cowboys, Gust, Clara Ellen, Golden Lady, Gold Mania, Looking Glass Eyes, Redwood Supreme topped in the generative productivity (4 to 6 pedicels); 22 varieties formed 1 to 3 pedicels. Varieties with 4–6 pedicels were recommended for the gardening assortment [7]. It is necessary to analyze their reproduction potential for a wide gardening use.

The clonal micro-propagation represents an innovative plant protection technique, which makes it possible to propagate valuable varieties, develop tissue banks in vitro and store hundreds of thousands of plant samples in small areas [4-6].

The research aims to develop a fast vegetative reproduction method of *Iris Spuria* species for the forest steppe in the West Siberian South.

2. Materials and Methods

2.1 Conditions

Regarding the climate, the forest steppe of the Altai Territory is characterized with prevailing winds and low winter temperatures, a sharp temperature change in spring and autumn and a varying rainfall, droughty May and June and a short vegetation period [8].

Weather conditions in the vegetation period. In 2017 the amount of rainfall exceeded a mean annual one, reaching 422.5 mm, in 2018 it was in line with a precipitation norm – 251.5 mm, and in 2019 – below the norm by 43 mm (212.6 mm). The heat sum 10° C, in 2019 it was above a value of 2017 and 2018. The hydrothermal coefficient is 1.8; 1.2; and 0.9, respectively.

A test area is located in a highland region of Barnaul, on the left bank of the Ob. The objects of interest are varieties of *Iris Spuria*. The nursery for primary estimation of varieties was founded in August 2015. The propagation was assessed for 3-4 year-old plants. A vegetative propagation coefficient was calculated as a ratio between a number of vegetative shoots and a period of growing. Weather conditions and a hydrothermal coefficient were estimated according to "Agro-climatic resources of the Altai Territory" provided by the weather station of the Scientific-Research Institute of Horticulture of Siberia [8].

2.2 Tissue culture

Unexpended flower buds of *Iris Spuria* (Art And Soul variety) provided by Z.V. Dolganova (M.A. Lisavenko Scientific-Research Institute of Horticulture of Siberia, Federal State Budgetary Scientific Institution, Federal Altai Research Center of Agro-Bio Technologies, Barnaul) were used as explants for introduction into the tissue culture. Regeneration plants and hydroponic plants were produced and grown in the Biotechnology Department of Altai State University according to the developed recommendations [9, 10].

Phytohormones of cytokinin type – 6-benzylaminopurine (BAP) Sigma, USA 1- 8 μmole, and auxin type – α-naphthaleneacetic acid (NAA) Sigma, USA 1-5 μmole, indole-3-butyric acid (IBA) Sigma, USA 0.1 μmole were added to the nutrient when introducing into the culture in vitro.

Phytohormones were added to the MS-based nutrient [11] in appropriate concentrations.

In the study saccharobiose (30 g/l) was applied as a main carbohydrate to cultivate organs and tissues. A value of pH of the nutrient was balanced by 0.1N HCl and 0.1 N KOH to a value of 5.0-5.8 before autoclaving (autoclaving conditions of the nutrients: pressure 1.0 atm., temperature 105-110 °C, 20 min.).

Explants were grown in a culture room at a temperature of 26-30 °C, 16-hour photoperiod, lightening intensity 2000–4000 lux. In 30 days shoots were sub-cultivated. Each experiment was repeated 10 times.

3. Results and Discussion

3.1 Vegetative propagation of *Iris Spuria* varieties

For the gardening a whole color spectrum of *Iris Spuria* varieties selected in Russia and abroad is used in the Altai Territory. The origin of varieties is different. Varieties by G.I. Rodionenko are selected from the population of species as follows: Lenkoran' from *Iris Klattii* Kem.-Nath, Moldova – *Iris monnierii* (DS), Frigia – cross-species hybrid *Iris orientalis* Miller × *Iris monnierii* [5]. Varieties,

which are not included in the gardening list, such as wine-red Countess Zeppelin and purple and brown Handsome Is enjoy a high degree of popularity.
The conditions of growing play the most important role for the productivity of varieties. The summer in the forest steppe of the Altai Territory is hot and sunny. At the latitude of Saint-Petersburg a sum of effective temperatures makes 1500–1900°C, a sunshine duration is 1700–1800 hours, and in the forest steppe of the Altai Territory – 2000-2200 °C and 2500 hours, respectively [1]. The climate in the forest steppe of the Altai Territory is more favorable for Iris *Spuria*, e.g. a 3-4-year-old Frigia develops 3–6 pedicels at the latitude of Saint-Petersburg [3], whereas 10–20 pedicels tend to develop in the Altai Territory. The vegetative productivity of 3-4-year-old plants is in a range: Lenkoran' 10±3 and 18±10, Frigia 15±5 and 25±6, Moldova 6±2 and 15±9, and Elegia 6±3 and 9±2. The productivity of two latter varieties is similar to that of sorts selected abroad (Table 1).

Table 1. Color of flowers and productivity of *Iris Spuria* varieties

Variety	A number of vegetative shoots		Author, year of the variety development*	Colors of perianth lobes**
	2018	2019		
Bay of Silk	4±2	5±1	B. Blyth, 1998/99, Australia	Upper lobes – blue-violet; lower lobes – white with a creamy spot
Belissinado	7±2	10±3	Corlew, 1988, USA	Upper lobes – purple; lower lobes – yellow with a violet mesh
Golden Lady	5±2	9±2	Combs, 1957, USA	Golden yellow
Gold Mania	8±2	10±2	Blyth, 1989/1990	Golden
Gust	4±1	6±2	Jenkins, 2001, USA	white, yellow signal
Clara Ellen	5±2	12±3	Jenkins, 1993	Upper lobes - purple; lower lobes - yellow, light-purple edges
Looking Glass Eyes	6±2	9±2	Niswonger, 1979, USA	Upper lobes – black and yellow; lower lobes – yellow, black and yellow edges
Redwood Supreme	5±3	7±2	Niswonger, 1979	Red and brown with a yellow spot
Wyoming Cowboys	5±1	7±2	Wickenkamp, 1994, USA	Golden yellow and brown
Lenkoran'	10±3	18±10	Rodionenko, Abrego, 1994 (1985), Russia	Thick violet and carmine
Frigia	15±5	25±6	Rodionenko 1985, (2001)	White, ochroid and yellow spot
Elegia	6±3	9±2	Rodionenko, hybrid	Blue, yellow signal
Moldova	6±2	15±9	Rodionenko, hybrid	White, yellow signal
Art and Soul	3±1	3±1	Blyth, 1998/99, Australia	Upper lobes – light-violet; lower lobes – white with a creamy spot

*Author and year of the variety development [5]. **upper and lower lobes of the perianth

The vegetative productivity of varieties selected abroad changed in 2018, as plants were 3 years old, and varied from 4±2 (Bay of Silk), to 8±2 (Golden Lady); and in 2019, as plants were 4 years old – from 6±2 (Looking Glass Eyes) to 12±3 (Clara Ellen).

A low vegetative propagation coefficient was recorded for *Iris Spuria* varieties: 1–4 for three-year-old plants, and 1–6 for four-year-old plants. If a mother plantation of 10 bushes is developed, it is possible to propagate quickly such Russian varieties as Frigia (60 plants for planting per year), Lenkoran' (45), Moldova (37), and varieties of foreign producers – Clara Ellen (30). The assortment may be increased if new propagation methods are introduced. The planting stock of varieties selected abroad with a low vegetative propagation coefficient should be grown via micro-cloning.

3.2 Micro-cloning propagation of Iris Spuria (variety Art and Soul)

Nutrients with 2.5, 5.0, 7.5, and 10.0, μmole BAP, as well as nutrients with the same amount of cytokinin and enriched with auxin compounds 1.0 μmole NAA and 0.1 μmole IBA, nine experimental variations, in total, were used when micro-cloning of *Iris Spuria*. For the purpose of control a nutrient with 1 μmole BAP was prepared. Plants fail to grow without phytohormones in the tissue culture. When cultivating *Iris spuria*, a number of micro-shoots and a height of the plant were measured. If compared with *Iris sibirica* [9], *Iris spuria* is less sensitive to the BAP concentration. A concentration of BAP of 2.5-5.0 μmole appeared to be the most appropriate. In the micro-cloning period it is necessary to alternate nutrients containing phytohormones and nutrients without hormone to evolve fully a morphogenetic potential of *Iris spuria*. Moreover, nutrients without hormones are to be enriched with L-glutamine and adenine sulphate in amount of 100 mg/l (MS+100mg/l L-glutamine+100mg/l adenine sulphate). Within this cultivation scheme regenerative plants develop adventive and auxiliary buds at high concentrations, and buds may evolve into shoots when reducing a hormone load in the next passage. Non-hormone growth stimulators L-glutamine and adenine sulphate in nutrients have a positive effect on the developing shoots and their rooting ability. It is strongly recommended to alternate the nutrients, otherwise excessive amounts of BAP are accumulated in tissues of regenerative plants, a propagation coefficient drops, shoots stop growing, and finally a plant dies (Table 2).

A nutrient containing 3 μmole NAA was used for rooting of *Iris spuria*. Then a plant was taken from the agar medium, rests of agar were washed in flow water and plants were put into trays into the aeroponics line for 30 days to adapt to non-sterile conditions. Plants were grown in the aeroponics line in the propagation area, equipped with a control climate block. Water and minerals were supplied to plants via periodic irrigation of a plant root system with a nutritive solution (under pressure of 3 atm.). Between irrigations roots were air drained. In 25-30 days plants were transplanted into pots with soil. The growing was completed within 30 days in greenhouses. 90% young plants produced from regeneration plants adapt and complete growing successfully. As growing was completed, young plants met a standard with a closed root system, being ready to go outside (Figure 1).

Calculating theoretically by a formula $B_n = B_1 \times G^{(n-1)}$ (where n – months of cultivation, B1 – number of micro-shoots, G – propagation coefficient), one mature plant may produce 2048 young plants of *Iris spuria* given that this cultivation technique is used for 12 months and a propagation coefficient is 2.0.

a) b) c)

Figure 1. Growing of the planting material *Iris spuria:* a) clonal micro-propagation, b, c) in aeroponics conditions

Table 2. The effect of hormone composition of the nutrients on the bud development in *Iris spuria* (Art And Soul)

BAP concentration µmole	Shoot length			Propagation coefficient			Total shoot length, mm
	Arithmetic mean, mm	Standard deviation	Confidence interval for a mean	Arithmetic mean, ps.	Standard deviation	Confidence interval for a mean	
1.0 (control)	106.25	36.8	18.1	1.13	0.48	0.23	120±1
2.5	81.5	33.9	16.6	1.95	1.19	0.57	157±2*
2.5+A	77.5	35.0	17.2	1.12	0.48	0.24	86±1
5.0	101.2	25.9	12.7	1.62	0.48	0.26	164±1*
5.0+A	110.0	27.7	19.2	1.75	0.82	0.40	189±2*
7.5	109.5	37.3	20.3	1.18	0.52	0.25	129±1
7.5+A	96.8	29.5	14.5	1.12	0.48	0.23	108±1
10.0	88.1	32.4	15.9	1.06	0.24	0.11	93±1
10.0+A	80.0	25.7	12.3	1.18	0.39	0.19	94±1

Note: A (auxins) - 1.0 µmole NAA+0.1µmole IBA; * - a difference from the control data is sound at $P \leq 0.05$

4. Conclusion

To sum up, an entire spectrum of flower colors of *Iris Spuria* species selected in Russia and abroad is used for gardening in the forest steppe of the Altai Territory. Nevertheless, a low vegetative reproduction coefficient of *Iris Spuria* is reported: 1–4 for three-year-old plants, and 1–6 for four-year-old plants. A suggested production method of *Iris Spuria* planting material involving the clonal micro-propagation and adaptation in aeroponics conditions is commercially applicable, furthers the growing high quality young plants and simplifies technological stages. The application of a tissue culture reduces labor and financial costs and production areas.

References

[1] Dolganova Z.V. 2019 Development specifics and germination of *Iris Spuria* seeds in the forest steppe of the Altai Territory *Bulletin of Altai Agricultural University* **12(182)** pp 71–76 – (*in Russian)* – Dolganova Z.V. 2019 Osobennosti razvitija i vshozhest' semjan sortov irisa spurija v uslovijah lesostepi Altajskogo kraja Vestnik AGAU **12(182)** pp 71–76

[2] Shawl A.S., Vishwapaul, Zaman A., Kalla A.K. 1984 Isoflavones of *Iris Spuria Phytochemistry* **23(10)** pp 2405–2406

[3] Alekseeva N.B. 2008 Genus *Iris* L. (Iridaceae) in Russia *Turczaninovia.* **11(2)** pp 5 – 68 – (*in Russian)* – Alekseeva N.B. 2008 Rod Iris L. (Iridaceae) v Rossii Turczaninovia. **11(2)** pp 5 – 68

[4] Rodionenko G.I. 2013 Penetrating into the secrets of nature *(My fate – irises)* SPb.: RIO GBOU SPO "SPbIPT" p 260– (*in Russian)* – Rodionenko G.I. 2013 Postigaja tajny prirody (Sud'ba moja – irisy) SPb.: RIO GBOU SPO "SPbIPT" p 260

[5] Rodionenko G.I. 1977 Iridaceae Juss. – Iris or Iridaceae *Decorative herbaceous plants* **1** pp 158–311 – (*in Russian)* – Rodionenko G.I. 1977 Iridaceae Juss. – Irisovye, ili Kasatikovye Dekorativnye travjanistye rastenija **1** pp 158–311

[6] Shidara H. 2005 Iris species and cultivars in the World *The Japan Iris Society* p 247

[7] American Iris Society, 2019. [Electronic resource], URL: http://wiki.irises.org/bin/view/Main

[8] Agro-climatic recourses of the Altai Territory 1971 L.: Hydro-meteo-publishing p 154 – (*in Russian)* – Agroklimaticheskie resursy Altajskogo kraja 1971 L.: Gidrometeoizdat p 154

[9] Tihomirova L.I., Bazarnova N.G., Il'icheva T.N., Martirosjan Ju.C. 2018 Biotechnological growing of planting material of *Iris sibirica* L. *Chemistry of plant materials.* **4** pp 235-245 DOI:https://doi.org/10.14258/jcprm.2018043887 (*in Russian)* – Tihomirova L.I., Bazarnova N.G., Il'icheva T.N., Martirosjan Ju.C. 2018 Poluchenie rastitel'nogo syr'ja irisa sibirskogo (Iris sibirica L.)

metodami biotehnologii Himija rastitel'nogo syr'ja. **4** pp 235-245 DOI:https://doi.org/10.14258/jcprm.2018043887

[10] Tikhomirova L. 2017 Morphogenesis and histology of cultures of *Iris ensata* Thunb. Generative organs *In Vitro Cellular and Developmental Biology - Plant* **43(7)** pp 270-273. DOI 10.1007/s11627-017-9817-6

[11] Murashige T., Skoog F. A 1962 Revised Medium for Rapid Growth and Bioassaya with Tobacco Tissue cultures *Physiol. Plant* **15(4)** p 473.

Research and Development of Process Conditions for Growing Young Plants *Iris sibirica* L. Aimed at Bio-Synthesis of Tannins, Flavonoids and Xanthones

L I Tikhomirova[1] and E V Malaeva[2,3]

[1]Altai State University, Lenina str., 61, 656049 Barnaul (Russia)
[2]State Budget Institution of the Volgograd region «Volgograd Regional Botanical Garden», Russia, c. Volgograd, set. Metallurgov, 68, e.malaeva@mail.ru
[3]Federal State Budget Financed Educational Institution of Higher Education «Volgograd State Socio & Pedagogical University», Volgograd, Russia

Email: L-tichomirova@yandex.ru

Abstract. The reported paper presents process conditions providing a maximal output of biomass and accumulated biologically active substances when growing young plants *Iris sibirica* (Sterh variety): growth medium composition according to Murashige and Skoog Media Formulation, growth medium pH 5.6-5.8, temperature of growing 24-26 ^{0}C, 16-hour photoperiod, illumination intensity (2000 – 4000 lux), and relative air humidity 70%.
It is recommended to use growth media containing 1.0-2.5 μM BAP and auxin compounds 1.0 μM NAA+0.1 μM IBA for an optimal build-up of biomass and phenol compounds (flavonoids, tannins, xanthones) in *Iris sibirica*.

1. Introduction

The production of secondary metabolites in the commercial crop farming suffers from certain limitations, such as a low output of biologically active substances, a dependence of biomass accumulation upon seasonal and geographical conditions of the environment. The growth of plant tissues and organs has contributed to the development of several cell and tissue culture techniques, which enable the production of valuable plant secondary metabolism products, such as glycosides, alkaloids, other biologically active substances broadly applied in medicaments, food colorants, flavor agents, etc. [1]. To date, a number of production companies use cell culture of *planta superior*: ESC genetic in the USA, Samyang Genex in Korea. However, many years of intensive work have hardly brought many successful examples. The most likely cause of it is a difficulty to obtain an efficient producer strain, and a solution to this issue is therefore a key direction in the development of present day cell technologies [2].
Biotechnology-based technologies, e.g. hydroponics, have a potential for large-scale plant cultivation and production of secondary metabolites. The micro-cloning makes it possible to grow a good planting material, i.a. perennial and difficult-reproducible species in a required quantity all the year round. A combination of these two techniques allows the developing a production biotechnology of medical herbs with a specified concentration of biologically active compounds [3].
A genus of Iris represents medical and ornamental perennial plants, which synthesize a variety of biologically active substances [4].
This study aims to determine optimal process conditions providing a maximal biomass and accumulation of biologically active substances under cultivation of *Iris sibirica* (Sterh variety).

IOP Conf. Series: Materials Science and Engineering **941** (2020) 012035 doi:10.1088/1757-899X/941/1/012035

2. Materials and Methods

Tissue culture

The study focused on micro-cloned (regenerative) plants and grown in aeroponics conditions at the Biotechnology Department of Altai State University was carried out using general methods of plant biotechnology [5]. A standard growth medium MS was used to explore *in vitro* a morphogenetic potential of organs and tissues [6].

Phytohormones, such as 6-benzylaminopurine (BAP) Sigma, and α-naphthaleneacetic acid (NAA) Sigma, indole-3-butyric acid (IBA) Sigma were added to a growth medium with the purpose to control the morphogenesis *in vitro* of Iris. In the study saccharobiose (30 g/l) was applied as a main carbohydrate to cultivate organs and tissues.

When micro-cloning, BAP (2.5-10.0 μM) was introduced into a growth medium, besides it we used several growth media containing the same amount of cytokinin and enriched with auxin compounds A (NAA 1.0 μM and IBA 0.1 μM). Eight growth media were prepared for the experiment. A concentration of BAP in the control growth medium was as high as 1.0 μM (irises don't grow in media without hormones). We calculated mean values of a propagation coefficient and plant height, these data were multiplied, and a general height of plants was obtained.

Containers with plants were kept in a culture room at a temperature of 24-26 °C, 16-hour photoperiod, a value of illumination intensity (2000–4000 lux), and relative air humidity 70%.

Plants were grown in the aeroponics line. A three-high aeroponics line designed in Federal State Funded Scientific Institution "All-Russia Research Institute of Agricultural Biotechnology" (Yu. Z. Martirosyan) was used in the study. This is a modular line, so it may be applied to the research into potato selection, as well as for propagation and growing other agricultural crops and medical herbs.

Methods of Research

A content of extractive materials was estimated as stated in the chapter of General Pharmacopeia (Russian ОФС.1.5.3.00006.15) [7]. To measure a content of flavonoids in extracts we used a method based on their capability to form a colored compound with $AlCl_3$ solution [8]. A content of tannins was evaluated via potassium permanganate oxidizing with indigo carmine according the chapter of General Pharmacopeia (Russian ОФС.1.5.3.00008.18) [9]. Magniferine and a sum of xanthones were calculated as T. A. Kukushkina and colleagues offered in their method (2011) [10].

3. Results and Discussion

Process parameters involve a number of conditions supporting a correct process operation with a maximal output of products. A strategy of producing metabolites from plant tissue is suggested to be a multiphase process: first, a high-yielding donor plant is selected with a focus on an individual substance or group of them. The second phase aims at a maximal output of biomass and accumulation of biologically active substances. Process parameters to be considered while a certain living period of plants in tissue culture and ensuring a maximal output of biologically active substances include a growth medium composition; a combination of nutritive salts and their amount per a consumed unit of a growth medium; pH of a medium; temperature of growing; a plant growing period; illumination and a spectral quality of light. A correlation between cytokinin and auxin phytohormones in the composition of a growth medium is important to the crop productivity.

To stimulate the fission *in vitro* auxins and cytokinins were added to the growth medium. It is reported auxins activate CDK-protein kinases of a cell cycle, and cytokinins – corresponding cyclins. CDK-cyclin complex is essential to start the fission [11]. A combination of various growth regulators has a high synergic effect. Using a combination of cytokinins and auxins, it is possible to produce either a lot of shoots or long shoots.

We conducted a set of experiments to explore a combined effect of different cytokinin concentrations (BAP) and auxins (1.0 μM NAA and 0.1 μM IBA) in the growth medium. The results of experiments indicate that a propagation coefficient goes up to 3.4 in auxin-containing growth media (2.7 in auxin-free media) and plants are a little higher. A correlation between cytokinin and auxin 5:1 (5.0 μM BAP+1.0 μM NAA and 0.1 μM IBA) is the most optimal ratio supporting a synergic effect of cytokinin and auxins, which in turn increases maximally a propagation coefficient of a crop. In this

case an important indicator is a total plant height. The introduction of auxins into growth media at 2.5 μM BAP increases this indicator by 5%, and at 7.5 μM BAP – by 13%, at 10.0 μM BAP – by 47%. If growth media containing 5.0 μM BAP are enriched with auxins, a plant is 70% higher (Table 1).
A maximal content of water-ejected BAP is possible in a growth medium with 2.5 μM BAP+1.0 μM NAA and 0.1 μM IBA. Other variants of BAP concentration studied experimentally have no significant effect on the accumulation of water-ejected biologically active substances (Table 1).

Table 1. Biomass development and accumulation of biologically active substances in planting stock of regenerative plants *I. sibirica* (Sterh variety) vs. hormone composition of growth media

Hormone composition of the growth medium, μM	Biomass development			Accumulation of biologically active substances
	Number of shoots, ps.	Height of shoots, mm	Total plant height, mm	
1.0 BAP	1.8±0.2	90±2	162+0.4	4.0±0.3
2.5 BAP	2.2±0.2	63.2±1.3	139±0.3	4.6±0.2
2.5 BAP+A	2.6±0.3	56.2±9.8	146.1±2.9	5.9±0.5*
5.0 BAP	2.7±0.2	61.6±3.1	166.3±0.6	4.7±0.1
5.0 BAP+A	3.4±0.4	83.3±2.3	283.2±0.9*	4.6±0.1
7.5 BAP	2.6±0.5	61.9±4.6	160.9±2.3	4.7±0.4
7.5 BAP+A	2.9±0.3	62.5±8.5	181.3±2.6*	4.8±0.6
10.0 BAP	2.2±0.2	43.3±6.6	95.3±1.3	4.3±0.2
10.0 BAP+A	2.7±0.4	52±7	140.4±2.8	4.4±0.2

Notes: A (auxins) – 1.0 μM NAA+0.1 μM IBA; * - a difference from the control data is sound at a $P \leq 0.05$; accumulation of biologically active substances – a sum total of water-ejected extractives, % absolute dry matter

Low concentrations of phytohormones initiate weak developmental processes in tissue and organ culture, whereas, excessively high concentrations can be toxic and stop the growth, even leading to the plant death. We know from the reference data the accumulation of flavonoids to correlate with the morphogenesis of plants. However, this correlation tends to have a negative coefficient. The lowest BAP concentration causes the forced accumulation of polyphenols, an increasing concentration reduces a content of metabolites, and in addition, high concentration of phytohormones impedes both morphogenesis and biosynthesis of biologically active substances [12].
A high output of biomass produced by *Iris sibirica* has a negative effect on the synthesis of secondary metabolites (flavonoids, tannins, and xanthones) (Fig. 1). An exception is the accumulation of total flavonoids in phytomass. An increase in BAP concentration above 5.0 μM caused a rise of flavonoids in samples. Since phytohormones and their synthetic analogues are expensive, it is recommended to use growth media with 1.0-2.5 μM BAP to accumulate optimally biomass and these groups of phenol compounds in *Iris sibirica*. To produce an actively proliferous crop a rotation of media with a high cytokinin concentration and those with 1 μM BAP+L-glutamine and adenine sulphate (100 mg/ml) is needed each second passage.
A principal factor regulating the synthesis of phenol compounds is an optimal combination of environmental conditions for a certain crop. This phenomenon was observed in Iris tissue culture. A growth medium with a balance of minerals and hormones MS facilitated the accumulating biomass and phenol compounds in Iris shoots. The study highlighted biosynthesis of phenol compounds is most intensive in young dynamically vegetative organs.
A high cytokinin concentration caused a stress situation, during its first phase a lot of adventive and auxiliary shoots formed, a raise of hormone concentration decreased a propagation coefficient, plants stopped reproducing, growing and died. In extreme conditions phenol compounds are consumed for protection, as a consequence, their total amount drops in tissues of *Iris sibirica*.

A relationship between cytokinin and auxin hormones in the growth medium influences a concentration of several phenol compounds in the planting stock. For instance, we found experimentally auxins have a positive effect on tannins: a 50% increase for 2.5 μM BAP, a 70% increase for 7.5 μM BAP, a 37% increase for 10.0 μM BAP, and even a slight reduction for 5.0 μM BAP. An opposite effect was observed for xanthones: auxins reduced the accumulation of these metabolites; a 92% decrease for 5.0 μM BAP, a 41% decrease for 7.5 μM BAP, a 57% decrease for 10.0 μM BAP, and no effect was recorded for 2.0 μM BAP.

Further we explored the accumulating total flavonoids calculated as quercetin. It is important, a sharply decreasing synthesis of flavonoids was observed in the growth medium with 5.0 μM BAP containing auxins as well. It appears, the more actively plants accumulate biomass; the less secondary metabolites are synthesized.

Data from several studies suggest a majority of plants build up a maximal quantity of flavonoids in an above-ground part of a plant when budding and flowering [13]. Within this ontogeny period a plant produces a maximum of phytomass and begins the fruiting – the other development phase. In the tissue culture regenerative plants being affected by phytohormones are constantly in the intensive growth phase. A researcher aims to find such a hormone composition of the growth medium, which provides a high propagation coefficient alongside with an optimal length of shoots. This is a principle of the micro-cloning technique. From data of the diagram (Fig. 1) it is apparent that as soon as a total height of shoots is 283 mm, a content of flavonoids in phytomass drops sharply. In this consequence a balance between biomass accumulation and content of flavonoids in *Iris sibirica* is possible provided that growth media with 2.5 or 10.05 μM BAP with auxins are applied. Media containing 2.5 μM BAP+auxins are less expensive.

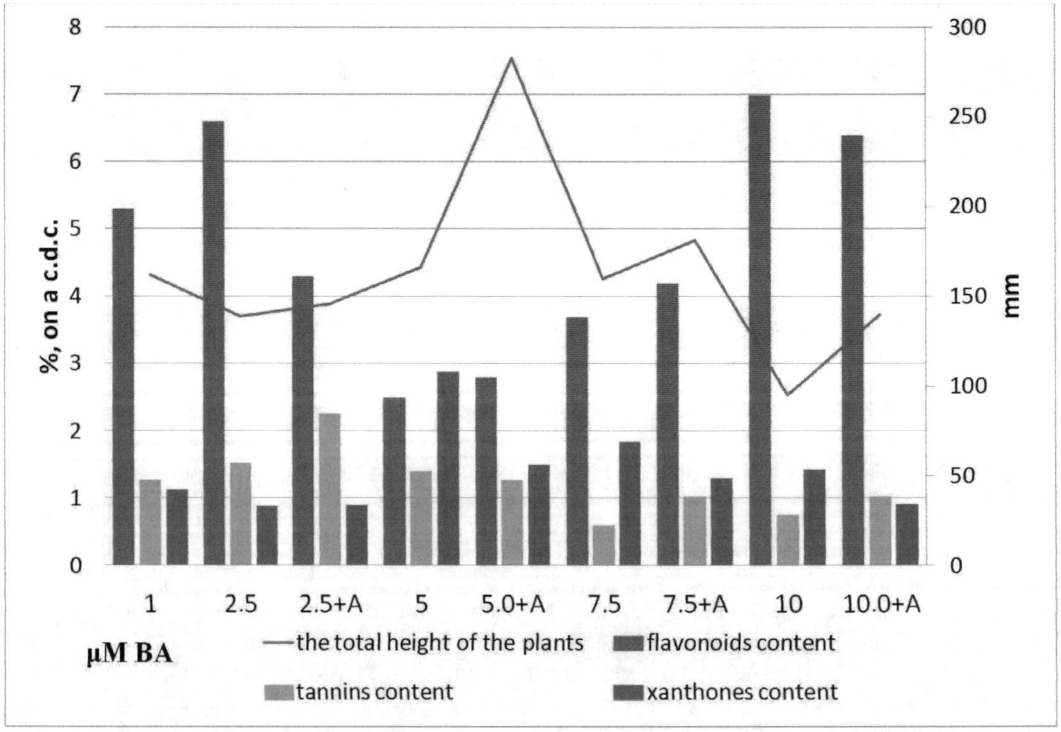

Figure. 1. The effect of a growth medium hormone composition on accumulating biomass and content of flavonoids, tannins, and xanthones in young regenerative plant *Iris sibirica* (Sterh variety)

Notes: c.d.c. - on a completely dry substance, μM BA – μM BAP, +A – a medium with auxins

The aeroponics growing of regenerative plants with the purpose of biomass output is a key factor in our production process of planting stock. Recently, this phase needs certain improvements. We obtained first outcomes of *Iris sibirica* (Sterh variety) planting stock. Over 6 months of cultivation in the aeroponics line a weight gain was 0.576 kg of dry product per $1m^2$. A content of phenol compounds was compared with that in a planting stock of native 6 year old plants, produced near Novoaltaysk, Altay Territory in 2015 (traditionally grown young plants).
In our study we measured several phenol substances. Numerical data on polyphenols depend on conditions where *Iris sibirica* grow. A researcher developing production biotechnologies of a planting stock aims to create the environment very similar to the natural one. Conditions of aeroponics line designed in Federal State Funded Scientific Institution "All-Russia Research Institute of Agricultural Biotechnology" (Yu. Z. Martirosyan) seem to meet these requirements. Numerical data on accumulated flavonoids, xanthones and tannins in the planting stock grown in aeroponics conditions are in line with traditionally output young plants within the experimental error. A hormone composition of growth media in a culture *in vitro* was important to the accumulating xanthones and tannins in regenerative plants. There was almost a three times difference between young plants grown in aeroponics and in a traditional way (Table 2).

Table 2. Quantitative content of some groups of polyphenols in young plants *Iris sibirica* (Sterh variety), % absolute dry matter

Biologically active substances	Regenerative plants	Young plants grown in aeroponics conditions		Young plants grown traditionally	
		Leaves	Rhizome with roots	Leaves	Rhizome with roots
Flavonoids (quercetin)	4.8±0.5	4.40±0.02	-	4.40±0.05	0.020±0.007
Flavonoids (rutin)	1.0±0.01	1.2±0.01	0.30±0.02	1.0	0.40±0.03
Tannins	1.2±0.4	0.49±0.08	1.1±0.1	0.53±0.07	0.73±0.07
Xanthones	1.4±0.2	0.50±0.02	0.43±0.03	0.44±0.01	0.40±0.01

4. Conclusion

It the paper process conditions providing a maximum of biomass production and accumulation of biologically active substances when growing young plants *Iris sibirica* (Sterh variety) are determined; a growth medium composition according to Murashige and Skoog Media Formulation, medium pH 5.6-5.8, temperature of growing 24-26 ^{0}C, 16-hour photoperiod, illumination intensity (2000 – 4000 lux), and relative air humidity 70%.
It is recommended to use growth media containing 1.0-2.5 μM BAP with auxin compounds 1.0 μM NAA+0.1 μM IBA for the optimal accumulating of biomass and phenol compounds (flavonoids, tannins, xanthones) in *Iris sibirica*.
A high cytokinin concentration (5.0 μM) caused a stress situation, during its first phase a lot of adventive and auxiliary shoots formed and biomass developed, a raise of hormone concentration decreased a propagation coefficient, plants stopped reproducing, growing and died. In extreme conditions phenol compounds are consumed for protection, as a consequence, their total amount drops in tissues of *Iris sibirica.*
The introduction of auxins neutralized a negative effect of cytokinin on the formation of phytomass and accumulating of biologically active substances.

References

[1] Grinkevich N.I. Medical herbs. 1991. 420 p. - (*in Russian*) – Grinkevich, N.I. Lekarstvennye rastenija. M.: Vyssh. shk., 1991. 420 p.
[2] Nosov A.M. Cell technologies: ate of the art and future // Prospects of phyto-biotechnology for life quality improvement in the North. Yakutsk. 2018. pp. 17-18. - (*in Russian*) – Nosov A.M. Kletochnye tehnologii: nastojashhee i perspektivy // Perspektivy fitobiotehnologii dlja uluchshenija kachestva zhizni na Severe. Jakutsk. 2018. pp. 17-18.
[3] Tihomirova L.I., Bazarnova N.G., Il'icheva T.N., Martirosjan Ju.C. 2018 Biotechnological growing of planting material of *Iris sibirica* L. *Chemistry of plant materials.* **4** pp 235-245 DOI:https://doi.org/10.14258/jcprm.2018043887 (*in Russian)* – Tihomirova L.I., Bazarnova N.G., Il'icheva T.N., Martirosjan Ju.C. 2018 Poluchenie rastitel'nogo syr'ja irisa sibirskogo (Iris sibirica L.) metodami biotehnologii Himija rastitel'nogo syr'ja. **4** pp 235-245 DOI:https://doi.org/10.14258/jcprm.2018043887
[4] Kaššák P. Screening of the chemical content of several Limniris group Irises //Journal of Pharmacognosy and Phytochemistry. 2014. № 3. Pp. 11-14. 2018. C. 1229-1233. DOI: 10.31255/978-5-94797-319-8-1229-1233
[5] Kalinin F. L., Sariatskaya V. V., Polishchuk V. E. Methods of tissue culture in physiology and biochemistry of plants. Kiev, 1980 - 488 p. – (*in Russian*) – Kalinin F. L., Sariackaja V. V., Polishhuk V. E. Metody kul'tury tkanej v fiziologii i biohimii rastenij / Kiev: Nauk. dumka, 1980 — 488 p.
[6] Murashige T., Skoog F. A Revised Medium for Rapid Growth and Bioassaya with Tobacco Tissue cultures // Physiol. Plant. 1962. Vol. 15. № 4. P. 473.
[7] General pharmacopeia chapter 1.5.3.00006.15. «Measuring a content of extractives in medical herbs and medicaments» // GF RF XIV. 2018. Vol. II. pp. 2356-2360. – (*in Russian*) – OFS.1.5.3.00006.15. «Opredelenie soderzhanija jekstraktivnyh veshhestv v lekarstvennom rastitel'nom syr'e i lekarstvennyh rastitel'nyh preparatah» // GF RF XIV. 2018. Tom II. S. 2356-2360.
[8] Muzychkina R.A. Production technology and analysis of phyto-medicaments. Almaty, 2011. – 360 p. – (*in Russian*) – Muzychkina R.A. Tehnologija proizvodstva i analiz fitopreparatov. Almaty, 2011. – 360 p.
[9] General pharmacopeia chapter 1.5.3.00008.18. «Measuring a content of tannins in medical herbs and medicaments » // GF RF XIV. 2018. Vol II. pp. 2365-2369. – (*in Russian*) – OFS.1.5.3.00008.18. «Opredelenie soderzhanija dubil'nyh veshhestv v lekarstvennom rastitel'nom syr'e i lekarstvennyh rastitel'nyh preparatah» // GF RF XIV. 2018. Tom II. pp. 2365-2369.
[10] Kukushkina T.A., Zinner N.S., Vysochina G.I., Sviridova T.P. Content of xanthones in the above-ground part of Hedysarum theinum Krasnob. and H. alpinum L. (fabaceae) when cultivationg in Siberian Botanical Garden (Tomsk) // Chemistry of plant materials. – 2011. №3. p. 113-116. – (*in Russian*) – Kukushkina T.A., Zinner N.S., Vysochina G.I., Sviridova T.P. Soderzhanie ksantonov v nadzemnoj chasti rastenij Hedysarum theinum Krasnob. i H. alpinum L. (fabaceae) pri vyrashhivanii v Sibirskom botanicheskom sadu (Tomsk) // Himija rastitel'nogo syr'ja. – 2011. №3. pp. 113-116.
[11] Chub V.A. Growth and development of plants. – 2003. – http: herba.msu.ru/russian. – (*in Russian*) – Chub V.A. Rost i razvitie rastenij. – 2003. – http: herba.msu.ru/russian
[12] Palacioa, L., Canterob J.J., Cusidoc R.M., Goleniowskia M.E. Phenolic compound production in relation to differentiation in cell and tissue cultures of Larrea divaricata (Cav.) // Plant Sci. 2012. P. 193-194:1-7. DOI:10.1016/j.plantsci.2012.05.007
[13] Korenskaja I.M., Ivanovskaja N.P., Izmalkova I.E. Medical herbs and medical plant raw materials with flavonoids. Voronezh State University Publishing. 2007. 81 p. – (*in Russian*) – Korenskaja I.M., Ivanovskaja N.P., Izmalkova I.E. Lekarstvennye rastenija i lekarstvennoe rastitel'noe syr'e, soderzhashhie flavonoidy. Izd-vo Voronezhskij gos. u-t. 2007. 81 p.

IOP Conf. Series: Materials Science and Engineering **941** (2020) 012036 doi:10.1088/1757-899X/941/1/012036

Study of Biological Characteristics of Leucinodes orbonalis Guenée which Destroyed Young Shoot and Fruit of Eggplant (Solanum melongena L.): A case in Trang Bom district, Dong Nai province, Vietnam

Vu Van Hung[1,2], Dang Viet Hung[1,2], Dang Thi Lan Anh[1]

[1]Department of Forestry, Vietnam National University of Forestry, Vietnam, Dong Nai Province, Trang Bom District, Trang Bom town, Tran Phu st., 54
[2]Department of Forestry, Saint-Petersburg State Forest Technical University, 5U, Instytutskiy, St. Petersburg 194021, Russian Federation

E-mail: vuhungln2@gmail.com

Abstract. Leucinodes orbonalis Guenée adult had white insect into dark brown colour. Wings were marked by three small triangular with a pinkish to bluish colour mixed pinkish orangish or dark brown. L. orbonalis female dimension were about 9.6 – 12.3mm, the dimension of L.orbonalis male were 8.1 – 9.5mm. Eggs were 1.0 – 1.2mm in length and oval or elliptical shape. The incubation time was around 5.0 – 6.0 days. Larvae are developed by six stages of age, larvae development period lasts 10 – 12 days. Pupae were pink to dark pink in colour and shiny with the length, width were 11.47 – 15.33 mm, 4.40 – 6.45mm respectively. Pupae developed about 8 – 9 days. The life cycle of L. orbonalis lasted 25.30 ± 0.92 days. L. orbonalis feeding in laboratory had high hatching rate, just 9.36% were not hatched. The early stage of larvae had high death-rate (27.98%) mainly due to dispersion. The older larvae had lower death-rate (10.12%). Pupae had high developed rate, only 4.10% were not development. There were equal rate between male and female adult of L. orbonalis. Laying eggs period of female adult lasted 3 days, in there the first day was the highest laying rate (113 eggs/female/day). L. orbonalis densities were highest on the young shoot and fruit of eggplant at the stage after planting 95 – 150 days and was reduced in a after growing.

1. Introduction

Eggplant (*Solanum melongena* L.) is a valuable vegetable plant. Eggplant used as a diuretic, communication circuit and precaution syndrome, atherosclerosis, diabetes, treat arthritis, reduce cholesterol in blood [1,2].

In Vietnam, eggplant is one of the most popular vegetables, the effect brought from the cultivation of eggplant is quite high, 1 ha average productivity of about 50 – 60 tons, harvested from 350 – 420 million VND/ha/crop. In recent years, eggplant is grown within pain processing in Dong Nai province. However, there are many types of dangerous insect damage reduces yield and quality in eggplant, which has deep fruit borers is harmful objects most serious [6,7,9]. This article provides results of research on *Leucinodes orbonalis* Guenée - an object cause dangerous insect on eggplant in Dong Nai province, Vietnam.

2. Materials and Methods

Biological characteristics of eggplant fruit and shoot borer – *L. orbonalis* Guenée was investigated by Borror, Delong (1981) [3,4]. Collecting field samples in Trang Bom district, Dong Nai province in the winter-spring crop of 2018 - 2019, creating the original source. When cocoons turn into the adult, collect each pair of identical ones on the same day, rearing them alone with honey 10% to make them lay eggs. When the eggs lay eggs, collect eggs laid on the same day to track. When eggs hatch the larvae on the same day are kept separately by eggplant and follow up until the larvae pupate. Pupae formed along this track continue. Number of pairs of larvae, pupae tracking 20 - 30 pairs, number of larvae tracking 50-60, numberof eggs tracking 300 eggs.

The monitoring criteria are the development time for each phase, each development stage of *L. orbonalis*, the color and body size of the development phases, the development completion rate of the development phases. These indicators are observed daily.

The investigation of the harmfulness of *L. orbonalis* was conducted in 3 gardens in Thanh Binh, Song Thao and Cay Gao communes (Trang Bom district). These surveyed gardens grow the same type of eggplant, each garden has an area of at least 2000 m^2. Investigate according to the method of 5 points evenly distributed on a row, translation without repetition, each point investigates 3 trees. Keep track of all the fruit and allshoots on plant. Conducting an investigation 5 days/time.

Monitoring indicators are fruit borers density (individuals/tree) and rate of shoot and fruit damage. These criteria density (MDS) and the rate of damaged shoots, damaged fruit (TLH) are calculated as follows (1) and (2) [5]:

$$\mathrm{MDS} = \frac{\sum A_s}{\sum A_o}, \quad (1)$$

Where A_s - number of insect on surveyed plants; A_o - number of surveyed plants.

$$\mathrm{TLH} = \frac{\sum D_d}{\sum D_s} \mathrm{x}100, \quad (2)$$

Where D_d - number of damaged shoot, damaged fruit; D_s - number of damagedshoots, damaged fruit of survey.

3. Results and discussion

3.1. Fluctuating levels of harmon eggplant

Monitoring the emergence and development of *L. orbonalis* in eggplant, studying changes in their damage in 3 gardens in Trang Bom district, Dong Nai province. The time of the eggplant research is from December 15, 2018 to July 7, 2019.

In Thanh Binh commune, it was noted that *L. orbonalis* first appeared on shoots 20 days after planting with a low density of 0.07 individuals/tree, the rate of damaged shootswas 2.38%. Within 30-95 days after planting, the density of *L. orbonalis* increased slowly and ranged from 0.13 to 1.53 individuals/tree, the rate of damaged fruit ranged from 3.57 to 9.29%, and the rate of damaged shootswas about 2.53-2.82% (fig. 1).

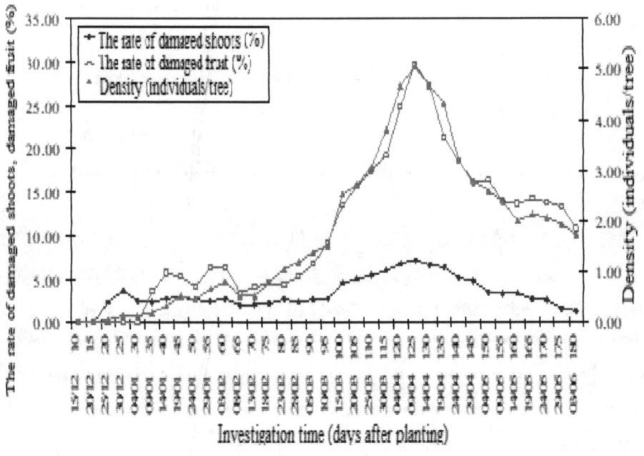

The period of 100-150 days after planting is a period of high density and long-term. During this period, high humidity creates favorable conditions for food sources, the environment helps the process of occurrence, damage and deep fluctuations 2.60-5.07 individuals /tree, the rate of damaged fruits was 13.46-29.79%, and the rate of damaged shoots was4.57 - 7.14%. Especially in the period of 125 days after planting, the density of harmful fruits and the intensity of damaged shoots increased the most (5.07 individuals /tree; 29.79% and 7.14%).

After 130-150 days after planting, the density of *L. orbonalis* tended to decrease, the density reached 2.60 individuals /tree, the rate of damaged shoots 3.50%, damaged fruit 16.38%.

Figure 1. Fluctuations in harmful levels of *L. orbonalis* in eggplants in Thanh Binh commune

In the period of 155 days after planting to the end of the crop (180 days after planting), old trees grow slowly, fruits and shootsare small, so the density of *L. orbonalis* decreases by 1.73 individuals /tree, the rate of damaged shoots is 1, 33% and damaged fruit 10.77%. To further determine the origin and development of *L.orbonalis*, we conducted a study of changes in their harm in Song Thao commune from January 14, 2019 to July 3, 2019. The results of the survey are shown in fig. 2.

In the study, *L. orbonalis* began to appear 25 days after planting, with an average density of 0.07 individuals /tree, and rate of damaged shoots 1.89%. During the period of 35-60 days after planting, the density of *L. orbonalis* gradually increased from 0.20 to 0.87 individuals / tree, especially rate of damaged shoots (4.19%) and damaged fruits (6.67%) take off. During the eggplant period from 65 to 90 days after planting, the density of *L. orbonalis* increased from 0.80 to 2.13 individuals/tree, the rate of damaged shoots from 3.19 to 3.30%, and rate of damaged fruit 3,45 - 9.09%. At this stage, the tree bears a lot of fruit, so it can create favorable conditions for the rapid growth of *L. orbonalis*.

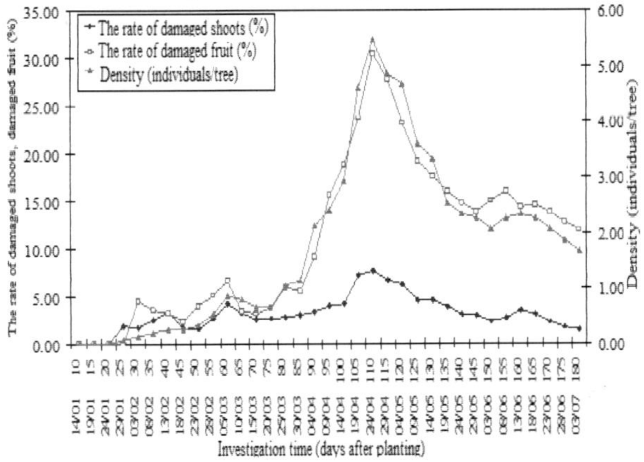

Figure 2. Fluctuationsin harmful levels of *L. orbonalis* in eggplants in Song Thao commune

At the eggplant stage, from 95 to 145 days after planting, the density of *L. orbonalis* increases rapidly and is harmful in the long term. 110 days after planting, the density of *L. orbonalis* reached a maximum of 5.47 individuals /tree, which corresponds to rate of damaged shoots of 7.67% and damaged fruit of 30.46%. 115-145 days after planting density of *L. orbonalis* had a gradual decline from 4.87 individuals / tree to 2.27 individuals / tree, which corresponds to a loss rate of damaged shoots of 6.67%. Down to 2.96% and damaged fruit from 27.81% to 13.79%.

In the period of 150-180 days after planting, the density of *L. orbonalis*, the rate of damaged shoots and damaged fruit began to decrease until the end of harvest, the density of *L. orbonalis* was 1.67 individuals / tree, the rate of damaged shoots of 1.45% and damaged fruit of 11.94%.

In Cay Gao commune, an area of 4,000 m^2 and was studied from January 17, 2019 to July 7, 2019, to further determine the occurrence and development of fruiting and shoots on eggplants. The results are shown in fig.3.

The study showed that *L. orbonalis* began to appear on the shoots 30 days after planting. In the period of 45 - 80 days after planting, the eggplant has strong branches, flowers and fruits. At this stage, the density of *L. orbonalis* was from 0.27 to 1.53 individuals / tree, the rate of damaged shoots varied from 2.24 to 3.19%, and the damaged fruit was about 2.25-7.79%. During the period from 80 to 110 days after planting, starting with the rainy season, humidity increases, eggplants thrive on stems, shoots, create favorable conditions for *L. orbonalis* grows and accumulates. 110 days after planting, the density of *L. orbonalis* increased dramatically and reached a maximum of 5.20 individuals/tree, equivalent to 6.93% of damaged shoots and 29.09% of damaged fruit.

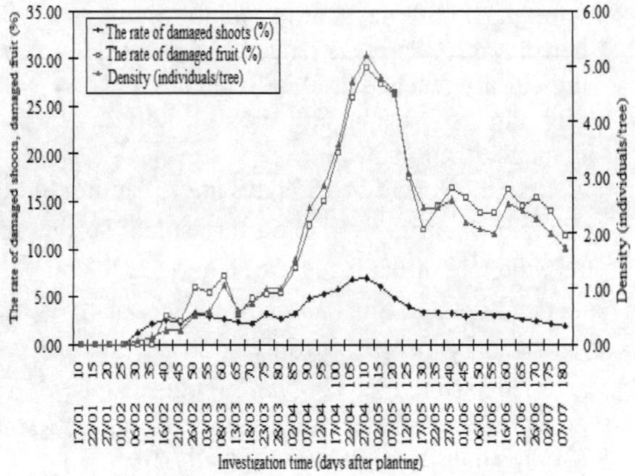

Figure 3. Fluctuations in harmful levels of *L. orbonalis* in eggplants in Cay Gao commune

During the period of 115-140 days after planting, the density of *L. orbonalis* tended to gradually increase, decreasing from 4.80 individuals / tree to 2.60 individuals / tree, the rate of damaged shoots from 6, 10 to 3.26%, and the rate of damaged fruit from 27.27 to 16.42%.

In the period from 145 to 180 days after planting, the density of *L. orbonalis*, the rate of damaged shoots and damaged fruit began to decrease before the end of harvest, the density of *L. orbonalis* was 1.73 individuals/tree, the rate damaged shoots of 1.92%, and the rate of damaged fruit of 11.36%.

3.2. Morphological characteristics of L. orbonalis

During the survey, some morphological characteristics of *L. orbonalis* were recorded, the results are presented in table 1.

Table 1. Dimensions of sexual phases of *L. orbonalis*

Survey period	Tracking criteria	Average Value ($\overline{X} \pm$ SD) (mm)	Variability	Number of observations
Female adult moth	Body length	11.00 ± 0.85	9.6 – 12.3	n = 20
	Body width	2.41 ± 0.08	2.3 – 2.5	
	Wingspan size	22.24 ± 1.71	19.9 – 25.2	
	Beard length	8.61 ± 0.58	7.6 – 9.7	
Male adult moth	Body length	8.66 ± 0.46	8.1 – 9.5	n = 20
	Body width	1.94 ± 0.13	1.7 – 2.1	
	Wingspan size	17.58 ± 1.24	15.2 – 19.4	
	Beard length	7.04 ± 0.42	6.2 – 7.7	
Eggs	Length	1.12 ± 0.06	1.0 – 1.2	n = 30
	Width	0.56 ± 0.04	0.48 – 0.60	
Larvae of age 6	Length	18.20 ± 2.74	14.85 – 22.50	n = 20
	Width	4.07 ± 0.37	3.44 – 4.55	
Pupae	Length	12.97 ± 0.90	11.45 – 15.33	n = 30
	Width	5.58 ± 0.50	4.40 – 6.45	

Note: $\overline{X}$ - The average value; SD - Standard deviation

* *Adult moth*

The adult moth of *L. orbonalis* is a small white butterfly with dark brown and gray-brown veins. The brown belly is covered with a milky layer. Two pairs of milky white wings. The forewing has a triangular shape, along the contour of the Costa there are brown stripes extending to the top of the wing, on the forewing there are 3 triangular dots with a slight pinkish greenish tint, orange-pink or dark brown. The root of the forewing has a spot of 1 spot, resembling a pink-brown triangle. The leading edge of the wing has many small feathers and 6 small black dots along the edge of the wing.

The rear wing is shorter and narrower than the front. There is a long feather border along the outer edge of the wing and there is a light pinkish-brown stripe near the outer edge. The hindwing has 2 black dots in the middle of the wing, 6 black dots and 1 black dot on the wings. When the wings are always spread, the belly is always arched, the front legs are pointed forward, and the beard is straightened back.

The abdomen has 10 segments, the last of which is sharper than the other. Female genitals have yellow hair, but men do not (fig. 4).

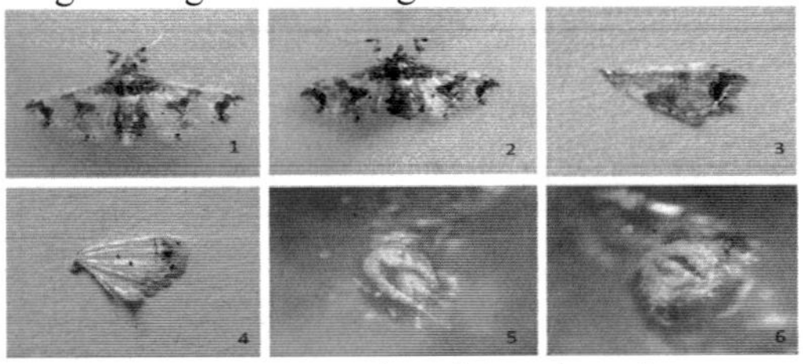

Figure 4. Maturity of *L. orbonalis*

1: Female adult moth 2: Male adult moth
3: Front wing 4: Rear wing
5: Female genital organs 6: Male genital organs

* *Eggs*

Female adult mothlay eggs on young parts of eggplants. Eggs are oval, ellipse. Eggs are laid scattered in separate fruits or in groups of 4-15 eggs, arranged in the form of tiled roofs on young leaves (usually the leaves of the second or fourth from top to bottom), shoots and flower shoots. When laying eggs, the eggs are slightly yellowish, milky white, after 2 days the eggs become dark yellow reddish, after hatching the eggs there is a white crust. The average egg size is 1.12 ± 0.06 mm, ranging from 1.0 to 1.2 mm in length and width by about 0.56 ± 0.04 mm with a deviation of 0.48 - 0.60 mm.

* *Larvae*

The survey reported that the caterpillar survived 6 years of age with changes in size and color (table 2).

Table 2. Size of larvae shells of *L. orbonalis*

Age of larvae	Head shell length (mm)		Width shell length (mm)		Number of observations
	($\overline{X}$ ± SD)	Variability	($\overline{X}$ ± SD)	Variability	
Age 1	0.31 ± 0.02	0.28 – 0.32	0.33 ± 0.03	0.28 – 0.36	15
Age 2	0.67± 0.04	0.56 – 0.72	0.68± 0.04	0.60 – 0.72	15
Age 3	0.99 ± 0.07	0.88 – 1.16	1.07 ± 0.09	0.96 – 1.20	15
Age 4	1.36 ± 0.05	1.24 – 1.44	1.38 ± 0.09	1.28 – 1.52	15
Age 5	1.77 ± 0.08	1.56 – 1.88	1.85 ± 0.06	1.72 – 1.92	15
Age 6	3.64 ± 0.18	3.36 – 3.92	3.95 ± 0.34	3.44 – 4.36	15

Note: $\overline{X}$ - The average value; SD - Standard deviation

Newly hatched larvae emerge from the shell and crawl out. Newly hatched larvae are very flexible and move quickly in search of food. The larvae undergo 6 molts. In preparation for molting, the larvae often release silk directly on a muddy road, they shrink, the color of the body changes and begins to shed. The molting process begins at the beginning, gradually molting down to the belly. After molting, they leave a dark brown shell.

Age 1: the body is pale transparent, the head is larger than the body, the head is dark brown, the body is hairy, and there are 2 small spines at the end of the abdomen. The caterpillar is agile, releasing the silk when sliced into fruit. The size of the head cap is 0.28–0.32 mm, the average length is 0.31 ± 0.02 mm, the width is approximately 0.28-0.36 mm, the average length is 0.33 ± 0.03 mm.

Age 2: the body is pale yellow, the head is yellowish-brown, with dark brown streaks on the front of the chest and 3 developed pectoral legs. The size of the head cover is 0.56 - 0.72 mm, the average length is 0.67 ± 0.04 mm, the width is approximately 0.60 - 0.72 mm, the average is 0.68 ± 0.04 mm.

Age 3: the body is yellowish brown, the head is dark brown, there is a horizontal dark brown stripe on the front of the chest, there are 3 small dark brown spots on the abdomen at the end of the abdomen, and the body is clearly lit. The size of the head cover is about 0.88-1.16 mm, the average length is 0.99 ± 0.07 mm, the width is 0.96 - 1.20 mm, the average is 1.07 ± 0.09 mm.

Age 4: the body becomes dark brown, the middle of the back has a pale pink stripe, the head is brown, the upper part of the head clearly looks like a "v", the front of the chest is noticeably transverse and there are many brown spots on each burn. The last abdominal segment has 2 spines. The size of the head cover is 1.24-1.44 mm, the average length is 1.36 ± 0.05 mm, the width is about 1.28 - 1.52mm, the average is 1.38 ± 0.09 mm.

Age 5: body becomes reddish-brown, middle of back has a pale pink stripe, head dark brown, upper part of the head is clearly visible "v" shape, and the front part of the chest noticeable horizontal black, head size is about 1,56 - of 1.88 mm in length, average length is 1.77 ± 0.08 mm, a width of about 1,72 - of 1.92 mm, and an average of 1.85 ± 0.06 mm.

Age 6: the body is reddish-brown, full strength 14.85-22.50 mm, average 18.20 ± 2.74 mm, width 3.44 - 4.55 mm, average 4.07 ± 0 37mm the body burns clearly. At the age of 6, the worms move to the left, the body squeezes its silk to prepare the pupae. The size of the head cap is 3.36-3.92 mm, average length 3.64 ± 0.18 mm, width 3.44-3.36 mm, average 3.95 ± 0.34 mm.

Descriptions of the color and size of the larvae of the above larvae correspond to the description of Dr. Daniela Lupi, 2008; Alam et al., 2003 [8,10].

* *Pupae*

By the end of six years of age, the larvae vigorously release silk pupae on tree tops, on old leaves, on stems, or under vegetation. Nymphs are located in 2 layers of thick cocoons, first the cocoon has a pinkish-brown color, on the 2nd and 3rd days gradually turns into a dark pink-brown iridescent, when it is going to become dark brown in color. The length of the cocoon varies from 11.47 to 15.33 mm, the average length is 12.97 ± 0.90 mm, the width is about 4.40 - 6.45 mm, the average is 5.58 ± 0.50 mm.

3.3. Biological characteristics of L. orbonalis

Study the biological characteristics of *L. orbonalis* in the laboratory to find out their potential danger. A short stage of development facilitates the development of *L. orbonalis* in many litters throughout the year, and a high rate of development at various stages contributes to a rapid increase in density. The development time of the larvae was controlled in the laboratory at room temperature of 26.4-30.0^{0}C and humidity of 71.8-84.3%. The larvae has 6 years, the development time at each age varies from 1-2 days. Time of development of larvae at the age of 1 year, an average of 1.70 ± 0.47 day, at the age of 2 years average of 1.83 ± 0.38 day, the age of 3 years by an average of 1.90 ± 0.31 day the age of 4 years average of 1.77 ± 0.43 day, the average age of 5 years average of 1.73 ± 0.45 day and age of 6 the average of 1.77 ± 0.43 day. The entire larvae phase varied from 10 to 12 days, with an average of 10.70 ± 0.75 days (table 3).

Table 3. Duration of development of the larvae of *L. orbonalis* in the laboratory of Vietnam National University of Forestry, Dong Nai Province

Phase phonetic	Phonetic time (days)		Number of observatios
	Average valuae ($\overline{X}$ ± SD)	Variability	
Phaseeggs	5.07 ± 0.25	5.0 – 6.0	30
Phase larvae	10.70 ± 0.75	10.0 – 12.0	30
Phase pupae	8.37 ± 0.49	8.0 – 9.0	30
Extra nutrition time	1.17 ± 0.38	1.0 – 2.0	30
Life cycle	25.30 ± 0.92	24.0 – 28.0	30
Average temperature (^{0}C)	28.16 ± 1.36	26.4 – 30.0	
Average humidity (%)	78.19 ± 5.54	71.8 – 84.3	

In laboratory conditions, the temperature of 26.4 - 30.0 and humidity of 71.8-84.3% recorded the life cycle of *L. orbonalis*, which passes through four stages: eggs, larvae, pupae and adultmoth.

The life cycle of *L. orbonalis* (fig. 5) ranges from 24 to 28 days, with an average of 25.30 ± 0.92 days, of which the eggs period lasts 5-6 days, with an average of 5.07 ± 0.25 days. The development time of the larvae ranges from 10 to 12 days, with an average of 10.70 ± 0.75 days. The pupae period lasts 8.0 - 9.0 days, with an average of 8.37 ± 0.49 days. Additional feeding time (time to spawning) of the adult moth varied from 1.0 to 2.0 days, averaging 1.17 ± 0.38 days.

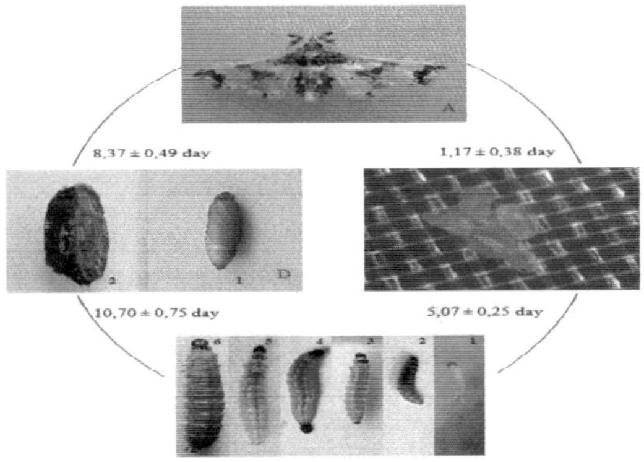

Figure 5. Life cycle of *L. orbonalis*
A - *L. orbonalis* Adult moth; B - Eggs *L. orbonalis*;
C - Larvae *L. orbonalis* age 1, 2, 3, 4, 5, 6;
D - Pupae (1) and cocoons (2) *L. orbonalis*

The survey showed that, theeggs of *L. orbonalis* has a high hatching rate, ranging from 52.09 - 96.64%, with an average of 85.92 ± 12.17%. The percentage of the larvae that survived to the pupae stage in three studies varied from 95.14 to 96.03%, with an average of 95.51 ± 0.47%. The pupae dance rate is about 96.95 - 98.35%, with an average of 97.70 ± 0.70%. The percentage of the adult moth lays eggs in the range from 85.45 to 90.80%, with an average of 88.71 ± 2.84% (table 4).

Table 4. Level of growth stage of *L. orbonalis* of Vietnam National University of Forestry

Tracking criteria	Average valuae ($\overline{X} \pm$ SD)	Variability	Number of observations
Percentage of egg hatches (%)	85.92 ± 12.17	52.09 – 96.64	13
Percentage of pupae larvae (%)	95.51 ± 0.47	95.14 – 96.03	3
Universe capsule rate (%)	97.70 ± 0.70	96.95 – 98.35	3
Adult spawning participation rate (%)	88.71 ± 2.84	85.45 – 90.80	3
Male adult maturation rate (%)	47.23 ± 1.71	45.28 – 48.51	3
Female adult maturation rate (%)	52.77 ± 1.71	51.49 – 54.72	3
Life cycle completion rate (%)	80.17	–	3

The survey results showed that, the life cycle completion rate reached an average of 80.17%. The rate of *L. orbonalis* maleranges from 45.28 to 48.51%, on average 47.23 ± 1.71%, the rate of *L. orbonalis* female ranges from 51.49 to 54.72%, on average 52.77 ± 1.71%. When conducting the survey 3 times, each time 126 - 172 individuals were interviewed, the number of *L. orbonalis* females for all 3 times showed that they are able to lay eggs.

The spawning period of adults of *L. orbonalis* ranged from 1.0 to 2.0 days, with an average of 1.17 ± 0.38 day. The spawning time lasts from 2 to 4 days, with an average of 3.2 ± 0.63 days. The time after spawning of females varied from 1-2 days, on average 1.2 ± 0.42 days. The average life expectancy of females is small 3-5 days, with an average of 4.4 ± 0.70 days. Life expectancy for male is shorter than for female and varies between 2-3 days, with an average of 2.4 ± 0.52 days and usually after the death of male. The ability of *L. orbonalis* lay eggs high, the number of eggs laid by a single female varies from 87 to 342 eggs, with an average of 249.5 ± 72.77 eggs. The number of unborn eggs varied from 0 to 116 eggs, with an average of 19.2 ± 36.23 eggs. This proves that the ability of *L. orbonalis* to lay eggs is not limited by artificial cultural environment (table 5).

Table 5. Period growth and fertility of adult *L. orbonalis*

Period	Phonetic time (days)		Number of observations
	Average value ($\bar{X} \pm$ SD)	Variability	
Extra nutrition time	1.17 ± 0.38	1.0 – 2.0	30
Time to lay eggs	3.2 ± 0.63	2 – 4	10
Time after spawning	1.2 ± 0.42	1 – 2	10
Longevity of female adult moth	4.4 ± 0.70	3 – 5	10
Longevity of male adult moth	2.4 ± 0.52	2 – 3	10
Ability to lay eggs / females (eggs)	249.5 ± 72.77	87 – 342	10
Number of unborn eggs (eggs)	19.2 ± 36.23	0 – 116	10
Average temperature (^{0}C)	28.16 ± 1.36	26.4 – 30.0	
Average humidity (%)	78.19 ± 5.54	71.8 – 84.3	

Observing the activity of *L. orbonalis* adults, they mate, lay eggs from evening to night (at 19-24 pm) and spawn at 21-22 pm.

4. Conclusion

The damage of *L. orbonalis* is mainly in the period of 95 - 150 days after planting and decreases until the end of harvest.

The adult moth of *L. orbonalis* is white with dark brown streaks. The front wing is triangular with 3 points of pink or dark brown. The rear wing is shorter and narrower than the front one. The femalesizes varies from 9.6 to 12.3 mm, the male ranges from 8.1 to 9.5 mm.

Eggs oval or ellipse. Egg sizes vary from 1.0 to 1.2mm long and 0.48 to 0.60mm wide. The larvae are 6 years old, strenuous to release silk pupae on the tree or under the vegetation and reproductive time is about 10 - 12 days. The pupae are iridescent dark brown nymphs, when they are about to turn into arms, they are dark brown in color, the length varies from 11.47 to 15.33 mm, the width is about 4.40 - 6.45 mm and the pupae lasts 8 - 9 days. The life cycle of *L. orbonalis* is about 25.30 ± 0.92 days. Average fertility of the female 249.5 ± 72.77 eggs, time to lay eggs at night.

Identify harmful levels, morphological and biological characteristics of *L. orbonalis* on eggplant in Trang Bom, Dong Nai.

References

[1] Alam SN., Rashid M.A., Rouf F.M.A., Jhala R.C., Patel J.R., Satpathy S., Shivalingaswamy T.M., Rai S., Wahundeniya I., Cork A., Ammaranan C. and Talkar N.S. 2003. *Development of integrated pest managenent strategy for eggplant fruit and shoot borer in South Asia*. Shanhua, Taiwan: AVRDC. Technical Bulletin No. 28. AVRDC Publication N. 03-548. 56 pp.

[2] Alam M.A. and Sana D.L., 1962. Biology of the brinal shoot and fruit borer, *Leucinodes orbonalis* G. (Pyralidae: Lepidoptera) in East Pakistan. *The Scientist*, 5:13-24.

[3] Aujla M.S., Thind H.S., Buttar G.S., 2007. Fruit yield and water use efficiency of eggplant *(Solanum melongema L.)* as influenced by different quantities of nitrogen and water applied through drip and furrow irrigation. *Scientia Horticulturae* 112: 142–148.

[4] Chen N. C. and Li H. M., 1995. *Cultivation and Breeding of eggplant*. Asian Vegetable Research and Development Center.

[5] Cork A., Alam S. N, Das A., Das C. S., Ghosh G. C., Farman D. I., Hall D. R., Maslen N. R., Vedham K., Phythian S. J., Rouf F. M. A. and Srinivan K., 2001. Female sex pheromone of Brinjal Fruit and Shoot borer, *Leucinodes orbonalis* blend optimization. *Journal of Chemical Ecology, 27 (9):*1867 – 1877.

[6] Daniela and Entomology, 2008. *Leucinodes orbonalis* (Lepidoptera: Pyralidae) Institute Università degli Studi di Milano.

[7] FAOSTAT data (2005) Agriculture: Agricultural Productions: Crops Primary. Last accessed 03/04/06. viewed on line at /faostat.fao.org/

[8] HossainM.M., ShahjahanM., SalamM.A. and BegumM.A., 2002. Screeuing of Some Brinjal Varieties and Lines Arinjal Shoot and Fruit Borer, *Leucinodes orbonalis* Guenée. *Pakitan Journal of Biological Sciences* 5 (10): 1032-1040.

[9] Kalloo G., 1993. Eggplant. In: G. Kalloo and B.O. Bergh (Eds.), Genetic Improvement of Vegetable Crops. Pergamon Press, Oxford, pp. 587–604.

[10] Mallik S. N., Kumar M., Sinha A. N. and Karn B. P., 1989. *Trathala flavoorbitalis* Cameron (Ichneumonidae) - aparasite of *Leucinodes orbonalis* Guen. From Bihar. *Current Science,* 58: 1098 - 1099.

IOP Conf. Series: Materials Science and Engineering **941** (2020) 012037 doi:10.1088/1757-899X/941/1/012037

Producing high-quality seeding material of Russian spring soft wheat varieties in Priobskaya zone of Altay forest steppes

S Zharkova[1], A Nechaeva[1], N Kiyan[1] and I Gefke[1]

[1] Altay State Agrarian University, 98, pr. Krasnoarmeysky, Barnaul, 656049 Russia
E-mail: ivgefke@mail.ru, stalina_zharkova@mail.ru

Abstract. A sufficient output of food and feed grain is one of priorities in Russian agribusiness. The reported study aims at the agrobiologic assessing spring soft wheat varieties in conditions of Priobskaya zone of Altay forest steppes. The experiments were carried out on production fields in 2018 – 2019. As an object of research 5 mid-ripening spring wheat varieties are taken. A maximum yield (2.1 t/ha) was produced when growing such varieties as Altayskaya zhnitsa and OmGAU 90. The variability of crops is reported to be low (Cv%<10). A high wheat gluten value is detected in the following varieties: Altayskaya zhnitsa (36.2%), Altayskaya stepnaya (36.9%), and OmGAU 90 (37.1%). A maximum of grain unit is given by Altayskaya zhnitsa (785 g/l). It is recommended to cultivate Altayskaya zhnitsa, OmGAU 90, Altayskaya 100 in Priobskaya zone of Altay forest steppes.

1. Introduction

To date, agricultural producers need to confront certain challenges caused by a difficult economic situation in the country. A current import substitution process is imposing its requirements on farmers. The output of agricultural products is to be based on domestically selected varieties and hybrids of cultivated crops.

For agricultural companies these targets are related to the searching for solutions to many problems. For instance, seeds used in production are to be of high quality and meet the State Russian Standard (GOST). High-quality seeding material is the basis for an efficient technological process and a high crop yield.

By nature, wheat is supposed to be an easily-grown crop, and plasticity makes its cultivation possible in many regions of Russia, irrespectively to soil and climate characteristics. The grain represents a principal product of wheat cultivation. The priority goal for agricultural producers in Russia remains a sufficient output of food and feed grain reserves.

In Siberian Federal District spring soft wheat is the most frequently cultivated crop with a total area of fields around 12 million ha. Wheat is a main grain crop of the region. The Altay Territory is a major granary in Siberia. Annually, an area of fields intended for wheat growing is around 2 million ha in the region [1].

Soil and climate conditions in the Altay Region are quite different. In view of these characteristics the territory is divided into 7 climatic zones with typical weather conditions and composition of soils. Generally, the climate in the Altay Region is considered as sharply continental.

Abiotic factors of the environment influence the growth and development of plants. Production use of domestically selected wheat seeds is seen as an element of agricultural technologies responsible for good harvest and high-quality grain. These varieties are adapted to cultivation conditions and have a positive response to all environmental characteristics of a zone [2].

Therefore, agricultural producers need varieties with a known response to cultivation conditions; moreover, it should be a positive one to provide a good and high-quality yield of a grown crop.

2. Materials and Methods

The study aims at the agrobiology assessing spring soft wheat varieties in Priobskaya zone of the Altay Territory. The objectives of the research are as follows: to determine a vegetation period for varieties, to evaluate a crop yield, to assess a grain quality in each variety, to find a variability range of crops using crop yield values, to detect factors influencing a crop yield in the zone of interest, and to give an economic assessment of the cultivating spring soft wheat varieties in Priobskaya zone of Altay forest steppes.

The experiment was carried out on production fields located in Priobskaya zone of Altay forest steppes in 2018 – 2019. Climate conditions in the region of experimenting are continental. A summer time is short with high temperatures; hot periods may turn into droughty ones unless there is no rain. Once we experimented, favorable weather conditions allowed evaluating objectively spring soft wheat varieties in Priobskaya zone of Altay forest steppes.

On the experimental territory a prevailing type of soils is regular slightly leached medium black-earth soils with a moderate concentration of loams and a humus horizon of 42 – 45 cm [3, 4].

As objects under study we took five mid-ripening varieties of spring soft wheat: Altayskaya 100, Altayskaya zhnitsa, Altayskaya stepnaya, OmGAU 90, and Svetlanka. Altayskaya 100 was accepted a standard variety. An area of a registration plot was 1 m^2, triple replication [5, 6]. A preceding crop was a grain one. A seeding rate amounted to 500 seeds per square meter. Soils were prepared for sowing in a way accepted in the region. Dates of sowing are as follows 16.05.2018 and 21.05. 2019. On days of sowing we carried out pre-seeding tillage. Within a vegetation period the weeding was done manually.

3. Results

A vegetation period represents a key characteristic of a variety. The crop phenology in the experiment was monitored over the entire plant vegetation period (Table 1).

Table 1. Development stage duration, days

Variety	Sowing – sprouts		Sprouts – tillering		Tillering – earing		Earing – waxy ripeness		Waxy ripeness – harvesting		Vegetation period, days		
	*	**	*	**	*	**	*	**	*	**	*	**	Average
Altayskaya 100	9	9	21	16	22	26	39	33	15	20	97	95	96.0
Altayskaya zhnitsa	9	9	21	16	23	28	38	32	15	19	97	95	96.0
Altayskaya stepnaya	9	9	21	16	24	26	41	35	12	17	97	94	95.5
OmGAU 90	9	9	21	16	22	26	40	34	14	19	98	95	96.5
Svetlanka	9	9	21	16	22	26	41	34	13	19	97	95	96.0
Average	-	-	-	-	-	-	-	-	-	-	97	95	96.0

*2018; ** 2019

A variety yield of all crops i.a. wheat is an integrated parameter highly sensitive to broadly ranging factors.

A variety yield is a sum total of several constituents, such as a number of productive stems per plant, and wheat spike characteristics: length, a number of spikelets per spike, a number of grains per spike and per plant, etc. Table 2 shows constituents of a yield formula.

Table 2. Yield formula of the varieties, 2018 – 2019

Variety/line	Number of productive stems pcs./pl.	Spike length, cm	Number of grains per		Grain weight per		Thousand kernel weight, g
			spike, pcs.	plant, pcs.	spike, g	plant, g	
Altayskaya 100	1.22	7.2	28.6	34.9	1.16	1.42	40.6
Altayskaya zhnitsa	1.41	7.8	32.7	46.1	1.26	1.78	38.5
Altayskaya stepnaya	1.35	6.8	26.7	36.0	1.05	1.42	39.3
OmGAU 90	1.46	7.3	28.0	40.9	1.18	1.72	42.1
Svetlanka	1.20	6.2	30.2	36.2	1.17	1.40	38.7
Average	1.33	7.1	29.2	38.8	1.2	1.42	40.6

Table 3 summarizes data on an experimental crop yield and its statistical processing.

Table 3. Crop yield of spring soft wheat, t/ha, 2018 – 2019

Variety	Year, t/ha						Average	
	2018	+/- vs. standard, t/ha	Cv%	2019	+/- vs. standard, t/ha	Cv%	2017-2018	Cv %
1*	1.97	-	5/9	1.73	-	8.8	1.9	9.5
2*	2.13	+0.1	5.4	2.00	+0.1	10.0	2.1	7.9
3*	2.23	0.0	6.8	1.77	+0.1	8.6	2.0	14.5
4*	2.17	+0.2	7.1	2.10	+0.3	4.8	2.1	5.7
5*	1.93	-0.2	3.0	1.67	-0.1	6.9	1.8	9.3
Average	2.09	-	-	1.85	-	-	2.0	
HCP_{05}, t/ha	0.22	-	-	0.27	-	-	0.22	-

*1 – Altayskaya 100; 2 – Altayskaya zhnitsa; 3 – Altayskaya stepnaya; 4 – OmGAU90; 5 – Svetlanka

To estimate how factors influence the variability of a crop yield we carried out a two-way analysis of variance (ANOVA) of yields, these varieties demonstrated over two years of experimenting (Figure 1).

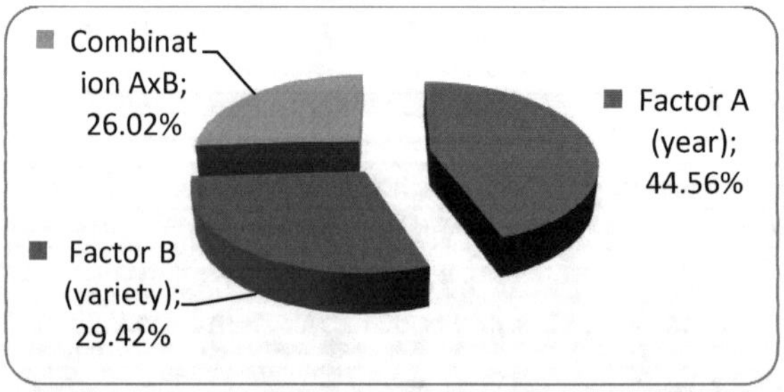

Figure 1. ANOVA outcomes based on a factor "crop yield" of spring soft wheat, 2018 – 2019

One of key quality characteristics relevant for baking properties of grain is gluten in endosperm (Table 4).

Table 4. Characteristics of varieties based on the grain quality, 2018 – 2019

Variety	Thousand kernel weight, g	Grain unit, g/l	Vitreousness, %	Gluten, %
Altayskaya 100	33.2	774	96.1	32.3
Altayskaya zhnitsa	35.1	785	96.1	36.2
Altayskaya stepnaya	35.4	769	98.3	36.9
OmGAU 90	34.4	765	98.4	37.1
Svetlanka	32.7	751	97.0	33.8

4. Discussion

A tillering – earing period of varieties was minimal both in 2018 and 2019, being 22 and 26 days, respectively (Table 1). The shortest earing – waxy ripeness period of 32 days was registered for a variety of Altayskaya zhnitsa in 2019 году. In 2019 this development stage of all varieties under study was shorter than in 2018 and ranged from 32 days (Altayskaya zhnitsa) to 35 days (Altayskaya stepnaya). On the contrary, this period was longer for all varieties in 2018, from 38 days (Altayskaya zhnitsa) to 41 days (Altayskaya stepnaya and Svetlanka). In our opinion, it was a result of weather conditions in the years of interest: in 2018 an amount of precipitation was bigger than in the droughty year 2019. The varieties were harvested simultaneously on the same day: 29.08 in 2018 and 02.09 in 2019.

A vegetation period of all crops was different, depending both on a variety and a year of consideration (Table 1). In the droughty 2019 the ripening of plants was quicker. In 2019 all varieties were more fast-growing than in 2018. An average vegetation period in 2019 was 2 days shorter in comparison with 2018, being 95 and 97 days, respectively.

In 2019 Altayskaya stepnaya demonstrated a shorter vegetation period (94 days), a standard – 95 days. In 2018 a vegetation period was 2 – 3 days longer than in 2019, approximately 97 – 98 days. Generally, all varieties under study belong to an early-ripening maturity group in view of a vegetation period.

A productive tillering capacity of a variety develops over the entire period of vegetation under the influence of hydrothermal conditions in the cultivation region. Varieties in the focus of our study developed low but good tillers (Table 2). A value of productive tillering ability was in a range from 1.20 pcs. per plant (Svetlanka) to 1.46 pcs. per plant (OmGAU 90), an accepted standard – 1.22 pcs. per plant. Therefore, a high average tillering ability within a period of consideration was formed by plants of Altayskaya zhnitsa – 1.41 pcs. per plant and OmGAU 90 – 1.46 pcs. per plant, an accepted standard – 1.22 pcs. per plant.

A spike length is an important parameter for many constituents of a yield crop since it is a spike where these elements form. In view of this characteristic Altayskaya zhnitsa and OmGAU were above the standard (7.2 cm), being 7.8 cm and 7.3 cm, respectively. In general, varieties formed a medium length spikes. The variability was recorded ± 0.5…1.6 cm.

A crop yield is closely related to such factor as a number of grains per spike. In our case, there were 29.2 pcs. per spike over the period of experimenting. The deviation from an average value shown by varieties in the experiment varied from 85.6% (OmGAU 90) to 113.1% (Svetlanka). Altayskaya zhnitsa with 32.7 pcs. per spike and Svetlanka with 30.2 pcs. per spike exceeded an average of 29.2 pcs. per plant; furthermore, a value of Altayskaya zhnitsa was maximal in the experiment. Thus, a number of grains per spike in tested varieties ranges from 26.7 pcs. per spike (Altayskaya stepnaya) to 32.7 pcs. per spike (Altayskaya zhnitsa).

A grain weight per spike is an integral characteristic depending on such constituents of a crop yield formula as a spike length, a number of grains per spike and plant, and a thousand kernel weight. An average grain weight per spike registered during 2 years of experimenting was as high as 1.0 g and varied from 1.05 g (Altayskaya stepnaya) to 1.26 g (Altayskaya zhnitsa). A grain weight per plant of Altayskaya zhnitsa was 1.26 g, being a maximal value in the experiment.

A parameter "thousand kernel weight" gives information on grain fineness and a concentration of substances a grain contains. The varieties formed well-filled grains. An average thousand kernel weight within a period of experiment was as high as 40.6 g. OmGAU 90 developed the heaviest grain (42.1 g). This parameter varied from 38.5 g (Altayskaya zhnitsa) to 42.1 g (OmGAU 90).

As seen in the data on a crop yield (Table 3), in 2018 conditions relevant for this parameter were better than in 2019. An average crop yield of varieties in 2018 was 2.09 t/ha, being 14.5% higher than the same characteristic in 2019 (1.85 t/ha).

In 2018 a crop yield varied from 1.93 t/ha (Svetlanka) to 2.23 t/ha (Altayskaya stepnaya). Altayskaya stepnaya produced a maximal crop yield in 2018 and was above the standard (1.97 t/ha) in view of this characteristics. At the standard there was OmGAU 90 (2.17 t/ha) and Altayskaya zhnitsa (2.13 t/ha). A variability of the parameter "crop yield" for all varieties in the experiment was low (Cv%<10), the most stable variety was determined to be Svetlanka (Cv% = 3.0%).

In 2019 a crop yield of varieties was 3-21% lower than in 2018. A crop yield of the standard variety in 2019 (1.73 t/ha) is 11.9% lower than in 2018 (1.97 t/ha). That year OmGAU 90 produced a high crop yield (2.10 t/ha), exceeding the standard (1.73 t/ha) in view of this parameter, the difference with a standard was as much as 0.3 t/ha. Four varieties of five under consideration demonstrated a low coefficient of variation (Cv%<10), they are Altayskaya 100, Altayskaya stepnaya, OmGAU 90, and Svetlanka. Altayskaya zhnitsa had a medium variation level 10>Cv%<20.

An average crop yield within two years of experiments was as high as 2.0 t/ha. No variety was above the standard crop yield (1.9 t/ha). Altayskaya zhnitsa and OmGAU 90 produced a maximal crop yield (2.1 t/ha). The variability of varieties is low (Cv%<10) except for Altayskaya stepnaya (Cv=14.5%).

When exploring an effect of factors on a crop yield of varieties certain trends were found (Figure 1). Such factor as "year" plays the most significant role for a crop yield – 44.56%, the second important factor is "variety" – 29.42%, and their combination results in 26.02%.

A level of gluten in all varieties was above 30% (Table 4). This factor varied from 32.3% (Altayskaya 100) to 37.1% (OmGAU 90). High values of gluten were found in Altayskaya zhnitsa (36.2%), Altayskaya stepnaya (36.9%), and OmGAU 90 (37.1%).

The vitreousness characterizes structural and mechanical properties of grain endosperm. A level of vitreousness in the study was approximately 96-98%. OmGAU 90 and Altayskaya stepnaya showed a maximal level of vitreousness 98.3 – 92.4%.

A grain unit of varieties was quite high and allows producing a good output of flour. Altayskaya zhnitsa had a maximal grain unit (785 g/l) by contrast to the standard (774 g/l). Therefore, Altayskaya zhnitsa and OmGAU 90 have better grain quality.

Conclusion

In general, all varieties under study are early-ripening in view of a vegetation period. A high tillering potential within a period of consideration was shown by plants of Altayskaya zhnitsa – 1.41 pcs. per plant and OmGAU 90 – 1.46 pcs. per plant, an accepted standard – 1.22 pcs. per plant. A number of grains per spike in tested varieties were 29.2 pcs. per spike within a period of experiment. OmGAU developed the heaviest grain (42.1 g). The value varied from 38.5 g (Altayskaya zhnitsa) to 42.1 g (OmGAU 90).

An average crop yield of varieties in the years of experiment was 2.0 t/ha. No variety was better than the standard (1.9 t/ha) according to this parameter. A crop yield of all varieties is at the standard. A maximal output was produced by Altayskaya zhnitsa and OmGAU 90 (2.1 t/ha). Except for Altayskaya stepnaya (Cv=14.5%) the variability of varieties is low Cv%<10. A crop yield is most sensitive to such factor as "year" – 44.56%, the second important factor is "variety" – 29.42%, and their combination gives 26.02%.

A level of gluten was above 30%. This factor varied from 32.3% (Altayskaya 100) to 37.1% (OmGAU 90). High values of gluten were found in Altayskaya zhnitsa (36.2%), Altayskaya stepnaya (36.9%) and OmGAU 90 (37.1%). A level of vitreousness in the study was approximately 96 – 98%.

OmGAU 90 and Altayskaya stepnaya showed a maximal level of vitreousness 98.3 – 92.4%. Altayskaya zhnitsa had a maximal grain unit (785 g/l) by contrast to the standard (774 g/l).

To summarize, such varieties as Altayskaya zhnitsa, OmGAU 90, and Altayskaya 100 are recommended for growing in Priobskaya zone of Altay forest steppes to provide the efficient production of high-quality spring soft wheat grain.

References

[1] P V Klyushin *et al* 2019 *IOP Conf. Ser.: Earth Environ. Sci.* **350** 012006
[2] N V Rygalova *et al* 2019 *IOP Conf. Ser.: Earth Environ. Sci.* **395** 012046
[3] A Kudryavcev *et al* 2019 *IOP Conf. Ser.: Earth Environ. Sci.* **395** 012016
[4] S Zavalishin and V Karelina 2019 *IOP Conf. Ser.: Earth Environ. Sci.* **395** 012014
[5] Dospekhov B A 1985 *Methods of field experience (with the basics of statistical processing of research results)* (Moscow, USSR: Agropromizdat)
[6] Methods of state testing agricultural crops. – M., 1985. – 257 p. (*in Russian*)

IOP Conf. Series: Materials Science and Engineering **941** (2020) 012038 doi:10.1088/1757-899X/941/1/012038

Increasing gross grain harvest in the southern forest-steppe zone of the Novosibirsk region

B D Dokin[1], O V Elkin[1], E A Balushkina[1] and A A Aletdinova[2]

[1]Siberian Research Institute of Mechanization and Electrification of Agriculture, Krasnoobsk, Krasnoobsk settlement, 630501, Russia
[2]Novosibirsk State Technical University, 20, Karl Marx Avenue, Novosibirsk, 630073, Russia

E-mail: sibime@ngs.ru, aletdinova@corp.nstu.ru

Abstract. The article analyzes the features of grain production in the Novosibirsk region of Russia. The authors analyzed the development of innovative technologies in farms. Unfortunately, extensive technology is still the most common. In the Novosibirsk region, 60% of agricultural enterprises use extensive technology and create 49% of the gross grain harvest with an average yield of 14.3 hundredweight/ha, 20% of farms create 26% of the gross grain harvest with a yield of 23.3 hundredweight/ha, 11% of farms create 21% of the gross grain harvest with a yield of 33.4 hundredweight/ha. The authors chose the Novomayskoye farm in the southern forest-steppe zone of the Novosibirsk region as a model. The simulation results were tested in practice, they showed that the transition to intensive technology with direct seeding (No-till) required an increase of 1.5 times the technical support and costs for mineral fertilizers and plant protection products, i.e. from 6 thousand rubles/ha of wheat crops to 9 thousand rubles/ha (in 2020, 1 USD≈65 Rub), but this allowed to reduce fuel consumption by 30 – 35% and the need for machine operators by 1.7 – 2 times.

1. Introduction
Production intensification is an economic process based on advanced achievements and more effective use of technical, material, labor, and information resources, as well as a higher level of production organization. Scientists show their interest in scientific and technical progress in crop production in many publications [1 –4]. The growth of crop yields and increase in their production is influenced by:
- selection of the technological map and the set of available resources [5–7];
- costs that a farmer can afford [8–11];
- management decisions, including the use of operational information [12, 13];
- formed machine and tractor fleet [14–17].

Thus, the change in cultivation technology leads to changes in resource requirements for production and production volumes. The purpose of this paper is to analyze this transition and justify the need for new wheat production technologies for the southern forest-steppe zone of the Novosibirsk region of Russia.

2. Materials and methods

2.1. Features of grain production in the Novosibirsk region of Russia

The Novosibirsk region is located in Russia in the South-East of the West Siberian plain. According to the Ministry of agriculture of the Novosibirsk region in 2019, the farms of the Novosibirsk region sown area was 2225.7 thousand hectares, including cereals and legumes sown 1416 thousand hectares. When sowing these crops, 271 thousand tons of seeds were used, their air-conditioning capacity was 97.3%. The share of original and elite seeds in the total volume of sown seeds – 8.2%, reproduction seeds – 47.5%. The gross harvest of grain in weight after completion in farms of all categories was 2431.5 thousand tons (97.5% compared to the level of 2018) with a yield of 17.2 hundredweight/ha [18].

In the southern forest-steppe zone, about 24% of agricultural enterprises have an area of grain crops from 6 to 16 thousand hectares, they specialize in the production of grain crops and dairy and meat cattle breeding. These are fairly large agricultural enterprises with a population of 2000 cattle, including a herd of 950 dairy cows. The size of the field is from 60 to 120 ha, the length of the rut is more than 1000 m, the distance of transportation is 33 km.

Extensive (herbicide-free) technologies are a system for obtaining grain with the maximum use of available soil fertility without restoring it. Normal technologies maintain a level of chemicals that allows the development of soil protection systems of agriculture, the average level of soil cultivation, they can eliminate the deficit of mineral nutrition elements that are at a critical minimum, and give a satisfactory product quality (i.e., a yield of 18–22 hundredweight/ha).

Intensive technologies are used to obtain the highest possible yield of high quality in the prevailing meteorological conditions in a given field. Their implementation requires continuous management of the production process of agricultural crops, which will ensure optimal mineral nutrition of plants and protection from harmful organisms and lodging. For intensive technologies, it is necessary to use intensive varieties and create conditions for more complete realization of their biological potential (then it is possible to yield more than 32 hundredweight/ha). These technologies can be based on dump plowing, on the basis of free-fall loosening, and on the basis of minimal and zero tillage. The element of intensive "No-till" technology is direct seeding.

The authors also highlight a haphazard technology that is used in some farms of the Novosibirsk region, it has no scientific justification and regularities in the stages.

Let's analyze the use of these technologies in the Novosibirsk region (table 1).

Table 1. Grouping of agricultural organizations and farms in the Novosibirsk region on the use of advanced technologies for the production of cereals and legumes.

Technologies	Agricultural organizations,		Farms		Application of technologies on acreage	
	number	%	number	%	tons of ha	%
Extensive	255	62	686	75.6	1277	59
Normal	72	17	108	11.8	402	18
Intensive	30	7	16	1.8	281	13
Intensive with elements of precision farming	3	0.07	0	0	24.8 (ha)	
Unsystematic	55	13	98	10.8	213	10
Total	412	100	908	100	2173	100
Of them, with the "No till"	32	8	20	2	126.0	5.7

IOP Conf. Series: Materials Science and Engineering **941** (2020) 012038 doi:10.1088/1757-899X/941/1/012038

In total, in the Novosibirsk region, most of the acreage is cultivated using extensive technologies. Although there is the use of intensive technologies with elements of precision farming.

2.2. Methods

The prediction method of the variations of annual field work complexes is the basis of the method for substantiating the perspective compositions of a machine and tractor fleet of model farms. The principle of forming annual field work complexes is described in the authors ' article [19].

Calculations can be performed for a model of agricultural enterprises that contains characteristics typical for most farms in the Novosibirsk region. The southern forest-steppe zone affects the choice of crop rotation, productivity and recommended technological maps of crop production.

Part of the values are given in the Russian Rubles (in 2020, 1 USD≈65 Rub).

3. Results and discussion

The Novomayskoye farm of the Krasnozersky district of the Novosibirsk region was taken as a model. The farm has the following structure of acreage: total crops – 20529 ha, including cereals – 15822 ha: spring wheat – 12493 ha, spring barley – 932 ha, spring oats – 953 ha, legumes (peas) – 1444 ha, annual grasses – 790 ha, perennial grasses – 1878 ha, pairs – 2039 ha. There are 4 John Deere tractors of the ninth series, one HewHolland and one K-744R tractors, 18 MTZ-80 and MTZ-82 tractors, 4 sowing complexes and 3 trailer sprayers in the machine-tractor fleet for grain cultivation. Over the past four years, the yield of wheat was from 17 to 24 hundredweight/ha. With such a yield, it is impossible to create after harvesting the necessary layer of crushed straw to cover the field to protect the soil from moisture loss.

The farm used normal technology, but it was necessary to switch to intensive technology with a yield of 32-34 hundredweight/ha, requiring other technical and technological support (timing of field work, nutrition and plant protection). The authors modeled the structure of the machine and tractor fleet based on the prediction method of the variants of annual field work complexes [2]. The technological support is based on the recommendations of the Institute of agriculture and chemization of the Siberian Federal scientific center for agrobiotechnologies of the Russian Academy of Sciences. The authors assessed the need for tractors and combines (table 2).

Table 2. Need for machines, tractors and combines.

Brands of equipment	Need, units
JohnDeere	6
NewHolland	1
MTZ -80 /82	5
John Deere seed complexes	6
John Deere 4730 self-propelled sprayer	1
OP-2000 sprayer	5
Hardi Navigator 300 sprayer	1
Amazone UX5200 sprayer	1

The cost of protective equipment amounted to 9063.83 rubles/ha, the remaining costs are shown in table 3.

Table 3. Operating costs for wheat cultivation using intensive technology with "No-till".

Types of costs	Cost, thousand rubles
Fuels and lubricants	7058.057
Salary	198.154
Depreciation and maintenance	17647.587
Purchase of protective equipment and fertilizers	113225.728
Buying seeds	39679.017
Total costs	177808.500
Cost per ha	14.232
Machine operators, people	12

Sowing complexes worked on satellite navigation in two shifts and sowed 200 hectares per day, while K-744R3 with SZP-3.6 seeders sowed only 50 ha per shift. The transition to intensive technology with direct seeding (No-till) required an increase of 1.5 times in technical support and costs for mineral fertilizers and plant protection products: from 6 thousand rubles per 1 ha of wheat crops to 9 thousand rubles per 1 ha. In 2019, Novomayskoe farm received a wheat crop of 24 hundredweight/ha. At the same time, the transition to intensive technology with direct seeding allowed to reduce fuel consumption by 30 – 35% and reduce the need for machine operators by 1.7 – 2 times. The neighboring farm of Alexander Weiss followed the example of the farm "Novomayskoe", it was able to get a wheat yield of 34 hundredweight /ha using intensive technology with direct seeding (No-till).

Thus, intensive technologies ensure maximum crop yield and maximum gross harvest. The use of unsystematic technologies, the lack of calculations of the composition of the machine and tractor fleet leads to losses. For example, in 2017, farms in the Novosibirsk region lost 6 hundredweight of crop per hectare of wheat due to a lack of sprayers.

4. Conclusion

The most common technology of grain crops in the Novosibirsk region is extensive. Scientists come to the aid of agricultural producers and calculate the optimal structure of crops and machine and tractor fleet. Calculations for the Novomayskoe farm as a model were used in production and tested in practice. To achieve maximum productivity, it was necessary to increase technical support, costs for mineral fertilizers and plant protection products.

When switching from normal grain cultivation technology to intensive technology with "No-till", it is necessary to create a layer of straw and cover the field with it after harvesting the grain. When switching to intensive technology, the cost of chemical treatment was 9-10 thousand rubles per 1 ha of wheat crops, and the load on one tractor is 2 thousand hectares of wheat crops for a 10-day sowing period.

References

[1] Reddy P P 2016 *Sustainable intensification of crop production* (Singapore: Springer Singapore) pp 241-252

[2] Pellegrini P and Fernández R J 2018 *Proceedings of the National Academy of Sciences* **115(10)** 2335-40

[3] Kipriyanov F A and Savinykh P A 2019 *EurAsian Journal of BioSciences* **13** 1651-58

[4] Rockström J, Williams J, Daily G, Noble A, Matthews N, Gordon L et al 2017 *Ambio* **46(1)** 4-17

[5] Shrestha J and Subedi S 2019 *South Asian Research Journal of Agriculture and Fisheries* **1(1)** 8-11

[6] Värnik R, Aste R and Ariva J 2017 *Precision Agriculture: Technology and Economic Perspectives* (Springer Cham) pp 201-21

[7] Kleijn D, Bommarco R, Fijen T P, Garibaldi L A, Potts S G and van der Putten W H 2019 *Trends in ecology & evolution* **34(2)** 154-66

[8] Lin M and Huang Q 2019 *Agriculture, ecosystems & environment* **274** 33-40

[9] Friedrich T and Kassam A 2016 *ASABE Annual International Meeting American Society of Agricultural and Biological Engineers* (Michigan: St. Joseph) p. 1

[10] Capellesso A J, Cazella A A, Schmitt Filho A L, Farley J and Martins D A 2016 *Agroecology and Sustainable Food Systems* **40(3)** 215-36

[11] Antonini C and Argilés-Bosch J M 2017 *Journal of cleaner production* **140** 796-803

[12] Surve U S, Dhonde A S, Kumbhar J S and Bhosale P U 2019 *IJCS* **7(3)** 399-402

[13] Kamilaris A, Kartakoullis A and Prenafeta-Boldú F X 2017 *Computers and Electronics in Agriculture* **143** 23-37

[14] Dokin B D, Aletdinova A A, Kravchenko M S and Tsybina T S 2018 *J. of Phys.: Conf. Ser.* **10015** 032005

[15] Korotchenya V M, Lichman G I and Smirnov I G 2019 *Agricultural Machinery and Technologies* **13(1)** 14-20

[16] Viesturs D and Kopiks N 2017 *Proc. Int. Conf. on Engineering for Rural Development* vol 16 (Latvia: Springer) pp 534-539

[17] Terentyev S E, Gnezdova J V and Semchenkova S V 2020 *E&ES* **459(6)** 062060

[18] Website of the Ministry of agriculture of the Novosibirsk region: http://mcx.nso.ru/page/444

[19] Dokin B, Elkin O and Aletdinova A 2019 *IOP Conference Series: Earth and Environmental Science* **341(1)** 012111

Intensification of secondary dampening of grain before milling

N S Protasov, I Y Fedorenko, S Y Buzoverov and V V Sadov
Altai State Agrarian University, 98 Krasnoarmeyskiy Prospekt, Barnaul, 656049, Russia
E-mail: protasow.n@mail.ru

Abstract. This article describes the problems encountered by grain processing industries. In this regard, we are looking for methods of intensive grain dampening, allowing to reduce the technological cycle of flour production by reducing the time of softening. The article also considers the process of secondary dampening of grain before milling by using pulsating vacuuming and subsequent pulsing pressure. An experimental unit for grain dampening before milling is presented and the basics of its operation are described. In addition, the paper provides the results of the experimental study of wheat grain dampening using pulsating vacuum and subsequent pulsating pressure, and provides control results of the experimental study of wheat grain dampening without vacuum and pressure. The results of the experimental study were analysed; according to said analysis it is evident the increase in moisture content of wheat grain in case of dampening with the use of pulsating vacuum and subsequent pulsating pressure is more intense than in case of dampening without the use of vacuum and pressure, which leads to a two-fold reduction of the softening time. The dependence between the dampening increment ΔW and the grain processing time t in the experimental dampening unit was determined. Having analysed this work, were made our conclusions that are presented in the article as well.

1. Introduction

The humidity of grain coming for processing is usually low, while the structural and mechanical properties of the endosperm and shells differ slightly. Therefore, it is difficult to separate them, and the grain gets poorly processed, i.e., the output and quality of the finished product do not meet the requirements of the "Rules for Organizing and Conducting the Technological Process at Mills" [1].

To change the technological properties of the grain, various methods of hydrothermal treatment (HTT) may be used. The first priority of the mills using HTT is to enhance the difference between the properties of the shells and the endosperm [2].

Intensity of grain dampening significantly affects the the effectiveness of HTT process. In this regard, we are looking for ways of intensive grain dampening that would allow us to reduce the technological cycle of flour production by reducing the time of softening, while maintaining the product yield and the quality of the produced flour at the same level [3].

2. Research material and methods

Based on the above, the aim of our work is "to increase the effectiveness of grain dampening before milling using variable pressure in the dampening chamber".

The device proposed by us (utility model patent No. 171424) improves the quality of dampening, while being small in size and requiring not much supply power [4].

This problem can be solved by using pulsating vacuuming of the grain before dampening, after which the actual dampening takes place and subsequent pulsating pressure in the dampening chamber occurs [5].

For the purpose of the experiments, the following experimental setup was developed and constructed; this setup is intended primarily for the study of the effect of variable vacuum before dampening and variable pressure after dampening on the intensity of moisture content increase in wheat (figure 1).

In order to reduce the loss of pressure and vacuum, in this experiment, we used an installation, in which grain was loaded and unloaded in portions; the loading and unloading device securely sealed the working area of the auger conveyor from the environment. In terms of process conditions, the dampening chamber can be divided into three parts. The first part generates pulsating vacuum, in which the grain is treated by pulsating vacuum; the second part dampens the grain at a constant vacuum, which prevents moisture entrainment by the vacuum pump; the third part of the chamber generates pulsating pressure and intensifies grain dampening. In this case, the grain is dampened constantly and not in portions.

In addition, we chose a dampening chamber of a small size in order to avoid the "receiver effect" that smoothens the pulsations of vacuum and pressure, which reduces the intensity of grain dampening and can make the proposed measures useless.

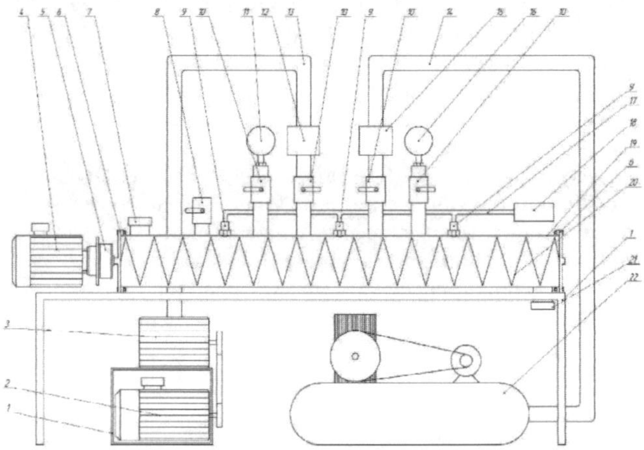

Figure 1. Experimental installation for grain dampening.

1 – frame, 2 – electric motor, 3 – vacuum pump, 4 – electric motor, 5 – reducer, 6 – flange, 7 – loading hole, 8 – air venting cock in the auger conveyor, 9 – nozzle, 10 – cock, 11 – vacuum meter, 12 – vacuum pulsator, 13 – vacuum pipeline, 14 – pressure pipeline, 15 – pressure pulsator, 16 – pressure gauge, 17 – water pipeline, 18 – dosing pump, 19 – housing, 20 – spiral auger conveyor, 21 – discharge hole, 22 – compressor.

The experimental installation works as follows. At the start of the experiment, we need to determine the moisture content of the grain that we need to moisten to 16.4% in order to understand, what amount of water is required for dampening. Based on the amount of water identified, we shall set the time for water injection into the dampening chamber. After that, we pour the grain into the dampening chamber and level it with the auger, then turn on the vacuum pump and treat the grain with pulsating vacuum using the vacuum pulsator. During the experiment, the auger constantly stir the grain in the chamber for better effect of vacuum, pressure and water mist during dampening of each grain. In order to avoid entrainment of a part of moisture through the vacuum pump, we need to set a permanent vacuum and dampen the grain by injecting water in the form of mist into the chamber through nozzles for a specified time. Next, we set the pulsating pressure in the chamber using the compressor and pressure pulsator; the set time must be observed as well. At the end of the experiment, we pour the grain with the auger conveyor into the dampening chamber and determine its moisture content [6].

3. Research results

The figure 2 shows the results of the experimental study of wheat grain dampening using pulsating vacuum and subsequent pulsating pressure, and provides control results of the experimental study of wheat grain dampening without vacuum and pressure. The numerical value of the grain moisture content obtained through wheat dampening without the use of pulsating vacuum and subsequent pulsating pressure is given in table 1.

Table 1. Grain dampening without the use of pulsating vacuum and subsequent pulsating pressure

Experiment	Wheat grain moisture content (%)		
number	Before dampening	After dampening	Note
1	12.44 ; 12.34	15.90 ; 15.90	Dampening time: 30 sec.
2	12.70 ; 12.98	15.99 ; 16.36	Dampening time: 1 min.
3	12.54 ; 12.48	16.03 ; 16.06	Dampening time: 2 min.
4	13.16 ; 13.25	16.04 ; 16.17	Dampening time: 3 min.
5	13.11 ; 13.27	16.25 ; 16.29	Dampening time: 4 min.
6	13.16 ; 13.38	16.19 ; 16.40	Dampening time: 5 min.
7	13.07 ; 13.19	16.21 ; 16.35	Dampening time: 6 min.
8	12.84 ; 13.05	16.31 ; 16.38	Dampening time: 7 min.

The numerical value of the grain moisture content obtained through wheat dampening with the use of pulsating vacuum and subsequent pulsating pressure is given in table 2.

Table 2. Effect of pulsating vacuum () up to -60kPa and subsequent pulsating pressure () up to 7kg/cm^2 on wheat grain moisture content

Experiment	Wheat grain moisture content (%)		
number	Before dampening	After dampening	Note
1	13.15 ; 13.29	16.92 ; 17.01	– 30 sec. ; – 30 sec.
2	13.51 ; 13.45	17.01 ; 16.64	– 1 min. ; – 1 min.
3	13.33 ; 12.91	17.16 ; 16.90	– 2 min. ; – 2 min.
4	13.27 ; 13.24	17.90 ; 17.71	– 3 min. ; – 3 min.
5	13.19 ; 12.94	17.72 ; 17.39	– 4 min. ; – 4 min.
6	13.42 ; 13.29	17.62 ; 17.52	– 5 min. ; – 5 min.
7	13.16 ; 13.31	17.80 ; 18.00	– 6 min. ; – 6 min.
8	12.52 ; 12.97	18.37 ; 18.18	– 7 min. ; – 7 min.

According to the analysis of the experimental study, generation of pulsating vacuum and subsequent pulsing pressure in the dampening auger conveyor significantly speeds up the process of wheat dampening, and, according to the graph, the grain reaches the optimum moisture content of 16.4% within the first 15-30 seconds, whereas when grain is dampened in the same auger conveyor without pulsating vacuum and pulsating pressure, wheat grains reach said parameters only by the eighth minute.

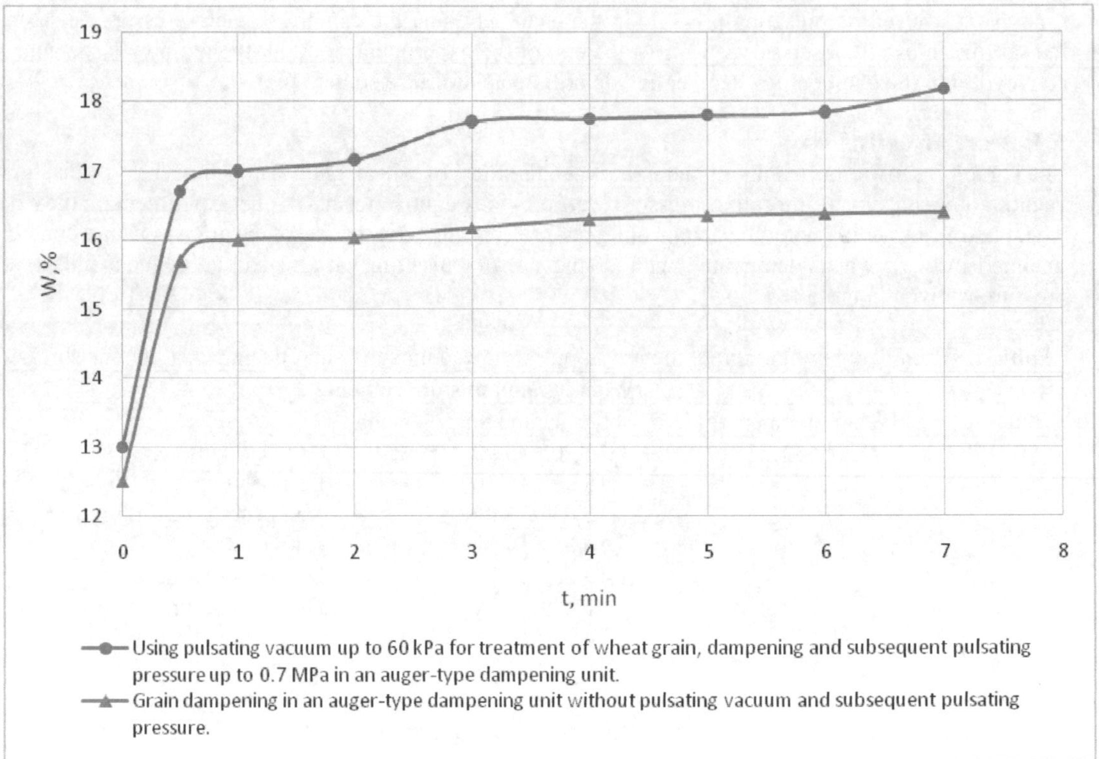

Figure 2. Graph of grain dampening at the experimental installation

According to the analysis of the experiment results, it is evident that the increase in moisture content of wheat grain in case of dampening with the use of pulsating vacuum and subsequent pulsing pressure is more intense than in case of dampening without the use of vacuum and pressure.

Let us determine the dependence between the dampening increment ΔW and the grain processing time t in the experimental dampening unit. The correlation and regression analysis shows that the following dependencies approximate the observation data quite well:

–linear dependence $y = 3.012x - 9.3984 (R^2 = 0.747)$

–exponential dependence $y = 0.024e^{1,094x}$ ($R^2 = 0.635$)

–second degree polynomial $y = 0.3516x^2 + 6.1407x - 16.211$ ($R^2 = 0.7552$)

Any of them can be used for forecasting, but the linear one is preferable (figure 3).

So, if the grain received for processing has the moisture content of 13.2%, and it needs to be moistened to 16.6%, then it will take $y = 3.012 * (16.6 - 13.2) - 9.398 = 0.84$ minutes to treat the grain.

The correlation coefficient is 0.865, which indicates a close dependence of the grain moisture change on the time of its processing in vacuum.

The dispersive analysis suggests that by using a linear dependence, the risk of getting an incorrect forecast is 0.002.

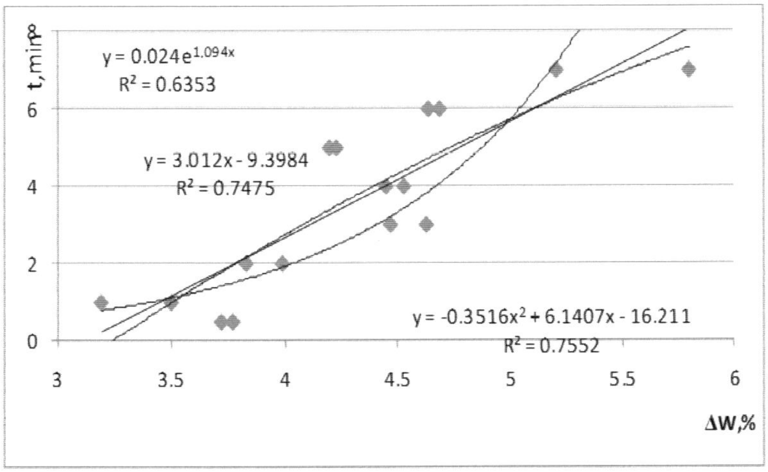

Figure 3. Dampening linear dependence

4. Conclusion

1. The experimental study has shown that the proposed dampening system reduces the time of grain dampening by 2 times.
2. The correlation and regression analysis shows that the following dependencies approximate the observation data quite well: linear dependencey $= 3.012x - 9.3984$ $(R^2 = 0.7475)$; exponential dependence $y = 0.024e^{1,094x}$ $(R^2 = 0.6353)$; second-degree polynomial $y = 0.3516x^2 + 6.1407x - 16.211$ $(R^2 = 0.7552)$.

References

[1] Rules for Organizing and Conducting the Technological Process at Mills Part 1 1991 Moscow: VNPO [All-Russian Research and Production Association] "ZERNOPRODUKT" [GRAIN PRODUCT] p. 74
[2] I A Naumov 1975 Improving Conditioning and Milling of Wheat and Rye Moscow: Kolos p. 76
[3] A Y Sokolov 1975 Technological Equipment for Grain Storage and Processing Enterprises. Moscow: Kolos p. 52
[4] Patent No. 171424. Grain Dampening Device Russian Federation IPC B02B 1/04 (2006.01) / Buzovcrov, S. Y., Protasov N. S.; applicant and patentee Protasov, N. S. No. 2016145712; application dd. 22/11/2016; publication on 31/05/2017, Information Sheet No. 16.
[5] S Y Buzoverov, N S Protasov 2015 On the Issue of Upgrading the Dampening Unit for Hydrothermal Grain Treatment *Bulletin of the Altai State Agrarian University* **12 (134)** 139-144
[6] S Y Buzoverov, V I Lobanov, N S 2018 Protasov Influence of the Degree of Grain Moisture in the Process of Hydrothermal Treatment on the Quality and Yield of Flour *Bulletin of the Altai State Agrarian University* **1 (159)** P. 172-176

IOP Conf. Series: Materials Science and Engineering **941** (2020) 012040 doi:10.1088/1757-899X/941/1/012040

The Technology of Dehydrated Soup Base from Poultry Meat and Bone Residue

S V Polyanskikh[1], **M M Danyliv**[1], **U R Dubrovina**[1], **O N Ozherelyeva**[1], **O A Vasilenko**[2]

[1]Voronezh state university of engineering technologies, 19, Revolyutsii Avenue, 394036, Voronezh, Russia
[2]Voronezh state agrarian university named after emperor Peter the Great, 1 Michurina Street, 394087, Voronezh, Russia

E-mail: max-dan@yandex.ru

Abstract. This research paper describes the technology for obtaining dehydrated soup base from the meat and bone residue obtained during the primary processing of poultry. The authors proposed a rational cutting of poultry and assessed the chemical composition of the meat and bone residue; during experimental studies it was found that a significant amount of fat predominates in the bone residue, where skin from the neck was used and makes up 21.3%, the protein remained at the level of 22.7 % Water-soluble proteins predominate in the meat and bone residue and make up 52.3 and 63.1% for the residue from the carcass and the residue from the carcass and skin, respectively, the mass fraction of alkali-soluble proteins is 23.5 and 17.6%, respectively. The total mass fraction of essential amino acids is lower than in minced meat, the limiting amino acids are lysine (71%), leucine (69%), and threonine (68%). To organize the convection drying, we used a dryer with a liquid vibro-boiling bed.

1. Introduction

Currently, significant volumes of poultry and poultry products are recorded in the Russian Federation, the USA, India, the EU countries, and many others, which undoubtedly leads to the achieving the food security in our and other countries. Such large Russian manufacturers as Prioskolye CJSC, BEZRK-Belgrankorm and many others continue to successfully develop in the market. At the same time, production growth leads to the formation of a significant amount of poultry slaughter products and other non-food waste from the poultry processing industry and other food industries, which can lead to certain environmental problems and pollution [1, 2, 3].

Assessing the research potential of the leading Russian scientists, we can state the fact that there are many technologies for processing waste obtained from poultry slaughtering, most of which are aimed at the manufactruring of B2B products, while the technologies that implement waste processing for food purposes are very insignificant.

According to experts of *businesstat*, in 2016-2020, the dynamics of sales of soups and broths in liquid form in the country will be multidirectional. Thus, in 2020, sales are expected to decrease by 14%, after which the market growth will begin. In 2020, the market capacity for soups and broths in liquid form in Russia will be 4.1 thousand tons, which is 21% higher than the similar indicator in 2019. The concept of the poultry farming development in the Russian Federation for the period of 2013-2020 (hereinafter referred to as the Concept) defines the main directions of the poultry farming development for the long term. Its implementation is aimed at satisfying the population's needs for poultry products to the level of the recommended consumption standards by increasing the production of poultry meat up to 4.5 million tons, and over 50 billion of eggs as well.

To improve the efficiency of production at the poultry processing companies, it is advisable to deeply process and introduce new progressive methods for cutting and deboning carcasses based on the rational use of poultry meat. The use of such technology helps to eliminate the negative effects of the defects in poultry farming, slaughter, and transportation. The most valuable parts of the carcass are sent to the semi-finished products and finished products manufacturing; the remaining parts with high bone share, less valuable in nutritional terms (bone carcasses, backs, necks, legs, wings) - for mechanical boning [4, 5, 6].

2. Materials and methods

As the objects of study, the meat and bone residue obtained during the cutting of poultry and the production of mechanically deboned poultry meat obtained in the conditions of the Russian poultry production was selected. When developing the technology for the production of dehydrated soup base, experimental studies were carried out under the conditions of the research laboratory of the Department of Animal Products Technology at VSUET. The cutting scheme was carried out according to GOST 31962-2013 "Chicken meat (carcasses of hens, chickens, broiler chickens, and their parts) and TU 9214-212-23476484-2010 "Natural semi-finished products from broiler chicken meat" during the production of semi-finished products. The determination of the total chemical composition of meat products was carried out according to generally accepted procedures: determination of the mass fraction of protein - by the Kjeldahl method according to GOST 25011-2017; determination of the mass fraction of fat - by the Soxhlet method according to GOST 23042-2015; determination of the mass fraction of moisture - according to GOST 33319-2015; determination of the mass fraction of total ash - according to GOST 31727-2012 (ISO 936: 1998); mass fraction of moisture - according to GOST 9793-2016. Amino acid composition was defined according to GOST 34132-2017, using the automatic amino acid analyzer AAA 400 according to the attached instructions by the method of classical ion exchange chromatography. Organoleptic assessment was performed according to GOST 9959-2015. The biological value of proteins was determined by the calculation method according to the recommendations of academician N.N.Lipatov using Biocen software.

3. Results

Currently, the modern poultry processing industry is characterized by considerable degree of overproduction, due to the high production development within the country. Improving the technology of poultry farming and processing involves the maximum use of all production capacities. Simultaneously with the increase in production of a wide range of poultry products including the key raw materials (natural semi-finished products, breasts, wings, fillets, hips, drumsticks, etc.), there is the need for processing the secondary slaughter products remaining after the cutting, i.e. organic waste.

Currently, innovative solutions for waste processing are met by the lack of comprehensive technological solutions and projects for their disposal, which has led to the accumulation of organic waste from the poultry processing industry in large companies. The waste accumulation near the companies will lead to serious environmental problems in the nearest future. To solve these problems, many operators apply rational poultry processing schemes, which provide for deep processing and maximum use of all parts of the poultry carcass. The technology of obtaining meat of mechanical deboning of poultry, carried out by processing substandard carcasses and skeletons on presses of mechanical deboning, was widely used in Russia. As a result of this technological process, the so-called mechanically deboned meat and meat and bone residue are obtained, the latter is subsequently disposed of or processed into semi-finished technical feed.

The problem of recycling the organic waste has been the subject of many scientific papers in all regions of the Russian Federation.

Professor Antipova L.V., associate professors Polyansky S.V. and Orekhov O.G., the scientists of Voronezh State University of Engineering Technology, noted that as a result of mechanical separation, secondary raw materials are accumulated - bone residue in the amount of 27–40% of the mass of raw materials, which is mainly used for the production of feed and, less commonly, food products. The authors of one of the developed technologies [Antipova L.V., Orekhov O.G., 2013] for processing the meat and bone residue indicate that accumulation of considerable volumes of such raw materials in the production conditions require comprehensive study of the composition and features of the meat and bone residue and the design of additional and more rational ways for its use including those applying the biotechnology methods.

To choose a poultry cutting scheme, GOST 31962-2013 "Chicken meat (carcasses of hens, chickens, broiler chickens and their parts). Specifications" used in the production of poultry carcasses was analyzed as well as TU 9214-212-23476484-2010 "Natural semi-finished products from broiler chicken meat" used for the production of semi-finished products.

According to the cutting scheme presented in the regulatory documentation, we have studied the ratio of carcass parts, as this is important for calculating the production cost and the calculation of technological output. The obtained experimental data for cutting the poultry are presented in table 1.

Table 1. Proportion of the poultry body parts in cutting according to GOST 31962-2013.

Semi-finished product	Average output values, %
Fillet	27.3
Wing	9.8
Hip	15.9
Drumstick	11.3
Offcuts	23.1
Skin from neck and breast	9.6
Process waste (cords from fillet, lungs, kidneys)	2.1
Process loss	0.9
Total	100.0

For the production of the dehydrated soup base, on the basis of studying the proportion of the poultry carcass parts, it is advisable to use the offcuts set, which is a skeleton with small cuts of meat left during boning of the carcass and the division of the semi-finished products. The meat of mechanical boning is produced on a separator for mechanical boning of poultry meat LIMA RM 350 S (France). The resulting meat and bone residue is pieces of bone from 1 to 3.5 cm in size and approximately 4% of muscle tissue cuts. The chemical composition of the meat and bone residue was evaluated for the mass obtained from the skeleton and skin and without skin, because chemical composition depends on the type of raw materials used and the technological regime of the separator. The experimental data are presented in table 2.

Table 2. Assessment of the chemistry of the meat and bone residue.

Semi-finished product	Mass fraction of, %			
	protein	fat	ash	moisture
Skeleton meat and bone residue	22.7	17.6	9.3	50.4
Skeleton and skin meat and bone residue	20.3	21.3	8.3	50.1
Ground meat of mechanical boning from skeleton	14.7	16.2	0.8	68.3
Ground meat of mechanical boning from skeleton and skin	11.8	24.2	0.9	63.1

During experimental studies, it was found that a significant amount of fat predominates in the bone residue, where, in addition to the skeleton, the skin from the neck was used, and amounts to 21.3%, which may indicate a possible oxidative damage to fat during storage of the finished product. Therefore, for the further research, it is preferable to use only carcass skeleton as the raw materials. Also, the meat and bone residue obtained from the skeleton contains higher share of protein at the level of 22.7%. The study of the fractional composition which we have performed is of no small importance for the further use of the meat and bone residue as a raw material for the production of dehydrated soup base. The experimental data of the study of fractional composition are presented in table 3.

Table 3. The fractional composition of the proteins of the meat and bone residue.

Raw material	Mass fraction of proteins, %		
	water-soluble	salt-soluble	alkali-soluble
Skeleton meat and bone residue	52.3	24.2	23.5
Skeleton and skin meat and bone residue	63.1	19.3	17.6

As follows from the data presented in table 3, water-soluble proteins predominate in the meat and bone residue and make up 52.3 and 63.1% for the residue from skeleton and the residue from skeleton and skin, respectively, the mass fraction of alkali-soluble proteins is 23.5 and 17.6% respectively. For food production, the amino acid composition is of utmost importance, so, for the further organization of the production of dehydrated soup base, we analyzed the amino acid composition of raw materials in order to study the planned nutritional and biological value of resulting soups and broths (Figure 1).

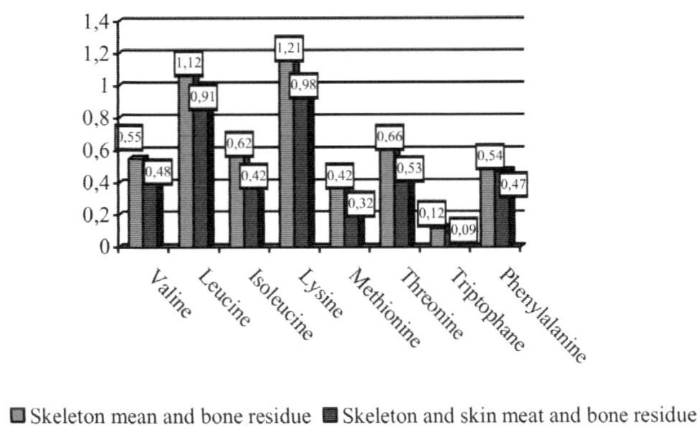

Figure 1. Comparative description of amino acid composition of meat and bone residue

The total mass fraction of essential amino acids is lower in the meat and bone residue than in the ground meat, the limiting amino acids are lysine (71%), leucine (69%) and threonine (68%).

Based on the literature review and patent search, to solve the production problem of producing the dehydrated soup base, we have compiled the main technological basic production operations: drying of bone residue, grinding, packaging and storage. The technological process for the production of the dehydrated soup base is presented in Figure 2.

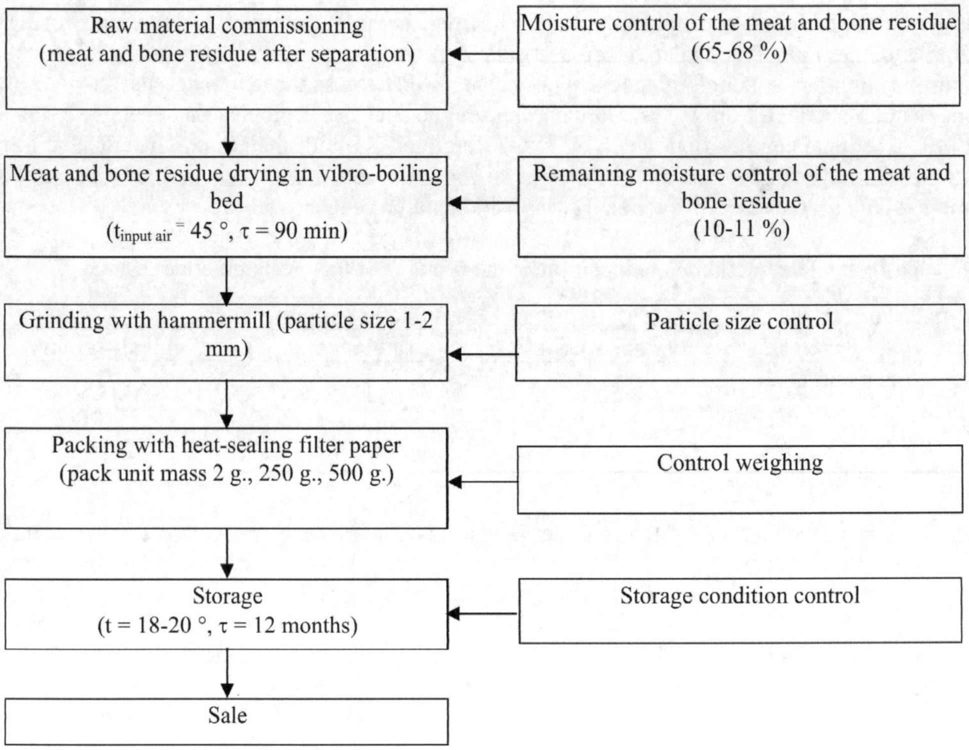

Figure 2. The scheme of the technology process of dehydrated soup base production.

To organize convective drying, we plan to use a liquid-bed dryer with the liquid bed, which is capable of drying many types of bulk materials, semi-finished products, and products in many industries, including food processing [8, 9]. The vibro-boiling dryer is the most energy-intensive and efficient solution for product drying. To achieve better drying results, the equipment uses vibration, which ensures high process efficiency, especially in products with high humidity, as is the case with the meat and bone residue under the study. To establish the optimal drying time, studies were carried out and the drying curve of the meat and bone residue was constructed (Figure 3). As it follows from the data presented in Figure 3, the skeleton meat and bone residue reaches the required residual moisture within 90 minutes. Later no decrease in humidity was observed, so, the process is advisable to be stopped. The skeleton and skin meat and bone residue reaches the specified values within 100 minutes, which is due to the higher mass fraction of fat. Subsequently, the drying of the meat and bone residue obtained from poultry cutting was carried out in a liquid boiling bed at the t_{air} = 45 °C, duration 90 minutes, air velocity 5 m/s, product layer height 120 ± 20 mm; drying is carried out till the residual moisture content of 10-11 % remained unchanged for 10 minutes. Based on the drying parameters and residual moisture in the product, the technological yield of the dehydrated base is 45-50%. During the experimental studies, it was found that the dehydrated base contains a significant amount of protein which is 54.7 and 54.9% for the

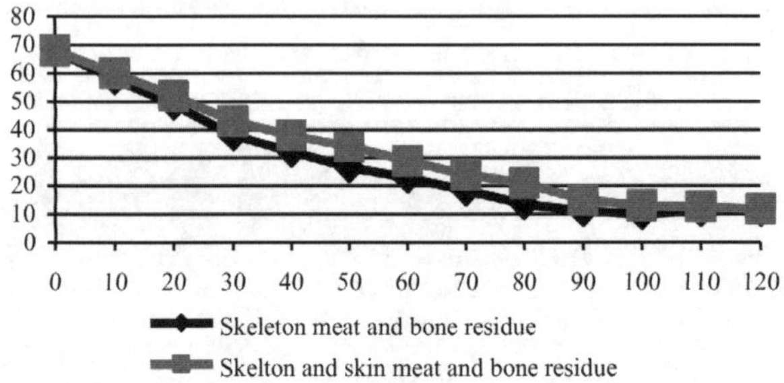

Figure 3. The drying curve for the skeleton meat and bone residue in the liquid-boiling bed

dehydrated base from the skeleton meat and bone residue and the skeleton and skin meat and bone residue, respectively. The total composition of the essential amino acids has not changed, while due to the increase in the mass fraction of protein and the decrease in moisture, the total content of amino acids has increased. The calculation of the indicators of biological value showed that the biological value of the developed dehydrated bases is at the level of 74%, KRAS 25%. The preparation of broths was carried out by brewing a package weighing 5 g for 10 minutes. To assess the organoleptic characteristics of the broth, it is poured into glass glasses, at least 50 cm^3 each, and determine the following features: appearance and color, smell (aroma), taste and richness (saturation with nitrogenous extractive substances). The assessment was carried out using the scale of organoleptic assessment. The organoleptic evaluation showed that the obtained broths from the dehydrated base are transparent with slight turbidity, the color is light yellow, lipid drops are absent; the consistency is liquid, the smell typical for chicken broth is moderate, the taste is typical for chicken broth as well as the smell is moderate. Overall rating is 4.8-4.9.

4. Conclusion

The developed technology provides for the rational use of the meat and bone residue obtained during poultry processing in order to obtain the dehydrated soup base, aimed at developing non-waste methods of poultry processing with maximum use of basic nutrients. The technology allows of expanding the range of poultry products, reducing the complexity of the process, increasing the shelf life of the product and economic efficiency of production.

References

[1] Martin-Rios, Carlos & Demen-Meier, Christine & Gössling, Stefan. 2018. *Food waste management innovations in the foodservice industry.* Waste Management. 79. 196. 10.1016/j.wasman.2018.07.033.

[2] Jayathilakan K, Sultana K, Radhakrishna K, Bawa AS. 2012 *Utilization of byproducts and waste materials from meat, poultry and fish processing industries: a review.* J Food Sci Technol. 49(3):278-293.

[3] Martynova, E.I. & Rusanova, G.Ye. 2019 *Food wastes and meat and poultry processing wastes treatment.* Poultry and Chicken Products. 21. 42-44.

[4] Orekhov O.G. 2013 *Dry broth-based concentrates from bone residue* Poultry and poultry products. No3, 2013. - Pp. 60-63.

[5] Antipova L.V., Polyansky S.V., Orekhov O.G., Sulina Ju.A. 2013 *Substantiation of the applied issues of the rational use of poultry bone residue* Bulletin of VSUET. No1 (55). pp. 110-115

[6] Song DH, Choi JH, Choi YS, et al. Effects of Mechanically Deboned Chicken Meat (MDCM) and Collagen on the Quality Characteristics of Semi-dried Chicken Jerky. *Korean J Food Sci Anim Resour.* 2014;34(6):727-735.

[7] Pavlova N.E., Danyliv M.M., Dvoryaninova O.P. 2013 *Development of the food and forage product technology using convection drying* Modern knowledge intensive technologies, No 8-2, 2013, pp. 214.

[8] Danyliv M.M., Dvoryaninova O.P., Astanina M.V. 2013 *Low waste food technology* Proc.conf. "Innovative technologies in the food industry: science, education, and production", Voronezh, VSUET, 2013. Pp. 563-564.

Seed drill used on complex configuration fields

V I Plyaka, N A Sergeeva, A I Panov and N A Yakovleva

Russian Timiryazev State Agrarian University, 49 Timiryazevskaya street, Moscow, 127550, Russia

plyaka@rgau-msha.ru

Abstract. The seed drill moves along the field in a shuttle manner. Non-productive time spent on the seed drill turn at the end of each run increases as the field area decreases. Consequently, the waste of time reaches the greatest value on small-scale fields and complex configuration areas, naturally encircled by forests, shrubs, ravines and swamps. This leads to a marked decrease in the seed drill performance and increased specific fuel consumption. In addition, the use of seed drills in complex areas has revealed a number of limitations associated with agrotechnical indicators. This problem was solved by developing a seed drill capable of uniform sowing, both in straight-line and curvilinear motion. Curvilinear sowing implies that different amount of grain should get into coulters moving with different radii. A fluted roller seed drill has two groups of sowing units, kinematically connected with the left and right support-driving wheels. The sowing units of different groups can be coupled. Tests of the seed drill with a coupled sowing unit showed that its performance when working on complex configuration fields in case of circular motion is 28... 79% higher and specific fuel consumption is 25... 41% less than that of a conventional shuttling seed drill.

1. Introduction

Non-productive waste of time increases when sowing on small-scale fields and complex configuration areas, naturally encircled by forests, shrubs, ravines and swamps. Growth reserves in the seed drill performance may be found in the reduction of shift time waste [1]. These indicators largely depend on the duration of the seed drill turn at the end of each run [2; 3]. Circular motion can eliminate non-productive waste of time typical for a shuttling seed drill. This problem can be solved by uniform sowing, both in straight-line and curvilinear motion [4].

Curvilinear sowing implies that different amount of grain should get into coulters moving with different radii. This can be done with two types of drills: a fluted roller drill featuring linear relationship between the roller working length and the amount of material introduced [5] or a drill with an electric dispenser [6]. The latter is much more expensive and more complicated.

2. Materials and methods

Such a seed drill (Figure 1) is developed based on the SZ-3.6 serial seed drill and includes an additional sowing unit, a driven reduction gear and self-aligning coulters.

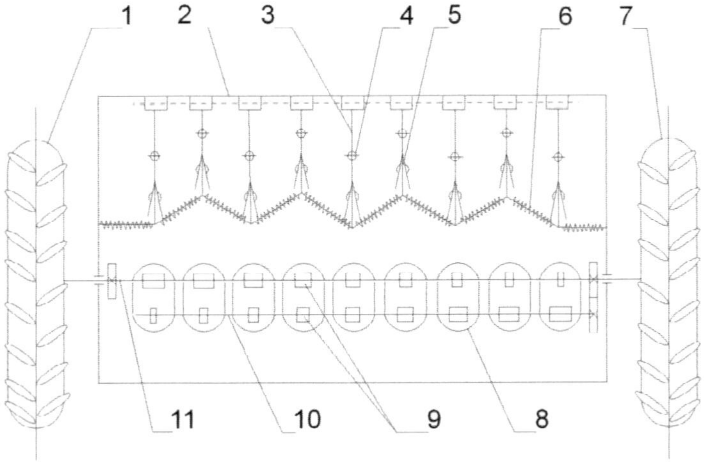

Figure 1. Design of a seed drill with a coupled sowing unit: 1 – left drive wheel; 2 – frame; 3 – joint; 4 – hinge; 5 – coulter; 6 – spring; 7 – right drive wheel; 8 – funnel; 9 – doubled sowing unit: 10 – first shaft; 11 – second shaft.

The seed drill countershaft consists of right and left parts [7]. Its left part transmits the left wheel torque to the sowing unit shaft through the reduction gear. Its right part transfers the right wheel torque to the shaft of the additional sowing unit through the additional reduction gear. The bottom of the grain box is expanded to house an additional sowing unit. Thus, the seed drill is equipped with two similar sowing units, their shafts being parallel to each other. The fluted rollers of the first shaft provide various flows of seeds introduced. Moreover, there is a linear relationship between these flows. The rollers of the second shaft provide seed flows, which also vary in a linear manner. The total operating length of coupled rollers for one funnel is the same for all coupled sowing units of the seed drill. The roller of the first shaft provides a minimum seed flow into the funnel, which is simultaneously filled from the second shaft roller having a maximum flow of material introduced. Thus, the seed flows formed by the rollers of one shaft differ from those of the other shaft but when combined they give the same total flows to each funnel. This is true for the straight-line motion of the seed drill.

When the seed drill moves along an arc, its support-driving wheels move at different speeds. The sowing unit shafts also rotate at different speeds. Therefore, the total seed flow to each funnel changes. Moreover, the minimum total seed flow enters the coulter, which moves along a minimum radius arc. The maximum flow enters the opposite extreme coulter, moving along a maximum radius arc. There is a linear relation between the seed flows supplied to the coulters as they move away from that with a minimum flow. Therefore, the same amount of seeds is sown per unit of distance travelled when different total seed flows are supplied to each coulter.

The joint hinge provides coulter mobility when turning the seed drill. Self-aligning coulters are fixed together by springs.

3. Results and discussion

The seed drill with a coupled sowing unit is developed to increase its shift performance when working on complex configuration fields. The coupled sowing unit and self-aligning coulter unit of the SZ-3.6 seed drill solve this problem with the minimum possible production and design changes of serial seed drills. Comparative tests of the seed drill with a coupled sowing unit and a serial one were carried out on complex configuration fields to determine the effect [8]. Moreover, the former moved in a circular manner along the field while the latter did it in a shuttle way. The gear ratio in the reduction gear was set I = 0.616. The set depth of seed planting was 60 mm. The field configuration complexity groups are shown in Table 1.

Table 1. Field configuration complexity groups.

Group No.	Complexity group	Main feature of the group
1	Rectangle-shaped fields, square-shaped ones or similar-form fields	Correct form
2.	Straight-line nonconvex fields	The smallest angle between adjacent sides is more than 27°
	Ellipse-shaped and semi-ellipse-shaped fields	The ratio of the major axis to the minor one is less than 2.1
3	Straight-line nonconvex fields	The smallest angle between adjacent sides is less than 27°
	Ellipse-shaped fields	The ratio of the major axis to the minor one is more than 2.1
	Fields with partial convex-concave and straight-line sides	
4	Convex-concave fields	The average value of half the ratio of the largest to the smallest in length and width ranges from 1.1 to 1.8
	Band-shaped fields, as well as with the beveled convex-concave and straight-line sides	
5	Convex-concave fields	The average value of half the ratio of the largest to the smallest in length and width is more than 1.8
		Fields must have at least three convexities or concavities

Studies show the coupled sowing unit meets the requirements of technical guidance documents [9] on the sowing rate unevenness of the units and on the overall sowing instability. As can be seen from Figure 2, the indicator of the sowing rate unevenness of the experimental seed drill units does not exceed 3% on all its trajectories. Whereas, the sowing rate unevenness of the units of the SZ-3.6 serial seed drill exceeds the above limit while moving along a trajectory with a radius R <60 m [10; 11; 12].

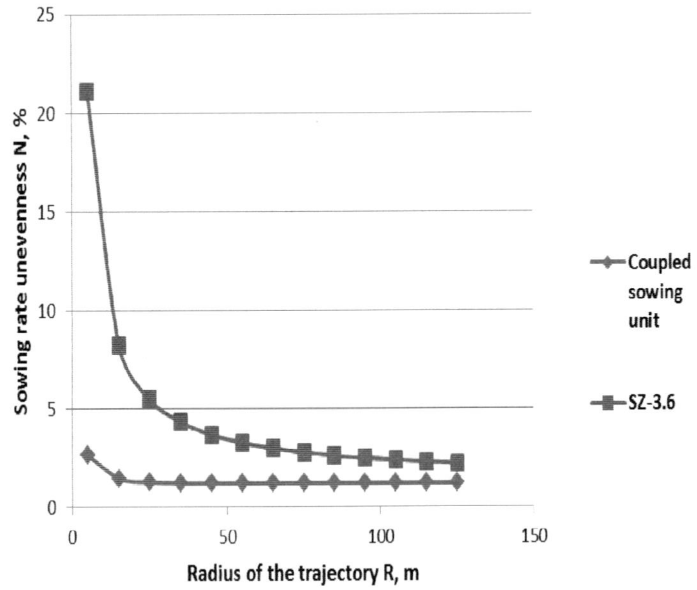

Figure 2. The relationship between the sowing rate unevenness N of the units and the radius R of the trajectory.

The operating speed of the sowing units was found to decrease when the field configuration gets complicated (Table 2). The sharp drop in speed has been noted for the serial seed drill, where the average operating speed decreased from 10.1 km/h to 7.9 km/h. The average operating speed reducing of the unit is explained by the decrease in the run length when the field configuration gets complicated. Therefore, the performance of the sowing units also decreases with the increase in the complexity group of the fields. The coupled sowing unit shows higher performance compared to the serial seed drill sowing unit.

Table 2. Seed drill study results.

	Indicator	Indicator value							
		Complexity group							
		Seed drill with a coupled sowing unit				SZ-3.6 seed drill			
		1	2	3	4	1	2	3	4
1	Unit composition	A tractor and a seed drill							
2	Type of work	Sowing							
3	Way of motion	Circular				Shuttle			
4	Working conditions: Field Area, ha	1.74	1.80	1.68	1.60	1.74	1.80	1.68	1.60
5	Operating mode: A) operating speed km/h	9.42	9.42	9.33	8.90	10.1	9.39	9.36	7.92
	B) working width m.	3.6	3.6	3.6	3.6	3.6	3.6	3.6	3.6
6	Performance, ha/h: A) Main	3.39	3.39	3.36	3.2	3.63	3.38	3.37	2.85
	Б) Shift	3.28	3.27	3.2	3.01	2.56	2.32	2.15	1.68
	Shift time performance increase, ha/h	0.72	0.95	1.05	1.33	0	0	0	0

	%	28.1	40.9	48.8	79.1	0	0	0	0
7	Specific fuel consumption kg/ha	1.51	1.53	1.55	1.61	2.01	2.18	2.3	2.73
	Reduction in specific fuel consumption kg/ha	0.5	0.65	0.75	1.12	0	0	0	0
	%	24.8	29.8	32.6	41.0	0	0	0	0

Studies of the seed drill with a coupled sowing unit have shown that its performance when working on complex configuration fields in case of circular motion is 28... 79% higher and specific fuel consumption is 25... 41% less than that of a conventional shuttling seed drill. At the same time, the crop yield remained equal for the seed drills compared.

4. Conclusions

1. There is linear relationship between the roller working length and the amount of material introduced.
2. The indicators of the sowing rate unevenness of the units and of the overall sowing instability when operating the seed drill with a coupled sowing unit do not exceed 3%.
3. The crop yield is almost equal for both the seed drill with a coupled sowing unit and the SZ-3.6 seed drill.
4. Shift performance of the seed drill with a coupled sowing unit when working on complex configuration fields is 28 ... 79% higher than that of the serial seed drill.
5. The seed drill with a coupled sowing unit when working on complex configuration fields in case of circular motion has its specific fuel consumption 25... 41% less than that of the serial shuttling seed drill.

References

[1] Khorunzhenko V E 1987 Technological grounds for designing sowing machines and prospects for seed drill development. *Tractors and agricultural machinery* **11**

[2] Klenin N I, Levshin A G and Kiselev S N 2008 *Agricultural Machinery* (Moscow: KolosS) p 816

[3] Lyubushko N I and Khorunzhenko V E 1986 Development trends of the technical level of seed drills. *Tractors and agricultural machinery* **8**

[4] Kornev A I 1958 Study of the influence of the working trajectory on the agricultural unit performance: PhD (Eng) thesis (Moscow)

[5] Semenov A N 1960 Theoretical grounds and methods of technological calculation of seed drill working units: DSc (Eng) thesis (Kishinev)

[6] Kamgar S, Noei-Khodabadi F and Shafaei S M 2015 Design, development and field assessment of a controlled seed metering unit to be used in grain drills for direct seeding of wheat. Information *Processing in Agriculture* **2** 169–76

[7] Plyaka V I and Vinogradov Yu A 1987 Copyright certificate of the USSR No.1299533, MKI A 01 C 7/16. Seed drill sowing unit / Plyaka V I and Vinogradov Yu A / B.I. **12**

[8] Udachin S A 1962 *Land / Land Use Essays* (Moscow)

[9] GOST 31345-2017 Agricultural machinery. Tractor seeders. Test methods INTERSTATE STANDARD

[10] Gurashvili G V 1970 Research of curvilinear motion influence on operational indicators of arable and sowing units: PhD (Eng) thesis (Tbilisi)

[11] Plyaka V I 1989 Working process background of a coupled sowing unit of the seed drill. "Technical means for providing intensive technologies for crop cultivation and harvesting." Collection of papers of Moscow Institute of Agricultural Engineers (Moscow)

[12] Plyaka V I 1989 Seed drill for complex configuration fields *Scientific and technical information digest* **3** (Moscow: AgroNIIETIITO)

The influence of the sowing coulters type on the seeding quality and the spring wheat yield

V I Belyaev[1], V V Vol'nov[1], D A Iakovlev[1] and L V Sokolova[2]

[1] Altai State Agricultural University, 98, Krasnoarmeyski ave., Barnaul, 656049, Russia
[2] Altai State University, 61, Lenina ave., Barnaul, 656049, Russia

E-mail: l.v.sokol@mail.ru

Abstract. The paper presents data on the influence of the sowing coulters type on the seeding quality and the spring wheat yield in the conditions of the steppe part of Altai Krai. For the grain sowing, seeders are used with the most common types of the sowing coulters such as V-shaped, chisel and disk openers. The wheat was sown after the fallows, the peas and the wheat forecrops. The chisel openers provided the highest sowing quality of wheat seeds. Also the maximum efficiency of the spring wheat cultivation in the extra dry steppe conditions (2012) was achieved by the chisel sowing coulters. In the years with the moderately dry conditions (2006–2011) in the steppe the seeders with the various types of the sowing coulters do not significantly affect the yield structure constituent elements; to a greater extent it depended on the forecrops, the maximal wheat yield was obtained after the fallow.

1. Introduction

The problem of rational use of agricultural land is currently acute [1-4]. In the future, obtaining the maximum possible yield of grain crops per unit of sown area allows to optimize the agriculture in the climate change conditions. In Altai Krai, the problem of getting the full value seedlings of the grain crops is directly related to the instability of the climatic conditions, pests and diseases, but to a greater extent with the failure of the agrotechnical requirements for the sowing. For the grain sowing, seeders are used with the most common types of the sowing coulters such as V-shaped, chisel and disk openers. Despite the fact that these sowing machines have been used in the Altai steppe for many years, the question of their influence on the soil moisture reserves, the soil nitrogen content, the weed infestation, the realisation of the agrotechnical requirements, the seed germination and the elements of the yield structure in the system plan are insufficiently studied [5].

The aim of the study is to investigate the influence of the sowing coulters type on the seeding quality and the spring wheat yield in the conditions of the steppe part of Altai Krai.

2. Materials and methods

In order to determine the most effective type of sowing coulter in 26 farms of the steppe part of Altai Krai, from 2005 to 2012, the studies were conducted with the double-disc coulters (SZP – 3.6A, John Deere 730), the V-shaped coulters (SZS-2.1, Agromaster and Morris) and the chisel coulters (Amazone Condor and DMC).

Assessment of the work of the seed openers was carried out in the field experiments by the main forecrops of the spring wheat (fallow, peas, wheat) in three repetitions according to the scheme (Table

1).

Table 1. The scheme of the experiments on the assessment of the sowing coulters, 2006-2012

Variant	Tillage, depth	Type of sowing machine, Russian or foreign-made	Sowing method	Type of sowing coulter
1	Autumn, 14-16 cm + presowing, 10-12 cm	single-function, Russian-made	row	double-disk
2	Autumn, 14-16 cm	combined, Russian-made	strip	V-shaped
3	No-tillage	direct seeder, foreign-made	row	chisel
4	No-tillage	combined, foreign-made	strip	V-shaped

The main concomitant observations were carried out in accordance with standard methods: moisture reserves before sowing in the meter soil layer; the content of the nitrate nitrogen in 0-40 cm of soil before sowing; the weed infestation of crops; the seeding depth; the field germination of seeds; the wheat yield structure.

Technological operations were carried out by the experimental scheme, the sowing of spring wheat was in the third part of May, taking into account generally accepted recommendations with the sowing rate of 3.0–4.0 million germinating grains per 1 ha, depending on the moisture content of the territories. The herbicidal treatment of wheat plants, seeded with the seeders with the double-disc sowing coulters, and with the V-shaped coulters was carried out 1 time against dicotyledonous weeds. In the variant of sowing wheat, seeded with the seeders with the chisel openers it had to be treated 5-10 days before seeding with the nonselective herbicides. The experiments were carried out without the use of the mineral fertilizers.

3. Results and Discussion

3.1. Soil moisture reserves

The main reason for the low and unstable productivity of spring wheat in the steppe part of the Altai Kray is the aridity of the climate. Crops lack moisture throughout the growing season. The intense water regime is exacerbated by the uneven precipitation and their low intensity. The amount of vegetation period' rainfall over the 6 year in the studied area amounted to the average of 168 mm (80% of the norm) with the fluctuations from 80 mm (2010) to 208 mm (2006). The year 2012 was exceptional in terms of aridity. So, in many areas of the Altai steppe there were in April only 2-6 mm of rainfall, in May 8-14 mm, in June 12-38 m, in July 9-18 mm and in August 21-27 mm. In general, in the all research area there was on average 83 mm (40% of the norm) precipitation during the vegetation period, which affected the soil moisture (Table 2, Figure 1a).

Table 2. The soil meter layer productive moisture reserves before sowing depending on the forecrop and the tillage, mm (2006-2012)

Variant	Moderately dry 2006-2011			Extra dry 2012		
	Forecrop			Forecrop		
	fallow	peas	wheat	fallow	peas	wheat
1	123	118	76	42	44	28
2	126	116	72	44	46	32
3	111	102	61	71	61	57
4	108	101	59	70	60	51
ẋ	117	109	67	57	53	42
σ	9	9	8	16	9	14
SEM	4	4	4	8	4	7
Cv, %	8	8	12	28	17	34

The soil water regime in the years with the average moisture supply developed more favorably in the variants with the autumn tillage (1 and 2 variants) after the fallow and peas forecrops. Productive soil moisture reserves were 14-17 mm higher than for backgrounds with no-tillage (3 and 4 variants). In the extra dry year 2012, the direct sowing with the chisel openers with the backgrounds of no-tillage (3 and 4 variants) contributed to an increase in the soil moisture reserves of 15-28 mm relative to another due to the better water preservation in the spring time.

3.2. Nitrogen supply and weed control

The plants supply with the nitrate nitrogen in the soil was the most dependent on the forecrops. Before sowing the content of the nitrate nitrogen in 0-40 cm of soil depending on the weather conditions of the particular year after the fallow ranged from 11.0 to 43.2 mg/kg, after the peas forecrops – 6.7-35.8 mg/kg, after the wheat it was 6.2-18.3 mg/kg. With the background of no-tillage the accumulation of the nitrate nitrogen in the soil did not differ in the experimental variants.

It is known that herbicides are an integral element of weed control in the cultivation of wheat using direct sowing technology. The use of herbicides in the experiments allowed to reduce the weed infestations of crops to a low level (up to 10% of the mass of weeds in the total plant mass from 1 m^2), but there is still a tendency to increase the weed infestations of crops on the variant sown by the seeder with the chisel seed openers, where nonselective herbicides were only used in time of 5-7 days before the sowing. Seeders with the V-shaped coulters had the advantage of weed control. They destroyed weeds during the sowing and the herbicidal treatment of crops in the period of seedling stage to tillering stage increased the efficiency of combating the weeds. The weed infestations of crops decreased in these variants to 3.8-6.0%.

3.3. Seeding depth

Among the agrotechnical methods (forecrops, tillage, etc.), the sowing is of no small importance in obtaining the full-fledged seedlings of grain. The task of sowing is to obtain the even sprouts through the optimal use of the method, the sowing time, the seeding rate, the seeding depth and the distribution of seeds, the row spacing. The field germination index is significantly affected by the sowing depth. When it smaller than 4 cm, the death of the sprouts in dry years reaches reaches 27.7%, and when the place of seeds is in the soil layer of 8-10 cm, the number of sprouts is reduced by 5-10%, so they need to be sown to a depth of 5-7 cm [6]. In our experiments, seeders with the disk coulters did not withstand the specified parameters of the seeding depth and sowed them at a smaller depth (Table 3).
In our experiments, seeders with the disk coulters did not withstand the specified parameters of the seeding depth and sowed them at a smaller depth (Table 3).

Table 3. The seeding depth depending on the type of sowing coulter and the forecrop, mm (2006-2012)

Variant	Type of sowing coulter	Forecrop		
		fallow	peas	wheat
1	double-disk	47.6	48.1	46.9
2	V-shaped	69.4	62.3	61.8
3	chisel	55.6	53.0	55.5
4	V-shaped	65.0	66.5	57.1
$\dot{x}$	-	59.4	57.5	55.3
σ	-	9.7	8.4	6.2
SEM	-	4.9	4.2	3.1
Cv, %	-	16.4	14.6	11.2

The depth of seeding with the V-shaped sowing coulters fits into the set parameters (50.0-70.0 mm), regardless of which seeders they were sown - Russian or foreign production. When sowing on plowed backgrounds (fallow, peas), the depth of seeding was higher by 9-7.3 mm than sowing on untreated stubble backgrounds. At the same time, a high uneven distribution of seeds was observed along the depth of their placement. The best indicators for the listed values were observed with sowing seeders with chisel-type sowing coulters. The depth of seeding in this case by the different forecrops ranged from 51.0 to 55.6 mm.

3.4. Field germimation

In the normal humidification years (2006–2011), the type of the sowing coulters did not have a significant effect on the differences in field germination index of the wheat seeds. Only in the extra dry year 2012 it was higher on the variants with the Condor and DMC seeders with the chisel openers. The difference in the field germination index relative to the other variants reached 4.8% after the fallow, and amounted to 7.4 and 12.0 % after the wheat and the peas (Table 4). The low level of the field germination index is explained by the influence of the hydrothermal regime of dry conditions during the sowing–sprouting period, as well as by the root rot damage, especially by the fallows (Figure 1b).

Table 4. The field germination index depending on the type of sowing coulter and the forecrop, % (2006-2012)

Variant	Type of sowing coulter	Moderately dry 2006-2011			Extra dry 2012		
		Forecrop			Forecrop		
		fallow	peas	wheat	fallow	peas	wheat
1	double-disk	67.6	70.8	69.1	57.2	55.9	53.3
2	V-shaped	62.6	69.3	62.1	58.0	53.2	52.1
3	chisel	68.1	75.0	70.5	62.0	65.2	59.5
4	V-shaped	66.2	74.4	68.8	57.4	55.2	53.3
$\dot{x}$	-	66.1	72.4	67.6	58.7	57.4	54.6
σ	-	2.5	2.8	3.8	2.3	5.3	3.3
SEM	-	1.2	1.4	1.9	1.1	2.7	1.7
Cv, %	-	3.8	3.8	5.6	3.9	9.3	6.1

3.5. Wheat yield

The potential yield of grain crops depends on the tilling capacity, the number of productive stems per 1 m^2, the number of the grains in the ear, and the mass of the 1000 grains. An analysis showed that in normal humidification years (2006–2011) in the dry steppe the seeders with the various types of the sowing coulters do not significantly affect the yield structure constituent elements, only the number of productive stems shows the advantage of the chisel openers. In this variant the number of productive wheat stems in the fields after the fallow was higher by 9.5%, after the peas by 10.8% and after the wheat 8.4% then in the other variants. In the extra dry year 2012 the positive influence of the chisel coulters affected all the elements of the yield structure after all forecrops. Thus, the number of the productive stems in the variant 3 was 55 pcs./m^2 higher than on the variant 1 with the double-disk coulters (188 and 242 pcs./m^2, respectively), the number of the grains in the ear was higher by 7.5 pcs. (10.4 and 17.9 pcs., respectively), and the weight of 1000 grains was higher by 4.0 g. The V-shaped openers occupied the intermediate position and did not have the large differences between the Russian-made sowing machines in the frames of the yield structure constituent elements.

The spring wheat productivity in the favorable moisture years did not differ significantly depending on the type of opener and ranged within the standard error of the mean for all forecrops. To a greater extent, it depended on the forecrop than on the type of sowing coulter (Table 5, Figure 1c). In the extra dry year 2012, the yield of the spring wheat in the variant 1 (double-disc coulter) was the lowest and consisted after the fallow 6.4 c/ha; after the peas 5.9 c/ha and after the wheat 2.5 c/ha. In this severe weather conditions high efficiency in the spring wheat cultivation was obtained on the fields sowing with the chisel openers (variant 3). Relative to the variant 1, the yield increase here after the fallow was 7.0 c/ha, after the peas 3.0 c/ha and after the wheat 4.5 c/ha. Seeders of Russian and foreign production with the V-shaped sowing coulters did not have the significant differences in the wheat yield (variants

2 and 4), but they exceeded the double-disc coulter seeders by 5.9 c/ha and 2.1 c/ha respectively for the best forecrops (fallow, peas). At the same time, the yield of wheat sown after the wheat with the double-disc coulters was lower than in the variant with the chisel openers.

Table 5. The wheat yield depending on the type of sowing coulter and the forecrop, c/ha (2006-2012)

Variant	Type of sowing coulter	Moderately dry 2006-2011 Forecrop			Extra dry 2012 Forecrop		
		fallow	peas	wheat	fallow	peas	wheat
1	double-disk	24.2	17.2	15.8	6.4	5.9	2.5
2	V-shaped	24.6	16.9	16.2	12.6	8.1	4.6
3	chisel	24.4	17.5	16.2	13.4	8.9	7.0
4	V-shaped	25.0	18.0	16.0	12.0	7.9	4.0
ẋ	-	24.6	17.4	16.1	11.1	7.7	4.5
σ	-	0.3	0.5	0.2	3.2	1.3	1.9
SEM	-	0.2	0.2	0.1	1.6	0.6	0.9
Cv, %	-	1.4	2.7	1.2	28.7	16.6	41.4

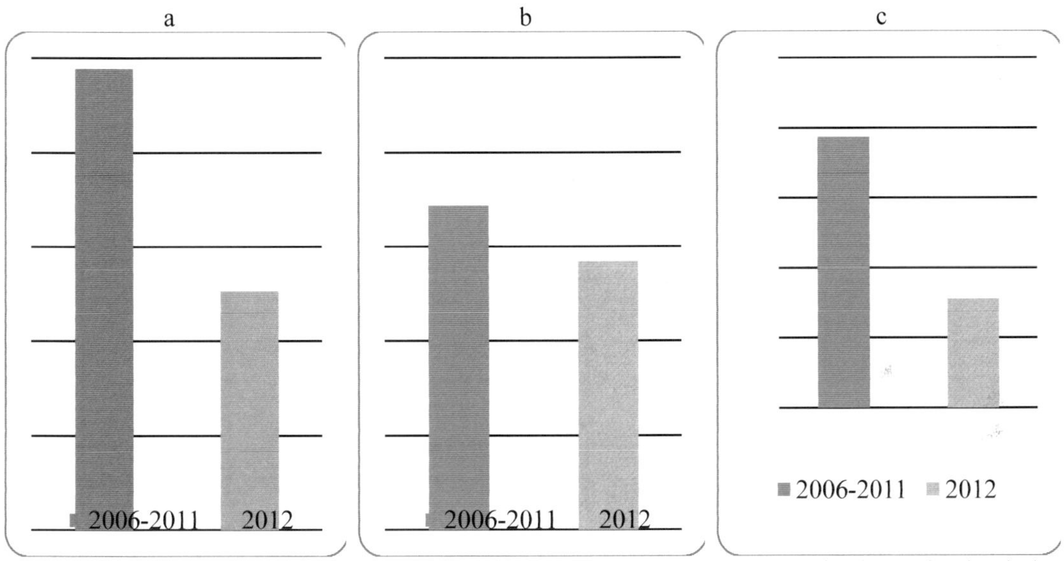

Figure 1. Comparison of the average soil productive moisture (a), the average field germination index (b) and the average wheat yield (c) in 2006-2011 and in extra-dry 2012.

4. Conclusion

The water regime of the soil in the moderately dry years was more dependent on the forecrops (fallow, peas) than on the type of the sowing coulters of the seeders. In an extremely dry year direct sowing with the seeders with the chisel openers contributed to the safety of the soil moisture reserves to the seeders with the double-disk openers of 17–28 mm. Provision of wheat plants with soil nitrogen between the experimental variants was almost the same and differed only in their forecrops. On the background of the use of herbicides, the weed infestation of wheat crops was low (up to 10%), although it increased on the variant with chisel openers to 8.0–9.1% with a contamination level of 3.8–6.0% on the variant with the V-shaped openers. The chisel openers provided the highest sowing quality of wheat seeds.

The maximum efficiency of the spring wheat cultivation in extra dry conditions (2012) was achieved by the sowing it with the seeders with the chisel sowing coulters. Due to the increase in the field germination index (by 4.8–12%) and other indicators of the crop structure, the yield of the spring wheat in the extra dry year was after the fallow 13.4 c/ha, after the peas 8.9 c/ha and after the wheat 7.0 c/ha with the yield in the variant with the double-disk openers 6.4, 5.9 and 2.5 c/ha, respectively by the forecrops. The seeders with the V-shaped openers held an intermediate position with the average wheat yield of 12.3, 8.0 and 4.3 c/ha, respectively by the forecrops. In the years with the moderately dry conditions (2006–2011) in the steppe the seeders with the various types of the sowing coulters do not significantly affect the yield structure constituent elements; to a greater extent it depended on the forecrops, the maximal wheat yield was obtained after the fallow.

References

[1] Guggenberger G, Bischoff N, Shibistova O, Müller C, Rolinski S, Puzanov A, Prishchepov A V, Schierhorn F and Mikutta R 2020 *KULUNDA: Climate Smart Agriculture, Innovationsin Landscape Research* ed M. Frühauf et al. (Switzerland: Springer Nature) pp 183-199

[2] Bischoff N, Mikutta R, Shibistova O, Puzanov A, Reichert E, Silanteva M, Grebennikova A, Schaarschmidt F, Heinicke S and Guggenberger G 2016 *Agriculture, Ecosystems & Environment* **235** 253-264

[3] Rockström J, Sheehan J, Siebert S, Tilman D and Zaks D 2011 *Nature* **478** 337–342

[4] Levykin S V, Chibilev A A , Gulyanov Yu A, Silantieva M M, Kazachkov G V and Yakovlev I G 2019 *Ukrainian Journal of Ecology* **9(3)** 371-375

[5] Belyaev V I, Silanteva M, Matsyura A V and Sokolova L V 2019 *Ukrainian Journal of Ecology* **9(4)** 723-728

[6] Belyaev V I, Rudev N V, Maynel T, Kozhanov S A, Sokolova L V and Matsyura A V 2017 *Ukrainian Journal of Ecology* **7(4)** 145–150

IOP Conf. Series: Materials Science and Engineering **941** (2020) 012043 doi:10.1088/1757-899X/941/1/012043

Combine head mover designing by modern software using

D N Pirozhkov and V V Sadov

Altai State Agricultural University, 98 Krasnoarmeyskiy Ave., Barnaul, 656049, Russia

E-mail: mms.asau@yandex.ru

Abstract. The development of technical solutions without using of modern computer-aided design systems is not possible at the present stage. The using of CAD and CAE systems made possible to reduce costs and save time for development of combine head mover. As a CAD and CAE systems were used KOMPAS-3D and APM WinMachine. As a result of a computational experiment, the minimum weight with the required structural strength and rigidity of the combine head mover design was obtained. Obtained data were the base of combine head mover prototype manufacturing and its successful test.

1. Introduction

The Altay region is one of annual leaders in grain production in Russian Federation. The average annual wheat production in the region is from 4 to 5 million tons. The sown area occupied by grain crops is about 3.5 million hectares. Acute problems of a timely harvesting campaign are inevitable, given such production volumes and relatively low grain yields. Besides unpredictable weather conditions worsen harvesting operations very often. So, depending on the weather, straight-combine or separate-combine method is used. Share of grain sown areas harvested by separate-combine method in the Altai region prevails. So, in some years up to 80% of grain crops are harvested by separate-combine method [1]. Separate-combine method involves using of both combine harvesters and self-propelled windrowers for mowing into windrows. In recent years, combine head movers for transporting headers used to quickly move harvesting equipment from field to field. The combine head mover is towed by a combine or self-propelled windrower and lets header transporting process in a more convenient way compared to header mounted on the harvester. In addition, the combine head mover solves the problem of harvesting equipment moving along public roads.

2. Results and discussion

The object of research is the designing process of combine head mover by using modern domestic computer-aided design and computer-aided engineering systems.

The research subject is the design of combine head mover for headers that have cutting width from 6 to 9 meters and weight of up to 2500 kg for using with Acros harvester and KSU self-propelled windrower produced by Rostselmash.

Engineering and scientific problems solution of developing new equipment and technologies is impossible without modern design and analysis systems today. Such systems called as SAPR in Russia or CAD and CAE in international practice. Computer-aided design and computer-aided engineering systems are considerable part of modern software that made it possible to carry out very significant volumes of structural design work in a short time [2-8]. Moreover, the developed design is obtained if not optimal, then near optimal according to selected optimization criterion. This work uses the same methods that we have successfully tested when optimizing the front-end loader design [9].

This paper presents designing of combine head mover on the basis of technical enquiry. As CAD system was used KOMPAS-3D [10] and CAE system was used APM WinMachine [11]. The technical enquiry requests to design a domestic combine head mover. The combine head mover design should surpass competitors' products in cost, structural strength and rigidity.

The first stage of project is blueprint of combine head mover on the basis of existing engineering solutions. The first stage takes into account dimensional specifications of hauling equipment and headers (figure 1).

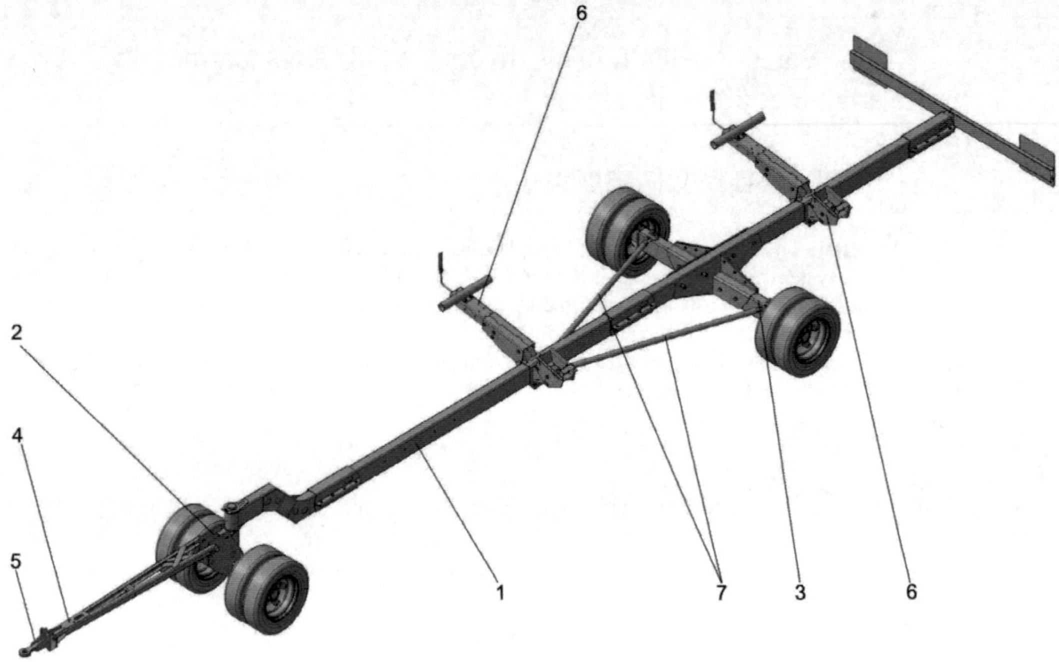

Figure 1. Blueprint of combine head mover: 1 – chassis frame; 2 – front axle; 3 – rear axle; 4 – towing beam; 5 – tow hook; 6 – header mounting bracket; 7 – pull rod.

The next stage of project is founding of design factors that affect weight, dimensional specifications, structural strength and rigidity. These design factors are rear axle pinning point, header mounting bracket pinning points and profiled bar of chassis frame (figure 1).

The third stage of project is creation of experimental design. The experimental design involved finding the dependence of structural strength on design factors using the second order regression equation. We used a central composite design that based on a two-level factorial design with the addition of 2k points between the axes plus repeat points at the centroid [12-14].

The fourth stage of project is computational experiment implementation. Stress strain and weight of combine head mover were calculated during experimental runs (figure 2).

The fifth stage of project is determination of design factors optimal values. As optimization criterion was used the weight of combine head mover. Optimal values of design factors were obtained by Mathcad software [15].

The final stage of project is combine head mover structural modification in light of optimal values of rear axle pinning point and header mounting bracket pinning points. Also production technology ability and transportation convenience in unassembled form were taken into account in the final stage.

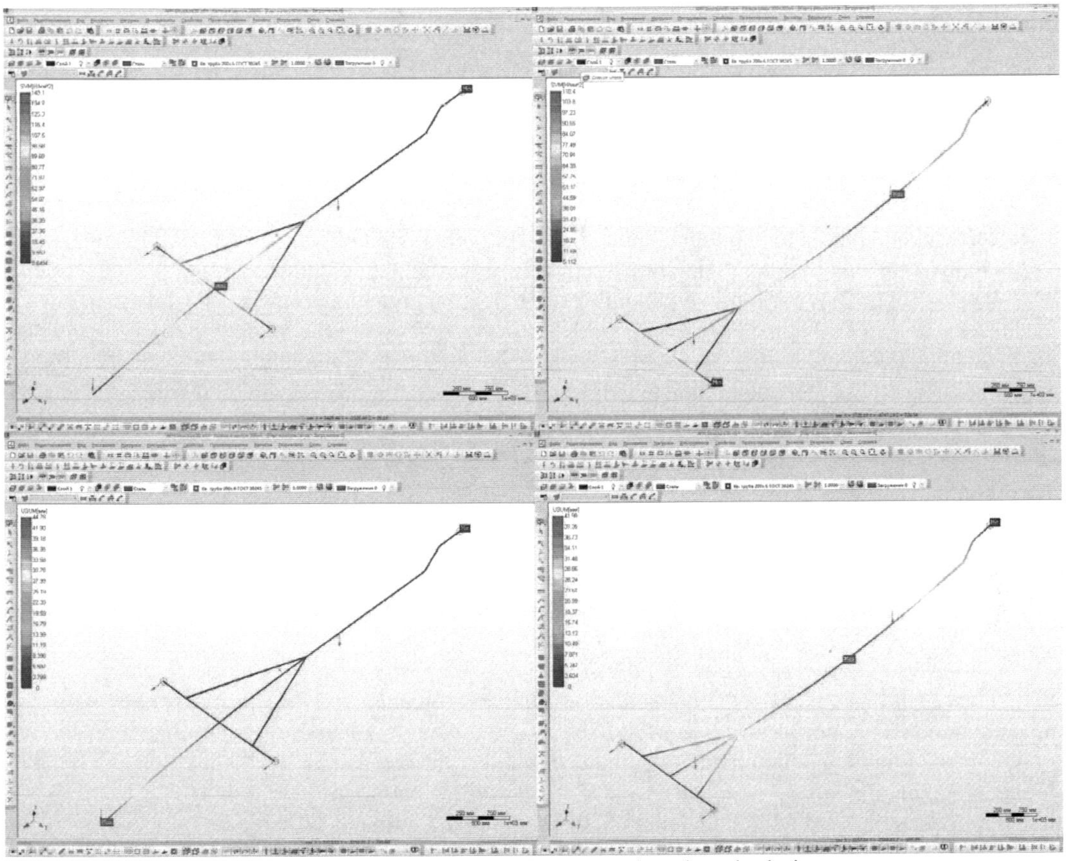

Figure 2. Some results of stress and strain calculations.

The preproduction model made according to the research results confirmed the declared characteristics of strength and rigidity (figure 3).

Figure 3. The preproduction model of combine head mover.

3. Conclusion

The following tasks were solved during the project implementation:

1. The design of combine head mover is developed for using with Acros harvester and KSU self-propelled windrower produced by Rostselmash.
2. The three most significant design factors are found that affect the strength and rigidity of the structure.
3. Optimal values of design factors were obtained that allowed getting the required parameters of strength and rigidity with minimum weight of combine head mover.

4. References

[1] Grain harvesting in Altai region https://regnum.ru/news/2310725.html
[2] Wu D, Rosen D W, Wang L, and Schaefer D 2015 *Computer-Aided Design* **59** 1-14.
[3] Zhang Z, Liu G, Jiang Z, and Chen Y 2015 *Journal of Manufacturing Science and Engineering* **137(4)** 040908.
[4] Farhan U H, Tolouei-Ra. M, and O'Brien S 2012 *International Scholarly and Scientific Research & Innovation* **6(12)** 365-8.
[5] Lad A C, Rao A S 2014 *International Journal of Emerging Engineering Research and Technology* vol 2 **2(7)** 157-167.
Jayakiran Reddy, E., Sridhar. C.N.V.. Pandu Rangadu, V.: Knowledge Based Engineering: Notion. Approaches and Future Trends. American Journal of Intelligent Systems 5(1). 1-17(2015)
[6] Jayakiran Reddy E, Sridhar C N V and Pandu Rangadu V 2016 *Advances in Intelligent Systems and Computing* 10.1007/978-3-319-23258-4_27.
[7] Liu W, Zhou X, Niu Q, and Ni Y 2014 *International Journal of Mechanical. Aerospace, industrial and Mechatronics Engineering* **8(12)** 1851-4.
[8] Titus A and Xuc Bin L 2013 *International Journal of Science and Research* **2(10)** 162-4.
[9] Pirozhkov D N Voronov Y V, Sorokin Sergey Anatolyevich, Bachurin Roman Nikolayevich. The use of modern software in agricultural machinery designing.
[10] CAD/AEC/PLM Solutions from ASCON. https://ascon.net/main/
[11] Optimum solutions in civil and mechanical engineering. https://apmwm.com/
[12] Harold F, Giles Jr John R and Wagner Jr 2013 *Extrusion 2nd Edition* (ISBN: 9781437734812. eBook ISBN: 9781437734829).
[13] El Hami A and Pougnet P 2015 *Embedded Mechatronic Systems 2 Analysis of Failures, Modeling, Simulation and Optimization* (ISBN : 9781785480140).
[14] Paulo Davim J 2012 *Mechatronics and Manufacturing Engineering. Research and Development* (ISBN: 978-0-85709-150-5).
[15] PTC Mathcad – Show Your Work! https://www.mathcad.com/

IOP Conf. Series: Materials Science and Engineering **941** (2020) 012044 doi:10.1088/1757-899X/941/1/012044

Study of the operation of a combine harvester cleaning system with a sieve screw separator in the conditions of operation on slopes

S F Sorochenko and V A Ugarov

Polzunov Altai State Technical University, Barnaul, 656038, Russia

E-mail: sorochenkosf@list.ru

Abstract. The article presents the results of studies of a cleaning system with a sieve screw separator in operation conditions of a combine harvester on slopes. The grain heap in the proposed cleaning system is moved by screws, which ensures uniform feeding of the grain heap to the upper sieve with longitudinal tilts of the combine. Autonomous operation of the screws reduces the uneven distribution of grain heaps across the width at the beginning of the upper sieve of the cleaning system. Preliminary separation of grain reduces the grain flow to the upper sieve by more than 60%. Laboratory studies have shown a decrease in grain loss with a transverse tilt of the combine of 8 degrees from 1.17% in the compared treatment system to 0.75% in the treatment system with a sieve screw separator. An additional reduction in grain losses was achieved by installing longitudinal partitions with a height of 0.13 m on the upper sieve, preventing the grain heap from shifting sideways on the upper sieve. The theoretical coefficient of variation in the grain heap layer thickness at the end of the upper sieve is 0.18. Grain losses in the proposed cleaning system, calculated by the mathematical model of separation, amounted to 0.14%, which is 8 times lower than the compared cleaning system.

1. Introduction.

The grain heap in the combine cleaning system is in a fluidized state due to sieve vibrations and exposure to airflow. When the combine moves up or down the slope, the speed of the grain heap increases or decreases, respectively. With a lateral tilt of the combine, the grain heap is shifted to the sidewall. This leads to deterioration of the cleaning system quality and the lateral tilt of the combine leads to increasing grain losses [1, 2]. Stabilization of the upper sieve position in two planes relative to the horizon [3, 4] provides a reduction in grain losses, while the uneven feeding of grain heaps to the upper sieve reduces the efficiency of the cleaning system. Therefore, the cleaning system of the hillside combine harvester should provide not only stabilization of the distribution of the grain heap on the upper sieve but also a uniform feed of the grain heap. One of the structural and technological solutions for the combine harvester cleaning system in the hillside operating conditions is the combination of preliminary separation, uniform feeding and distribution of grain heaps on the upper sieve of the cleaning system.
The article aims to study the operation of a cleaning system with a sieve screw separator in the conditions of operating a combine harvester on slopes.

2. Object of study.

The object of research is the separation of grain heaps in the cleaning system, including a sieve screw separator [4] and an air-sieve cleaning system.

The sieve screw separator (figure 1) is intended for preliminary separation of a grain heap. The sieve screw separator is installed instead of the shaking board and contains augers 1 with blades, a wavelike sieve 2 and a shortened shaking board 3. The centrifugal fan 4 contains an additional nozzle. The separation of the grain heap in the considered cleaning system occurs due to the influence of the airflow and the blades when moving the grain heap on the sieve.

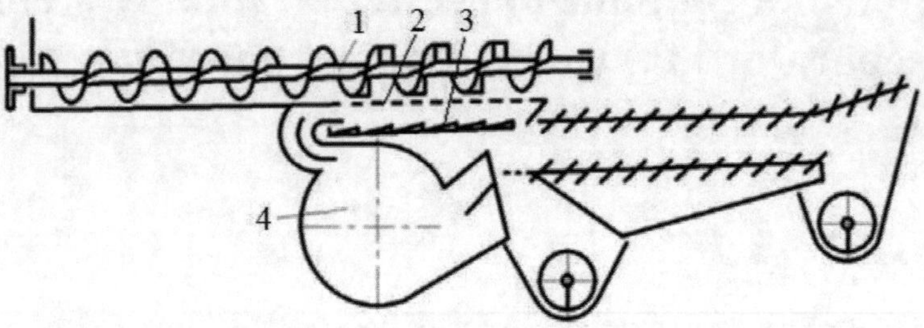

Figure 1. The cleaning system with a sieve screw separator.

The proposed cleaning system provides a uniform feed of grain heaps onto the upper sieve, evenly distributing it both across the width of the upper sieve and in time.

3. Research methods.

The study of grain separation in the proposed cleaning system was performed using comparative laboratory tests and mathematical modeling.

In theoretical studies, the grain loss after the cleaning system (P,%) was determined by the separation model [5], which for the studied cleaning system has the following form:

$$P=100\cdot\prod_{i=0}^{k}\left[\exp\left[-\mu_P\cdot\Delta x\cdot\exp\left(\Delta V_h\cdot i-V_{hn}\right)\right]\right]. \quad (1)$$

Where i is the number of the current section of the grain heap along the length of the upper sieve; k is the number of Δx -long sections along the length of the upper sieve; μ_P is the separation coefficient of the grain heap; V_{hn} is coefficient of variation of grain heap layer thickness at the beginning of the sieve; ΔV_h is the step of changing of layer thickness variation coefficient along the sieve length.

The parameter ΔV_h was determined by the formula:

$$\Delta V_h=\frac{V_{hn}-V_{hk}}{k}, \quad (2)$$

V_{hk} is the coefficient of variation of the layer thickness at the end of the sieve.

The separation coefficient depends on the separation coefficient in the basic cleaning system and on the thickness of the grain heap layer in the cleaning under examination. In the adopted separation model, the parameter is set equal to the shift of the grain heap on the upper sieve in one oscillation [5]. For one oscillation of the sieve, the grain heap moves over a distance Δx, and the coefficient of variation changes by value.

The coefficient of variation at the beginning of the upper sieve depends on the operation of devices installed in front of the upper sieve of the cleaning system. In the studied cleaning system it depends on the operation of the sieve screw separator. The heap variation coefficients were determined experimentally using grain heap samplers and theoretically according to the developed method [5].

The experiments were carried out in a laboratory setup (figure 2). When examining the operation of experimental cleaning, a sieve screw separator was installed instead of the shaking board, containing four augers with right hand and left hand screws, installed alternately.

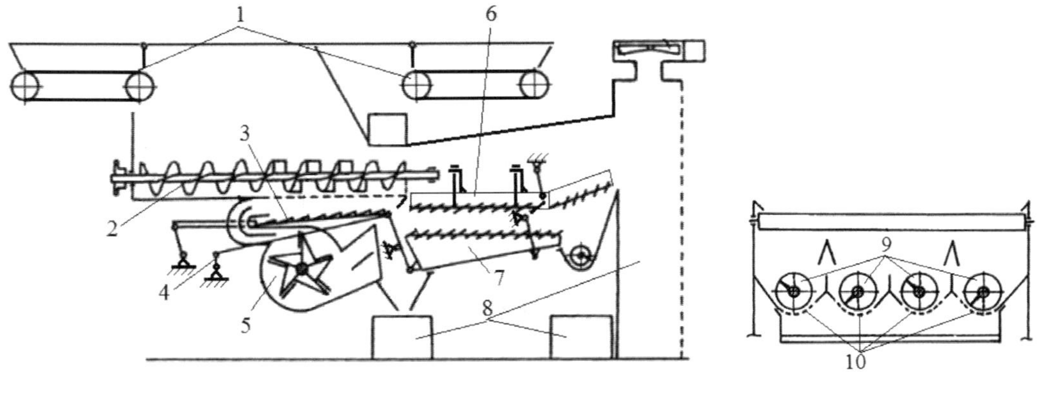

Figure 2. Scheme of the laboratory setup: side view (a); cross section of a sieve screw separator (b); conveyor feeders (1); sieve screw separator (2); shaking board (3); oscillatory drive mechanism (4); fan (5); upper sieve with longitudinal partitions (6); lower sieve (7); containers for collecting separated materials (8); auger (9); sieve (10).

The comparison was carried out with a basic cleaning system containing a shaking board, an upper sieve, a lower sieve and a centrifugal fan. The experiments were carried out on wheat heap with a moisture content of 10 to 12% and a content of straw impurities of 30%. Design and technological parameters of a laboratory setup with a sieve screw separator: pitch of sieve screw separator blades - 100 mm; blade width - 60 mm; screw rotation speed - 330 rpm, sieve holes diameter - 12 mm; fan impeller rotation speed - 690 rpm; the gaps between the fins of the upper and lower sieves are 13.5 mm and 9 mm, respectively. Qualitative indicators of the operation of the cleaning system were determined — grain loss (P, %) and the content of impurities in grain (Z, %).

4. Results and Discussion.

The displacement of the grain heap on the upper sieve with the lateral tilt of the combine leads to a change in the distribution of the grain heap on the upper sieve. Basic cleaning parameters: two longitudinal partitions 0.05 m high were installed on the shaking board; three partitions 0.05 m high are installed on the upper sieve; coefficient of variation at the beginning of the upper sieve $V_{hn} = 0.46$; at the end of the upper sieve $V_{hk} = 0.52$. The sideways shift of the grain heap on the upper sieve leads to an increase in grain losses (figure 3).

It was found that with a transverse roll of 8° (α), grain losses of 0.5% were observed when feeding grain heaps of 1.6 kg / s (q), and when feeding 2.9 kg / s, grain losses amounted to 1.17%. The theoretical value of grain losses calculated by the formula (1) equals P=1.25%.

In a cleaning system with a sieve screw separator, screw conveyors operate autonomously, transporting and separating grain heaps. This leads, firstly, to a decrease in the load on the upper sieve (up to 40% of the total grain weight comes onto the sieve), and secondly, the distribution of the grain heap is improved - the coefficient of variation at the beginning of the upper sieve is $V_{hn} = 0.24$. However, in the cleaning system of the scheme under consideration, the grain heap is shifted sideways on the upper sieve towards the slope, which causes an increase in the coefficient of variation by the end of the sieve to 0.52, which explains the high level of grain loss. When feeding a grain heap of 2.9 kg / s, the grain loss was 0.75% (figure 4).

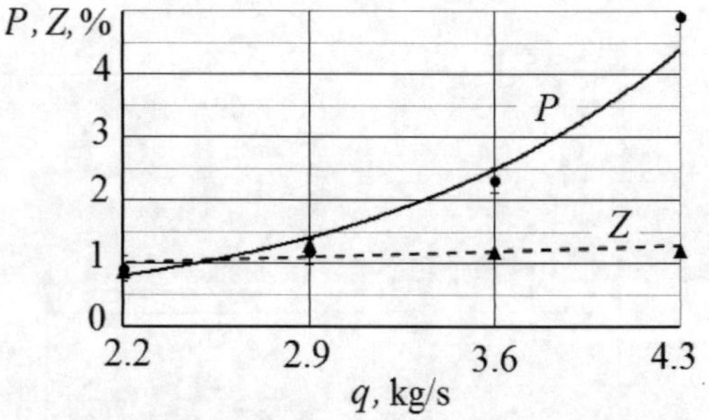

Figure 3. Qualitative indicators of the basic cleaning system.

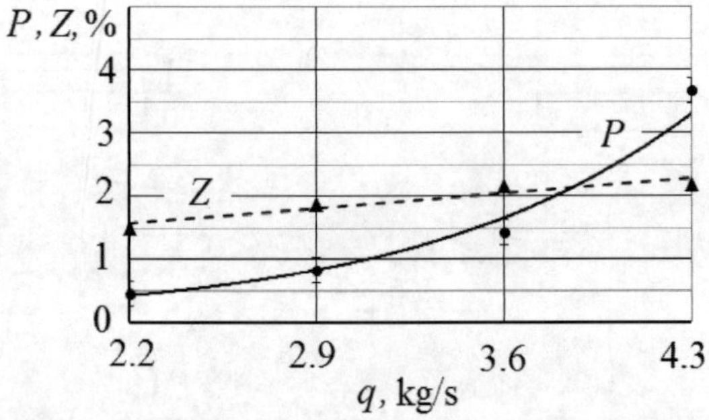

Figure 4. Qualitative indicators of the cleaning system with a sieve screw separator.

The shift of the grain heap sideways towards the slope can be prevented by installing longitudinal partitions on the upper sieve. Since the sieve screw separator consists of four screws, it is advisable to install three longitudinal partitions on the upper sieve, dividing the upper sieve into four parts. Figure 5 shows the results of calculating the grain heap distribution at the end of the upper sieve, provided that at the beginning of the upper sieve the grain heap is fed uniformly across width. The data were obtained according to the developed method [6]. To avoid the redistribution of grain heaps on the upper sieve, the height of the partitions should be at least 0.13 m.

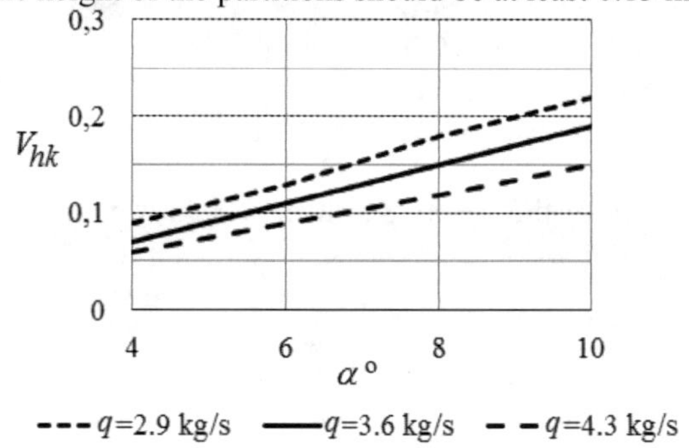

Figure 5.

Therefore, one of the technical solutions of a combine harvester with a sieve screw separator is the use of four longitudinal partitions 0.13 m high on the upper sieve. To feed the grain heap to the upper sieve at a rate of 2.9 kg / s and with a lateral tilt angle of 8 °, the coefficient of variation at the beginning of the sieve is 0.24, and at the end of the sieve 0.18.

The grain loss of the proposed cleaning system was determined by the formula (1). The calculation was carried out with a length of the upper sieve of 1.3 m, a grain separation coefficient of 5.1 m^{-1}. The estimated value of grain losses is 0.14%, which is 5 times lower than the cleaning system with a sieve screw separator and 8 times lower than the basic cleaning system.

5. Conclusion

The cleaning system with a sieve screw separator containing four screws and three longitudinal partitions 0.13 m high mounted on the upper sieve is proposed. Loss of grain in the proposed cleaning system is reduced eightfold in comparison with the basic cleaning system.

References

[1] Boettinger S and Fliege L 2011 *Conference: Agricultural Engineering: Land.Technik AgEng.* **2124** 63-8.
[2] Berdyshev V Ye and Lomakin S G 2017 *Bull. Krasnoyarsk State Agrar. Univ.* **7** 186-90.
[3] Kotov A V and Chaus V P 2010 *Bull. Gomel State Tech. Univ.* **2 (41)** 3-10.
[4] Sitnikov A A, Sorochenko S F and Dryuk V A 2017 *Achievements of Science and Technology in Agro-Industrial Complex* **3(31)** 35-8.
[5] Sorochenko S F 2017 *Bull. Altai State Agrar. Univ.* **12 (158)** 134-40.
[6] Sorochenko S F 2014 *Polzunov Bull.* **4/2** 64-9.

Pneumatic boom applicator for dispersion of mineral fertilizers

Nikolay Byshov[1], Valentin Makarov[1], Olga Zhravleva[2], Alexander Shemyakin[1], Oksana Starovoitova[3] and Aleksandra Manokhina[4,5]

[1]Ryazan State Agrotechnological University Named after P.A. Kostychev, Ryazan, Russian Federation
[2]Ryazan State Institute for Technical Support of Agriculture, Branch of Federal State Budgetary Institution "Federal State Research Center VIM", Ryazan, Russian Federation
[3]Lorch Potato Research Institute, Moscow region, Russian Federation
[4]Russian State Agrarian University - Moscow Agricultural Academy named after K.A. Timiryazev, Moscow, Russian Federation
[5]Author to whom any correspondence should be addressed.

Corresponding author's e-mail: alexman80@list.ru

Abstract. Almost 90% of machines are used with throw-type seeding devices. Serious disadvantages of such machines include uneven application of 25% or more, depending on air humidity and wind force and can reach about 50%. The best seed-fertilizer seeders have the best performance of about 10% irregularity of application, but they have a small width of application and productivity, and in practice they are not used to fertilize the soil when preparing for sowing crops. The solution to the problem of increasing productivity and improving the uniformity of fertilizer application can be found by creating pneumatic boom applicators with a greater width than that of grain-seeders.It was found that the test machines in the field had 8% non-uniformity of the introduction of ammonium nitrate that was lower than in a case of machines equipped with throwing working bodies and approved by agricultural requirements for them. The obtained non-uniformity of the fertilizer application allows to evaluate its indicators even at the design stage of the machines, depending on the quality of dosing and the distance between the sowing ejectors.

1. Introduction

Modern science is in constant search for improving the uniformity of the application of solid mineral fertilizers in the system of agriculture and the production of high-quality products.

When determining the quality indicators of boom applicators for solid mineral fertilizers, special attention is paid to the uniformity of dispersing their particles on the field surface [1, 2, 3, 4, 5]. In this regard, granulometric composition of fertilizers, the uniformity of their supply by roller feeds, air humidity, wind direction and speed relative to the axis of movement of the unit across the field are important for machines with throwing working bodies.

The purpose of the study is the creation of a pneumatic boom applicator that allows increasing the uniformity of fertilizer dispersion when making the main dose and top dressing on vegetating plants.

2. Research objective

The development of a pneumatic boom applicator with a working width of 12 meters and irregularity of application within 10%, regardless of temperature fluctuations, humidity and wind direction.

3. Materials and methods

In the second half of the 20th century, boom applicators for dispersing mineral fertilizers were widely used in the world practice. Legalized in Russia agrotechnical requirements for machines for fertilization found that the uneven application for machines of a throwing type should not exceed 25% of the width of the application and no more than 10% in the direction of movement. For boom applicators these figures should not be more than 10% and 5%, respectively. The technological scheme of the machine, the shape and dimensions of metering and distribution devices (as design parameters), the seeding rate of the fertilizer, the distance between the holes in the boom, air pressure and the machine speed (operational parameters) can be considered as factors influencing the quality of dispersion in pneumatic machines [6].

The uniformity of distribution of fertilizers on the area will depend on many factors. First, it will depend on the even distribution of fertilizers in individual boom ejectors, caused primarily by the main parameters of the working bodies of the centralized seeding system.

Such factors as the seeding rate, air pressure, speed of the machine and parameters of the outlet section have no less influence on the uniformity of distribution of fertilizers. The required width of the sowing strip with one ejector outlet in the boom and the quality of fertilizers distribution over the width are directly related to their flight range when they hit a reflecting scattering surface, which, in turn, is determined by the initial velocity of fertilizer particles when they arrive at a distribution reflecting surface.

The schematic diagram of the boom applicator for dispersion of solid mineral fertilizers is shown in Figure 1.

To solve the problem of increasing the uniformity of the surface dispersion of solid mineral fertilizers and eliminating the shortcomings of existing spreaders, it is possible to solve the problem by using pneumatic boom applicators (Figure 2).

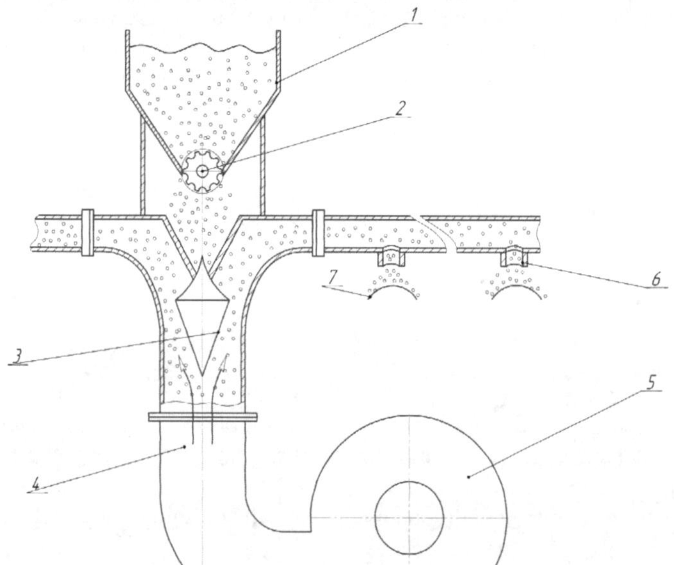

1 – hopper; 2 – dispenser; 3 – flow divider; 4 – air supply direction,
5 – fan; 6 – pipeline with ejectors; 7 – spheres of dispersion

Figure 1. The schematic diagram of the device for dispersion of mineral fertilizers

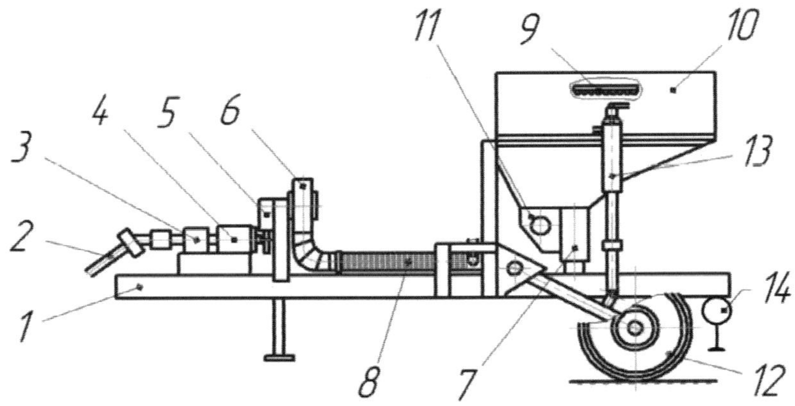

1 – frame; 2 – driveshaft; 3 – overrunning clutch; 4 - elastic coupling; 5 - belt drive; 6 - fan; 7 – mixing chamber; 8 – duct; 9 – fertilizer spreader; 10 – bunker; 11 – sowing unit; 12 – support wheel; 13 – hydraulic cylinder; 14 – boom with ejector

Figure 2. The schematic diagram of a pneumatic boom applicator for dispersing solid mineral fertilizers

4. Results and discussion

The quality of fertilizers dispersion by a pneumatic boom applicator with holes along the length of the boom is mainly influenced by the uniformity of fertilizers supply by roller feeds into the air stream created by the fan and into the boom with holes (let's call them ejectors) and the nature of the distribution of fertilizers depending on the distance between these ejectors (Figure 3).

Achieving this goal was carried out by theoretical and experimental studies. A theoretical study is devoted to obtaining dependencies that allow obtaining rational structural and kinematic parameters of the device for dosing and distributing fertilizers through ejectors and scattering surfaces according to laws and methods of classical mechanics and mathematics [7, 8].

If the irregularity of fertilizers supplies to the air flow to the sowing ejectors, the nature of their distribution and the supply of the air-mineral mixture to the ejectors are known, then the quality of the distribution of fertilizer particles depending on the distance between the ejectors can be assessed even at the design stage of boom applicators and timely necessary adjustments made in their design [4, 5, 9, 10]. The uneven dispersion of fertilizers into ejectors is characterized by average supply m_y and standard deviation σ_y found in the experiment.

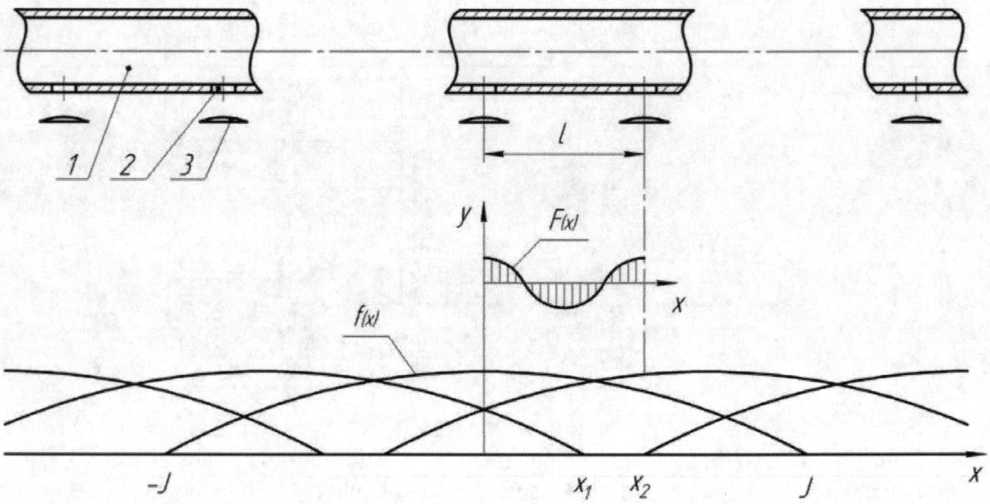

Figure 3. The scheme for determining the non-uniformity of the dispersion of fertilizers according to the width: 1 – boom; 2 – seed ejectors; 3 – diffusers

The distribution of fertilizers by ejectors can be described by a functional dependence of the form *y = f (x)*, which allows determining the uneven distribution when changing the distance between the ejectors. Then, in the area between two adjacent ejectors, the distribution of fertilizers can be described by the following dependence:

$$F_{(x)} = \begin{cases} f(x) + f(x+l) + (x-l) & when\ 0 \le x \le x_1; \\ f(x) + f(x-l) & when\ x_1 \le x \le x_2; \\ f(x) + (x-1) + f(x-2l)| & when\ x_2 \le x \le l \end{cases} \qquad (1)$$

where l – is the distance between adjacent ejectors, m.

The average value of function F (x) in this section is represented as

$$\bar{y} = \frac{1}{l}\int_0^l F(x)dx \qquad (2)$$

Since a certain integral of function *F (x)* on interval l is equal to a certain integral of function *f (x)* on interval *[-j, j]*, the average value of function *F (x)* can be calculated as:

$$\bar{y} = \frac{1}{l}\int_{-j}^{j} f(x)dx \qquad (3)$$

Here $\int_{-j}^{l} f(x)dx$ corresponds to average supply $\overline{m}_D$ of fertilizer into the ejector, so $\bar{y} = m_D / l$.

The root-mean-square deviation of the distribution of fertilizers by ejectors over the working width is presented in the following form:

$$\sigma_p = \sqrt{\frac{1}{l}\int_0^l [F(x) - y]^2 dx} \qquad (4)$$

The standard deviation of the distribution of fertilizers over the field surface, taking into account the error in their supply into ejectors, can be determined by the following dependence:

$$\sigma = \sqrt{\sigma_{dl}^2 + \sigma_P^2 + 2r_{dp}\,\sigma_{dl}\sigma_p} \tag{5}$$

where r_{dp} is the correlation coefficient of fertilizers supply into ejectors and their distribution σ_{dl} is standard deviation of fertilizers supply into ejectors.

Thus, the uneven distribution of fertilizers on the surface of the field along the width of the machine will be represented as the following dependence:

$$N_P = \sqrt{\sigma_d l + \sigma_p^2 + 2r_{dl}\sigma_{dl}\,\sigma_{dl}\sigma_p} \times 100/\bar{y} \tag{6}$$

That is, the non-uniformity of fertilizers distribution along the field surface is equal to the sum of the metering unevenness in ejectors.

For rough calculations, one can take r_dp=1. Then formula (6) takes the following form:

$$N_P = \frac{\sigma_{dl} + \sigma_p}{\bar{y}} 100 = N_d + N_p$$

That is, the non-uniformity of fertilizers distribution along the field surface is equal to the sum of the metering unevenness in ejectors.

So the uneven distribution of fertilizers on the field surface is equal to the sum of uneven metering in ejectors N_p and their distribution by ejectors N_d.

Thus, the proposed method for determining the irregularity of fertilizers distribution can be used when designing a pneumatic boom applicator with ejectors along the length of the boom.

The dispersion unevenness of ammonium nitrate and ammophos in laboratory studies was 6-7% (with a total boom length of 12 m and a spreading rate of 600 kg/ha), which is somewhat lower for machines equipped with throwing working bodies and approved by agro requirements.

The nature of the distribution of ammonium nitrate over the capture width is established by the results of experiments.

When checking the pneumatic boom applicator under production conditions, the dispersion non-uniformity of ammonium nitrate amounted to 7.5-8.0% over the length of the boom (with a total boom width of 12 m and application doses of 200 and 600 kg/ha), which indicates satisfactory match results.

5. Conclusion

Thus, determining the non-uniformity of solid mineral fertilizers distribution over the field, taking into account the non-uniformity of their supply to the sowing ejectors, makes it possible to evaluate the quality indicators of the machine as a whole.

References

[1] Makarov V.A. and Khodakova T.A. 2011 On some areas of applied research and development to improve the field of chemicalization of agriculture in the region. Bulletin of the Russian State Logistics University. 2 19–21.

[2] Kuleshov M.S., Makarov V.A. and Marchenko N.M. 2015 Study of the distribution of the air-mineral mixture in channels of boom applicators for fertilizers. Agricultural machines and technologies. 2 34–36.

[3] Izmailov A.Yu. and Makarov V.A. 2016 To the question of the feasibility study of agricultural machines and equipment. Agricultural machines and technologies. 3 17–19.

[4] Pakma G. and Calil C. 2005 Propriedades fisicas de alguns produtos armazendos. Rer. Brasil. Armazenmebto. Vol. 30. pp. 65–72.

[5] Fronlich G., Broker K.-H. and Rodel G. 2007 Computer Parcel Dispatcher with computer spreader for Exaktersuche. Agricultural Engineering. Vol. 62. 3 Pp. 150–151.

[6] Andreev K.P., Terentyev V.V. and Shemyakin A.V. 2019 Technological process of application of mineral fertilizers by self-loading machine. IOP Conference Series: Earth and Environmental Science XII International Scientific Conference on Agricultural Machinery Industry. Don State Technical University. 012180.

[7] Lachuga Yu.F. 2015 Scientific and methodological support of the development of basic and

IOP Conf. Series: Materials Science and Engineering **941** (2020) 012045 doi:10.1088/1757-899X/941/1/012045

exploratory research in the field of agriculture. AIC: Economics, Management. 2 3–11.
[8] Lachuga Yu.F. and Ksendzov V.A. 2005 Theoretical mechanics. Moscow. Kolos. p. 576.
[9] Izmailov A.Yu. and Lobachevskiy Ya.P. 2013 The system of machines and technologies for the integrated mechanization and automation of agricultural production for the period up to 2020. Agricultural machines and technologies. 6 6–10.
[10] Stepuk L.Ya. and Zheshko A.A. 2012 Construction of machines for chemicalization of agriculture. Minsk: RUE NPC NAN of Belarus on Agricultural Mechanization. p. 442.

IOP Conf. Series: Materials Science and Engineering **941** (2020) 012046 doi:10.1088/1757-899X/941/1/012046

Modeling Elastic and Dissipative Properties of Feed Grain Subjected to Vibration

I Ya Fedorenko[1], A M Levin[2] and A V Tabaev[3]

[1]Altai State Agricultural University, 98, Krasnoarmeysky Avenue, Barnaul, 656049, Russian Federation
[2]LLC STK, 20, Malo-Tobolskaya Street, Barnaul, Russian Federation
[3]LLC Promyshlennoye resheniye, 7, Selektsionnaya St.reet, Stavropol, Russian Federation

E-mail: ijfedorenko@mail.ru

Abstract. Appropriate parameters of feed grain crushers are to be selected relying on the in-depth knowledge of fragmentation regularities in single grains and clear understanding of their deformation rheology. Mathematically, the vibration deformation is a considerably complex process related with originating and propagating cracks in a grain body. Therefore, mechanical analogues and adequate models are needed. The paper demonstrates a suggested rheological model – a combination of Ishinsky, Hook, and Newton bodies correctly demonstrates mechanical processes taking part in a being deformed grain when crushing. Analytically, the loading and unloading curve of a grain is reduced to an appropriate form of the Davidenkov mathematical model.

1. Introduction

To date, the mechanical crushing has been extensively used in the production. However, the intensifying of this manufacturing process is needed, what may be achieved via application of vibratory machinery [1-3], requiring the in-depth knowledge of its operation principles and specifics of the equipment configuration, as well as characteristics of raw materials. In particular, basic parameters of feed grain crushers are selected appropriately provided that the particular attention is paid to fragmentation regularities of single grains and their deformation rheology [4].

2. Materials and methods

A prior analysis [4] of experimental data suggests a rheological model for a single grain, which covers its deformation behavior and provides an explanation of energy dissipation if cyclically loaded.
To begin with, we find out whether an inertia element is to be included into the being developed rheological model. In most cases, the loading of bodies in vibratory crushers is referred to as vibratory crushing. We consider if it is correct to use this term for vibratory grain crushing. For this purpose dynamic operation modes of vibratory and hammer-type crushers are compared.
In vibratory crushers an angular frequency ω working tools affect a material is always below 300 s^{-1} and amplitude A – 0.005 m. Therefore, a maximal velocity of material deformation amounts to $\upsilon = A\omega = 0.005*300 = 1.5$ m/s. This is considerably lower than that of hammer-type crushers (up to 100 m/s).
This difference is even more serious under consideration of a dimensionless criterion used in the impact theory to characterize impact stiffness and called Johnson's damage number [5]. When referring to crushers this criterion is understood to mean an impact stiffness coefficient

$$\pi = \frac{\rho \upsilon^2}{\sigma_T}, \qquad (1)$$

where ρ – grain density;
σ_T – dynamic yield stress of a grain material.
The ratio of these criteria for hammer-type and vibratory crushers is as much as $100^2/1.5^2 \approx 4444$.

Thus, the deformation of grains in the vibratory crusher can be defined quasi-static, and the inertia element can be rightly omitted in the rheological model.

3. Indings The proposed rheological model comprises two blocks (Figure 1).

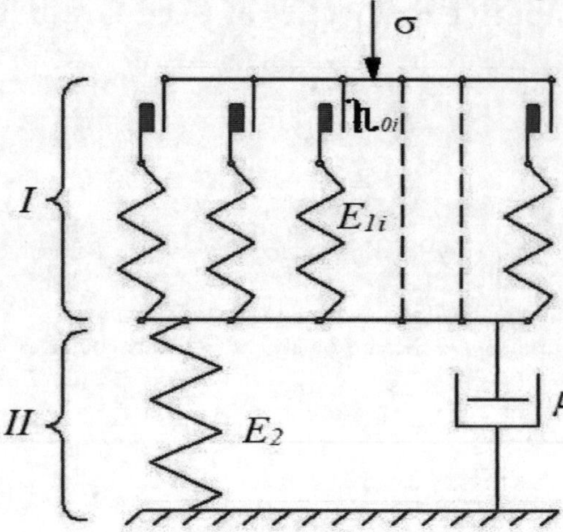

Figure 1. The rheological model of a grain

Block I presents a set of dry friction type elements h_{0i} (general Saint-Venant body), each of them is sequentially joined with its elastic element E_{1i}. As known, a Saint-Venant body is used to model plastic properties of a material. It is not malleable if loaded below the yield stress and tends to deform unlimitedly above it. An equation of state for a shear in a lever *i* can be expressed:

$hi < h_{0i}$ – no deformation; $hi > h_{0i}$ – flow (sliding). (2)

Importantly, within this model a dimensionless microscopic yield stress of a lever *i* is used.

$$h_{0i} = \tau_{0i}/E_{1i}, \quad (3)$$

where τ_{0i} – yield stress of a lever *i*.

Since the model includes a set of Saint-Venant bodies, their yield stresses are different ($h_{01} < h_{02} < .. h_{0i} < h_\infty$, ($i=1...\infty$, see Figure 1). A general Saint-Venant body, therefore, emphasizes the heterogeneity of strength properties throughout a grain.

For an insignificant load below h_{01} a general Saint-Venant body is not malleable. Given that $h_{02} > h > h_{01}$, a sliding motion appears in the first element only. The increased loading makes other elements operate.

This sliding motion is possible if each plastic element is sequentially connected with a Hooke element.

Assume the rigidity modulus of all elastic elements is similar and equals E_1, and a set of microscopic yield stresses has a continuous spectrum with a distribution density *p(h).* Here block I in our model turns into the Ishlinsky model [6].

Summing forces in all levers of the model with a probability *p(h)* of any value of *h,* stresses in block *I* are formulated [6]

$$\sigma_1 = E_1\varepsilon_1 - E_1\int_0^\infty \varepsilon_h\, p(h)dh; \quad (4)$$

$$h\, sgn(\dot{\varepsilon}_h) + \varepsilon_h = \varepsilon_1, \quad (5)$$

where ε_1 – full deformation of a random lever, equal to all levers;

ε_h – plastic deformation in a lever *h*.

The function $sgn(\dot{\varepsilon}_h)$ takes a value of +1 at a positive deformation rate and –1 for a negative deformation rate. For $\dot{\varepsilon}_h=0$ this function may take any value from a range of (+1,-1).

Several cracking types and surface sliding of a crack are reported [7]: I –normal fracture (abrupture); II – cross shear (or simply shear) III – longitudinal shear (cut off). It is apparent block I in the rheological model (see Figure 1) demonstrates the second cracking type, and it is supposed to be a key one in a grain.

Hence, in the proposed model Saint-Venant elements refer not only to plastic deformations but also to the crack propagation under application of a compressive force. Plastic deformations are known to cause the origination and propagation of cracks [7]. However, not all cracks open up when primary crushing a grain, some of them remain in fragments.

Further, we select a distribution density of yield stresses $p(h)$. Since fatigue fracture defects are attributed to Saint-Venant elements, it is quite reasonable to pay attention to an experimental relation of the restoration coefficient $R=f(\upsilon)$ formulated as a function of a grain impact velocity [4]. Obviously, the lower R the more defects arise in a being deformed grain.

In terms of this relationship the pattern of defects $F(h)$ is stated

$$F(h)=1-exp(-\alpha h), \quad (6)$$

where α – coefficient estimated experimentally.

This correlation can be approximated by a less complex exponential formula

$$F(h)\approx h^{\beta}, \quad (7)$$

where β – coefficient (a positive constant).

The coefficient β can be expressed with the help of the experimentally determined coefficient α. In fact, the equating of right parts in expressions (6) and (7) results in

$$1-exp(-\alpha h) = h^{\beta}.$$

Taking the logarithm of right and left parts in the last-mentioned expression, it is found

$$\beta = \frac{\ln\{1-\exp(-\alpha h)\}}{lnh}. \quad (8)$$

To calculate an averaged value ß of this coefficient the right part is integrated in a range from 0 to h

$$ß = \int_0^h \frac{ln\{1-exp(-\alpha h)\}}{lnh} dh.$$

It is impossible to determine analytically this integral. So, we use the numerical method and MathCAD system. We apply probable integration limits from 0 to 0.3 and a value of $\alpha = 0.037$ to wheat. The estimation of the integral is $ß = 0.82$.

To sum up, it is quite reasonably to use a simplified formula (7). Differentiating this expression with respect to h, a probability density for an integral (4) is determined

$$p(h)= ß\, h^{ß-1}. \quad (9)$$

Substituting this expression into the mentioned integral and estimating it, we obtain

$$\sigma = E(\varepsilon - \frac{\varepsilon^{ß+1}}{\varepsilon+1}). \quad (10)$$

The expression obtained complies with the N.N. Davidenkov formula [6], which includes both loading and unloading curves of a material in the cycle (Figure 2).

Comparing this correlation with the available experimental data [8], their sufficient qualitative congruence is revealed. Therefore, our assumptions appear to be true.

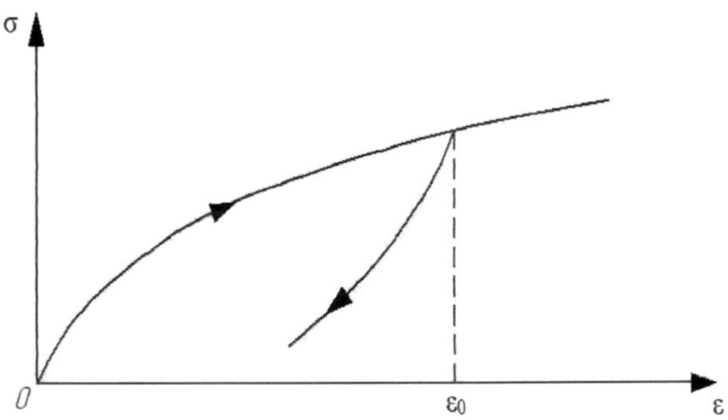

Figure 2. The hysteresis loop curve stated in the formula (10)

The rheological model of a grain includes an elastic E (Hook body) and viscous (Newton body) μ element (block II, Figure 1). However, the experimental data [8] show phenomena of creep, inverse creep, and relaxation of stresses are poorly developed in a grain and arise if mechanically loaded in a continuous manner (up to 4-5 minutes). The pulsed loading in vibratory crushing is faster.

Consequently, viscous properties of a grain are not essential to power consumption when it is subjected to deformation in the vibratory crushing.

We turn now to general reactions of the rheological model to cyclic loads. Thus, if a load is applied to the model, springs E_1 get compressed immediately (see Figure 1), and a spring E_2 compresses gradually, as well as a piston μ moves. Given that a load is below the yield stress the weakest part in the general Saint-Venant body has, the model behaves almost elastically (for times complying with the vibratory crushing). An increase in the external loading causes the sliding in elements of the general Saint-Venant body; furthermore, the higher a load, the more elements are involved into sliding. It imitates the origination of microcracks in a grain body.

IOP Conf. Series: Materials Science and Engineering **941** (2020) 012046 doi:10.1088/1757-899X/941/1/012046

When unloading, the springs E_1 extend rapidly, and their potential energy returns to its source. The spring E_2 extends slowly due to the piston μ. If a preliminary load is high and the sliding is observed in elements of the general Saint-Venant body, the reverse sliding might occur when unloading. It simulates the micro-plasticity and overhead energy consumption to be minimized.
Some Saint-Venant friction couples demonstrate no sliding. It is the reason for the incongruent loading and unloading curves of a grain (see Figure 2). Here, microstresses may remain in the grain body.

4. Conclusion

In summary, the proposed rheological model – a combination of Ishlinsky, Hook, and Newton bodies – represents correctly mechanical processes taking part in a being deformed grain when crushing. The analytical expression for a loading-unloading curve of a grain is formulated as the Davidenkov mathematical model.

References

[1] Veisberg L A 2004 Vibrating crushers. Principles of calculation, design and technological application (SPb.: ARRIGS publishing house) p. 306.
[2] Crusher of grain material. Fedorenko I.Ya., Levin A.M., Tabaev A.V. Patent of invention RU 2688424 C1, 21.05.2019. Application № 2018105815 received 15.02.2018 (in Russian) Izmel'chitel' zernovogo materiala. Fedorenko I.Ja., Levin A.M., Tabaev A.V. Patent na izobretenie RU 2688424 C1, 21.05.2019. Zajavka № 2018105815 ot 15.02.2018
[3] Fedorenko, I & Levin, A & Tabaev, A. (2019). Dynamic properties of vibration crusher of feed grain taking into account technological loading. IOP Conference Series: Earth and Environmental Science. 341. 012115. 10.1088/1755.
[4] Rheological model of a single. Fedorenko I.Ya., Levin A.M., Tabaev A.V. In the collective book: Theory and practice of the present day agrarian science. Collective papers of the III national (all-Russian) scientific conference with international participation. Novosibirsk State Agrarian University. 2020. – p. 476-479. (in Russian) Reologicheskaja model' edinichnoj chasticy furazhnogo zerna. Fedorenko I.Ja., Levin A.M., Tabaev A.V. V sbornike: Teorija i praktika sovremennoj agrarnoj nauki. Sbornik III nacional'noj (vserossijskoj) nauchnoj konferencii s mezhdunarodnym uchastiem. Novosibirskij gosudarstvennyj agrarnyj universitet. 2020. – p. 476-479.
[5] Kolmogorov V.L. Johnson U. Rid S.R. Korbert G.G. Impact loading and crushing of solids: review and new theory. Transl. from English by E.E. Verstakova // Edited by V.L. Kolmogorov. – Yekaterinburg: UrB RAS, 2006. – 321 p. (in Russian) Kolmogorov V.L. Dzhonson U. Rid S.R. Korbett G.G. Udarnoe nagruzhenie i razrushenie tverdyh tel: obzor i novaja teorija. Per. s angl. E.E. Verstakovoj // Pod red. V.L. Kolmogorova. – Ekaterinburg: UrO RAN, 2006. – 321 p.
[6] Pal'mov V. A. Oscillations of elastic-plastic bodies. M.: Nauka, 1976. – 328 p. (in Russian) Pal'mov V. A. Kolebanija uprugo-plasticheskih tel. M.: Nauka, 1976. – 328 p.
[7] Parton V.Z. Mechanics of elastic-plastic fracture. M.: Nauka, 1985. – 504 p. (in Russian) Parton V. Z. Mehanika uprugoplasticheskogo razrushenija. M.: Nauka, 1985. – 504 p.
[8] Naumov I.A. Improvement of conditioning and crushing of wheat and rye M.: Kolos, 1975. – 176 p. (in Russian) Naumov, I.A. Sovershenstvovanie kondicionirovanija i izmel'chenija pshenicy i rzhi. – M.: Kolos, 1975. – 176 p.

Stress-strain modelling state and fatigue of the working body of the tillage unit for weed removal

A V Shindelov[1], N M Ivanov[2], A E Nemtsev[2], B B Vakhrushev[2] and A D Nikolaev[1]

[1] Novosibirsk State Agricultural University, 160, st.Dobrolyubova, Novosibirsk, Russia, 630039
[2] Siberian Research Institute of Mechanization and Electrification of Agriculture, SFNZA RASpos. Krasnoobsk, Novosibirsky district, Novosibirsk region, Russian Federation, 630501

E-mail: dr.schindelov@ngs.ru

Abstact. It is possible to increase the efficiency of crop cultivation technology through the using of multifunctional equipment for soil cultivation. The analysis of tillage equipment showed that the development of a multi-functional tillage unit capable of performing the main tillage with weed removal and a set of operations for pre-sowing tillage is relevant. To quickly adapt the soil cultivating unit to changing operating conditions, the method has been developed for regulating the angle of attack of the combing machine, the width of the capture unit and the angle of entry of the working body into the substrate [1]. A method for calculating the fatigue life of the working body of a tillage unit is proposed.

1. Introduction

Repeated application of loads on the working bodies of tillage machines, in particular on the working body of the combing unit, causes their destruction at stresses significantly lower than in the case of their single loading. With a large number of repeated loads, the fracture stress can be not only less than the tensile strength and yield strength, but even less than the elastic limit [1,2]. The destruction of parts as a result of the action of a large number of alternating stresses is a consequence of the fatigue of the metal of the working body of the combing unit. Since the working body is operated under load conditions of variable magnitude, taking into account the peculiarities of their action is necessary, since its destruction will lead to failure of the unit and disruption of the technological process of pre-sowing cultivation of the soil. The most appropriate technique in assessing fatigue life and conducting full-scale experiment is the use of computer technologies for engineering analysis of stress-strain state (SSS) and numerical simulation of structures, which include software packages MSC. Patran and MSC. Fatigue.

2. Materials and methods

To assess the SSS and fatigue of the working body, we consider the general case of its contact with an elastic-viscous substrate (soil). The load on the working body of the gun is determined as:

$$F = -m\omega \frac{1}{tg(2\alpha)} \quad (1)$$

where m is the mass of the working body, kg;

ω - acceleration of the working body, $\frac{m}{s^2}$;
$tg(2\alpha)$ is the angle of entry into the working body in the substrate, deg.
The kinematic parameters and device design for weed removal are shown in Figure 1.

By varying the basic structural parameters of the working body, it is possible to create the necessary SSS with a characteristic ratio of the main stresses in its design. In particular, it does not seem difficult to model the SSS of the working body in the design of the tillage unit for removing weeds in the area of its attachment to the disk. The estimated fatigue life of such a design, at present, has low accuracy, since the nature of the deformation of the working body and its attachment point from operational loads are not taken into account.

Modeling was carried out in two stages. At the first stage, a static assessment of the SSS of the working body in the MSC. Patran environment was carried out. The geometric model of the working body is built in the MSC. Patran preprocessor. The finite element mesh (CE) was generated by volumetric eight-node elements in the form of an isoperimetric hexahedron.

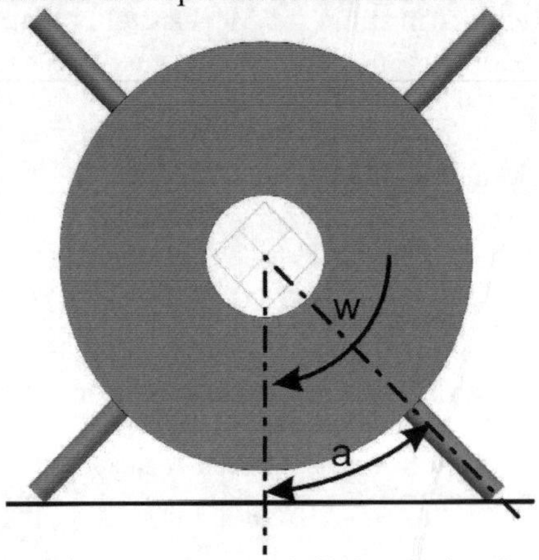

Figure 1. Kinematic parameters and device design for weed removal

The following boundary conditions are accepted:
- in the center of the disk with the working bodies - rotation around the Y axis is allowed;
- on the working body along the Y axis - the rated load of 365, 588 and 750 Newton is applied. The load is determined based on the formula 1 for the acceleration of rotation of the working body 5, 6 and 10 $\frac{m}{s^2}$.

The material of the working body is Steel 65G according to GOST 1050-88 with the following mechanical characteristics:
- modulus of longitudinal elasticity $E = 2 \times 10^5$ MPa,
- Poisson's ratio $\mu = 0.3$,
- tensile strength $\sigma_B = 2200$ MPa,
- yield strength $\sigma_T = 790$ MPa [4,5].

In the calculation model, the boundary conditions of fastening for the disk were taken as rigidly fixed along the supporting surface with the possibility of rotation around the Y axis. A force was applied to the working body along the entire length of the rod. The static solution of the computational model was carried out in the MSC.Patran postprocessor.

3. Results and discussion

The graphical interpretation of the results of the solution is shown in Figure 2. From the graphs of the stress isolines it follows that the maximum stresses are concentrated in the area of sealing the combing bar in the disk. So, the maximum stresses are observed during the rotation of the combing apparatus with an acceleration of 10 $\frac{m}{s^2}$ and reach 2/3 of the yield strength of the material. The durability of the

IOP Conf. Series: Materials Science and Engineering **941** (2020) 012047 doi:10.1088/1757-899X/941/1/012047

working body in terms of fatigue strength with the identified type of SSS was estimated taking into account the linear accumulation of damage in the MSC. Fatigue software environment.

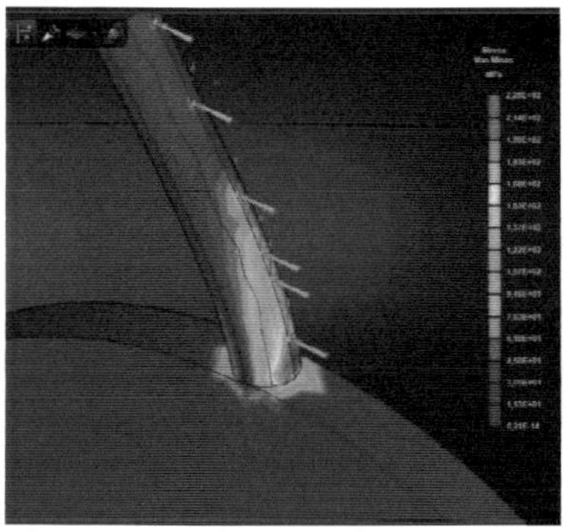

The load on the working body - 365 Newton (corresponds to the acceleration of rotation - 5 $\frac{m}{s^2}$, the angle of entry into the working body in the substrate - 10 ^{0}

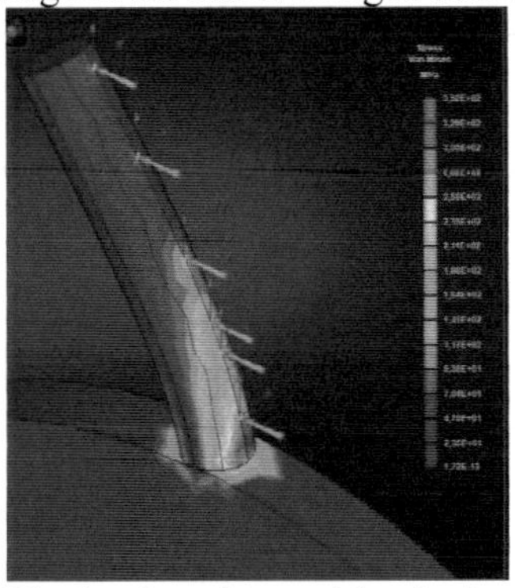

The load on the working body - 588 Newton (corresponds to the acceleration of rotation - 6$\frac{m}{s^2}$, the angle of entry into the working body in the substrate - 10 ^{0}

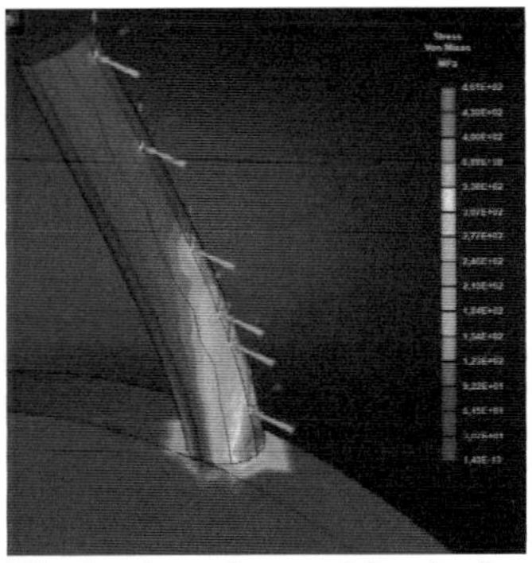

The load on the working body - 750 Newton (corresponds to the acceleration of rotation - 10$\frac{m}{s^2}$, the angle of entry into the working body in the substrate - 10 ^{0}

Figure 2. Isolines of stresses of the working body of the apparatus for removing weeds

The fatigue life analysis procedure included the following steps:

- the history of loading of the working body was set, taking into account the static load, which varies in time;
- the curve of material fatigue and the probability of destruction of the working body were set.

The history of loading is given in the amplitudes of the stresses of the working body, based on its stress-strain state, in connection with this, the constructive resource (by fatigue strength) in the area of attachment to the disk is determined. Since the data on the structural fatigue curve are not known, the fatigue life was determined taking into account the existing Weler curve for Steel 65 G.

Consider an example with the conditions: the load on the working body is 365 N (corresponds to an acceleration of rotation of 5 $\frac{m}{s^2}$, the angle of entry into the working body into the substrate is 10^0 According to the calculated data (Figure 2), the applied force on the working body creates equivalent stresses in the seal corresponding to an amplitude of 299 MPa, while the cycle shape in calculating fatigue life with zero asymmetry is 0.5. Fatigue resistance characteristic Steel 65 G was determined based on the work [5, 6].

Since MSC.Fatigue accepts a linear model of the accumulation of Palmgren - Miner fatigue injuries, and also uses the zero cycle of the working body loading mode to determine durability, the effect of average stresses according to the Goodman hypothesis is taken into account in the numerical experiment. Taking into account the peculiarities of the MSC.Fatigue software package, the probability of sample destruction in the computational experiment is taken to be 50% [7].

As a result of calculating the fatigue life of the working body under complex stress conditions, taking into account the first hypothesis of strength, taking into account the maximum normal stresses in absolute value, failure will occur when:

$\sigma = 299$ MPa after $5.32 \cdot 10^7$ loading cycles;

$\sigma = 352$ MPa after $2.73 \cdot 10^5$ loading cycles;

$\sigma = 461$ MPa after $4.16 \cdot 10^4$ loading cycles.

4. Conclusion

IOP Conf. Series: Materials Science and Engineering **941** (2020) 012047 doi:10.1088/1757-899X/941/1/012047

The analysis of the calculated data allows us to state that the fracture zone will be located in the zone of maximum equivalent stresses located at the attachment point of the working body and the disk. The value of the SSS determined during the static tests does not significantly affect the location of the fracture site, but the value of the maximum equivalent stresses increases significantly from 299 to 461 MPa. At the maximum speed of movement of the working body, the calculated fatigue life does not exceed, in equivalent, the operating time of 1000 m traveled by the track unit. From what it becomes obvious, the need to improve the design of the working body and conduct fatigue tests of field samples.

References

[1] A V Shindelov, N M Ivanov *2019 Inzhenerno-ekologicheskoye obespecheniye bor'by s sornoy rastitel'nost'yu Sel'skokhozyaystvennye mashiny i tekhnologii* **13 (3)** -23 DOI 10.22314/2073-7599-2019-13-3-18-23

[2] A V Shindelov, L N Korobova, A V Tanatova 2014 Tekhnologicheskaya predraspolozhennost' i ekologicheskoye obosnovaniye marshrutizatsii polevykh agregatov *Dostizheniya nauki i tekhniki APK* **11** 22-26

[3] A V Shindelov, N M Ivanov, V M 2018 Medvedchikov *Pochvoobrabatyvayushchiy posevnoy agregat* Patent RU 184739 U1,. Zayavka № 2018109848

[4] A V Shindelov, V M Medvedchikov, N M Ivanov 2018 *Sposob poseva sel'skokhozyaystvennykh kul'tur i ustroystvo dlya yego osushchestvleniya* Patent na izobreteniye RU 2675492 C1 Zayavka № 2018104167

[5] V Bolotin 1990 *Resurs mashin i konstruktsiy* M.: Mashinostroyeniye 448

[6] Rukovodstvo pol'zovatelya MSC. Patran

[7] V A Zhilkin, V V Vakhrushev 2008 Obosnovaniye tekhnologii remonta posadochnykh otverstiy pod podshipniki kacheniya *Vestnik CHGAU* **53** 47-63

IOP Conf. Series: Materials Science and Engineering **941** (2020) 012048 doi:10.1088/1757-899X/941/1/012048

Role of metrology inaccuracy for variance of test results under reproducibility conditions

V.G. Zakhakhatnov, V.M. Popov, V.A. Afonkina and V.N. Levinsky

Federal State Budgetary Educational Institution of Higher Education «South Ural State Agrarian University», 13 ul. Gagarina, Troitsk, Chelyabinsk Region 457100, Russia

zahvg@inbox.ru, ntc-es@mail.ru, afva82@mail.ru, lv_74rus@mail.ru

Abstract. A source of inaccuracy in regression models is supposed to be the statistical nature of a model and metrology inaccuracy related to the measuring explanatory and response variables. A number of approaches to the assessing inaccuracy of regression models are available. However, their use in searching for a solution to engineering problems concerning the assessing multiple regression inaccuracy is restricted by insufficiently developed mathematical tools. When explanatory variables assigned appropriately the inaccuracy of a regression depends on the reproducibility of test results. The reported simulation study investigates the importance of a measurement error to the variance of a normally distributed random variable for small sample sizes between 2 and 30 and coefficients of variation 1, 2, 10, 20, 40, 60. The effect of inaccuracy is analyzed relying on a relative confidence interval of variances estimated as a difference between variance intervals for a variable measured inaccurately and accurately divided by a confidence interval of the variance for an accurately measured variable. The outcomes to emerge from the study are quantity-related estimates for the effect of metrology inaccuracy in the measuring random variables with various coefficients of variation for small sample sizes and different relative measurement errors. For the assessing we use 10^6 retests and a maximal variance for the same number of repeated tests. The results make possible to evaluate an effect of metrology inaccuracy for diverse coefficients of test results variation, as well as determine an appropriate extent of testing for assigned model inaccuracies and a known measurement error. In addition, practical application of obtained results is shown.

1. Introduction

Without analytical treatment of a process, regression models have been broadly used to analyze a relationship between factors related to a value in focus of a researcher. An experiment is carried out; its outcome represents numerical values of a response variable for different values of an explanatory variable. In most cases the method of least squares (LS method) is used to find an unknown regression relationship. In this case it is postulated that a response variable is a normally distributed random variable, whereas explanatory variables are not random with precisely assigned values. In this perspective, the inaccuracy of a regression model depends on an averaged sample obtained according to a variance freedom degree, i.e. a variance of test results under repeatability conditions. A statistical error calculated in this way seems not to involve components related to measurement uncertainty of explanatory and response variables, what is normally typical for practice. Furthermore, a number of tests are usually limited, and it is difficult to control whether the distribution of obtained values complies with the normal law of distribution.

2. Literature Review

Generally, an error is estimated, which is a product of measurement inaccuracy of factor values under condition they are independent and a distribution law of measurement errors "Equation 1" [1].

$$\sigma[y(x1,x2)] = \left[\sum_{i=1}^{n} c_i^2\, \sigma(x_i)^{\,2}\right]^{1/2} \tag{1}$$

here $c_i = {\partial y}/{\partial x_i}$ – sensitivity coeffcients of response function for each factor, which characterize the inaccuracy of a response variable related to each factor. Given that a regression model is adapted, inaccuracy calculated as in (1) includes no component attributed to the statistical nature of a model, which depends on a variancy of response function predicted values [2].

Measurement errors of a response function and factors are taken into account due to the so-called "Deming regression" [3]; it was introduced first for a dual regression and suggests the minimizing an error sum of squares for explanatory and response variables, differing in the variance in order to estimate regression coefficients. Based on this assumption in studies [4, 5] an idea is developed and a numerical method for building a multiple regression is proposed.

Another approach to assess the inaccuracy of regression models with the uncertainty of factors is formulated with the help of mathematical tools within an interval analysis suggested first in [6] and further developed by A.P. Voshchinin [8…10] and A.I. Orlov [11…13] for solving interval systems of equations [7] and determine the regression uncertainty. In view of this theory we can take into consideration uncertainty related to the statistical nature of the model and metrology inaccuracy caused by a measurement error of a response variable. The study [14] generalizes this method for a case there is a measurement error both of explanatory and response variables. The interval method makes it possible to estimate an upper and lower boundary of a response variable with no regard to a distribution law of measurement errors.

I.M. Sobol' [15] suggests a notion of global sensitivity indices to estimate an influence of a variable or a group of variables on the output of mathematical models. The method has been broadly employed in a variety of research domains to assess the sensitivity of models with a high number of variables to factors and their groups [16…20]. However, its use suffers from certain restrictions for it is a complex issue to find multidimensional integrals within a range of varying factors.

The application of approaches mentioned above to the solution of engineering problems, such as the assessment of multiple regression inaccuracy, is complicated by insufficiently developed mathematical tools normally intended for a double regression or a significant number of complex calculations.

Since the presented approaches are somehow attributed to the computing variances, this study aims to provide a quality-related assessment how a measurement error of a random value influences its variance for small size samples and diverse coefficients of variation; it represents an important issue for practical use of regression models.

3. Materials and Methods

A variance of a predicted value for a response function can be determined by "Equation 2" [2].

$$s^2_{\{\hat{y}_i\}} = X_i (X^T X)^{-1} s^2_{\{y\}} X_i^T \tag{2}$$

here X_i^T – a transposed matrix of conditions for i experiment; X – a matrix of experiment conditions (plan matrix); $s^2_{\{y\}}$ - variance of reproducibility.

It is an obvious fact that a measurement error is relevant for the reproducibility variance. The statistics of interval data shows a measurement error of a random value extends a confidence range of a variance calculated on the base of these measurements [12].

$$(s^2(y) - A;\ s^2(y) + A)$$

here $s^2(y)$ – variance without a measurement error.

$$A = \frac{u(\gamma)}{\sqrt{n(n-1)}} \left[\sum_{i=1}^{n} \left(y_i^2 - \frac{1}{n}\sum_{j=1}^{n} y_j^2\right)^2\right]^{1/2} + \frac{2\Delta}{n-1}\sum_{i=1}^{n} |y_i - \bar{y}| \tag{3}$$

here $u(\gamma)$ – a quantile of a normal standard distribution; Δ - an absolute maximal measurement error.

"Equation 3" assumes the expansion of a confidence range for certain experimental results depends on a significance point, measurement error, and a sample size. In this connection, it is necessary to determine an influence trend of inaccuracy and sample size on a variance of a random value, in our case – a variance of test results under conditions of repeatability.

It was a purpose of a numerical experiment under following assumptions:

- measurement results of a response variable represent a normally distributed random value;
- metrology inaccuracy contains no regular component;
- metrology inaccuracy is a normally distributed random value.

The matter of an experiment is that a sample with $n \in [2\ldots30]$ values is modeled from a general population distributed according to a normal law, which matches with values of a random value measured accurately. Values beyond the scope of an interval equal to a significance point of 0.05 are eliminated from the sample. Further, a sample is formed, which corresponds to measured values with relative errors of δ =1%, 2.5%, and 5%. For this purpose each value of a sample from the general population is used as a distribution center, and a mean-square deviation is assigned through a probable "measurement error". Values out of the interval with a significance point of 0.05 are also ignored. Confidence intervals for variance and mean-square deviation are determined by "Equation 4"and "Equation 5"

$$\frac{sd^2(n-1)}{\chi^2_{1-\frac{\alpha}{2},n-1}} < \sigma^2 < \frac{sd^2(n-1)}{\chi^2_{\frac{\alpha}{2},n-1}} \quad (4)$$

$$sd\sqrt{\frac{(n-1)}{\chi^2_{1-\frac{\alpha}{2},n-1}}} < \sigma < sd\sqrt{\frac{(n-1)}{\chi^2_{\frac{\alpha}{2},n-1}}} \quad (5)$$

here $\chi^2_{1-\frac{\alpha}{2},n-1}$ and $\chi^2_{\frac{\alpha}{2},n-1}$ – critical values of a distribution χ^2 for significance points of 0.975 and 0.025, respectively, sd^2 – a sampling variance, σ^2 – a general population variance, n – a sample size. To determine an interval fitting the variance of "measured accurately" samples, sd^2 is calculated as an averaged in view of variance degrees according to 10^6 iterations. To find an interval matching with a variance of samples "measured inaccurately", a maximal value of sd^2 from 10^6 iterations was selected for substituting into "Equation 3", what, to the point, is in compliance with the methods of interval analysis when an interval is assigned through a distribution quantile.

To compare samples "measured accurately" and "measured inaccurately" we sought for a relative confidence interval in form of a difference between intervals determined for a variance calculated with and without inaccuracy divided by an interval for a variance "without inaccuracy".

A distribution center of a modeled general population is accepted to be 100. We calculated for coefficients of variation $\upsilon = \frac{\sigma}{\bar{y}}100$ equal to 1, 2, 10, 20, 40, and 60.

4. Results of Research

Table 1 provides data on a fractional increase in a confidence interval of the variance determined for an asymptotic approximation at $N \to \infty$ ($N = 10^6$) and a sample size $n \in [2\ldots30]$, a significance point is accepted 0.05, coefficients of variation are different.

Table 1. A fractional increase in a confidence interval for variances, $N \to \infty$ ($N = 10^6$), and $n \in [2\ldots30]$ for a point of significance 0.05.

	υ =1	υ =2	υ =10	υ =20	υ =40	υ =60
δ=1%	0.96	0.24	0.01	0.0	0.0	0.0
δ=2.5%	6.22	1.5	0.06	0.01	0.0	0.0
δ=5%	24.97	6.20	0.24	0.06	0.02	0.01

Table 2. A fractional increase in a confidence interval for variances determined on the base of maximal variances from 10^6 iterations, a point of significance is 0.05.

	n	2	3	4	5	6	7	8	9	10	15	20	30
$\upsilon=1$	δ=1%	57.4	23.6	20.5	13.3	11.4	9.0	8.3	8.1	7.4	5.9	5.0	4.4
	δ=2.5%	171.0	78.5	55.8	44.3	40.2	36.4	30.4	29.5	26.7	23.2	18.9	16.0
	δ=5%	477.5	228.6	184.3	147.9	127.5	118.4	108.7	107.1	99.3	80.5	68.9	56.9
$\upsilon=2$	δ=1%	28.9	12.5	10.0	7.3	6.9	6.7	4.7	4.3	3.8	2.9	2.515	2.4
	δ=2.5%	65.5	27.9	21.2	19.0	13.8	13.2	11.5	10.3	9.6	7.5	7.1	5.2
	δ=5%	156.4	83.8	60.4	48.4	41.8	36.7	31.6	28.7	29.4	21.0	19.2	15.6
$\upsilon=10$	δ=1%	16.2	7.9	6.5	4.1	3.9	3.3	3.0	2.8	2.5	1.8	1.6	1.3
	δ=2.5%	19.9	9.3	7.07	5.0	4.9	4.38	3.5	3.2	2.9	2.1	1.7	1.6
	δ=5%	26.7	12.6	8.4	6.9	6.4	5.8	4.6	4.1	3.8	2.7	2.5	1.9
$\upsilon=20$	δ=1%	15.7	7.3	5.4	4.9	3.7	3.5	3.1	2.7	2.5	1.7	1.6	1.26
	δ=2.5%	17.9	7.7	6.4	4.6	3.6	3.4	2.8	3.0	2.7	2.0	1.8	1.3
	δ=5%	20.5	9.1	7.1	5.1	4.6	4.3	3.1	3.0	2.7	2.4	1.6	1.6
$\upsilon=40$	δ=1%	15.0	7.2	5.6	4.7	3.7	3.5	2.9	2.8	2.6	2.0	1.5	1.3
	δ=2.5%	15.7	7.4	5.6	4.4	4.0	3.5	2.9	2.8	2.6	2.1	1.6	1.2
	δ=5%	17.2	8.0	6.1	4.7	4.4	3.5	3.3	3.0	2.9	2.1	1.6	1.2
$\upsilon=60$	δ=1%	14.7	6.9	5.6	47	4.3	3.5	3.2	2.9	2.6	1.8	1.6	1.2
	δ=2.5%	15.3	7.1	5.6	4.6	3.7	3.5	3.0	2.8	2.6	1.9	1.7	1.2
	δ=5%	16.3	7.5	5.8	5.3	4.1	3.7	2.9	2.6	2.5	2.1	1.7	1.2

5. Discussion and Conclusions

The data presented in Table 1 and obtained for the asymptotic approximation as a number of iterations tends to infinity (N→∞) demonstrate a 96% increase in a confidence interval for υ=1 and inaccuracy of 1%, as well as a 24.97 increase for 5%. Given that υ=2, an influence of a measurement error becomes less significant and a 6.22 times increase is recoded for inaccuracy of 5%. A measurement error is below 1% for δ=1% at υ >10, for δ =2.5% at υ >20, for δ =5% at υ >60. It evidences confidence intervals for a variance at $N\to\infty$ are independent on a sample size but are related to a metrology uncertainty of a measurement, as concluded in [14].

A maximal increase of a variance confidence interval ever possible for a selective assessment (simple sampling) correlates closely with a sample size (Table 2). The worst instance is detected for a minimal coefficient of variation v =1, and a maximal measurement error of 5%. If a sample size n=2, an error-related (5%) increase of the interval under selective assessment with a probability of 95% may be 477.5 times more expanded than a variance confidence interval for a value "measured accurately". When increasing a sample size and a coefficient of variation to 60 an influence of a measurement error becomes far less important, being, however, significant for $n < 30$. A relationship between a relative confidence interval of a variance and a sample size for different coefficients of variation v is provided in Figure 1.

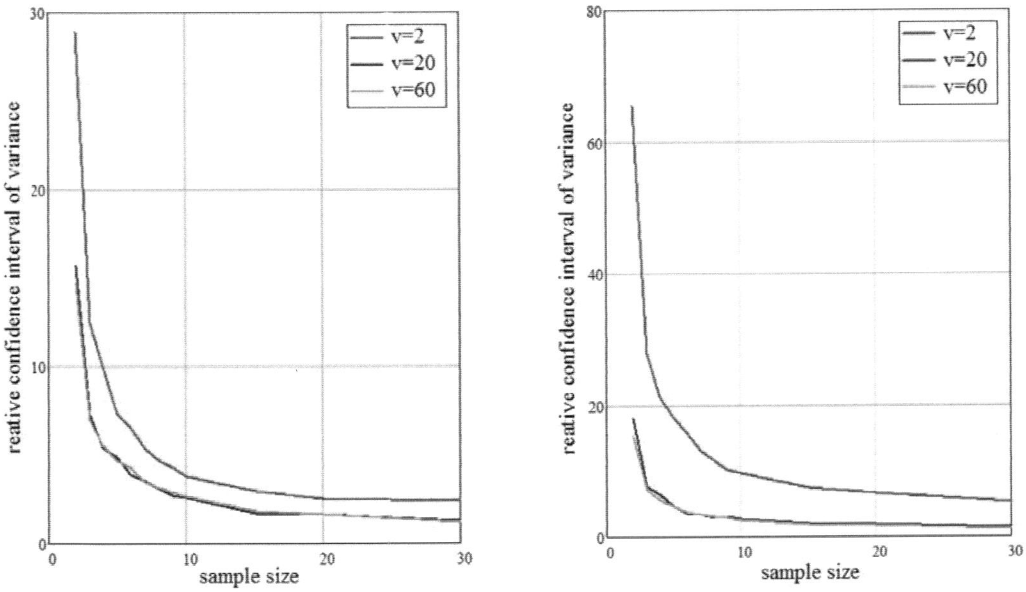

a b

Figure 1. Relative confidence interval of variance vs. sample size for different coefficients of variation *v*; a – a relative error δ=1%, b – δ=2.5%.

We use the method given above to evaluate the uncertainty of a model suggested in [21]. We implemented a full-factorial experiment 2^3. In view of results obtained we suggested output models of a grain dryer unloader for such construction and process factors as grain moisturex_3, operation gapx_2, and vibration frequency of the machine x_1 "Equation 6".

$$\hat{y} = 1046 + 453.8x_1 + 205.3x_2 - 306.4x_3 + 77.1x_1x_2 - 140.8x_1x_3 - 41.6x_2x_3 \qquad (6)$$

Below we provide a plan matrix and outcomes of the experiment carried out in four parallel trials for each combination of factors.

$$Y := \begin{pmatrix} 338 & 335 & 319 & 330.7 \\ 915 & 934 & 965 & 958 \\ 1688 & 1609 & 1603 & 1633 \\ 1322 & 1317 & 1272 & 1304 \\ 586 & 588 & 562 & 578.7 \\ 513 & 491 & 562 & 503.7 \\ 782 & 809 & 816 & 802.3 \\ 2239 & 2275 & 2271 & 2262 \end{pmatrix} \qquad X := \begin{pmatrix} 1 & -1 & -1 & 1 & 1 & -1 & -1 \\ 1 & -1 & 1 & -1 & -1 & 1 & -1 \\ 1 & 1 & -1 & -1 & -1 & -1 & 1 \\ 1 & 1 & 1 & 1 & 1 & 1 & 1 \\ 1 & -1 & -1 & -1 & 1 & 1 & 1 \\ 1 & -1 & 1 & 1 & -1 & -1 & 1 \\ 1 & 1 & -1 & 1 & -1 & 1 & -1 \\ 1 & 1 & 1 & -1 & 1 & -1 & -1 \end{pmatrix}$$

Figure 2. *Y* – outcome matrix of a plan, *X* – plan matrix in coded units

A unit of measurement for unloader output (response values) is accepted "gram" and it was determined with inaccuracy of 1 g. Therefore, a maximal relative error is possible when measuring a minimal value, being $\delta = \left({}^{1}/_{319}\right)100 = 0.3\%$ A coefficient of variation $v = \left({}^{sd}/_{\bar{y}}\right)100\% = 619.254 * 100/1046$=59, a sample size is as high as 32. Here, two approaches to the interpretation of imitation modeling results are possible. In the optimistic perspective, the averaged expansion of a confidence interval to the nearest hundreds according to 10^6 iterations for a variance with a probability of 0.95 equals to zero. Thus, measurement errors can be ignored and measurement results can be referred to as a part of a general population. In the pessimistic perspective, the maximally possible expansion of a confidence interval determined as proposed in the above methods for these conditions of a variance and related to a measurement error is as high as 1.906. For a mean-square deviation a calculated expansion is 0.705. Therefore, it is quite expectable a measurement error may be a reason that a confidence interval of a reproducibility variance expands significantly in the selective assessment, in our case – almost twice.

We calculate regression inaccuracy for both cases. A reproducibility variance is sought as in "Equation 7":

$$s_{\{y\}}^2 = \frac{\sum_{i=1}^{N}\sum_{q=1}^{n}\left(y_{iq}+\bar{y}_i\right)^2}{N(n-1)} = 783.464 \quad (7)$$

Here $s_{\{y\}}^2$– a reproducibility variance calculated on the base of input *X* and *Y*; *N*=8 – a number of lines in a plan, *n*=4 – a number of retests. A mean-square deviation found with the help of (2) $s_{\{\hat{y}\}} = 27.99$. Ignoring a measurement error (optimistic perspective), a confidence interval of a mean-square deviation determined as assigned in (5) for α=0.05 and a sample size *n*-1=31, $\chi^2_{1-\frac{\alpha}{2},n-1} = 48.232$, $\chi^2_{\frac{\alpha}{2},n-1} = 17.539$ equals to [37.2; 22.4]. A confidence interval for a mean-square deviation calculated for a maximally possible deviation of the selective assessment as the above methods propose is [68.6; 47.1], what represent the worst case.

Comparing the outcomes, we can conclude a measurement error affects the model inaccuracy; a confidence interval becomes more expanded and tends to be more significant. Since a regression of interest was obtained as a result of full factorial planning, its inaccuracy is the same on boundaries of a factor space.

6. Summary

A measurement error of a response value is relevant for the reproducibility of test results, so it affects the accuracy of a regression model. A significance point depends on a value of metrology inaccuracy, a sample size, and a coefficient of variation for the response value. Given that a number of tests are increased infinitely, there is almost no correlation between a reproducibility variance and a sample size, a measurement error becomes less important as a coefficient of variation increases. The influence of a measurement error is below 1% for δ=1% at υ >10, for δ =2.5% at υ >20, for δ =5% at υ >60.

In simple sampling a confidence interval is closely related to a sample size and a coefficient of variation, it demonstrates a sharp increase for a sample size less than 10. A higher coefficient of variance decreases a confidence interval, which is 120% for a sample size of 30 at υ >40.

The reported example of the assessing a relative measurement error (0.3%) for a response value in the output model for a grain dryer unloader can increase almost twice the uncertainty of a regression model when simple sapling. The outcomes obtained point at the influence of metrology inaccuracy on a random value distribution variance and make it possible to find an appropriate sample size provided that measurement errors, model inaccuracy, and confidence probability of results are known.

7. References

[1] Guidliness EVRAHIM / SITAK CG 4 QUAM: 2012. *Quantaty-related description of uncertanity in analytical measuring. 3d edition, edit. by* S. L. R. Ellison A. Williams p 156 (*in Russian*)
[2] Adler Yu P, Markova E V and Granovsky Yu V 1976 *Experiment planning when searching for optimal conditions* (M.: Nauka) p 280 (*in Russian*)
[3] Deming W E *Statistical adjustment of data* 2011 (New York, *Dover Publications*) p 288
[4] Basilevsky M P 2016 *Mathematical modeling and numerical methods.* **2(10)** 104-116 (*in Russian*)
[5] Basilevsky M P 2019 *Scientific journal. Series: Economics. Informatics.* V 46 **1** 117-129 (*in Russian*)

[6] Moore R E 1967 *Journal of the Franklin Institute.* 284; Issue **2**:148–149
[7] Shary C P 2020 Finite-dimensional interval analysis. (Novosibirsk, Institute of Computational Technologies SB RAS, XYZ) p 635 (*in Russian*)
[8] Voshchinin A P 1987 Method of optimization objects using interval models of response function. (M.: MEI) p 109 (*in Russian*)
[9] Voshchinin A P and Sotirov G R 1989 Optimization under uncertainty conditions (M.: MEI – Sofia: Tekhnika) p 224 (*in Russian*)
[10] Voshchinin A P 2002 *Plant laboratory.* V 68 **1** 118-126 (*in Russian*)
[11] Orlov A I 2015 *Plant laboratory.* V 81 **3** 51-65 (*in Russian*)
[12] Orlov A I 2004 Non-numerical statistics (M.: MZ-Press) p 513 (*in Russian*)
[13] Orlov A I 2002 Econometrics (M.: Ekzamen) P 576 (*in Russian*)
[14] Gus'kova E A and Orlov A I 2004. *Plant laboratory.* V 70 **11** 57 (*in Russian*)
[15] Sobol' I M 1990 *Mathematical modeling. V*2 **1** 112-118 (*in Russian*)
[16] I M Sobol' 2001. *Maths and Computers in Simulation* **55(1-3)** 271-280
[17] I M Sobol' and Yu L Levitan 1999 *Computer Physics Communications* **117(1-21)** 52–61
[18] Yicheng Zhou, Zhenzhou Lu, Kai Cheng and Wanying Yun A 2019 *Mechanical Systems and Signal Processing* **11715** 498-516
[19] S Kucherenko and S Song 2017 *Reliability Engineering & System Safety* **165** 222-238
[20] O V Klymenko, S Kucherenko and N Shah 2017 *Computer Aided Chemical Engineering* **40** 151-156
[21] Zakhakhatnov V G, Dylenya V K and Kashin P V 2018 *Bulletin of KrasGAU* **4** 99-103 (*in Russian*)

Information and search space for monitoring the machine and tractor fleet

V. D. Chervenchuk[1], G. V. Redreev[1], I. V. Chervenchuk[2], A.V. Shimokhin[1], A. A. Luchinovich[1]

[1]Omsk State Agrarian University named after P. A. Stolypin, Omsk, Russia
[2]Omsk State Technical University, Omsk, Russia
E-mail: gv.redreev@omgau.org

Abstract. The machine and tractor fleet of agricultural enterprises is a set of objects of agricultural machinery linked in technological chains, spread over the area of agricultural land. For effective use of the park, it is necessary to monitor it in order to maintain or restore its working condition in a timely manner through the implementation of repair and maintenance actions. The applied means of maintenance and repair are spread among various objects of the repair and maintenance base, being an integral part of it. However, for monitoring purposes, it is convenient to separate maintenance and repair facilities from the objects of the repair and maintenance base. This technique, which expands the theoretical possibilities of representing the monitoring process, allows you to get practically significant consequences. First, the degree of optimal equipment placement on the objects of the repair and maintenance base is revealed and it becomes possible to solve the problem of improving this placement. Secondly, the formation of a system of element interaction of «park – equipment – base objects» sets creates prerequisites for the development of a computer monitoring system.

1. Introduction

Under the monitoring of the machine and tractor fleet (MMTF), we will understand the system of constant monitoring of the operation, maintenance and repair of the entire set of machines necessary for the work mechanization on the cultivation of crops [1].

As part of this monitoring system, the statistical data is processed to evaluate, control and justify management decisions to ensure the economic efficiency of the machine and tractor fleet (MTF), the reliability and safety of equipment, and the safety of service personnel [2].

One of the most important tasks of creating the MMTF is to develop a data storage model that reflects the specific features of technical information and allows further statistical research. First of all, the specific features of the information that MMTF operates with are:

1) the data structure is based on a system of indicators that have an accepted classification;
2) the data is linked to an object;
3) the data is time-bound.

2. Research methods

IOP Conf. Series: Materials Science and Engineering **941** (2020) 012049 doi:10.1088/1757-899X/941/1/012049

The information core of the statistical information processing system is a hierarchical database. The hierarchy of objects and classification of indicators underlying the proposed model implies the use of a hierarchical structure.

The scientific concept of hierarchical structuring of this information system is based on some ideas of information algebra, which was developed by the CODASYL committee group [3–4].

The information algebra at its core contains two fundamental concepts: the essence and the property of the entity. The entity is physically existing objects, elements of a real system that has a complex hierarchical structure, and the properties of an entity are its logical and numerical characteristics that this entity possesses. For the proposed MMTF, as a system of continuous monitoring, the entity is transformed into the concept of object and property into the indicator concept. This terminology is more typical for monitoring. It was also used in environmental monitoring in Omsk [5-6]. Here the world of objects is a special case of the world of entities, and the world of indicators is a specific feature of the world of properties.

The object will be called the characteristic carrier (logical or numeric). For example, an object can be a machine and tractor fleet as a set of machines necessary for the work mechanization on cultivation of agricultural crops. Such an object consists of the following groups of objects:

- tractors as a universal energy tool;
- agricultural machines connected to it in a common unit (plows, seeders, harrows, cultivators, mowers, various harvesting machines, non-self-propelled, and others);
- self-operating harvesting machines;
- stationary machines with individual or group drive of working bodies;
- transport vehicles, etc.

The agricultural machines included in the MTF are combined into complexes (in accordance with the requirements of complex mechanization), taking into account the production peculiarities. The structure of these machine complexes depends on the farm specialization, production technology, weather, soil and other factors that affect the choice of machines.

3. Results

The aggregates and complexes of machines form a set of objects of technical support for the cultivation of agricultural crops. Let's introduce a notation for this set. It has a hierarchical structure in the form of a root tree (Fig. 1).

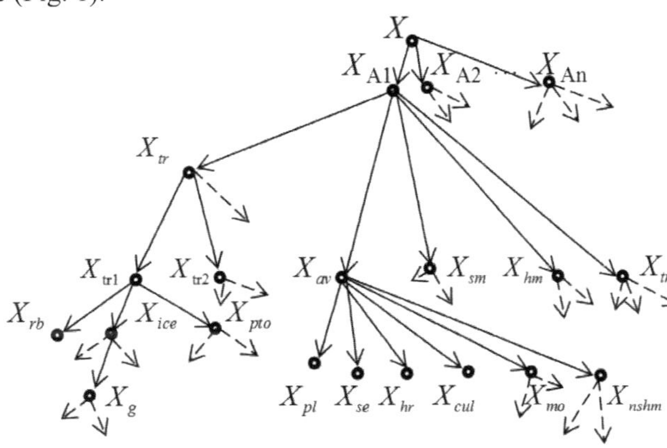

Figure 1. MTF technical support tree

X - tree root; $X_{A1}, X_{A2} \ldots X_{An}$ – aggregates for cultivation of various crops; X_{tr} – tractors; X_{av} – agricultural vehicles (X_{pl} – plough, X_{se} – seeder, X_{hr} – harrow, X_{cul} – cultivator, X_{mo} – mower, X_{nshm} – various non-self-propelled harvesting machines); X_{sm} – stationary machines with individual or group drive; X_{hm} – self-operating harvesting machines; X_{tm} – transport machine; X_{rb} – rechargeable battery; X_{ice} – internal combustion engine; X_{pto} – PTO; X_g – generator.

Another group of objects is a tool for technical maintenance and repair of objects of the first group. This set of objects Y includes machines, test benches, diagnostic scanners, various tools and measuring devices. The structure Y is the same as in X (Fig. 2).

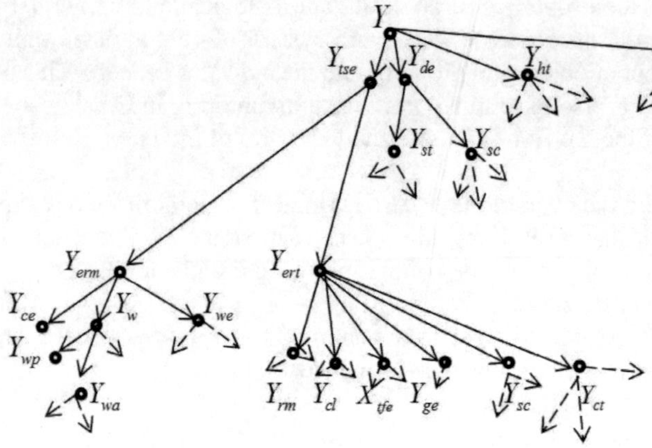

Figure 2. MMTF tool tree

Y – tree root; Y_{tse} – technical service equipment; Y_{de} – diagnostic equipment; Y_{ht} – hand tools; Y_{mi} – measurement instrumentation; Y_{st} – stands; Y_{sc} – scanners and testers; Y_{erm} – equipment for repairing tractors and agricultural machinery; Y_{ert} – equipment for repair of trucks; Y_{ce} – compressor equipment; Y_w – washings;
(Y_{wp} – washing of parts, Y_{wa} – washing of aggregates); Y_{we} – welding equipment; Y_{rm} – riveting machine for brake pads; Y_{cl} – car lifts; X_{tfe} – tire fitting equipment; Y_{ge} – garage equipment; Y_{sc} – start-up chargers; Y_{ct} – current transducers

The garages, hangars, warehouses and other specially equipped premises (battery, generator, electromechanical, mechanical workshops, etc.) intended for storage, maintenance and repair of vehicles and removed onboard equipment form the third group of objects Z. These entities determine the location of objects in the first and second groups. A specific MTF is linked to a certain territory. This territory can be considered as the Z root object. On this territory there are buildings for various purposes, inside which there are rooms premises for the MTF work. Therefore, the objects of the set Z have the same tree structure (Fig. 3).

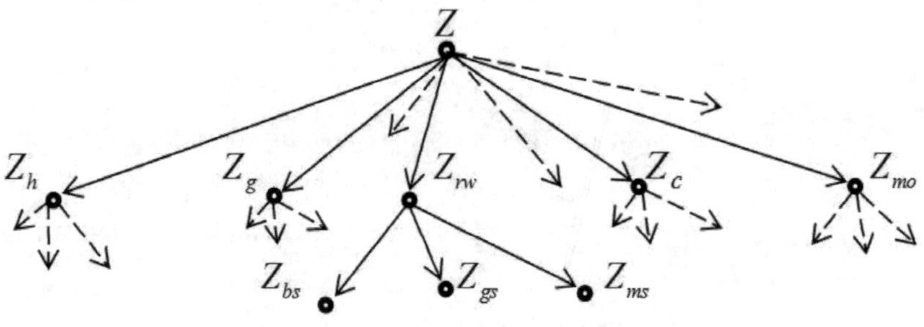

Figure 3. Tree of the MTF production premises

Z_h – hangars; Z_g – garages; Z_{rw} – repair workshops; Z_c – cellarage; Z_{mo} – management offices; Z_{bs} – battery shop; Z_{gs} – generator shop; Z_{ms} – mechanical shop.

The binding objects of the first two groups (X and Y) to its location (group objects Z) is set by matching

$$q_Z = (X \cup Y, Z, Q_Z), \tag{1}$$

where $X \cup Y$ – conformity dispatch area q_Z; Z – conformity arrival area q_Z; Q_Z – law of conformity q_Z, then

$$Q_Z \subseteq (X \cup Y) \times Z. \tag{2}$$

The scope for determining this compliance represents the law projection Q_Z on the first component

$$\Pi p_1 Q_Z \subseteq X \cup Y. \tag{3}$$

This set includes only those items from X and Y for which their exact location on the territory of the MTF is determined. These objects can have indicators (numeric or logical values). During the operation of the MTF, the indicators of objects may change over time. Therefore, all objects must be linked not only by location, but also by time. Time is usually set by date (YYYY. MM.DD) and time of day (hh.mm.ss), which is also represented as the T root tree: year consists of months, month of days, day of hours, hours of minutes, minute of seconds. For a specific point in time, an object's indicator is a specific number or logical value. For a time interval, the indicator will be represented by an array of values.

The binding objects $X \cup Y$ to T time is set by matching

$$q_T = (X \cup Y, T, Q_T), \tag{4}$$

where $X \cup Y$ – conformity dispatch area q_T; T – conformity arrival area q_T; Q_T – law of conformity q_T, then

$$Q_T \subseteq (X \cup Y) \times T. \tag{5}$$

The correspondences (1) and (4) determine the relation of objects $X \cup Y$ to the territorial-temporal space $Z \times T$. This relation is determined by the match

$$q = (X \cup Y, Z \times T, Q), \tag{6}$$

where $X \cup Y$ – conformity dispatch area q; $Z \times T$ – conformity arrival area q ; Q – law of conformity q , then

$$Q \subseteq (X \cup Y) \times Z \times T. \tag{7}$$

Definition 3. The set of triples $I \equiv (X \cup Y) \times Z \times T \supseteq Q$ will be called the MMTF information search space (or simply PE space).

Not all points $(x,z,t) \in I$ where $x \in X \cup Y$, $z \in Z$ и $t \in T$, here will have indicators. Some points in the PE space are logically contradictory (for example, a tractor cannot be located in the accounting department of a management office). Accessing these points to determine its indicators will be accompanied by the message «request error». Some logically consistent points in the PE space may also have no indicators due to underloading of the MMTF database. In this case, when accessing them, the message «there is no data for this object» will be displayed.

In general, each point of the PE space can correspond from zero to different n-indicators (passport data of the technical object, service life, statistics of failures and its elimination, etc.) [7]. With the help of statistical data processing for the same type of objects, the reliability indicators of these technical means, the probability of failure-free operation are calculated, and the terms of their maintenance, preventive and capital repairs are compiled. Here there is a prospect of merging PE space, which provides search and collection of statistical data for specified types of objects, with mathematical support for statistical processing of large information amounts [8–9].

The set of «MTP personnel» also has a tree-like hierarchical structure and each specialist constantly monitors those indicators and achieves their most optimal value, for which he is directly responsible in accordance with the profile of his job responsibilities. The access to the necessary information is provided by hierarchical menus that mirror the tree structures of object search X (Fig. 1), Y (Fig. 2) and Z (Fig. 3). For example, the main menu of the MTF monitoring system can be represented as follows:

The main menu for monitoring the machine and tractor fleet:

1. Service personnel
2. Aggregates and complexes
3. Repair equipment
4. Office premises
5. Exit from the main menu.

When you choose the second point «Aggregates and complexes», a menu consisting of items corresponding to the second tier of the X tree will appear on the computer screen.

Aggregates and complexes for:

1. Grain crops
2. Forage crops
3. Roots
4. ...
5. Return to the previous menu
6. Exit from the main menu.

When you choose the second point «Aggregates and complexes», a menu consisting of items corresponding to the third tier of the X tree will appear on the computer screen.

Aggregates and complexes for grain crops:

1. Tractors
2. Agricultural machinery and tractor trailers
3. Self-operating harvesting machines
4. Stationary machines with individual or group drive
5. Transport vehicles
6. ...
7. Return to the previous menu
8. Exit from the main menu

When you choose the first point «Tractors», the menu will display a list of tractors with the corresponding garage number. When you select any of it, a new menu will appear. For example:

Tractor Belarus, garage no. 08

1. Indicators
2. DVS
3. 6STS-50EMS Battery
4. PTO
5. Chassis
6. ...
7. Return to the previous menu
8. Exit from the main menu.

This object already has indicators (numeric and/or logical values). When you choose the first point «Indicators», a menu appears on the monitor screen with a list of indicators corresponding to this tractor.

Thus, the hierarchical structuring here is based on a given hierarchy of objects, on the basis of which, taking into account the semantics of its description, a classification of indicators and accumulated statistics of its changing values is designed. This concept, in addition to the obvious practical benefits obtained in the development of specialized monitoring (in this case, MMTF), is also an important scientific interest in terms of theoretical research of the properties of the PE space and related information search root trees (great-trees). The hierarchy of objects identified at the system design stage is then incorporated into the development of the user interface. This interface is represented here by the system of hierarchical menus described above. Its use allows the user to easily describe a subject area or complex object in terms of objects and indicators of the PE space.

The software implementation of the user interface is supposed to be implemented as an add-on to the Microsoft Access DBMS using the Visual Basic for Applications (VBA) programming language built into it.

To evaluate the effectiveness of applying the hierarchy to the semantic description of structured objects, we introduce the following numerical characteristics.

The compactness characteristic of the X tree of objects is defined as:

$$\lambda(X) = \frac{l(X)+n}{m+n}, \tag{8}$$

where $n = |X|$ - the number of indicators (X hanging tree nodes),

m - X number of internal tree vertexes,

$l(X)$ - X total length of the tree description is defined as the total length of X tree branches.

The compactness of the indicator tree shows how much computer memory costs are reduced while describing the semantics of system indicators due to the hierarchical structuring of this description.

The characteristic of a priori performance of the search for indicator in X is defined as:

$$\tau(X) = \frac{(1+n)n}{2} \Big/ l(X). \tag{9}$$

This characteristic shows how many times the speed of indicator search improves due to structuring.

Enter the average length of the object tree:

$$\bar{l}(X) = l(X)/n. \tag{10}$$

It is obvious that between the characteristics $\bar{l}(X)$, $\lambda(X)$, $\tau(X)$ there is a certain functional relation. From (9) and (10)

$$n = 2\bar{l}(X)\tau(X) - 1, \tag{11}$$

and from (8) and (11) we get

$$\lambda(X) = \frac{(2\bar{l}(X)\tau(X)-1)(\bar{l}(X)+1)}{m+2\bar{l}(X)\tau(X)-1} \tag{12}$$

On the basis of two criteria $\lambda(X)$ and $\tau(T_X)$, the task of two-criteria optimization is set to find effective structures for searching for indicators, meaning by effective structures the Pareto optimal structures for the criteria $\lambda(X)$ and $\tau(X)$.

If it is impossible to vary n and m, the problem is solved trivially (any structure is effective in this sense). To solve this problem, we introduce a certain class of trees (a subclass of root trees), which are called ST-trees – trees with a trunk. ST-trees have a trunk – i.e., an open sequence of vertices $<d_1,d_2,..,d_r>$, connected by edges; each such vertex d_i can only have edges leading to hanging vertices or to a vertex d_{i+1}. The case $r=1$ is possible. The set of hanging vertexes coming from the vertex d_i forms a tier of the ST-tree (here, for convenience, we have slightly departed from the traditional concept of «tier», we believe that the root has the 0th tier, not the 1st). The examples of ST-trees are shown in Figure 4.

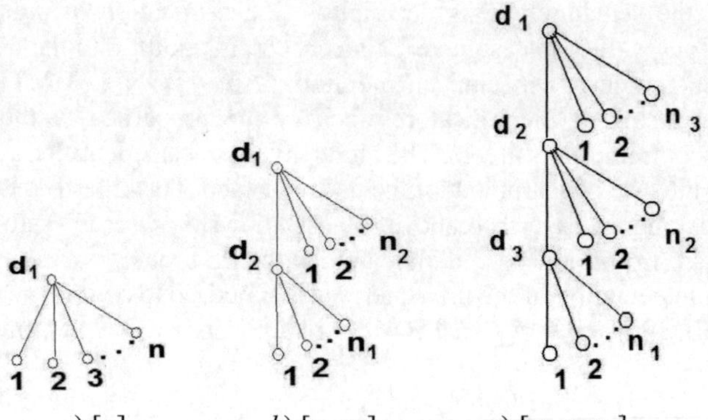

a) [n] *b*) $[n_1,n_2]$ *c*) $[n_1,n_2,n_3]$

Figure 4. Examples of ST-trees

Conclusion

The specialized databases (statistical information processing system) developed on the basis of the proposed tools for optimizing the search for the MTF indicators will allow a specialist in the subject area to perform statistical analysis of a large volume of source information with a multi-level structure [10]. Combining the statistical analysis tools and data manipulation tools into one integrated system makes it possible to dramatically increase the volume of processed information and significantly reduce the time for making managerial decisions.

References

[1] G V Redreev, A A Luchinovich, E I Ustiyantsev and A S Laskin 2018 Information System of Machines and Tractors Fleet Technical Service Journal of Physics: Conference Series Volume 1059 https://doi.org/10.1088/1742-6596/1059/1/012003

[2] G V Redreev, A V Shimokhin and P V Kiyko 2018 Perfection of Technical Service of Manufacturing Plants Equipment Journal of Physics: Conference Series 1059 https://doi.org/10.1088/1742-6596/1059/1/012008

[3] History of the CODASYL committee http://chernykh.net/content/view/208/220/

[4] Joe Celko 2012 Trees and Hierarchies in SQL for Smarties 2nd Edition. – Morgan Kaufmann p. 296

[5] I V Chervenchuk, V D Chervenchuk 2013 Some means of information support for environmental monitoring Current state and potential of tourism development in Russia: materials of scientific and practical conference, Omsk: Publishing House of the Omsk State Institute of Service 215-217

[6] I V Chervenchuk 2012 Automation of the preliminary processing of statistical information Mathematical and software systems in the industrial and social spheres: international collection of scientific works - Magnitogorsk: Publishing House of Magnitogorsk State Technical University named after Nosov 52–57

[7] D Christopher Manning, Prabhakar Raghavan, Heinrich Schutze 2011 Introduction to information search M.: Williams p. 528.

[8] Takase S, Okazaki N, Inui K rapid and large – scale uncontrolled extraction of relations - 2015.

[9] D A Devyatkin, R E Suvorov, I V Sochenkov 2016 «Architecture of the search and analytical system and research of the information space associated with the Arctic zone» **01** *INTELLIGENT SYSTEMS AND TECHNOLOGIES* 37-46.

[10] Hoffman R. et al. 2011 Knowledge-based weak supervision of information extraction from overlapping relations proceedings of the 49th annual meeting of the Association for computational linguistics: human language technologies **1** *Association for computational linguistics* 541-550.

Influence of inclination angle of piezoelectric receiver of ultrasonic sensor on the error in measurement of the average fiber diameter

A A Bagaev[1], L V Kulikova[2], and R A Kunitsyn[1]

[1] Altai State Agrarian University, 98 Krasnoarmeysky Avenue, Barnaul, 656049, Russian Federation
[2] Altai State Technical University named after I. Polzunov, 46 Lenin Avenue, Barnaul, 656038, Russian Federation

E-mail: bagaev710@mail.ru

Abstract. *During experimental studies, an ultrasonic device was used which in terms of design ensured maximum acoustic pressure on the receiving element, minimized the effect of external disturbances (environmental condition parameters) on the measurement result; decreased spatial variations of the source and receiver of ultrasonic waves during measurement, optimized the design and spatial positioning of ultrasonic vibration receiver.*
The measuring channel settings accepted during experimental studies are as following: ak – design constant of measuring channel; ak = 14, where a is half-width of the channel, a = 5 mm; k – wave number, k = 2.8. Experimental evaluation had confirmed theoretically substantiated effect of inclination angle of the receiving element of the ultrasonic sensor on the error in measurement of wool fineness in a disarrayed sample. At the inclination angle of the ultrasonic sensor receiver of $\beta = 48°$ *relative to the longitudinal axis of the measuring channel, the error does not exceed ±1.5%, and at the angle of* $\beta = 0°$ *it does not exceed ±5%.*
The reason for more accurate measurement of wool fineness at $\beta = 48°$ *in accordance with the developed theoretical provisions is that the inclination of the receiving piezoelectric element of the sensor relative to the longitudinal axis of the measuring channel eliminates the need for receiver installation at the angle of* $\beta = 0°$ *strictly in so-called ultrasonic wave antinode, which poses a significant technical problem at ultrasonic wave frequency of 150 kHz. Besides, the receiver arrangement at inclination relative to the longitudinal axis of the measuring channel allows increasing the spots of contact of ultrasonic wave with the surface of the receiving element irrespective of wave phase, and increases the average acoustic pressure and decreases the measurement error.*

Checking of fiber diameter of synthetic and organic nature in a continuous process for the production of artificial, synthetic, natural fibers and cords is an actual scientific and technical problem.

Yarns and fibers, located in the disarrayed sample, represent a special class of objects, characterized by a significant anisotropy of physical properties.

In the case, when the method and technical means measurement of the fiber diameter allows to obtain the fiber diameter value in a disarrayed sample with the required error, the sensor is able to estimate the fiber diameter with the required assurance in other conditions.

A disarrayed sample of sheep wool was selected as an universal object of research (physical generalized model). In this case, the revealed regularities can be extended to the processes of assessing the fibers diameter of a different physical nature subject taking into account certain restrictions.

This work continues the researches started by the authors in papers [1-5] and is associated with the development of methods and the improvement of technical means of wool fineness measurement of fibers in a disarrayed (irregular) sample.

The result of the theoretical studies is the following hypothesis: inclination angle of the receiving piezoelectric element of the ultrasonic sensor effects on the error in measurement of **the average fibers diameter** in a disarrayed sample.

It should be borne in mind a number of preset conditions. The maximum average pressure remains constant in a wide range of variation of the setting $ak = (7...25)$, where a is the half-width of the acoustic channel, mm; $k=2\pi f/c$ - wave number; f –ultrasonic wave frequency (f= 150 kHz); c – sound speed. In particular, when ak=14 the inclination angle of acoustic vibrations receiver is ≈48, and the half-width of the channel a=5 mm. The geometric parameters of the acoustic channel for a given frequency can be determined for other values of the coefficient ak. The specified frequency of the ultrasonic wave, the inclination angle of the receiver and the half-width of the channel are used in the sensor design.

The hypothesis of the influence of inclination angle of the receiving piezoelectric element of the ultrasonic sensor on the measurement of the average wool fineness requires empirical support.

The aim of this work is the experimental evaluation of the effect of inclination angle of the receiving element of the ultrasonic sensor on the error in measurement of the **average fiber diameter** in a disarrayed wool sample.

The task of the study is to conduct experimental studies, mathematical processing and interpretation of the measurement of the average wool fibers diameter at the inclination angles of the receiving element of the sensor $\beta=0^0$ and $\beta=48^0$.

1. Experimental installation

Figure 1 shows the structure-functional scheme of the device for ultrasonic express control of the average wool fibers diameter. The scheme contains a reference frequency generator 1, an integrating amplifier 2, a piezoelectric emitter 3, acoustic vibration receivers 4 and 5, a standard wool sample with a known fiber diameter 11, a sample of test fiber material 12, amplifiers 6 and 9, a reference signal source 7, an AGC signal conditioner 8, recording device 10.

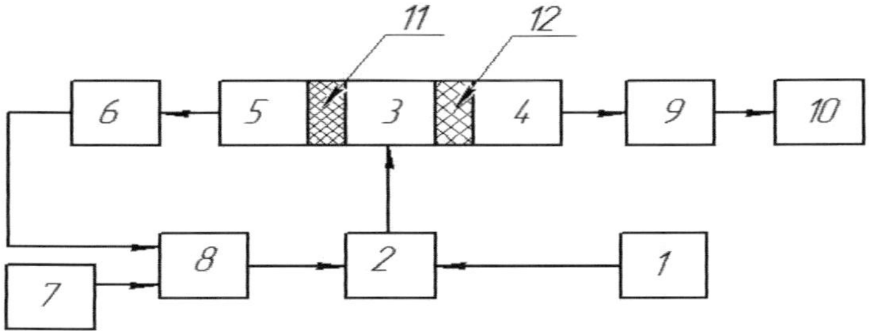

Figure 1. Structure-functional scheme of the device for ultrasonic control of the wool fibers diameter.

The main elements of the device are the sensors (blocks 1,2,3,4) and the registering device 10. The device is characterized by measuring (blocks 4,9,10) and reference (blocks 5,6,7,8) channels.

Acoustic vibrations in the measuring channel of the sensor are attenuated as a function of the fiber material settings 12 and converted in the receiver 4 into electrical signals, which are amplified in block 9 and are fed to the input of the digital recorder 10.

In the reference channel (blocks 5-8) acoustic vibrations varying in amplitude and phase under the environment condition are perceived by the receiver 5, converted by the amplifier 6 and fed to the input of the unit phase-locked loop 8 and adjusted by the signal generator 2 to compensate for the influence of external disturbances. Thus, a constant intensity of acoustic vibration emission in the sensor channel is kept. This allows to exclude the influence of such environmental factors as temperature, atmospheric pressure, humidity and gas composition on the measurement result.

The sensor shown in Figure 2 contains a split housing 1, into which the emitter 2, receiver 3 and additional receiver of the reference channel 4 are embedded.

IOP Conf. Series: Materials Science and Engineering **941** (2020) 012050 doi:10.1088/1757-899X/941/1/012050

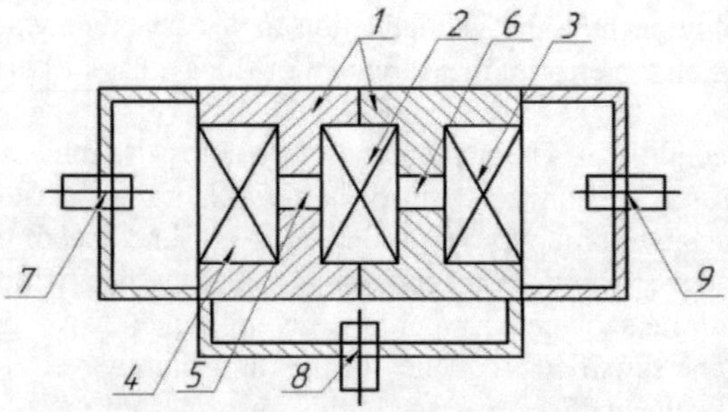

Figure 2. The scheme of ultrasonic sensor: 1 – housing; 2 – emitter; 3- receiver; 4 - additional receiver; 5 - reference channel; 6 – measuring channel; 7 - electrical connectors.

Emitter 2 is a piezoelectric transducer of two-way emission. The receiver 3 is made as a one-sided piezoelectric transducer. Measuring 6 and reference channels 5 connect the emitter 2 and receiver 3 and an additional receiver 4.

The piezoelectric ceramics of the lead zirconate-titanate type -19 is used as a material of the source and transducer receiver.

The emitter design has been developed (Figure 3) to increase the temperature stability and spatial stabilization of the emitter and the receiver. A distinctive feature of the design is the fixation of the piezoelectric element 1 in the foam rubber housing 2, which improves the accuracy of the installation of the piezoelectric elements source and receiver relative to the longitudinal axis of the sensor, reduces the oscillation of the device during operation, expands the region of resonant frequencies and increases the temperature stability of the emitter, improves the piezoelectric elements alignment.

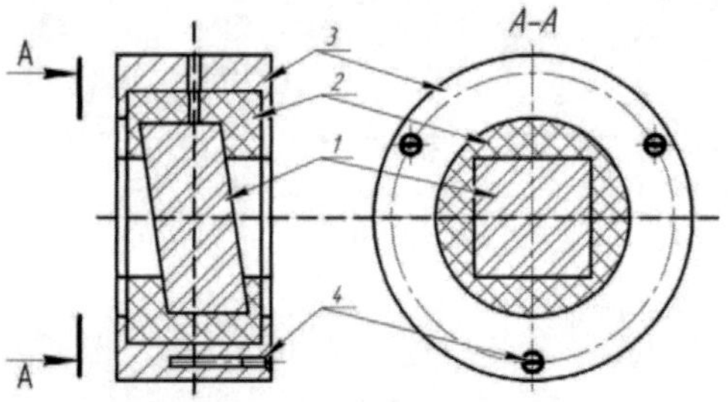

Figure 3. Receiver design with inclined piezoelectric element: 1 – piezoelectric element; 2 - rubber housing; 3 – brass case; 4 – screw connection.

An acoustically transparent tray-holder (Figure 4) was developed to fix the test material sample. It allows to research the sample with acoustic waves from four sides, which gives an average attenuation independent of the fibers distribution in the sample.

Figure 5 shows the external view of the device for express-control of the average diameter of the fiber material.

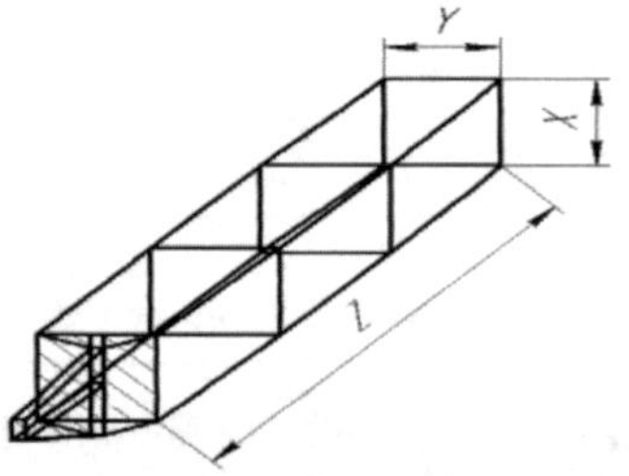

Figure 4. Tray – holder design of the test sample: X - height, Y - width, L - length of the tray.

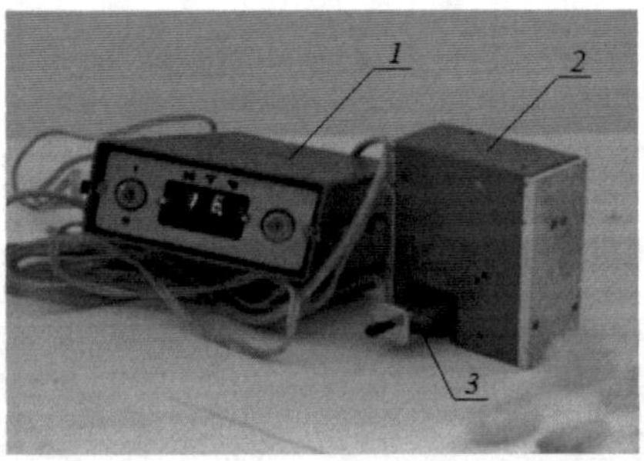

Figure 5 - External view of the installation of ultrasonic monitoring of the average fiber diameter: 1- digital signal processing and indication unit; 2- ultrasonic sensor; 3 - tray holder.

2. Research methodology

The authors have proposed a method for determining the average fiber diameter in a disarrayed sample [8], which consists in the researching of the test material sample by an ultrasonic wave with simultaneous measurement of the amplitude of acoustic vibrations transmitted through the sample. It is distinguished by the fact that, in order to increase the accuracy, the sample is subjected to a compressive load, placed into an acoustically transparent holder (see Figure 4), kept up to complete relaxation of the elastic component, placed into the measuring channel of the sensor (see Figure 2) and sequentially researched with acoustic waves from four longitudinal edges of the sample perpendicular to the compression. The sample fiber diameter is calculated as the arithmetic average found from the results of measuring the amplitudes of the ultrasonic wave along the four sample faces.

The method is implemented as follows. The device (Figure 5) is switched on to the network, heated for 10 minutes, and the zero point is set by the control hinge of the fibers.

A standard sample with a known average fibers diameter with a mass of 0.2 g is placed into the holder tray (see Figure 4), maintained until the relaxation state and placed into the reference channel 5 of the device (see Figure 2).

The test sample is treated in the same way, but it is put into the channel 6 of the ultrasonic sensor (see Figure 2).

Passed through the sample an acoustic signal is fixed. The measurement results are displayed; automatically recording on electronic media is possible.

Then the fiber holder with the test sample is removed from channel 6 and researched with acoustic waves from the side of the three remaining faces, consistently turning. At the same time rates of acoustic attenuation are fixed at a constant level of the emitter signal. The average fiber diameter is calculated as the arithmetic average of four measurements.

Main content pre-processing of measurement results is to check the normality of the distribution.

2. Processing methods of measurement results

It is known [9], that the main content of preliminary measurement results is to test the normality of distribution with the evaluation of such sample characteristics as the average value of the observable characteristic $\bar{x}$ and standard deviation S. the Distribution of errors of measurement result in most cases submits to the normal law of the Gaussian distribution [10]. For small samples (n<120) is known [11] a simple check of normality of distribution associated with the calculation of the average absolute deviation (CAO).

$$CAO = \sum|x_i - \bar{x}|/n \qquad (1)$$

For the sample n=10 with a normal distribution should occur the following condition

$$\left|\frac{CAO}{S} - 0.7979\right| < \frac{0.4}{\sqrt{n}} = 0.12 \qquad (2)$$

Confidence interval for true value of fineness a [12]

$$\bar{x} - \frac{t_\alpha S}{\sqrt{n}} < a < \bar{x} + \frac{t_\alpha S}{\sqrt{n}} \qquad (3)$$

where t_α is the limiting value. For the probability α = (1-P) / 2 = 0.005, the accepted sample is n = 10 and the number of degrees of freedom is k = n-1 = 9, the limit is t_α=3.25 [12].

Confidence limit for the variance S^2 [12]

$$\frac{(n-1)S^2}{\chi^2_{\alpha_1}} < S^2 < \frac{(n-1)S^2}{\chi^2_{\alpha_2}} \qquad (4)$$

where α_1, α_2 are probabilities, for a confidence level 0.95 α_1=(1-0.95)/2=0.025, α_2=1-(1-0.95)/2=0.975; $\chi^2_{\alpha_1}$, $\chi^2_{\alpha_2}$ - values of the upper and lower limits of the interval, respectively, for the number of degrees of freedom k=n-1=9$\chi^2_{\alpha_1}$=19.023, $\chi^2_{\alpha_2}$=2.7 [12].

The absolute Δ, μm, and relative δ, %, errors are determined by known methods [10,12]. Merino wool with a fiber diameter of 19.84 μm, defined in accordance with GOST 17514-93 is used as a standard or measure.

3. Results and discussion

Table 1 shows the results of statistical processing of 10 samples using the above method with the inclination angle of the receiving element of the ultrasonic sensor of β =0^0 and β =48^0 relative to the longitudinal axis of the measuring channel.

Table 1. Preliminary processing results of n-measurements of the average wool fineness in a disarrayed wool sample at different inclination angles of piezoelectric receiver *в*.

$\bar{x}$, μm	$\sum(x_i - \bar{x})$	$\sum(x_i - \bar{x})^2$	S^2	S	CAO, μm	$\left\|\frac{CAO}{S} - 0.7979\right\|$
β=0^0						
20.805	8.8	10.126	1.01	1	0.885	0.087
β=48^0						
20.055	6.57	5.173	0,517	0.719	0.657	0.11

Analysis of table 1 using condition (2) allows for the following conclusion: the hypothesis of the normal distribution of the data sample is confirmed in both cases.

Table 2 presents the results of estimating the confidence intervals of the fiber diameter *a*, the standard deviation S, the absolute Δ, and the relative δ errors at different inclination angles of the piezoelectric receiver, β The analysis shows that the mathematical expectation of the fiber diameter, found as an arithmetic average of the measurement results on 4 faces, when the receiving element of the sensor is at an angle β=48^0, is closest to the reference value compared to the angle β =0^0; at the inclination angle β =48^0, the error does not exceeds ± 1.5%, and at the angle β =0^0 - ±5 %.

Table 2. Results of evaluation of confidence intervals of fiber diameter a, standard deviation S, absolute Д and relative д errors at different inclination angles of the piezoelectric receiver в (n = 10).

confidence $interval\ for\ a$, μm	$\chi^2_{\alpha_1}$	$\chi^2_{\alpha_2}$	$interval\ for\ S$	Δ, μm	δ, %,
β=0^0					
20.805±1.08	19.023	2.7	0.69<*S*<1.98	0.965	±5
β=48^0					
20.055±0.778	19.023	2.7	0.49<*S*<1.313	0.215	±1.5

The reason for more accurate measurement of fiber diameter at β = 48° in accordance with the developed theoretical provisions is that the inclination of the receiving piezoelectric element of the sensor relative to the longitudinal axis of the measuring channel eliminates the need for receiver installation at the angle of β = 0° strictly in so-called ultrasonic wave antinode, which poses a significant technical problem at ultrasonic wave frequency of 150 kHz. Besides, the receiver arrangement at angle allows increasing the spots of contact of ultrasonic wave with the surface of the receiving element irrespective of wave phase, and increases the average acoustic pressure and decreases the measurement error. The obtained results are in qualitative agreement with the scientifical attitude outlined in the works [13, 14]. At the same time, the analytical and experimental base proposed by the authors makes it possible to increase the accuracy in measuring of average fiber diameter on the basis of previously unknown revealed regularities.

4. Conclusions. The measuring channel settings accepted during experimental studies are as following: ak = 14, where a- is half-width of the channel, a = 5 mm; f–ultrasonic wave frequency, f= 150 kHz; k – wave number, $k = 2\pi f/c$, c – sound speed. Experimental evaluation had confirmed theoretically substantiated effect of inclination angle of the receiving element of the ultrasonic sensor on the error in measurement of average fiber diameter in a disarrayed sample. At the inclination angle of the ultrasonic sensor receiver of β = 48° relative to the longitudinal axis of the measuring channel, the error does not exceed ±1.5%, and at the angle of β = 0° it does not exceed ±5%.

The established regularities make it possible to extend the results of the present studies to the processes of determining the fibers diameter in a disarrayed sample of materials of a different physical nature and geometric orientation in space. The main provisions of the theory of similarity are analytical and experimental base for further research are.

References

[1] Bagaev A A Mathematical model of ultrasonic analyzer of Merino wool quality /A. A. Bagaev, TS. I. Kalinin, R. A. Kunitsyn//Bulletin of the Altai agrarian University.- Barnaul: publishing house of agau, 2010.-№3 (65).-pp. 72-75.
[2] Bagaev A A Ultrasonic device for research of Merino wool/A. A. Bagaev, TS. I. Kalinin, R. A. Kunitsyn//Polzunovskii Herald.- Barnaul: publishing house of AltSTU named after I. I. Polzunov, 2010. - no. 2. - P. 57-59.
[3] Bagaev A A Improving the accuracy of volume density control of disordered fiber media using ultrasonic sensors/A. A. Bagaev, R. A. Kunitsyn, TS. I. Kalinin//Polzunovskii Herald.- Barnaul: publishing house of AltSTU named after I. I. Polzunov, 2011.-№2\2.-pp. 8-12.
[4] Bagaev A A Analysis of the influence of the parameters of the measuring channel and the ultrasonic wave on the value of acoustic pressure when measuring the Tonina of fiber agricultural materials/A. A. Bagaev, V. G. Rezinov, R. A. Kunitsyn// Bulletin of the Altai agrarian University.- Barnaul: publishing house of agau, 2013.-№11(109).-pp. 107-111.
[5] Bagaev A A Results of research on the dependence of acoustic pressure of ultrasonic vibrations on the mass of a sample of disordered fiber material and the diameter of fibers/A. A. Bagaev, TS. I. Kalinin, V. G. Rezinov, R. A. Kunitsyn// Bulletin of the Altai state agrarian University.- Barnaul: publishing house of agau, 2015.-№3(125).pp. 117-122.
[6] Allain, D., and R. G. Thébault.1996. A rapid method for measuring cross section characteristics of animal fibres. Pages 99–110 in Metrology and Identification of Speciality Animal Fibres Ocas. Publ. no. 4. J. P. Laker and F. J. Wurtmann, ed. Eur. Fine Fibre Netw., Aachen, Germany.
[7] Qi, K., C. J. Lupton, F. A. Pfeiffer, and D. L. Minikhiem.1994. Evaluation of the optical fibre diameter analyser (OFDA) for measuring fiber diameter parameters of sheep and goats. J. Anim. Sci. 72:1675.–1679.
[8] Method for determining the fiber tone. Patent no. 2465582, G 01 N29/00, 2006 Russian Federation/TS. I. Kalinin, R. A. Kunitsyn, A. A. Bagaev, Federal state educational institution of higher education Altai state agrarian University. Publ.27.10.12.
[9] Lvovsky E N Statistical methods of constructing empirical formulas: textbook/E. N. Lvovsky. - M.: Higher school, 1982. -224 p.
[10] Semenov B A Engineering experiment in industrial heat engineering, heat power engineering and heat technologies: Textbook/B. A. Semenov.- Saint Petersburg: LAN, 2013. -400 p.
[11] Lothar Z. Statistical evaluation/Z.Lothar. - M.: Statistics, 1976.
[12] Kulikovsky K L Methods and means of measurement: Textbook/K. L. Kulikovsky, V. ya. Kuper. - M.: Energoatomizdat, 1986. -448 p.
[13] Ilivanov V. M. Investigation of acoustic wave propagation in solid bodies and air media: monograph/V. M. Ilivanov, Yu. V. Kandrin, V. A. tsymbalist. - Barnaul: ASAU Publishing house, 2003. -124 p.
[14] Ilivanov V M, Physical acoustics: monograph/V. M. Ilivanov, Yu. V. Kandrin, V. A. tsymbalist 2nd ed., add. Barnaul: Izd-voagau, 2004. – 158 p.

IOP Conf. Series: Materials Science and Engineering **941** (2020) 012051 doi:10.1088/1757-899X/941/1/012051

Analysis of discriminatory characteristics of a video sensor with an exponential correlation function

V P Gerasimov[1], V D Kovalev[1], A Yu Darzhaniya[1], R A Magomedov[1], E V Sokolova[1], O V Klimenko[1]

North Caucasus Federal University, 1 Pushkin, Stavropol, 355000, Russia

E-mail: ram0707@mail.ru

Abstract. This article presents an analysis of the characteristics of the video sensors functioning both in the visible and "thermal" regions of the spectrum. The influence of errors caused by the non-linearity of the discriminatory characteristic (DC) in the idealized model is studied. The article is devoted to the refinement of the model, taking into account the constant exponent parameter and introducing its value into the algorithm implemented by the onboard video sensor.

1. Introduction

In autumn 2015, presidents and heads of government of nearly one and a half hundred states gathered in Paris at an environmental conference to discuss climate change caused by human activities. One of the manifestations of this influence on nature turned out to be fires that destroy not only forests, but also settlements on almost all continents. The leaders considered the strategic issues and tactical decisions are made by engineers - in this area can offer the use of robotic systems - unmanned aerial vehicles (remotely piloted air system (RPAS)) and fire-fighting equipment (RPAS FFE) for searching, automatic detection, tracking and liquidation of fires. RPAS and RPAS FFE will be able to perform the indicated functions with the help of on-board video sensors functioning both in the visible and "thermal" spectral regions [1].

On-board video sensors that automatically accompany the tracking objects - vizors are video sensors that calculate the mismatch by measuring the correlation functions (CF) of the reference and current images of the foci of ignition [2]. Such video sensors are characterized by high noise immunity, and discriminators that implement the differential method for calculating the mismatch between the reference and current directions of the line of sight (LP) are the least demanding on computer resources [3].It is known that the CF of the image of the underlying surface formed by the onboard video system of the RPAS and RPAS FFE, for example, television [4], is exponential (Figure 1).

$$\mathbf{K(X)} = \mathbf{e}^{-\alpha|\mathbf{X}|}, \quad (1)$$

where α – is the exponent constant, depending on the conditions of use of the onboard video sensor, for example, Figure 1 illustrates the CF of various types of the underlying surface, characterized by parameters α_1, α_2, α_3.

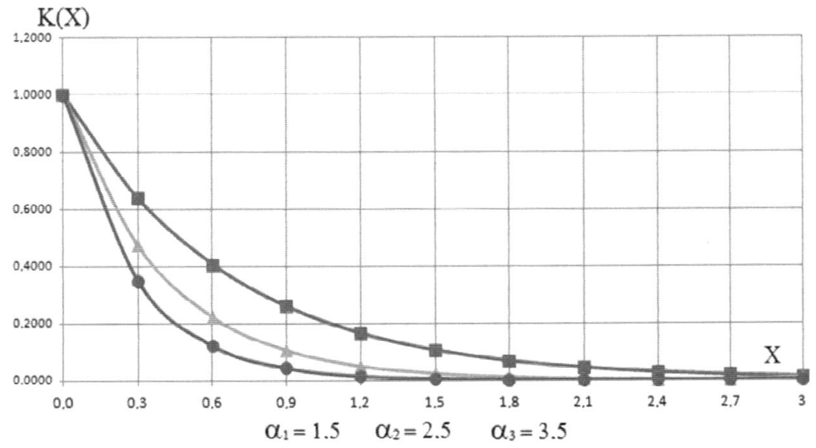

Figure 1. The exponential nature of the CF television system PRAS.

This feature should be taken into account when developing on-board video sensors and, in particular, for PRAS and PRAS FFE, since in automatic systems the requirements for accuracy of calculations for the meter are incomparably higher than in automated systems, where the operator can correct errors of measurement of mismatches [5].

The resources available on-board control computers - controllers, place great demands on the algorithms they implement: in terms of memory footprint, speed, etc. [6]. These requirements are met by the differential method of calculating the mismatch between the reference and the required positions of the line of sight.

Taking into account (1), the reference exponential CF K (X – Δ), K (X + Δ) and the discriminating characteristic (DX) D(X) they form when implementing the differential method are analytically and graphically described as follows (Figure 2):

$$K(X-\Delta) = \mathbf{e}^{-\alpha|X-\Delta|},$$
$$K(X+\Delta) = \mathbf{e}^{-\alpha|X+\Delta|}, \qquad (2)$$
$$D(X) = K(X-\Delta) - K(X+\Delta) = \mathbf{e}^{-\alpha|X-\Delta|} - \mathbf{e}^{-\alpha|X+\Delta|},$$

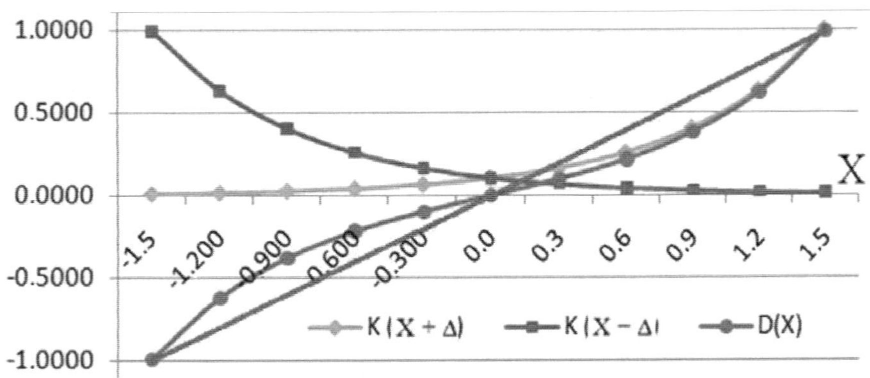

Figure 2. Exponential CF - K (X + Δ), K (X – Δ), nonlinear and linear discriminatory characteristics DC (X) of the PRAS video sensor

which, after normalizing the argument, up to a factor of **2/e**$^{\alpha}$ is described by a hyperbolic sine:

$$\mathbf{D}\left(\frac{\mathbf{X}}{\Delta}\right)=\frac{\mathbf{2\,Sh}\left(\frac{\mathbf{X}}{\Delta}\right)}{\mathbf{e}^{\alpha}} \tag{3}$$

Usually, when constructing DCs, they are linearized and the mismatch in such video sensors is calculated under the hypothesis of its linearity. From (4) it is easy to obtain an analytical expression for the error in determining the mismatch by a discriminator tuned to a linear DX and caused by neglecting the exponential nature of the correlation function of the onboard controller:

When determining the characteristics of the discriminatory characteristics of a one-dimensional differential video sensor with an exponential correlation function, we analyzed the discriminatory characteristic D (α, X, Δ) [4] with the parameters:

- exponent constant $\alpha = 1$ (in the future, the results obtained can be compared with the solutions obtained for $\alpha < 1$ or $\alpha > 1$);
- mismatch (argument) Z $\in$ {-3 … 3}, since D (1, X≥3, 0) ≤ 0,05, which with a 5% error can be considered a zero value and the behavior of the exponential function beyond the specified limits for the problem being studied is not of interest;
- samples i $\in$ {0 … L}, where L – is the maximum number of the sample;
- a fixed shift $\Delta = 1{,}5\ \alpha^{-1}$ (in this case, the aperture of the discriminatory characteristic on the segment [$-\Delta$ ... $+\Delta$] is equal to $2\Delta = 3\alpha^{-1}$).

2. Results and Discussion

The algorithm for constructing the discriminatory characteristics of a one-dimensional differential correlator with an exponential correlation function in the MS Excel spreadsheet processor is as follows:

- mathematical description of the correlation function,
- setting the values of the parameters α, Δ and the number of samples n,
- calculation of the **i**-values of the argument X,
- calculation of the values of the correlation function shifted by $+\Delta$,
- calculation of the values of the correlation function shifted by $-\Delta$,
- the calculation of the values of the discriminatory characteristics Di (X_i, Δ),
- calculation of the linearized discriminatory characteristics D_{lin} (X_i, Δ),
- rationing of correlation functions and discriminatory characteristics.

An analysis of the results and a comparison of the discriminatory characteristics formed by the difference of the correlation functions shifted by $\pm\Delta$ with the linearized DC (see Figure 2) leads to the conclusion that the working section of the real discriminatory characteristic is non-linear at X $\in$ [−1,5 … 1,5].

Therefore, to construct a correlator that is tolerant to changes in the conditions of use, expressed in a change in the parameter α, the following problems are of interest:

- formalization of the work site of discriminatory characteristics;
- clarification of the influence of the parameters α and Δ on the accuracy of determining the mismatch of the correlator;
- minimization of the error in calculating the mismatch by optimizing the ratio of the parameters α and Δ.

A computer analysis of the error in calculating the mismatch under various conditions (5) made it possible to propose an adaptive algorithm for determining the mismatch by the differential correlator in the PRAS and PRAS FFE video sensor:

1. Depending on the conditions of use of the onboard video sensor of the PRAS and PRAS FFE, the coefficient α is calculated.
2. The value of δ is determined.

3. The maximum value of the discriminatory characteristic in the given conditions of combat use is calculated by – $\mathbf{D_{max}}\left(\frac{Z}{\Delta}\right)$, if $\frac{Z}{\Delta} = 1$.

4. To build a discriminatory characteristic, normalization is performed

$$\mathbf{d}\left(\frac{\mathbf{X}}{\Delta}\right) = \frac{\mathbf{D}\left(\frac{\mathbf{X}}{\Delta}\right)}{\mathbf{D_{max}}\left(\frac{\mathbf{X}}{\Delta}\right)}$$

5. After constructing the discriminatory characteristics, the value of the mismatch is calculated as follows:

$$\mathbf{x} = \Delta\mathbf{d}^{-1}\left(\frac{\mathbf{X}}{\Delta}\right) \quad (4)$$

where d^{-1} – inverse function $\mathbf{d}\left(\frac{X}{\Delta}\right)$.

To verify the mathematical model of the video sensor and analyze the influence of the parameters of the real discriminatory characteristics on the error of calculation of the mismatch, it is proposed to develop and create a collimator - a simulator of visual environment (IVO), which allows you to create images of different foci of ignition taken into account by parameter α, the structural diagram of which when combined with the on-board video sensor is shown in Figure 3.

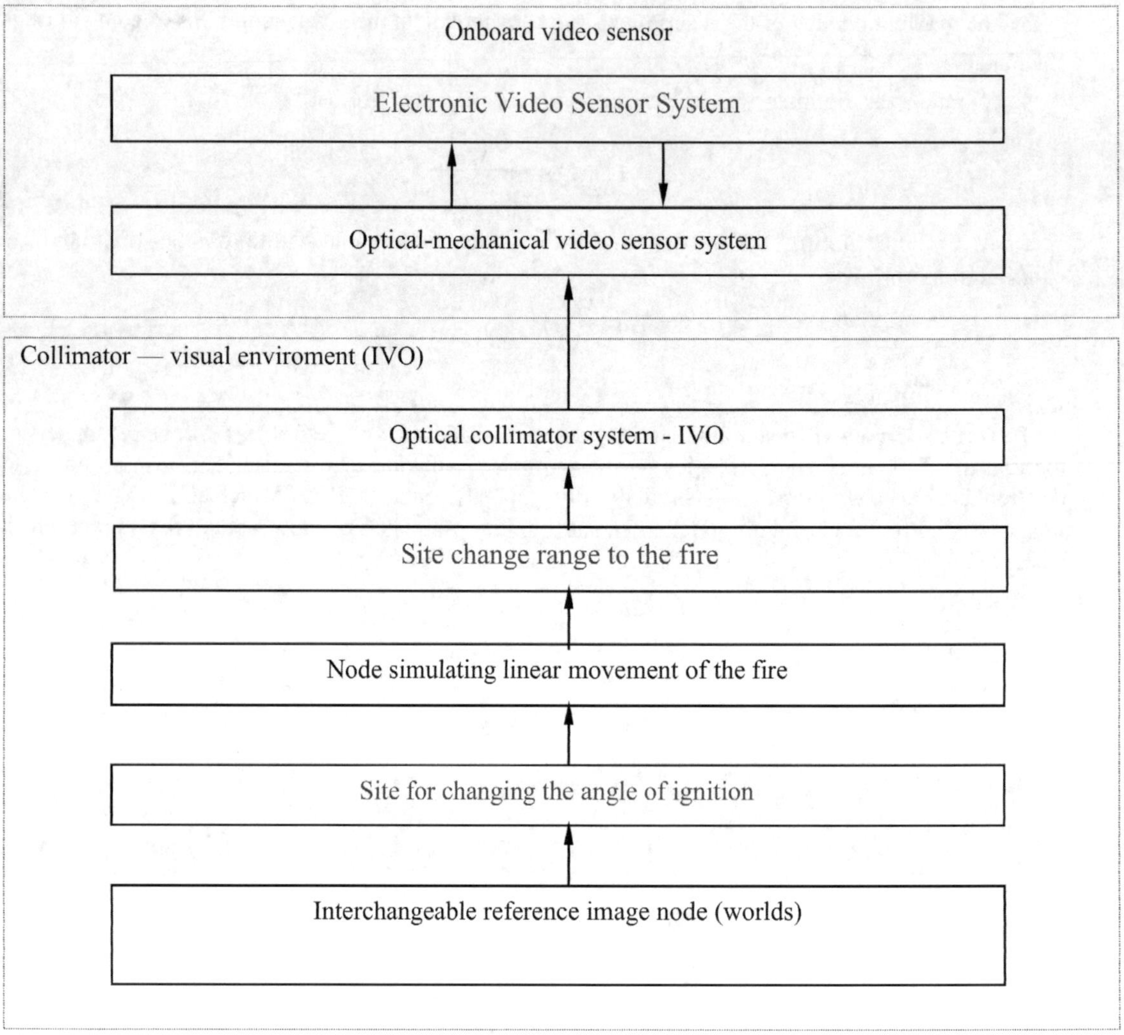

Figure 3. The block diagram of the pairing of the video sensor and the collimator - a simulator of visual environment

3. Conclusion

Therefore, the study of the influence of errors caused by neglecting the nonlinearity of the HX in an idealized model is an urgent task. The quality of the video sensor tracking of the source of fire will improve if we compensate for the error in the calculation of the mismatch during its processing during the calculation of the mismatch. Achieving this goal is possible by refining the model, taking into account the parameter a and introducing its value into the algorithm implemented by the onboard video sensor.

References

[1] Lecuyer M, Ducoffe G, Lan F, Papancea A, Petsios T, Spahn R, Chaintreau A, and Geambasu R 2014 *XRay: Enhancing the Web's Transparency with Differential Correlation* (23rd USENIX Security Symposium. August 20–22 San Diego, CA) 59–63

[2] Liu C, Chakraborty S, Mitta P. *Dependence Makes You Vulnerable*: Differential Privacy Under Dependent Tuples - www.princeton.edu

IOP Conf. Series: Materials Science and Engineering **941** (2020) 012051 doi:10.1088/1757-899X/941/1/012051

[3] H A Bruck, S R McNeill, M A Sutton, W H Peters 1989. *Digital image correlation using Newton-Raphson method of partial differential correction* / Experimental Mechanics, September, **29** (3) 261–267

[4] A T McKenzie, I Katsyv, W M Song, M Wang, B Zhang 2016 DGCA – *A comprehensive R package for Differential Gene Correlation Analysis.* https://bmcsystbiol.biomedcentral.com/articles/10.1186/s12918-016-0349-1

[5] K Kechris, C Siska 2017 *Differential correlation for sequencing data* BMC Research **56**

[6] I Bilogrevic, K Huguenin, S Mihaila, R Shokri, J-P Hubaux 2015 *Predicting users motivations behind location check-ins and utility implications of privacy protection mechanisms*, in Proceedings of ISOC, NDSS.

Research of the process of vegetable oil extracting under the influence of a high frequency wave field

V Yu Ovsyannikov, A A Berestovoy, N N Lobacheva, V V Toroptsev and S A Trunov

Federal State Budget Educational Institution of Higher Education "Voronezh State University of Engineering Technologies", Revolution Av., 19, Voronezh, 394036, Russia

E-mail: ows2003@mail.ru

Abstract. Research in a laboratory press equipped with an ultrasonic vibrator was performed to study the nature of a high frequency wave field influence on the process of vegetable oil extracting from safflower seeds. The initial moisture content of safflower seeds, the rotational speed of the press screw, the frequency and amplitude of ultrasonic vibrations have the greatest influence on the oil yield and the cake residual oil content. The analysis of the dependences obtained showed that the parameters of the safflower seed pressing process, the values of which could be varied during the experiments, affect the oil yield identically, namely, with an increase in the rotational speed of the press screw, the oil yield decreases and, consequently, the cake oil content increases. It was found out that moisture content raising increases the oil yield and reduces the residual oil content. Moreover, the most rational value of moisture content in terms of absolutely dry matter is 8.7...11.2%. It was revealed that an increase in the pressure given to the product in the press occurs with a decrease in the effective viscosity inside the surface layer of the mass being pressed due to ultrasonic vibrations imposition. The following parameters should be focused on as rational parameters for safflower seeds pressing providing the minimum residual oil content in the cake: the ultrasonic waves frequency is 22...28 kHz, the amplitude is more than $40 \cdot 10^{-3}$ m and the pressure value in the press is from 12...13 MPa. These values obviously depend directly on the grain behavior and its movement in the press during exposure to high-frequency ultrasonic vibrations.

1. Introduction

Being one of the most important branches of the food industry, the production of vegetable oils is continuously developing and improving. Both food and medicinal drug oil products extracted from crops of various origins are used in modern life. Nowadays, oils obtained from unconventional crops are gaining popularity. Safflower oil belongs to one of these types of vegetable oils by right. The specified oil is extracted from safflower seeds, which are mainly grown in the southern arid regions of Russia, as well as in Central Asia. The taste of safflower oil is in many ways similar to the taste of sunflower oil. At the same time, it found its application in the production of margarines, the preparation of cold snacks and salads, and the high ignition temperature allows it to be used for deep-frying culinary products [1, 2].

Traditionally, vegetable oils are extracted from seeds by screw or hydraulic presses. However, a significant amount of oil is contained in the cake remaining after oil extraction. Therefore, the search for ways to increase the yield of vegetable oil from seeds, as well as to reduce its residual content in the cake is very important. To achieve this goal, we propose to combine the process of safflower seeds pressing with the physical effect of a high-frequency wave field on the cake removed.

The application of ultrasonic vibrations is known to provide great opportunities for the intensification of technological processes, automation of operations, improving product quality and the overall production culture. The greatest effect is achieved with ultrasonic vibrations applying in viscous, liquid and gaseous media, which allows its use in vegetable cakes processing [3-5]. This research was carried out to study the prospects and influence of ultrasonic vibrations on the nature of the process of safflower seeds oil obtaining.

2. Materials and methods.

The study of the process of safflower seeds pressing under the influence of ultrasonic waves was carried out on an experimental unit - a single-screw press (Figure 1).

The press included a frame 1, a driving mechanism 2, an oil press chamber 5 with a loading hopper 4, a screw 7 (Figure 2), a grain chamber 6 with a device for regulating the cake exit gap 19.

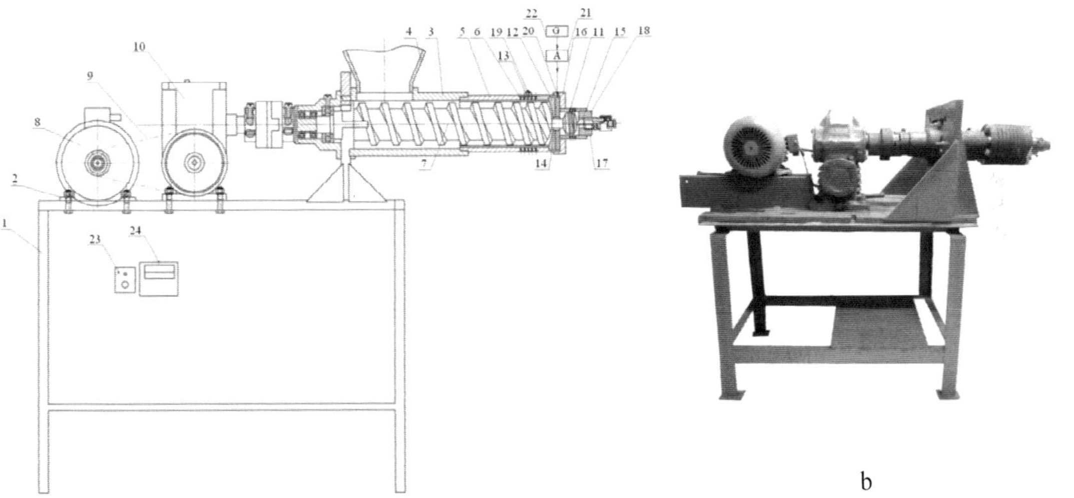

a b

Figure 1. Single screw press:
a - scheme of the press; b - general view of the press:
1- frame; 2- driving mechanism; 3- unit housing; 4- loading hopper; 5- operating camera; 6-11 - oil outlet holes; 7-screw; 8 - engine; 9 - chain transmission; 10 - gear; 12 additional oil pressing chamber; 13-grooves; 14- central hole; 15- forcing screw element ; 16-nut; 17- muft; 18 - cake outlet hole; 19- regulatory devices; 20 - piezoceramic ultrasonic emitter; 21- power amplifier (A);22-generator (G); 23 - start button; 24 – temperature gauge.

Figure 2. Press screw

The press contained the following sensors: for taking temperature readings at different points simultaneously; to measure the pressure in the grain cylinder, which are connected to a system of monitoring and controlling of the pressing process properties.

The press driving mechanism was equipped with an alternating current electric motor 8, a worm gearbox 10 and a chain gear 9. The screw shaft was connected to the output shaft of the gearbox through a muft. The design of the driving mechanism allowed to adjust the screw rotation frequency by means of a set of sprockets of various diameters, allowing to control the speed step by step in the range of 15 ... 25 min^{-1}. The press was turned on and turned off with button 11. The power of the press driving mechanism electric motor was 3 kW.

Measurements of the operating parameters of the pressing process were carried out with accepted instruments. The system of sensors - pressure transducers of the KRT-ST-25-0.5 brand was installed to measure prematrix chamber pressure. These sensors made it possible to provide pressure control in the prematrix area of the unit up to 25 MPa and provided pressure readability of ± 1 kPa. The temperature field of the mass processed along the length of the press was measured with a two-channel meter, which included a secondary device of the OVEN TPM 200 brand with thermocouples. The temperature measurement accuracy was about 1°C. Pressure sensors were connected directly to the PC.

The press working element was the screw, shown in figure 2 with a length of 0.33 m and a diameter of 0.075 m. The screw thread had a constant pitch of 0.02 m, the turns had a constant thickness of 0.008 m, the depth of the pitch of the thread was also constant. The grain chamber was attached to the housing from the end.

A device for the annular gap width regulating, which could vary between (0.5 ... $8 \cdot 10^{-3}$) m, was used to study the degree of influence of the channel living area for the cake molding on the pressure indicator in the chamber.

Previously, the safflower seeds were cleaned of impurities and weighed. Heating of the press housing 3 and the grain chamber with the screw 15 was carried out before pressing. Then the sawdust weighing 0.3 kg, with the moisture content of 12 - 18%, was charged into the press hopper. The electric motor was turned on, and the rotational speed of the working screw was equal to 25 min^{-1}, all the chamber outlet holes being closed. After the temperature in the chamber and housing approached 328 ... 333 K, safflower seeds weighing 5 kg were loaded into the press funnel. The seeds moisture content, the rotational speed of the working screw, the length of the grain chamber and the area of the cake output hole were changed during the study. A set of chain drive sprockets with specific diameters was used to change the screw speed. The yield of oil and cake was controlled by changing the cross-sectional area of the hole with devices 7 and 14. The grain chamber length was varied by changing the depth of its screwing into the press.

After emptying the loading hopper to push the remaining product out, the sawdust in the amount of 0.1 kg was put in the hopper again and the process was repeated. The pressure values in the grain chamber and the temperature of the grain mass were recorded in each subsequent experiment. The measurement of the volume and mass of the oil obtained and the oil yield in relation to the weight of the loaded seeds was carried out after pressing of the measured portion of safflower seeds.

The first series of experiments was carried out without applying ultrasound. This was necessary to identify the range in which safflower seeds oil was extracted to the fullest extent possible. In the second series of experiments, an ultrasonic generator with an amplifier was turned on. The influence of the amplitude and frequency of the ultrasonic field on the safflower seeds oil yield and the residual oil content of the removed cake was investigated.

The experimental data obtained made it possible to evaluate the processes that take place when pressing the safflower seeds initial mixture with a single screw press.

3. Results and their discussion.

Preliminary studies of safflower seeds pressing with the moisture content of 4.2 ... 16.3% in terms of absolutely dry matter showed the following. The moisture content of the initial seeds has a great effect on the oil yield. With the moisture content increase, the oil yield increases as well, and the residual oil content of the cake decreases to a certain interval corresponding to 8.7 ... 11.2%, after which it increases nonlinearly. This is explained as follows. At a seed moisture content of less than 8.7%, the cellular structure of the membrane destroyed by the screw adsorbs a certain amount of oil and does not allow it to be separated at appropriate pressures in the grain chamber. When the seed moisture content is more than 11.5%, a sufficiently moistened pulp gets plastic properties and comes back and rotates together with the shaft to a greater extent, due to the gap between the screw shaft and the inner surface of the grain cylinder [6, 7].

It was found out that with an increase in the rotational speed of the press screw, the oil yield decreases and, consequently, the cake oil content increases. This is explained by the fact that oil extracting from seeds, as a dynamic process, occurs in time and the more time the cake is pressed, the longer the oil will drain and, consequently, the residual oil content will decrease [8].

The gap width of the annular hole for the cake output has a significant impact on the oil yield as well. With an increase in the gap width of the outlet annular hole over $2.5 \cdot 10^{-3}$ m, the oil yield decreases significantly, and the cake amount increases, which correlates with the physical picture of the process. Therefore, it was this value of the gap width for the cake exit that was chosen by us in further experiments. The change in pressure P, MPa along the press chamber length L, m is shown in figure 3.

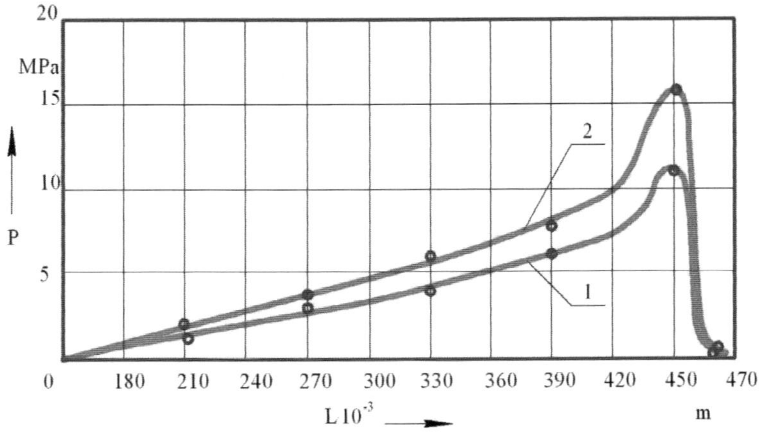

Figure 3. The change in pressure P, MPa along the press chamber length L, m with the safflower seeds moisture content of 9% and the gap width of the output annular hole: 1 – $1.5 \cdot 10^{-3}$ m; 2 – $1.5 \cdot 10^{-3}$ m

At first the pressure increases smoothly and then sharply in the grain chamber, as can be seen from the graphs. This is due to the fact that when the gap for oil outlet is reduced, the outlet pressure value increases sharply, while the pressure does not increase significantly at the final turn, since the cake is a rather plastic product, and the pressure gradient is very large in it. The distribution of pressure is also hindered by the screw last turn. It follows from the graphs that the oil outlet is the most difficult in the grain chamber due to the increase in pressure, and clogging of the product pores occurs as a result. It is in the grain chamber that it is necessary to install an ultrasonic emitter to create vibration in the layer and to form channels for additional oil output.

Figures 4...7 show graphs of changes in the frequency response of the kHz frequency and ultrasound impact amplitude. After analyzing the dependence, one can distinguish the characteristic inflection points at which the minimum values are minimal at various amplitude values and ultrasound impact frequency.

It was found out that an increase in pressure is the result of a decrease in viscosity within the surface layer of the system.

IOP Conf. Series: Materials Science and Engineering **941** (2020) 012052 doi:10.1088/1757-899X/941/1/012052

A decrease in the coefficient of external friction against the chamber wall concerning the vibrating product layer contributes to an increase in the degree of permeability and equal density of safflower cake. The observed decrease in oil content is explained by higher indices of vibration parameters than the resonance mass range of the pressed material, which leads to the destruction of the boundary layer [9, 10].

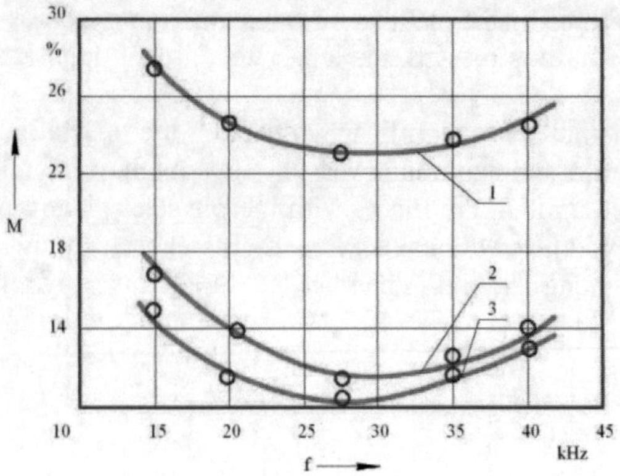

Figure 4. The influence of the emitter vibration frequency on the cake oil content M,% with the amplitude of ultrasonic vibrations Ah = 40 · 10^{-3} m and the press screw rotational speed n, min $^{-1}$: 1 – 15; 2 – 20; 3 – 25

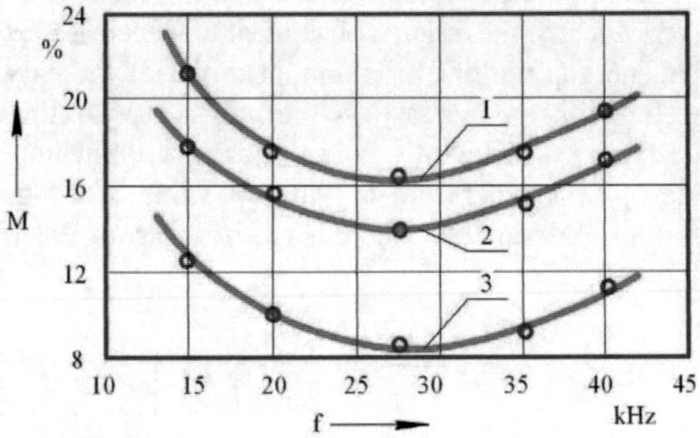

Figure 5. The influence of the emitter vibration frequency on the cake oil content M,% at the press screw rotational speed n = 20 min^{-1} and the ultrasonic vibrations amplitude Ah · 10^{-3} m: 1 – 30; 2 –40; 3 – 50

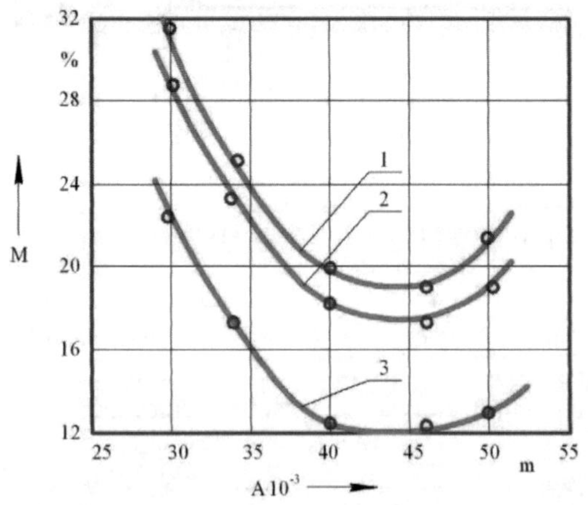

Figure 6. The influence of the emitter vibration amplitude on the cake oil content at a rotational speed of the press screw n = 20 min^{-1} and the frequency of ultrasonic vibrations f, kHz: 1 – 15; 2 – 25; 3 – 40

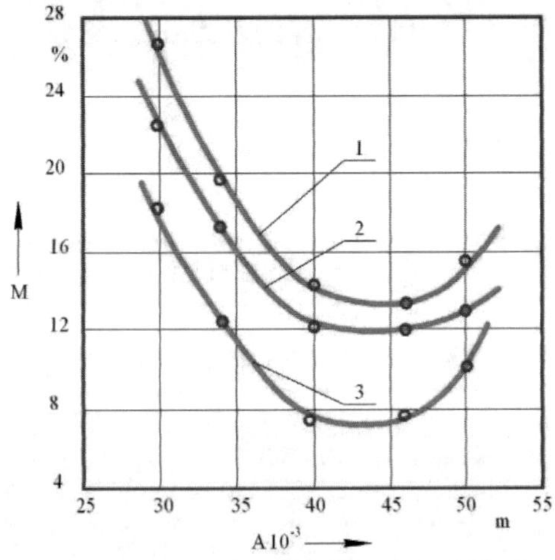

Figure 7. The influence of the emitter vibration amplitude on the finished product oil content on the screw speed n, min at the frequency of ultrasonic vibrations f = 40 kHz: 1 – 15; 2 – 20; 3 – 25

4. Conclusions

An analysis of the experimental data made it possible to find out the following. The greatest influence on the oil yield and residual cake oil content is made by the initial safflower seeds moisture content, the rotational speed of the press screw, the frequency and amplitude of ultrasonic vibrations passed to the cake removal zone. The observed loss of the adhesive relationship of the polydisperse system (safflower cake) with the vibrating surface exists for a certain range of vibrations, which is localized in a small surface layer. Due to the available energy concentration in the boundary layer, its transition to a highly elastic state can be observed. The product layer examined acquires adhesion-friction characteristics that differ from the mixture total mass. The decrease in near-wall movement is explained by the migration of the binder component - vegetable oil - into the outer layers.

The characteristic features of the process of safflower seeds pressing, providing the maximum oil yield and allowing to evaluate the physical picture of the process under study are identified. It should also be noted that carrying out the process of press production of safflower oil with the identified rational parameters is most effective and expedient.

The most effective values of the safflower seeds pressing parameters providing the minimum oil content in the removed cake are: the frequency of ultrasonic waves in the range of 22 ... 28 kHz, the amplitudes of more than $40 \cdot 10^{-3}$ m and the pressure in the press of 12 ... 13 MPa. These values obviously depend directly on the behavior of safflower grain and its movement in the press when oil extracting under the influence of high-frequency ultrasonic vibrations.

References

[1] Aydeniz B, Guneser O and Yilmaz E 2014 Physico-chemical, Sensory and Aromatic Properties of Cold Press Produced Safflower Oil. *Journal of American Oil Chemist Society.* **91** 99-110

[2] Galavi M, Ramroudi M and Tavassoli A 2012 Effect of micronutrients foliar application on yield and seed oil content of safflower (Carthamus tinctorius). *Afr. J. Agric. Res.* **7** 482-486

[3] Rozina E Y 2002 Effect of Pulsed Ultrasonic Field on the Filling of a Capillary with a Liquid. *Colloid Journal.* **64** 359–363

[4] Dridi W, Henry D and Ben Hadid H 2010 Stability of Buoyant Convection in a Layer Submitted to Acoustic Streaming. *Phys. Rev.* **81 056309**

[5] Moudjed B, Botton V, Henry D, Ben Hadid H and Garandet J P 2014 Scaling and Dimensional Analysis of Acoustic Streaming Jet. *Phys. Fluids.* **26 093602**

[6] Dentry M B, Yeo L Y, Friend J R 2014 Frequency Effects on the Scale and Behavior of Acoustic Streaming. *Phys. Rev.* **89 013203**

[7] Brand A E, Vershinina S V, Vengerov A A and Mostovaya N A 2015 Applying the technology of hydrodynamic cavitation treatment of high-viscosity oils to increase the efficiency of transportation *IOP Conf. Ser.: Mater. Sci. Eng.* **93 012005**

[8] Vengerov A A and Brand A E 2014 Oil cavitation treatment to prevent formation of paraffin deposits J. *Natur. and intellect. resource. of Siberia.* **15** 36-37

[9] Hossein hamidi, Mohammadian E, Junin R and Rafati R 2014 The Effect of Ultrasonic Waves on Oil Viscosity. *Petroleum Science and Technology* **32(19)** 2387–2395

[10] Mohammadian E, Rahmani O, Junin R and Idris F K 2012 Effects of sonication radiation on oil recovery by ultrasonic waves stimulated water-flooding. *Ultrasonics* **53(2)**

IOP Conf. Series: Materials Science and Engineering **941** (2020) 012053 doi:10.1088/1757-899X/941/1/012053

Electrically Energized Comfort Heating System Based on Electric Film Heaters and intended for Post-Weaned Pigs

V M Popov[1], V A Afonkina[1], E N Epishkov[1]

[1]South Ural State Agrarian University, 75 Lenin Avenue, Chelyabinsk, 454080 Russia

E-mail: afva82@mail.ru

Abstract: The reported paper summarizes outcomes of the thesis "Electrically energized comfort heating system intended for pre-nursery pigs kept in an unheated sow house" [2]. This alternative method of keeping warm pre-nursery pigs has been proven to be efficient both in the pilot phase and in pilot batch production. The further research aimed at the application of this system to the keeping warm pre-weaned pigs. The paper describes and presents a calculation procedure for parameters of a comfort heating system based on film heaters and intended for pre-weaned pigs.

1. Introduction

An analysis of mortality rates in various age groups of pigs has shown a sum percentage vs. an aggregate livestock ranges up to 14% within the suckling period and after weaning [Russian standard] (Fig. 1). Such a high die-off value is related to the violation of sanitary and hygienic standards in keeping animals, and a thermal regime is one of the vital characteristics to be observed.

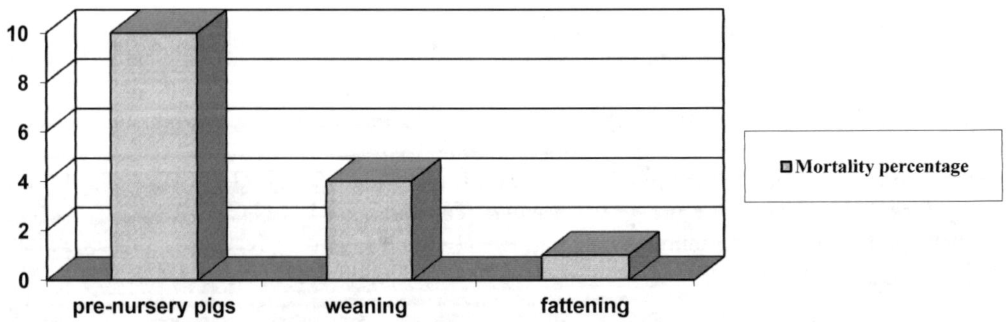

Figure 1. Mortality percentage of animals in different age groups vs. livestock sum

The development of a comfort heating system planned for the suckling period was in focus of the thesis "Electrically energized comfort heating system intended for pre-nursery pigs kept in an unheated sow house" [1] Further studies deal with the same issues during the weaning phase [2].

As a core in the comfort heating system for post-weaned pigs we considered a heat generator [3] used for keeping warm pre-nursery pigs and built on the basis of an electric radiant film heater (Figure 2). The heater is constructed like a blanket consisting of two layers of a polyethylene terephthalate film. Between these layers there is a meander resistant band produced of precision steel 20НХГ. This construction has been granted a useful model patent [3].

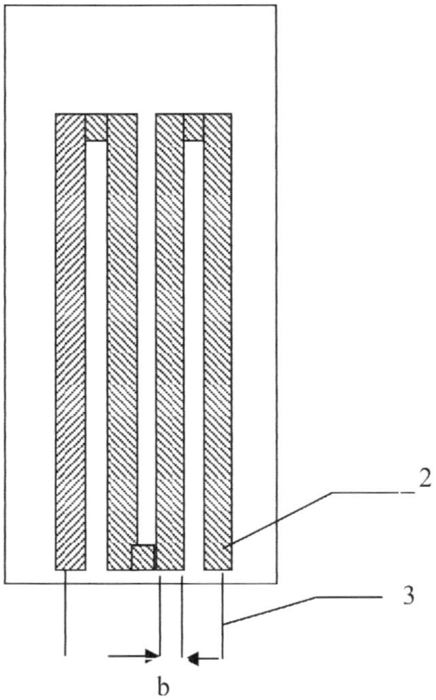

1- Two-layer blanket made of polyethylene terephthalate film;
2- Resistant elements;
3- Electric current conducting wires;
a – space between resistant channels;
b – band width of a resistant element.

Figure 2. General configuration of an electric radiant film heater

An electric radiant film heater has been broadly used in the drying agricultural products [4], heating plants in greenhouses [5], and heating commercial and residential premises [6].

To create a targeted radiation flux with minimal dissipation outside a being heated zone an electric film heater is mounted into a heat generator. It is made as a set of heat insulating elements and reflectors. Figure 3 presents its construction.

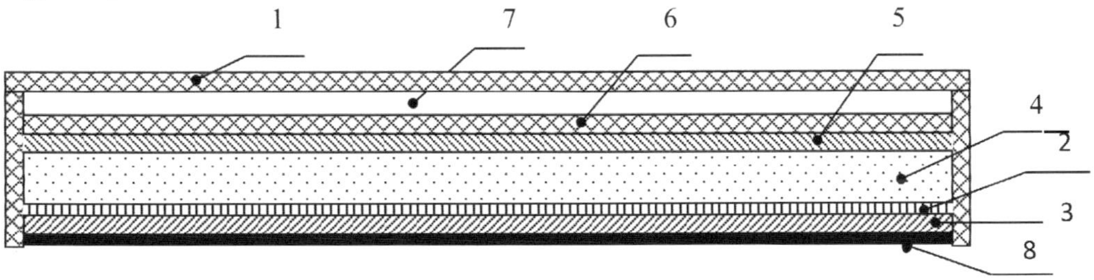

Figure 3. A construction diagram of the heat generator: 1 – frame of a kennel lid; 2 – electric radiant film heater; 3 – ply-wood layer to fix an electric heater; 4 – heat insulating layer; 5 – metallic reflector; 6 – ply-wood layer to fix a reflector; 7 – air layer; 8 – water-resistant film.

2. Materials and Methods

Using diverse fixtures, a heat generator was suspended over a zone to be heated. This construction was built with side curtains to reduce draught-caused heat losses. The prior research [2] has been focused on a system of equations describing a heat exchanging process in a zone to be heated. There is a vast body of literature where Russian and international experts explore and make attempts to find solutions to this problem. A majority of works [7-16] suggest a solution to this system within the vorticity-stream function approach and artificial-compressibility method [17], and by the marker-and-cell method [18]. A temperature pattern on the floor surface in a zone to be heated is formulated in a differential equation:

$$\frac{\partial^2 T}{\partial x^2} - \frac{\sigma*\varepsilon1*c*\rho}{\lambda*h}*T^4 - \frac{\alpha*c*\rho}{\lambda*h}*T = -\frac{\sigma*\varepsilon2*c*\rho}{\lambda*h}*Tucm^4 - \frac{\alpha*c*\rho}{\lambda*h}*T0 \tag{1}$$

where σ – the Stefan-Boltzmann constant; $T_{ист}$ – temperature of radiation source; λ – heat conductivity of the floor in a heated zone; c – specific heat capacity of the floor in a heated zone; ε1 – emissivity factor of the floor in a heated zone; ρ – material density of the floor in a heated zone; ε2 – emissivity factor of the radiation source; T0 – ambient temperature.

When calculating parameters of an electrically energized comfort heating system a solution to this equation is to be sought. Therefore, it is necessary to find its parameters. One of the parameters is ε1 (emissivity factor of the floor surface).

In our consideration we used the Fresnel law [18] for a regular electromagnetic wave coming to the dielectric (magnetic permeability of media is 1)

$$\varepsilon1 = \frac{4*\sqrt{\varepsilon}}{(\sqrt{\varepsilon}+1)^2} \tag{2}$$

where ε – relative dielectric permeability of medium receiving the radiation; ε1 – emissivity factor.

This equation for a normal wave incidence is not related to the position of a wave polarization plane.

It is seen in the equation an emissivity factor is a dielectric permittivity function of the medium.

The difficulty in the estimating dielectric permittivity of a material the floor is made of is that a material represents an inhomogeneous medium consisting of a straw layer, sawdust, and animal waste products.

The study [19] has determined both the dielectric permittivity and an emissivity factor. The recent research has been seeking for other parameters constituting the equation (1). In our further calculations we used these parameters as accepted for the chernozemic soil, since they are most similar to the floor material in a heated zone.

It is difficult to find an analytical solution to the equation (1). A temperature pattern on the floor in a heated zone found when seeking a solution to this equation is adapted as a boundary condition in further calculations.

It is suggested [2] a temperature pattern on a heater surface is homogenous. Moreover, this study has proven an optimal temperature of the heater surface is as high as 40^0C. This temperature is used as a boundary condition for the upper part of a heated zone.

Systems of differential equations 3 – 6 were solved numerically for three power values of a heat generator:

$$\frac{\partial u}{\partial x} + \frac{\partial \vartheta}{\partial y} = 0 \tag{3}$$

$$\frac{\partial u}{\partial t} + u\frac{\partial u}{\partial x} + \frac{\partial u}{\partial y} = -\frac{1}{\rho}\cdot\frac{\partial \rho}{\partial x} + v\left[\frac{\partial^2 u}{\partial x^2} + \frac{\partial^2 u}{\partial y^2}\right] \tag{4}$$

$$\frac{\partial \vartheta}{\partial t} + u\frac{\partial \vartheta}{\partial x} + \frac{\partial \vartheta}{\partial y} = -\frac{1}{\rho}\cdot\frac{\partial \rho}{\partial y} + v\left[\frac{\partial^2 \vartheta}{\partial x^2} + \frac{\partial^2 \vartheta}{\partial y^2}\right] - \beta g T' \tag{5}$$

$$\frac{\partial T}{\partial t} + u\frac{\partial T}{\partial x} + \vartheta\frac{\partial T}{\partial y} = a\left[\frac{\partial^2 T}{\partial x^2} + \frac{\partial^2 T}{\partial y^2}\right] \tag{6}$$

where *x*, *y*, *z* – coordinates of points inside the kennel, where values of temperature are estimated; *T* – temperature, K; *t* – period of time from the start of a heating process inside the kennel; *u* – velocity component of air circulation along the axis *X*; ϑ – velocity component of air circulation along the axis *Y*; ν – air kinematic viscosity coefficient; ρ – air density; β – volume expansion coefficient of air; *g* – gravitational acceleration; *a* – coefficient of air temperature conductivity; *p* – pressure deviation from the mean on in the kennel volume.

In the other study [20] this system is confined to a dimensionless form, its solution is sought using the elimination method:

$$\frac{1}{Sh}\frac{\partial \Omega}{\partial \tau} + U\frac{\partial \Omega}{\partial X} + V\frac{\partial \Omega}{\partial Y} = \frac{1}{\sqrt{Gr}}\Delta\Omega + \frac{1}{2}\frac{\partial \Omega}{\partial X} \quad (7)$$

$$\frac{1}{Sh}\frac{\partial \theta}{\partial \tau} + U\frac{\partial \theta}{\partial X} + V\frac{\partial \theta}{\partial Y} = \frac{1}{Pr\sqrt{Gr}}\Delta\theta \quad (8)$$

$$\frac{\partial^2 \Upsilon}{\partial X^2} + \frac{\partial^2 \Upsilon}{\partial Y^2} = -2\Omega \quad (9)$$

where: x, y – Cartesian coordinates; X, Y – dimensionless coordinates corresponding to coordinates x, y; l – length of a region in the considered solution zone along the axis x; t – time; t_0 – time scale; τ – dimensionless time; u, v – velocity components in the projection to axes x, y, respectively; U, V – dimensionless velocities, corresponding to velocities u, v; V_0 – velocity scale (velocity of convection); θ – dimensionless temperature; ψ_0 – stream function scale; Υ – dimensionless analogue of a stream function; ω_0 – vortex vector scale; Ω – dimensionless analogue of a vortex vector.

3. Results

A numerical solution to the equation (1) is given in the curve (Figure 4):

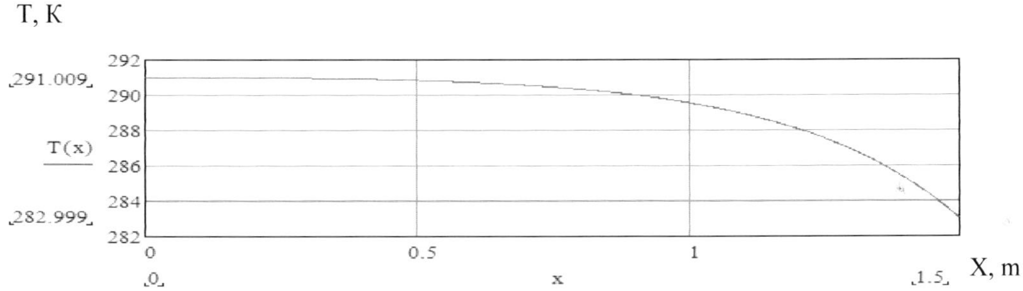

Figure 4. Temperature pattern (Kelvin degree) on the floor surface in a heated zone

This pattern is used as a lower boundary condition for temperature. As an upper boundary condition we use a surface temperature of the electric film heater (40°C). As side boundary conditions we adapt a background temperature of the pigsty (10-12°C). A temperature pattern inside a heated zone is shown graphically:

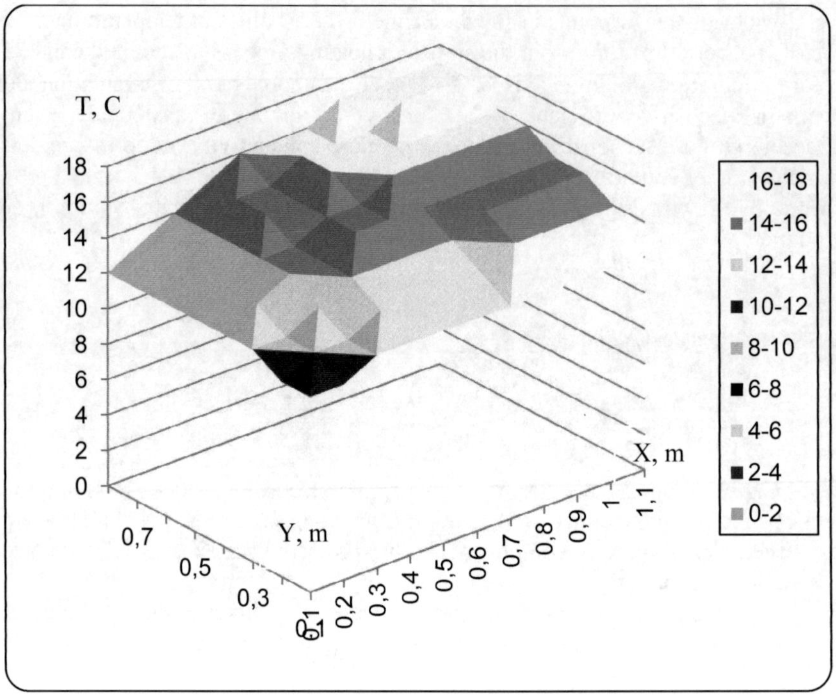

Figure 5. Temperature pattern (Celsius degree) inside a heated zone at a height of 10 cm

4. Conclusions:

1. In the research appropriate dimensions of a heated zone were found, as well as a power value of a heat generator 250 W (a background temperature of the pigsty 10-12^{0}C). These parameters support a temperature pattern inside a heated zone, which satisfies with veterinary and zootechnical standards in keeping young animals.
2. When using a pilot sample of this heating system a necessity was revealed to improve a hermetic state of the heat generator and reduce its mass and dimension parameters.
3. This heating system allowed a considerable mortality reduction among young animals because of cold-related diseases and increased their weight till the end of a weaning period.

References

[1] Tamvakidis S, Firfiris V, Martzopoulou A, Fragos V, Kotsopoulos T 2015 *Energy and Buildings* **97** pp 162-174

[2] Epishkov E, Zagidullin R 2017 *Proceedings of the International Scientific Conferences "Energy to agricultural complex in Russia"* (South Ural State Agrarian University) pp 67-74

[3] Epishkov E, Epishkov N, Deaf S 2006 *Pat. on useful. Model 76764 Russian Federation. Film electric heater No. 2006110752*

[4] Afonkina V, Zakhakhatnov V, Mayerov V, Popov V 2016 *Mordovia University Bulletin* **26** 1 pp 32-39

[5] Kalinina T, Polyakova V, Kichin K 2016 *Young scientist* **29** 133 pp 81-86

[6] Blednic V 2013 Vestnik BSAU **4** 28 pp 81-83

[7] Wu W, Sun Q, Sun M, Dehmer M, Chen Z 2019 *Applied Mathematical Modelling* **70** pp 378-401

[8] Mohammed H, Belkacem A, Noureddine K, Elhadj B 2017 *Energy Procedia* **139** pp 511-516

[9] Sai Ravi Gupta Polasanapalli, Kameswararao Anupindi 2019 *Computers & Fluids* **181** pp 259-282

[10] Guo B, Xie B, Zeng L 2019 *Journal of Differential Equations*

[11] Cui H, Xu F, Saha S C, Liu Q 2019 *International Journal of Thermal Sciences* **139** pp 282-291

IOP Conf. Series: Materials Science and Engineering **941** (2020) 012053 doi:10.1088/1757-899X/941/1/012053

[12] Gao M, Zhang L-Sh, Zhang D, Zhang L-X 2019 *Experimental Thermal and Fluid Science* **104** pp 9-14
[13] Guha A, Jain A, Pradhan K 2019 *International Journal of Heat and Mass Transfer* **135** pp 1130-1151
[14] AL-Bayati S A, Wrobel L 2019 *Engineering Analysis with Boundary Elements* **101** pp 89-101
[15] Kooij G, Botchev M, Frederix E, Geurts B, Verzicco R 2018 *Computers & Fluids* **166** pp 1-8
[16] Necati Mahir, Zekeriya Altac 2019 *International Journal of Heat and Fluid Flow* **75** pp 103-121
[17] Ganapathy R, Mohan A 2018 *Alexandria Engineering Journal* **57** 3 pp 2035-2042
[18] Sidney A 2017 *Studies in History and Philosophy of Science Part A* **64** pp 38-52
[19] Popov V *et al* 2019 XII International Scientific Conference on Agricultural Machinery Industry *IOP Conf. Ser.: Earth Environ. Sci.* **403** 012007
[20] Epishkov E 2020 Proceedings of the national scientific conference «Actual issues of agroengineering in the field of energy of the agroindustrial complex: theory and practice » (South Ural State Agrarian University) pp 50-59

IOP Conf. Series: Materials Science and Engineering **941** (2020) 012054 doi:10.1088/1757-899X/941/1/012054

Use of low-temperature plasma for stabilizing a combustion process under the utilization of waste products with high moisture content

V N Delyagin[1], N M Ivanov[1], A N Karzanov[1], V N Bocharov[1] and A A Kondrat'ev[2]

[1] Siberian Federal Scientific Centre of Agro-BioTechnologies of the Russian Academy of Sciences, 2B Central'naya ul., Krasnoobsk, Novosibirsk Region, 630501 Russia
[2] Novosibirsk State Agrarian University, 160 Dobrolyubova ul., Novosibirsk, 630039 Russia

E-Mail: valdel@ngs.ru

Abstract. Thermal recycling and disinfection of waste products in different physical states represent a comprehensive solution to the problem of waste utilization. The utilization of waste materials with high moisture content (more 70...75%) is associated with a need to evaporate a significant volume of water and causes, as a consequence, substantial additional expenses on an energy source. To cut costs related to the utilization of wastes with a high amount of water it is suggested using a coal cleaning product – a coal-water slurry fuel. The direct combustion of incompletely mechanically dehydrated (to 70-75%) wastes with high water content (cake) is an economically efficient process. However, it poses a severe problem of combustion stabilizing. The paper presents outcomes of the research into the stabilization of thermal recycling waste materials with a high level of moisture by means of low-temperature plasma in the adiabatic vortex combustion chamber. To control a burning process in the technology adapter low-temperature plasma is supplied into the base of flame. An output of a plasma generator (power supply source) is estimated as 1 kW power input per 13 kg/h suspension fuel. A dynamic range of heat output for a coal-water slurry fuel and wastes is determined to be 70-100% when generating thermal energy and utilizing wastes with high moisture content by means of low-temperature plasma as a stabilizer of a combustion process. Experimental studies carried out on a machine simulator have proven the efficiency of low-temperature plasma for utilization and disinfection of waste products in commercial farming.

1. Introduction

Large-scale livestock breeding complexes and poultry factories are main pollution sources of the biological environment around metropolitan areas. As a rule, waste products are in different physical states – solid, liquid, and gaseous.

The utilization of multicomponent wastes via thermal processing leads to a significant cost increase. One of the possible solutions to this problem seems to be the use of a coal-water fuel – a product of coal cleaning [1]. In this case, a cost price of thermal energy is 3...4 times lower. The future for this utilization procedure is related to stabilizing the combustion of wastes with a high amount of water in a material to be recycled and in a fuel as well.

IOP Conf. Series: Materials Science and Engineering **941** (2020) 012054 doi:10.1088/1757-899X/941/1/012054

Recently, a solution to the aforementioned problem, i.e. to stabilizing combustion processes is reported to be found in the use of electrical and magnetic fields, electrical discharges, ozone and low-temperature plasma [2-8].
The study aims to evaluate the efficiency of utilizing waste materials with high moisture content by means of a coal-water slurry fuel and low-temperature plasma used as a stabilizer of thermal processes when recycling wastes.

2. Materials and Methods

A power technological adapter including a coronation electrical unit and a plasma generator (Figure 1) was mounted on an experimental unit (Siberian Research Institute of Mechanization and Electrification in Agriculture, Siberian Federal Research Center for Agrobiotechnologies, Russian Academy of Sciences) equipped with an adiabatic vortex combustion chamber (output 500 kW), where a material to be utilized is supplied directly into the furnace. The adapter was installed to provide both separate or tandem arrangement of air blasting unit, and a unit supplying low-temperature plasma into the flame.

Figure 1 A power technological adapter

As a low-temperature plasma supply source we used a high-frequency streamer discharge. An output of a power source was as high as 3.5 kW. A voltage of a power source (no-load operation) was set 12 kV. For power production in the adiabatic vortex combustion chamber intended for the utilization of wastes we adapted a coal-water suspension mixed of coal cleaning wastes and pulverized coal fuel.
We modeled the low-temperature plasma supply on the example of coal-water slurry fuel (Figure 2 a) and pulverized coal fuel (Figure 2 b).
The electrical and physical effect of charged particles on the behavior of combustion was assessed relying on the dynamics of temperature in the furnace (a response of the system). Parameters to be measured – a number of ion pairs (millions pairs/cm3), a volume of supplied ionized air (m3/min), input current and power of a plasma generator for charged particles (A, W), and a heat flow output (W).
Parameters to be controlled included an amount of air supplied into the furnace, fuel consumption, and a temperature field of a burner flame.
Parameters were measured and recorded using analytical-recording software. We used such measuring devices as temperature sensors, a differential pressure gauge, an ion number (MAC 01) gauge, a radiometer with a spherical probe, a thermal imaging camera Testo 885-2, pyrometers Testo 845 and "Calvin", digital multiple-purpose meters ACTACOM, and a measuring device for electrical and magnetic fields PZ-70. Data were processed in the mathematical package "Statistica".

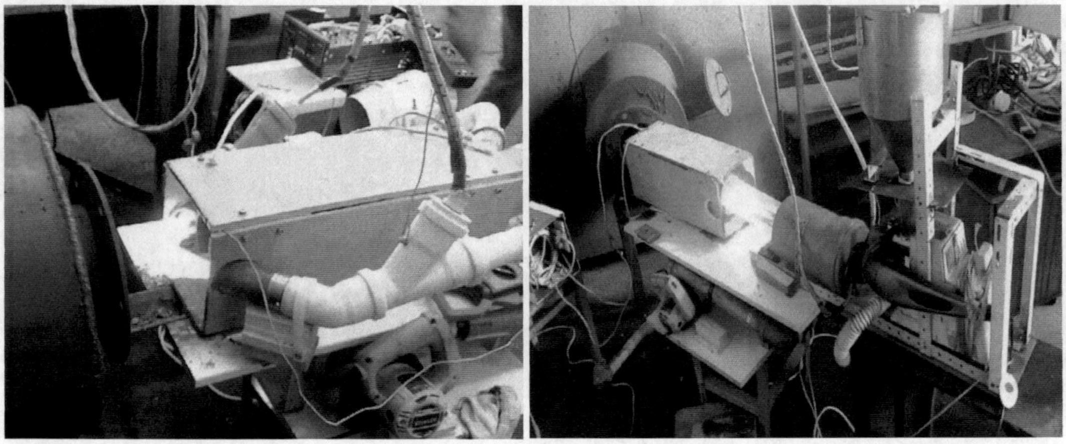

a b

Figure 2. Low-temperature plasma supply when using: a – a coal-water slurry fuel, b – a pulverized coal fuel.

3. Results and Discussion

Table 1 summarizes outcomes of experimental studies.

A value of the reduced electric field was determined as E/N, where E – electric-field strength, N – concentration of neutral particles 800...1200 Td.

A number of negative air ions was varied from 1.28*105 to 8*105, that of positive ions was in a range from 1.47*105 to 1.00*106.

Plasma source parameters: source output 3 kW, fuel consumption 12...40 kg/h, volume of supplied initiated air 80...110 kg/h.

Table 1 – Operation parameters of a technology adapter

Parameter	Symbol	Value
Temperature in the combustion zone inside the furnace (C^0)	tc	960-1050 ^{0}C
Temperature of an applied mixture (C^0)	tm	10/105I
Output in view of coal-water slurry fuel (kg/h)	Vw	12-40
A number of ions -positive (N+/cm^3) -negative (N-/cm^3)	N	1.28*105 ...8*105 1.47*105 ...1*106
Air supply (kg/h)	V	80-110
Reduced electric field, Td		800-1200
Recombination time of air-ions (s)	T	0.067...0.081
Voltage on electrons of a plasma generator (kW)	U	12/4
Discharge current (A)	I	1-2
Moisture content in wastes (%)	W	72-75
Dynamic range	%	75-100
Specific output of a plasma generator, kW/(kg*h)	Vp	0.61
Ratio coal-water slurry fuel: wastes		3:1

When experimenting with the combustion initialization of coal-water slurry and pulverized fuel mixtures, an impact of low-temperature plasma on the stabilization of a burning process appeared to be far significant than 3 g/s ozone supply into the flame base with a similar number of air-ions.

The combustion of wastes was stabilized more efficiently by means of a pulverized fuel than a coal-water slurry fuel.

An explanation is supposed to be that electrons under the given electric field frequency gain additional energy in a range 2...4 eV, therefore, plasma can be considered non-equilibrium. High-energy electrons committing with fuel and gas molecules form chemically active radicals CH and OH, which initiate the development of chain combustion reactions. Non-equilibrium plasma stimulates the strong inflammation with a great number of fire spots.

To stabilize a combustion process we suggest using non-equilibrium low-temperature plasma in the technology adapter, which is generated by a high-frequency spark discharge (energy of electrons to 4 eV) and supplied into the flame base. An output of plasma generator (power source) is determined as 1 kV input network power per 13 kg/h fuel (a ratio between a fuel and wastes kept at the humidity of 72...75% is 3:1).

The dynamic range of heat output in the running system "adapter – boiler unit" (thermal energy production and utilization of wastes with high moisture content) varies 70 to 100%.

Non-equilibrium plasma furthers the combustion stabilizing of fuel components with high moisture content in the process of utilization and disinfection of waste materials.

References

[1] Delaygin V.N., Murko V.I., Ivanov N.M., Revyakin E.L. Use of coal-water slurry fuel for power supply in agriculture M.: FSBSI "Rosinformagrotech", 2013. – 92 p. (*in Russian*)

[2] Research and developments of Siberian Branch Russian Academy of Sciences in the field of energy efficient technologies / edit. by S.V. Alekseenko: SBRAS Press, issue 20, 2009. – 405 p. (*in Russian*)

[3] Malinovsky A.E., Lavrov F.A. On the effect of electric field on combustion processes in gases // Technical Physics. – 1931. – V.2, Issues 3-4. – p. 530-534 (*in Russian*)

[4] Wang F, Liu J B, Sinibaldi J, Brophy C, Kuthi A, Jiang C, Ronney P and Gundersen M A. IEEE Trans. Plasma Sci. 33 844–9 (2005)

[5] Pancheshnyi S V, Lacoste D A, Bourdon A and Laux C O. IEEE Trans. Plasma Sci. 34 2478–87(2006)

[6] Bao A, Utkin Y G, Keshav S, Lou G and Adamovich I V IEEE Trans. Plasma Sci. 35 1628–38(2007)

[7] Kosarev I N, Aleksandrov N L, Kindysheva S V, Starikovskaia S M and Starikovskii A Yu Combust. Flame 154 569–86(2008)

[8] Heywood J.B., Internal combustion engine fundamentals, McGraw-Hill international editions (1988).

[9] Sosnin E.A. Source of an atmosphere pressure plasma jet formed in the air or nitrogen actuated by a barrier discharge / E.A. Sosnin, et al. // Technical Physics. – 2016. – V. 86. – Issue 5. – № 4. – p. 151-153 (*in Russian*)

IOP Conf. Series: Materials Science and Engineering **941** (2020) 012055 doi:10.1088/1757-899X/941/1/012055

The project solution of the refrigeration machine scheme with the heat utilization of condensation

E N Neverov, I A Korotkiy, M Y Mokrushin, I A Prib, D I Goleshov

Kemerovo State University, 6 Krasnaya Street, Kemerovo, 650000, Russia

E-mail: neverov42@mail.ru

Abstract. Refrigeration machines are widely used in various industries to produce artificial cold. In the process of maintaining low temperatures, the refrigeration machine generates a lot of heat to compress a refrigerant, which is not used in the future, and during the condensation it is released into the environment in the form of water vapor. This negatively affects to the environment because of increasing the concentration of greenhouse gases in the atmosphere. And the use of chlorine-containing refrigerants causes the destruction of the ozone layer. To reduce environmental damage, it is necessary to increase the energy efficiency of refrigeration machines and develop schemes that allow the use of low-grade heat, which is released during operation. In the work, a scheme of a refrigeration machine was developed, which allows utilizing the heat of condensation, using it for various consumers of thermal energy, therefore allowing to reduce emissions of harmful substances into the Earth's atmosphere and increase technological capabilities of cooling.

1. Introduction

At present, refrigeration equipment is used in many industries, is an integral part of production. The world operates a large number of installations for the production of cold. The most widely used cold in industry and commerce, used for long-storage products, cooling different objects, making a comfortable life support. People used to use cold for their needs, but not everyone thinks that using artificial cold is harmful to the planet and is mainly manifested in the destruction of the ozone layer and the greenhouse effect.

To operate the refrigeration system refrigerant is used, which is compressed in the compressor, cooled in the condenser, it expands in the throttling device and evaporates in the evaporator. In order to obtain compressed gas encountered with selection of a large amount of heat. The gas temperature after compression in the compressor can be in wide limits and will reach 210 ° C and above. This heat is usually transferred from the compressed refrigerant to the cooling medium, which is water or air. Due to heat exchange during condensation, water vapor is formed, which is released into the Earth's atmosphere [1,2].

Also, compression requires a lot of energy, and as you know, the main method of its production is thermal power stations. In the process of obtaining energy, a lot of carbon dioxide is released on them, which increases the concentration in the atmosphere. The release of greenhouse gases into the environment, such as carbon dioxide, methane or water vapor, detains infrared radiation from the Earth, which can ultimately lead to climate change on Earth. [3,4,5].

Freons are used as refrigerants in refrigeration systems, which lead to destruction ozone layer as it is composed of chlorine, which in the process of reaction, converts ozone into oxygen under atmospheric conditions [6].

Based on this, specialists working in the field of refrigeration technology ask questions related to the environmental situation and assist that it is necessary to be more careful about the environment, to create installations that will compliance with the standards, but not every invention can be called environmentally safe. Almost all refrigeration systems in the process of the work produce a significant amount of heat, which is often not used, but is released into the environment. However, the use of this heat has tremendous energy potential; therefore, the utilization of emitted thermal energy is the most important economic task, the solution of which is carried out in all developed countries of the world. [7,8,9].

In recent years, more and more important is the development of techniques to decrease the environmental impact and the environment. The use of refrigerators with heat recovery is very important, because the use of heat recovery can reduce heat emissions into the atmosphere and turn this heat to use in other processes, which ultimately improves energy efficiency by reducing energy consumption for these needs. It is important to note that the heat recovery system can be installed not only on new installations, but also implemented on already deserted [10].

1.1. Objectives of the Study

This article proposes a scheme of a refrigeration system, which in the course of its work minimizes the negative impact on the environment. The creation of a system for the use of utilized heat for other needs at the facility is the basis of this development. Using the allocated heat, the efficiency of the refrigerating machine as a whole increases, and this also leads to a decrease in the load on the condenser of the machine, which helps to reduce the greenhouse effect by reducing excess emissions of thermal energy into the environment.

2. Materials and methods

At Kemerovo State University, a scheme compression refrigeration system was developed for the efficient utilization of the heat generated during its operation, with subsequent use for technological needs. The tasks laid down in the proposed scheme are to ensure that the heat of condensation is utilized, which will reduce the load on the condenser, transfer heat to the required distance for future use, and increase the efficiency of the chiller as a whole.

To solve these problems, to ensure heat recovery and reduce the load on the condenser, two or more intermediate heat exchangers are installed in the area between the compressor and the condenser in the circuit of the plant, depending on the technological solution in the installation. It is most preferable to use prepared water as the heat carrier, since it is economically more profitable than any other heat carrier. High potential heat is removed in the intermediate heat exchangers due to the thermodynamic properties of the media used. The heat removed by the heat carrier is used to heat pipeline water to consumers, as well as for further heating, for example, soil, snow melting pits, and process water. Improving the efficiency of the chiller is achieved by transferring part of the condensation heat in the heat exchanger to the coolant, as well as by the ability to control the condensation temperature using three-way valves, which allow one, two or both plate heat exchangers to be switched on or off from the refrigeration machine. Also, this feature allows you to use the refrigeration machine in climatic areas with different outdoor temperatures. And the possibility of cooling the coolant is achieved by installing a plate heat exchanger on the low pressure line between the throttling device and the compressor. Further, the cooled coolant is used to remove heat from ice fields or chambers located at a considerable distance from the refrigeration machine. Propylene glycol is recommended to be used as a coolant, since it is an environmentally friendly, fire- and explosion-proof solution. In order to save budget use a solution of propylene glycol and water in a percentage ratio of 40% to 60%. Depending, on the geographical location of the installation, on the maximum calculated winter temperature and the temperature of the coolant [11].

Figure 1 shows a diagram of a refrigeration machine with utilization of the heat of condensation.

The refrigeration machine has a rotary-screw compressor (1), a high pressure switch KP (2), a shutdown valve Rotolock (3, 16, 19, 25), a relief valve SFV (4, 17), an oil separator (5), through which oil is separated from refrigerant, check valve SCA (6), condensation pressure regulator KVR (7), pressure regulator in the receiver KVD (8), pilot valve CVP (9), pilot-operated relief valve with pilot control ICS (10), ball valve GBC (11), plate heat exchangers (12, 23), three-way valve (13, 30), air-cooled condenser (14), non-return valve NVR (1 5), linear receiver (18), filter drier DCR (20), sight glass (21), temperature control valve TE (22), low pressure switch KR (24), oil temperature regulator ORV (26), oil cooler (27) , oil filter (28), oil flow switch (29).

The refrigeration machine operates as follows.

When the compressor motor (1) is turned on, the refrigerant is compressed, after which it is supplied to the oil separator (5), where the refrigerant vapor is separated from oil droplets that pass through the oil temperature regulator, and then enter the oil cooler (27), where they can be forced cooled by a fan installed on the oil cooler, and then enter the compressor crankcase. Forced cooling of the oil prevents it from boiling and reaching the flash point of the oil. Also, oil can pass through a temperature controller (26), and then immediately enter the compressor crankcase. And refrigerant vapors, passing through a three-way valve (13), enter the plate heat exchanger of the first stage (12), where high potential heat is removed to heat the water, and then to the second stage heat exchanger, where the average potential and remaining heat is removed for heating soil, floors and pits. In this case, the three-way valve (30) disconnects the condenser (14) from the system and the refrigerant enters immediately into the linear receiver (18). Also, heat exchangers (12) can be used separately for heating the ground, floors, snow melting pits, piped water or other needs of the enterprise. In this case, the three-way valve (30) includes a condenser (14) in the system, in which the process of complete condensation occurs. The number of heat exchangers included in the system from 0 to 2, depending on the selected mode of heat consumption and the position of three-way valves (13). Steam after heat exchangers, or passing them, enters the condenser (14), where it condenses and is supplied to the linear receiver (18), where it enters the thermostatic valve (22) through the filter dryer (20). Next, the refrigerant is throttled and fed into the plate heat exchanger (23), where the cooling medium is cooled, which cools the ice field or chambers remote from the refrigeration machine. After the plate heat exchanger (23), the refrigerant passes through the filter dryer (20), enters the compressor (1) and the cycle repeats.

3. Results and Discussion

Refrigerating machine consisting of shutdown and control valves connected in series a compressor, condenser, throttling device, an evaporator, a piping connecting between the compressor and the condenser, two heat exchangers removing heat of condensation in order to remove high-potential heat from the refrigerant. Removing the heat load from the condenser, the cooling capacity of the installation increases, and it also becomes possible to use the removed heat for other purposes (ground heating, snow melting pits, process water). Also, the refrigerating machine is equipped with the possibility to disconnect the condenser from the system by the three-way valve, which allows to minimize the thermal energy emissions. The refrigeration machine allows three-way valves to switch the heat exchangers to provide different condensation temperatures in the condenser, thereby allowing you to adjust the cooling capacity of the plant.

The refrigerating machine is equipped with a plate heat exchanger, which allows cooling the coolant in order to use it to remove heat from ice fields or chambers located at a considerable distance from the chiller.

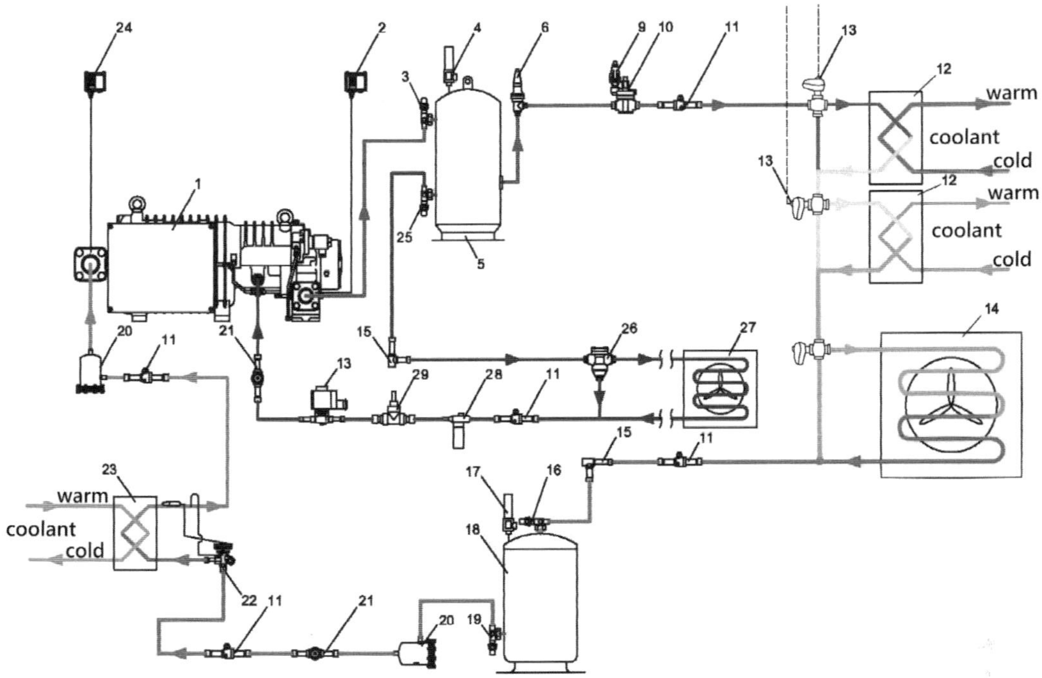

Specification

1 Screw type compressor
2 High pressure switch KR
3 Shut-off valve Rotolock
4 Safety valve SFV
5 Oil separator
6 Check valve for port SCA
7 Condensing pressure regulator KVR
8 Pressure regulator in the KVD receiver
9 Valve pilot CVP
10 Pilot operated servo valve ICS
11 Ball valve GBC
12 Plate heat exchanger
13 Three-way valve
14 Condenser air cooling
15 Check valve NRV
16 Shut-off valve by Rotolock
17 Safety valve SFV
18 The receiver is linear
19 Shut-off valve by Rotolock
20 Desiccant filter DCR
21 Viewing glass
22 Thermostatic valve TE
23 Plate heat exchanger
24 Low pressure switch
25 Shut-off valve Rotolock
26 Oil temperature regulator ORV
27 Oil cooler
28 Oil filter
29 Oil shut off relay
30 Three-way valve

Figure 1. The scheme of refrigeration machine with utilization of heat condensation

4. Results and Discussion

Refrigerating machine consisting of shutdown and control valves connected in series a compressor, condenser, throttling device, an evaporator, a piping connecting between the compressor and the condenser, two heat exchangers removing heat of condensation in order to remove high-potential heat from the refrigerant. Removing the heat load from the condenser, the cooling capacity of the installation increases, and it also becomes possible to use the removed heat for other purposes (ground heating, snow melting pits, process water). Also, the refrigerating machine is equipped with the possibility to disconnect the condenser from the system by the three-way valve, which allows to minimize the thermal energy emissions. The refrigeration machine allows three-way valves to switch the heat exchangers to provide different condensation temperatures in the condenser, thereby allowing you to adjust the cooling capacity of the plant.

The refrigerating machine is equipped with a plate heat exchanger, which allows cooling the coolant in order to use it to remove heat from ice fields or chambers located at a considerable distance from the chiller.

5. Conclusion

The proposed refrigeration machine provides utilization and transportation of heat to various consumers of thermal energy, which increases the energy efficiency of the refrigeration machine by removing heat from heat exchangers, and also reduces environmental pollution during operation. Also, the installation allows you to cool objects at a considerable distance from them, which helps to increase technical features or eliminates the possibility of harmful substances entering the product that is being cooled.

The refrigerating machine runs on ozone-safe refrigerant that do not contain chlorine atoms, which minimizes the potential of ODP and GWP. This plant is an environmentally safe, complies all modern requirements and standards, reduces the concentration of heat emissions and helps to minimize the destruction of the ozone layer and the greenhouse effect. And the use of non-toxic refrigerant and coolants allows not only to preserve nature and the environment, but also to preserve the quality of refrigerated products. The installation provides an object not only of artificial cold, but also allows the use of extracted heat for different technological needs of the enterprise, which leads to significant savings in general.

Acknowledgment

The main authors are very grateful to the student of the Kemerovo State University Radzhaboi Rustamov, for his help in translating the article into English.

References

[1] Shakirova G G 2015 *Modern Innovations: Fundamental and Applied Research* vol 1 no 1, ed S V Valtsev, A S Kotlova et al (Moskva: OOO «Olimp») pp 6-7

[2] Denisov-Vinsky N D 2011 *MegaPascal* **2** 8-16

[3] Avdeenko O A and Lazareva L P 2009 *Vologda Readings* **75** 5-8

[4] Belsky Y D 2012 *Modern Society, Education and Science* (Tambov: OOO "Consulting company Ucom") pp 16-7

[5] Lebedev S V and Penkov M V 2019 *Structural Transformations of Territory Economics: in Search of Social and Economic Equilibrium* T 3 ed A R Khalikov (Ufa: OOO "NIC Bulletin of Science ") pp 136-45

[6] Israphilov D I and Mukhamatdinov I A 2017 *Science Yesterday, Today, Tomorrow* vol 8 no 42, ed M A Vasinovich, N V Dmitrieva et al (Novosibirsk: ANS «SibAk») pp 58-62

[7] Buchin S and Smagin S 2010 *Climate World* **62** 74-7

[8] Neverov E N, Korotky I A, Raschepkin A N and Ibragimov M I 2020 *Bulletin of the International Academy of Refrigeration* **1(74)** pp 22-6

[9] Neverov E N, Korotkiy I A and Korotkaya E V 2019 IOP *Conf. Series: Earth and Environmental Science* **395(1)** 012066

[10] Neverov E N, Korotkiy I A, Korotkih P S, Lifenceva L V et al 2019 IOP *Conf. Series: Earth Environ. Science* **224** 012039

[11] Baranenko A V, Bukharin N N, Pekarev V I, Sakun I A and Timofeevsky L S 1997 *Refrigeration Machines* (St. Petersburg: Polytechnic) p 499

Using the organic Rankine cycle for heat supply of greenhouses at agricultural enterprises

K V Osintsev

South Ural State University, 76, Lenina Ave., Chelyabinsk, 454080, Russia

E-mail: pte2017pte@mail.ru

Abstract. The greenhouses are used at agricultural enterprises producing crop products. It is proposed to utilize the heat of the ground in order to increase the high-quality energy supply of these facilities. Organic Rankine cycle is needed to get heat in the form of hot water, as well as electricity. Hot water can heat air in heaters and maintain the required air temperature in the greenhouses themselves. Electricity production is possible due to its generation by a micro-turbine or an expander. The most promising area is the conversion of the kinetic energy of freon into mechanical work in Tesla's radial turbines. The article proposes a variant of the heat supply scheme of the greenhouse. The most effective technological solutions of putting into practice the organic Rankine cycle are shown.

1. Introduction

Cultivated constructions of sheltered soil are production facilities that are protected from the external environment by a translucent roof and walls. They are intended for growing vegetables and flowers. The main elements of such structures are the soil, where plants are grown, and engineering networks for heating the room. Small-sized constructions are simple constructions designed for growing early vegetables. They consist of a transparent polymer film, frame. Large-sized structures are, in fact, greenhouses, which will be discussed in the article. In the production of products in sheltered ground, the main expenditure of energy resources is accounted for by irradiation of plants and heating of the greenhouse in the winter.

There are several types of heating greenhouses. The first of these is electric heating. In agriculture, electrical energy is used to heat various objects. Electric heating plants, widely used in agriculture, include electric water heaters and steam generators, electric heaters and plants for heating the soil in greenhouses, greenhouses and floors in livestock buildings. Water heating is carried out using elemental and electrode water heaters. The working body of elemental water heaters is a tubular electric heater. Electrical heating installations are used for air heating and are used in heating and ventilation systems of livestock and poultry farms, greenhouses, domestic and industrial premises, at grain cleaning drying centers. Their advantage is that heating and ventilation functions are combined in one unit. They have simple flow control and the amount of air supplied. The heating elements are divided into independently adjustable sections. For agriculture, electric air heaters with centrifugal fans and axial fans are produced. To heat the soil in greenhouses, a wire is used that is designed for a voltage of 220 V or 380 V. It is laid on the bottom of the pit in two sections and covered with a cement screed, as well as a metal mesh. Sometimes the heating wire is laid in a monolithic asphalt concrete block, laid under the soil layer. The second type of heating is the construction of its own boiler room. The boiler room can be hot water or steam. Steam boilers are often used. In this case, it is necessary to additionally construct a condensate drainage system back to the boiler room. During the construction of a boiler house, it is necessary to build heating networks. The main element of engineering networks is a heater. Air is heated with hot water in this heat exchanger. Air can be additionally forced by a fan.

2. The principle of operation of the technological scheme and scientific novelty

The optimal composition is understood as the selection of thermal units of this type and quantitative ratio, which make it possible to obtain products in agriculture at the lowest cost of labor and funds per unit of output. The selection and justification of the optimal composition of thermal units and utilities is a very responsible task [1]. Its complexity is because agricultural enterprises are diversified farms cultivating various crops, and the conditions for making production decisions are diverse. Almost every type of thermal aggregates and devices in their technical capabilities can be most effectively used only in a certain range of natural-production conditions for which it is designed. Therefore, from the wide variety of units, only the minimum necessary number is selected, which will ensure that all work is performed with the greatest economic effect. It is advisable to start the choice with energy resources, which account for the largest share of cost and operating costs. These options include the use of the organic Rankine cycle. In this case, fuel costs are sharply reduced. Fuel is the main type of cost in any industrial enterprise [2, 3]. The principle of operation of the proposed technological complex is as follows. The scheme was developed taking into account the latest research by the authors [4, 5]. As the evaporator, soil heat exchange tubes 7 are used, in which freon boils. These pipes are in the soil 8. Freon steam is compressed by the compressor 9. The compressed steam is sent to the condenser 3. In the condenser, the freon steam gives its heat to water, which is pumped by a circulation pump 4. A radial micro turbine 5 is used as an expander, which transfers mechanical work to the electric generator 6 An important factor in heat transfer in a greenhouse is air exchange. Cold air 10 is pumped by the fan 11. Next, the air is sent to the air heater (it is not shown in the figure). Air circulates in the air supply system 12. Heat exchangers 2, ventilation 12 and air conditioning systems are widely used in industry [6, 7].

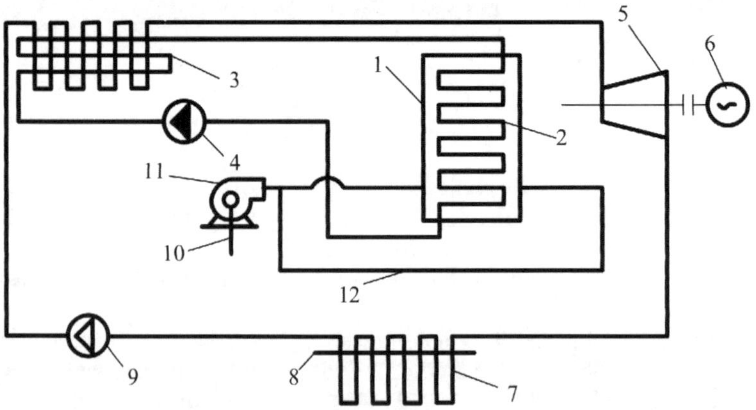

Figure 1. The application scheme of the organic Rankine cycle: 1 - greenhouse, 2 - heat transfer surfaces, 3 - condenser, 4 - circulation pump, 5 - turbine, 6 - electric generator, 7 - ground heat transfer pipes, 8 - ground, 9 - compressor, 10 - cold air, 11 - fan, 12 - air duct

3. Theoretical part of using organic Rankine cycle

Cogenerators are easy to transport and install. They allow solving the acute problem of uneven daily electricity consumption, insoluble for large generating plants. Indeed, for a cogenerator, a linear dependence of fuel consumption takes place starting from 15–20% of the rated power. By partitioning the total power operating in parallel, it becomes possible to work from 1.5–4.0% to 100% of the rated load with the estimated specific fuel consumption. Boiler units can be used as a source of thermal energy [8, 9]. Boiler units burn various fuels. At the moment, there are studies regarding water-coal mixtures [10, 11]. Using trigeneration technology allows you to maintain high efficiency year-round. For example, in summer, heating is not required, but it is necessary to air-condition residential premises, offices, hospitals [12]. Cold water and cold are widely used in industry. The disadvantage of cogenerators is only limited power up to 3 MW for one machine. The average industrial consumer has an installed capacity of 1–2 MW. If necessary, several parallel cogenerators can be installed. Technological complexes operating on the principle of the organic Rankine cycle should also be attributed to cogeneration plants. Heat sources for the Rankine organic cycle can be heat from the combustion of natural gas, propane, blast furnace gas, pyrolysis gas, landfill gas, shale gas, associated gas, biomass. In addition, it is possible to use waste heat from geothermal energy, waste hot water of various industries, steam from steam boilers or turbines, heat from the exhaust of gas turbines, waste heat from boiler plants. For effective use in a system based on the organic Rankine cycle, the optimal working environment must satisfy the following requirements. It is necessary that there is a low freezing temperature, and it must be below the lowest temperature of the cycle and below the ambient temperature. High heat of evaporation and density are required for more efficient absorption of energy from a heat source.

4. Practical part. Application

In steam cycles, the pressure to heat ratio of the turbine is very high. As a result, turbines with several expansion stages are typically used. In the organic Rankine cycle, the reduction in heat content is lower and usually one-stage or two-stage turbines are used, which entail a reduction in cost. Additional effects of lower heat reductions in organic fluids include lower rotational speeds. A lower rotation speed allows the direct drive of an electric generator without a reduction mechanism (this is especially beneficial for plants with a low power range); while a low speed reduces the voltage on the turbine blades and simplifies their design. The efficiency of the current high temperature cycle of the organic Rankine cycle does not exceed 24%. A typical steam cycle has a thermal efficiency of more than 30%, but with a more complex design cycle. The Rankine organic cycle is best suited to operate in the low and medium power ranges, as small power plants cannot afford a local operator. The cycle also has a relatively simple structure and does not require components difficult to manufacture. Consequently, it is more adapted to decentralized electricity production. For high power ranges, the steam cycle is more suitable, with the exception of low-temperature heat sources.

5. Conclusion

Thus, the general range of increasing of heat supply efficiency is possible in case energy and technological combination. The main advantages of the Rankine organic cycle are as follows. As noted earlier, organic fluids usually remain superheated at the end of expansion. Thus, there is no need for overheating in the Rankine organic cycle, in contrast to steam cycles. The absence of condensation also reduces the risk of corrosion on the turbine blades, and extends the service life to 30 years, and for steam turbines, the service life is 15-20 years. Another plus is the low heat recovery temperature. Due to the lower precise boiling point and due to properly selected organic working fluids, high temperature can be restored at much lower temperatures. In the organic Rankine cycle, it is possible to use once-through boilers, which avoids steam drums and recycling. This is due to the relatively small difference in density between the vapor and the liquid. In contrast, the low vapor density in steam boilers can produce a completely different heat transfer due to the characteristics of the pressure drop between the liquid and the steam. In order to avoid droplet formation during expansion, in the Rankine steam cycle, due to overheating, temperatures above 450 K are required at the turbine inlet. This leads to higher thermal stresses in the boiler and on the turbine blades, and increases its cost. The pump flow rate is proportional to the volumetric flow rate of the liquid and the pressure drop between the inlet and the outlet. This can be expressed as the ratio of pump operation to turbine operation, which is defined as the pump flow rate related to turbine power. In the Rankine steam cycle, water consumption is relatively low. For a high-temperature organic Rankine cycle using toluene, the ratio is 2-3%. For the low-temperature organic Rankine cycle using tetrafluoroethane (HFC-134a), the values exceed 10%. In the OCR organic Rankine cycle, the pressure does not exceed 30 bar (3000 kPa). In addition, the working fluid does not evaporate directly, not at the expense of a heat source, but through an intermediate heat exchange circuit. This simplifies heat recovery. To avoid air entering the cycle, it is advisable that the condensation pressure be higher than atmospheric pressure. Low temperature organic liquids such as tetrafluoroethane (HFC-134a), pentane fluoropropane (HFC-245fa) or 2,2-dichloro-1, 1,1-trifluoroethane (HCFC-123) meet this requirement. Water is one of the best working fluids. Its main advantages are low cost and high availability, non-toxicity, incombustibility, environmentally friendly (low global warming potential and zero ozone-depleting ability), chemical stability (no deterioration of the working fluid), also low viscosity, leading to lower friction losses, and higher heat transfer coefficient. However, the steam cycles are not very tight: water is lost as a result of leaks, drainage or purging of the boiler. Thus, the water treatment system must be combined with the power plant in order to enrich the cycle with high purity deionized water. A degasser must also be included to avoid corrosion of metal parts due to the presence of oxygen in the cycle.

Acknowledgments
The work was supported by Act 211 Government of the Russian Federation, contract №02.A03.21.0011.

References

[1] Sadhu S 2011 Handbook of mechanical engineering (S.Chand & Company Ltd) p.2080
[2] Osintsev K V, Tregubova M V, Kuskarbekova S I and Prikhodko I S 2019 Shifting the boiler units of mining enterprises and processing plants to a new scheme of heat supply *Earth and Environmental Science* **378** 012049 1-4
[3] Osintsev K V, Tregubova M V, Kuskarbekova S I and Prikhodko I S 2019 The use of drive turbine cooling systems based on reneawble energy sources *Earth and Environmental Science* **378** 012048 1-4
[4] Bukowska M, Nowak K, Proszak-Miąsik D and Rabczak S 2017 Concept of heat recovery from exhaust gases *Materials Science and Engineering* **245** 052057 23-25
[5] Shatskikh Y V, Sharapov A I, Byankin I G 2017 Analysis of deep heat recovery from flue gases *Journal of Physics* **891** 012188 1-5
[6] Pashchenko D I, Semenov B A, Shchelokov A I, Naplekov I S and Mustafin R M 2018 Thermodynamic analysis of thermochemical recuperation of high-temperature flue gas heat *Journal of Physics* **1111** 012009 6-10
[7] Osintsev K V, Prikhodko I S and Zavyalova M I 2018 Methods for improving energy efficiency of air handling unit using factor analysis of data *Earth and Environmental Science* **194** 052019 7-12
[8] Batrakov P A, Mikhailov A G and Ignatov V Yu 2018 Fire-tube boiler optimization criteria and efficiency indicators rational values defining *Journal of Physics* **944** 012009
[9] Khavanov P, Chulenyov A 2017 Boiler water regime *Earth and Environmental Science* **90** 012199 8-13
[10] Vershinina K Y, Glushkov D O and Strizhak P A 2017 Characteristics of the ignition of the drops of organic coal–water fuels based on waste oils and industrial oils *Solid Fuel Chemistry* **51(3)** 188-194
[11] Nyashina G S, Vershinina K Y, Shlegel N E and Strizhak P A 2019 Effective incineration of fuel-waste slurries from several related industries *Environmental Research* **176** 108559 238-242
[12] Thulukkanam K 2013 Heat exchanger desing handbook 2nd ed. (CRC Press) p. 1245

IOP Conf. Series: Materials Science and Engineering **941** (2020) 012057 doi:10.1088/1757-899X/941/1/012057

Using absorption principles of water-ammonia mixture for energy supply of agricultural enterprises

K V Osintsev

South Ural State University, 76, Lenina Ave., Chelyabinsk, 454080, Russia

E-mail: pte2017pte@mail.ru

Abstract. The most promising developments in the field of energy supply for industrial enterprises are research in the field of alternative energy sources. As experience in the construction of the first geothermal power plants operating along the Kalina cycle shows, the efficiency of heat use increases sharply due to the utilization of the heat of low-potential energy sources. As such, energy sources use ground, in particular, geothermal water. In regions where there are enough geothermal sources, power plants based on the Kalina cycle can be used. The generated heat and electricity can be used for agro-industrial complexes. This is possible, as theoretical studies show, for enterprises in the field of crop production, animal husbandry and processing of agricultural products. In addition, the article describes the principles of using absorption refrigeration machines in agriculture.

1. Introduction

Crop production is a leading branch of agricultural production, the most important source of human food resources. Green plants, binding the energy of sunlight, create in the process of photosynthesis with inorganic low-energy substances, for example, carbon dioxide, air, water and mineral compounds of the soil, as well as various organic substances. The main goal of crop production is the creation of optimal technological and ecological prerequisites for the production of the required amount of high-quality crop products based on intensive photosynthesis in crops of field crops while maintaining or improving soil fertility. The main tasks of the plant growing industry at the present stage of development are the production of high quality, environmentally friendly products with minimal energy and labor costs with a maximum output per unit time per unit area, which requires the widespread introduction of varietal, intensive, energy- and resource-saving environmentally appropriate technologies. Vegetable growing is divided into vegetable growing of open and protected soil. In the first case, crops are cultivated in the fields, in the second case; they are cultivated on a specially designated area or in rooms where the necessary microclimate is artificially created. Growing vegetable seeds is part of the task of seed production. Each branch of vegetable growing is specific and requires deep knowledge of plant biology, based on which agricultural requirements for mechanized work on their cultivation and harvesting are developed. Technical requirements for soil preparation, sowing and care coincide with those presented to previously considered row crops [1]. Currently, there are various ways to use crop waste in agriculture [2, 3].

Agro-industrial complexes need heat supply and electricity supply, which can be ensured by building the first geothermal power plants in our country, operating on the Kalina cycle, the efficiency of heat use increases sharply due to the use of heat from low-potential energy sources.

2. The principle of operation of the technological scheme and scientific novelty

Now, the first geothermal power plants operating on Freon are existed. These power plants are built in the USA and Iceland. The principle of their work is to use low-potential energy sources. The Kalina cycle uses a mixture of ammonia and water.

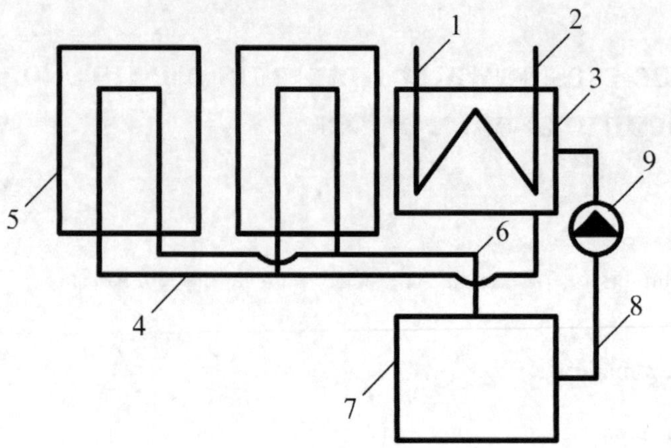

Figure 1. The heat supply scheme of greenhouses using geothermal water: 1 - a supply pipe for geothermal water, 2 - a return pipe for geothermal water, 3 - a heat exchanger, 4 - a supply pipe for circulation water, 5 - a greenhouse, 6 - a return pipe for circulation water, 7 - an expansion tank, 8 - recirculation pipe, 9 - circulation pump

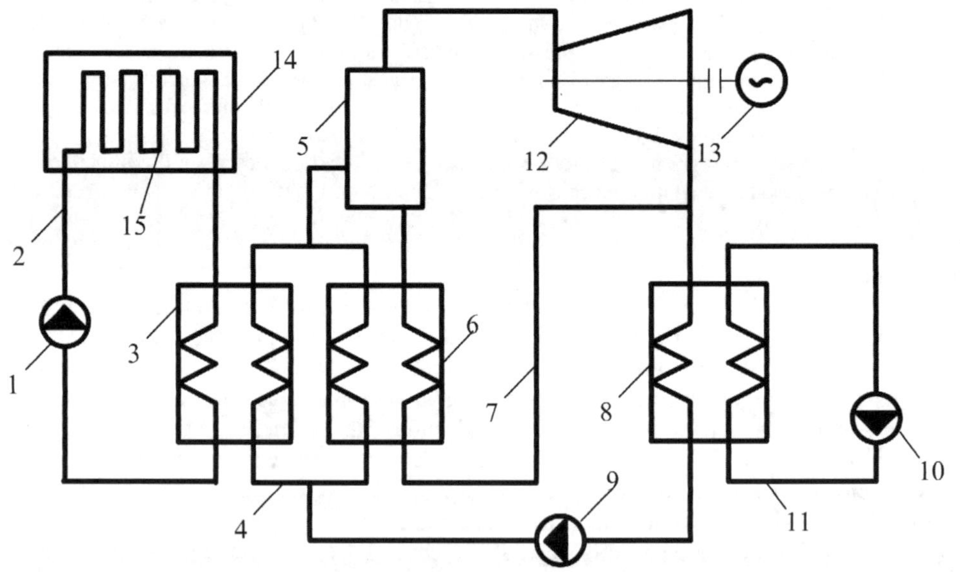

Figure 2. The application scheme of the Kalina cycle: 1 - hot water pump, 2 - hot water circulation pipe, 3 - evaporator, 4 - ammonia-water mixture circulation pipe, 5 - separator, 6 - intermediate heat exchanger, 7 - ammonia-water mixture recirculation pipeline, 8 - condenser, 9 - ammonia-water mixture recirculation pump, 10 - cold water pump, 11 - cold water circulation pipeline, 12 - turbine, 13 - electric generator, 14 - greenhouse, 15 - heat exchange surfaces

At the same time, methods for heating rooms are known, in particular at agricultural enterprises in which the flow of geothermal water is the heating medium. To heat greenhouses, geothermal water is supplied directly to the greenhouse heating system. If water has a high salt content, that it is necessary to use more complex water treatment systems and additional heat exchangers. The authors' studies were carried out in the field of heat transfer [4, 5]. Three electricity production schemes have been developed. There is a direct scheme that uses water vapor. The scheme is used where there is direct access to geothermal pairs. An indirect scheme uses not steam, but water. It is fed to the evaporator, converted to steam by the technical method and sent to the turbine. Water requires additional purification, because it contains aggressive compounds that can destroy working mechanisms. Waste but not cooled steam is suitable for heating needs. There is also a binary circuit. In this cycle, water replaces fuel, which heats another liquid with higher heat dissipation and it drives a turbine.

The principle of operation of the circuit in Figure 1 is as follows. Geothermal water flows through the supply pipe 1 to the heat exchanger 3. This water transfers heat to the water located in the annulus of the heat exchanger 3. Then the geothermal water returns through the pipe 2 to the well of the geothermal source. The circulation water heated in the heat exchanger is sent through the supply pipe 4 to the greenhouses 5, where it gives it is heat to heat the agricultural building. After which the circulating water is directed through the return pipe 6 to the expansion tank 7. From the expansion tank, the circulating water through the pipe 8 is pumped by the pump 9 again to the heat exchanger 3.

The principle of operation on the Kalina cycle in Figure 2 is as follows. Heated water is pumped through pipeline 2 by pump 1. Part of the heat is supplied to the heat supply of greenhouses 14 through heat exchange surfaces 15. Next, hot water is sent to the evaporator 3. In the evaporator 3, a mixture of ammonia and water is heated and evaporated. The vapors of the ammonia-water mixture rise into the separator 5. The dry saturated steam is sent to the turbine 12 to generate electricity in the electric generator 13. In the separator, not only dry saturated steam is generated, but also the liquid phase of the mixture of ammonia and water, which is first sent through recirculation line 7 to the intermediate heat exchanger 6. Then the cooled mixture is sent to the condenser 8 along with the condensate after the turbine 12. The cooled mixture in the condenser 12 is additionally cooled with cold water, which is pumped by the pump 10 through cold water pipeline 11.

3. Theoretical part of using absorption machine

The use of geothermal energy is repelled from the initial temperature. The temperature +30 °C and above is suitable for heating without additional transformation. Water, high temperature steam is used to generate electricity. The principle of operation of a thermal power plant is similar to a power plant device. The heat source at standard thermal power stations is a steam generator [6, 7]. The working element in both cases is heated steam [8, 9]. However, the heating methods differ. In thermal power plants, water is converted into steam using coal, fuel oil or natural gas for heating. Thermal installations and coolant are taken ready. The upper layers of the soil warm up or freeze naturally under the influence of solar heat or in its absence. Other external factors play a role. A little deeper the temperature is kept at the same level regardless of solar activity. Geothermal energy is based on an increase in the temperature of the Earth as it dives inside. The temperature increases on average by 2.5 °C every 100 meters. In mining mines, it is hot; the temperature is kept within 300 °C. Sometimes the problem of geothermal energy production is solved by nature itself. Heated water or steam is a natural coolant. They come to the surface or lie at a shallow depth. At the same time, their temperature is at least not much, but higher than the surrounding air.

4. Practical part. Application of the absorption machine in agriculture

Geothermal energy has a direct geographical dependence and is concentrated in zones with tectonic cracks in mountain ranges and seismic activity. There are various methods for analyzing the effectiveness of thermodynamic cycles [10, 11]. Therefore, its share in the total mass of energy is only 1%, and in some regions it rises to 25-30%. Technologically, the production of geothermal energy is much simpler than the generation of wind and solar electricity. Further, it will spread and grow, as it has high indicators of accessibility and environmental friendliness. This despite the fact that alternative sources of traditional energy are steadily becoming more expensive, sooner or later they will be exhausted and there simply will be no other choice. Absorption is the absorption of gas by a liquid. The operation of absorption refrigeration units is because the solubility of gases in a liquid increases with decreasing temperature, and with increasing temperature, the solubility decreases. Binary solutions, that is, solutions consisting of two components with different boiling points at the same pressure, serve as the working fluid of absorption chillers. One of these two substances, boiling at a lower temperature, is a refrigerant; the other substance with a higher boiling point serves as an absorber, or absorbent. Substances used, as absorbents should not enter into an irreversible chemical reaction with an absorbed substance called a refrigerant. Despite the large number of binary solutions, only a water-ammonia solution and a solution of lithium bromide in water received industrial use. According to its thermodynamic properties, ammonia is one of the best refrigerants actively absorbed by water: at 0 °C, one liter of water absorbs up to 1140 liters of ammonia. Like a compression chiller, an absorption chiller has an evaporator, a condenser and a control valve. During operation, periodically do analyze of the coolant and water for the presence of ammonia in them. To protect the equipment of water-ammonia absorption refrigeration units from corrosion, distilled water is used in the solution, to which potassium dichromate or ammonium dichromate is added in an amount of 0.5% by weight of the solution. Absorption chillers are widely introduced in the chemical, metallurgical and other industries. They allow the use of secondary energy resources: heat of chemical reactions, heat of cooling of molten metal's, flue gases and products of combustion of chemical wastes, low-pressure steam from waste heat boilers, and so on. The amount of these heat resources is growing from year to year; using them to get cold dramatically increases the efficiency of refrigeration units.

5. Conclusion

Thus, the general range of increasing of heat supply efficiency is possible in case energy and technological combination. Currently, only water-ammonia absorption chillers and continuous bromide lithium absorption chillers of a single-stage type are widely used in industry, using steam, hot and superheated water, hot gases and gas-vapor mixtures as a heat carrier. When describing the design features of individual devices, information is given on the main heat and mass transfer processes that occur in them. In such cases, it is necessary to check the appropriateness of the use of water-ammonia absorption refrigeration machines, in the absorbers of which water can be heated to the required temperature. The disadvantages of absorption machines compared with compressor machines are lower efficiency when using expensive heat sources, high metal consumption, cumbersomeness, increased cooling water consumption. Nevertheless, the experience of using absorption machines at artificial fiber plants for the joint production of cold and heat, as well as at a number of chemical and petrochemical enterprises, confirmed the possibility of their economical use in the above conditions. Other advantages of absorption machines are the ability to place them in open areas, low complexity of equipment and reliable operation. Water-ammonia absorption machines are more often used in the low-temperature range, while lithium bromide is used to produce cold water. Lithium bromide machines in a comparable temperature range have a higher thermal coefficient than water-ammonia ones. Part of the exhaust steam from the steam-powered unit is sent to a water-ammonia absorption refrigeration machine, the rest of it enters the hot water boiler. The water passing through the absorber, condenser and reflux condenser of the absorption chiller is heated. Then it enters the boiler, where it is heated by the exhaust steam to 80°C. However, their low cooling capacity necessitated the installation of several machines at each of the plants. The specific metal consumption for them was much higher, and, in addition, a much larger area was required for the placement of machines, and they could not satisfy the growing demand for artificial cold. The processes of absorption chillers are carried out using working substances, namely solutions, consisting, as a rule, of two components: refrigerant and absorbent. When the machine is operating, the refrigerant is periodically in both liquid and vapor states, and the absorbent is only in the liquid state. In recent years, in order to increase the efficiency of the absorption chiller, a series of studies of cold processes has been carried out using solutions such as choline chloride and water, monomethylamine and glycerin, methanol and lithium bromide, lithium chloride, lithium nitrate and water. It is advisable to check the effectiveness of a particular solution. Since in some cases it is necessary to exclude the rectification of steam of the working substance in the generator (in comparison with a water-ammonia solution), increase the solubility of the binary lithium chloride-water system, and reduce the corrosivity of the aqueous solution of lithium bromide in relation to the structural materials of absorption refrigeration machine or achieve other indicators. In general, the energy efficiency of the investigated alternative solutions turned out to be approximately the same with the corresponding known solutions, however, according to a number of other indicators and under certain conditions, they can be recommended for use in an absorption refrigeration machine. If a steam generator already exists at an agricultural enterprise, it must be used as an additional source of thermal energy. A water-carbon mixture can be recommended as a fuel [12, 13]. As experience in the construction of the first geothermal power plants operating along the Kalina cycle shows, the efficiency of heat use increases sharply due to the utilization of the heat of low-potential energy sources. As such, energy sources use ground, in particular, geothermal water. In regions where there are enough geothermal sources, power plants based on the Kalina cycle can be used. The generated heat and electricity can be used for agro-industrial complexes.

Acknowledgments
The work was supported by Act 211 Government of the Russian Federation, contract №02.A03.21.0011.

References

[1] Miranda T, Román S, Montero I, Nogales-Delgado S, Arranz J, Rojas C and González J 2012 Study of the emissions and kinetic parameters during combustion of grape pomace: Dilution as an effective way to reduce pollution Fuel Processing Technology **103** 160-165
[2] Wang Y and Cai J 2017 Renewable energy development in North West China *Sustainable Renewable Energy* **4(2)** 56-62
[3] Sadhu S 2011 Handbook of Mechanical Engineering (S.Chand & Company Ltd) p.2080
[4] Osintsev K V, Tregubova M V, Kuskarbekova S I and Prikhodko I S 2019 Shifting the boiler units of mining enterprises and processing plants to a new scheme of heat supply *Earth and Environmental Science* **378** 012049 1-4
[5] Osintsev K V and Prikhodko I S 2019 Application of mathematical modeling to optimize the operation of a tubular oil heating furnace *Earth and Environmental Science* **560** 012081
[6] Thulukkanam K 2013 Heat exchanger desing handbook 2nd ed. (CRC Press) p. 1245
[7] Bukowska M, Nowak K, Proszak-Miąsik D and Rabczak S 2017 Concept of heat recovery from exhaust gases *Materials Science and Engineering* **245** 052057 23-25
[8] Batrakov P A, Mikhailov A G and Ignatov V Yu 2018 Fire-tube boiler optimization criteria and efficiency indicators rational values defining *Journal of Physics* **944** 012009
[9] Shatskikh Y V, Sharapov A I, Byankin I G 2017 Analysis of deep heat recovery from flue gases *Journal of Physics* **891** 012188 1-5
[10] Pashchenko D I, Semenov B A, Shchelokov A I, Naplekov I S and Mustafin R M 2018 Thermodynamic analysis of thermochemical recuperation of high-temperature flue gas heat *Journal of Physics* **1111** 012009 6-10
[11] Osintsev K V, Prikhodko I S and Zavyalova M I 2018 Methods for improving energy efficiency of air handling unit using factor analysis of data *Earth and Environmental Science* **194** 052019 7-12
[12] Vershinina K Y, Glushkov D O and Strizhak P A 2017 Characteristics of the ignition of the drops of organic coal–water fuels based on waste oils and industrial oils *Solid Fuel Chemistry* **51(3)** 188-194
[13] Nyashina G S, Vershinina K Y, Shlegel N E and Strizhak P A 2019 Effective incineration of fuel-waste slurries from several related industries *Environmental Research* **176** 108559 238-242

IOP Conf. Series: Materials Science and Engineering **941** (2020) 012058 doi:10.1088/1757-899X/941/1/012058

The Use Of Combined Composites Of Non-Traditional Raw Materials Of Plant Origin In The Production Of Functional Bakery Products

I D Konovalova[1], V N Orobinskaya,[2] E N Kholodova[2], O N Pisarenko[2], E V Galdin[2]

[1]Pyatigorsk medical and pharmaceutical Institute (Affiliate), Volgograd state medical University, Russia, Pyatigorsk;
[2]Institute of Service, Tourism and Design (Affiliate), North-Caucasus Federal University, Pyatigorsk;

e-mail: orobinskaya.val@yandex.ru

Abstract: The development of new types of food products using non-traditional plant materials rich in useful substances necessary for the human body is an actual direction of the STI of the Russian Federation “FoodNet”. Bread and bakery products occupy a high share in the consumer basket of the population of the Russian Federation, research in the field of designing recipe compositions and technologies for new types of bakery products is of significant importance.
The disadvantage of the main share of bakery products is the low biological and physiological value and high calorie content. Modern production of bread and bakery products is aimed at their enrichment with vitamins, minerals to increase biological value and reduce calorie content.

1. Introduction

Issues of a healthy lifestyle and nutrition in developed countries of the European community are elevated to the rank of state.

This issue is relevant for Russia. The Government, the State Duma of the Russian Federation, the Ministry of Health and Social Development, set modern scientists with tasks related to the organization of a healthy diet for the Russian population. For example, the “Fundamentals of State Policy in the Field of Healthy Nutrition of the Population of the Russian Federation for the Period Until 2020” define as a priority the increase in the production of nutrient-enriched mass-consumption foods, incl. wheat flour of the highest and first grades, bakery products [1,2,4].

The development of new types of food products using non-traditional plant materials rich in nutrients necessary for the human body is an actual direction of the STI of the Russian Federation “FoodNet”. Bread and bakery products occupy a high share in the consumer basket of the population of the Russian Federation, research in the field of designing recipe compositions and technologies for new types of bakery products is essential.

In 1995, a law was passed in China in charge of food quality and safety, on the basis of which administrative measures were developed to regulate food products, which included the concepts of “functional foods” and “nutraceuticals”. The main direction of improving the technology of functional products is the use of food components and biologically active additives, not only increasing their nutritional value, but also giving the products targeted therapeutic and preventive properties [1,2,4,7,14]. Functional food products are divided into 3 groups (Fig. 1).

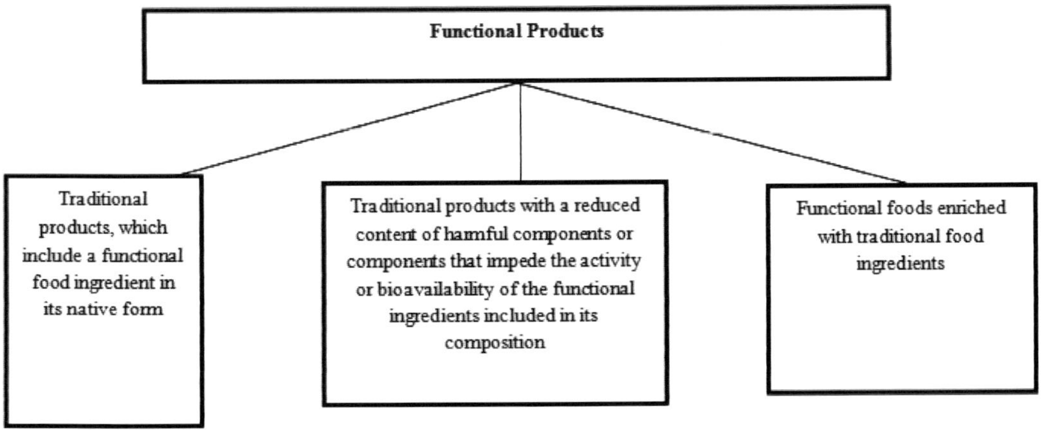

Figure 1. Classification of functional food products of therapeutic and prophylactic orientation

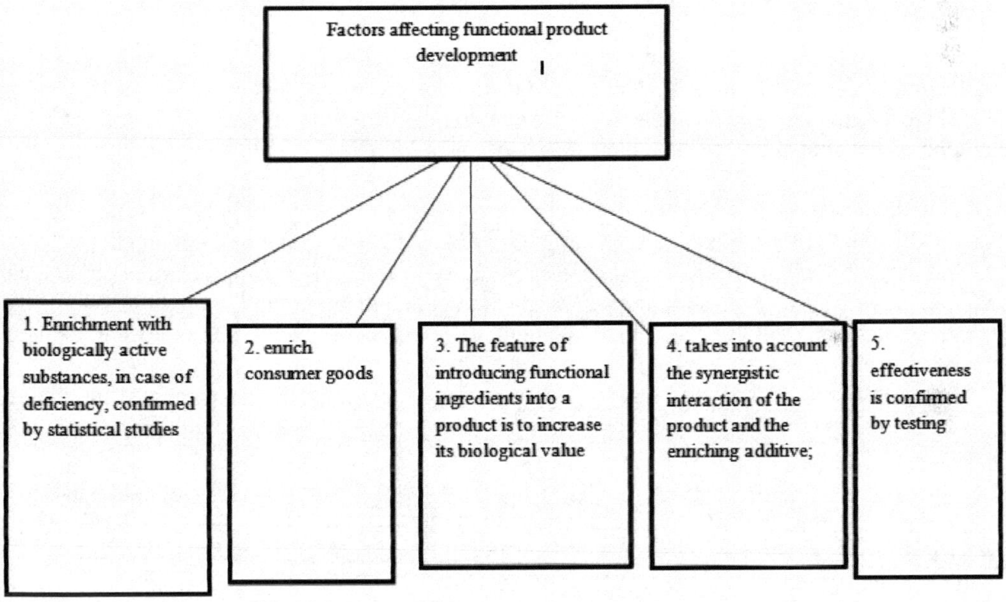

Figure 2. Classification of functional food products

In Russia, bakery products are traditional foods that are consumed daily. But recently, the assortment of bakery products has changed dramatically, the consumption of products from premium wheat flour has increased, where there are very few physiologically active substances - vitamins and minerals [1,2,4,7,11,14].

According to Intesco Research Group agency, bakery products are divided into 1) bakery products of long-term storage; 2) bakery products of non-durable storage% of the market ratio is presented in Fig. 3.

Figure 3. Market structure of bread and bakery products as estimated by Intesco Research Group

The disadvantage of the main share of bakery products is the low biological and physiological value and high calorie content. Modern production of bread and bakery products is aimed at their enrichment with vitamins, minerals to increase biological value and reduce calorie content.

One of the possible ways to obtain functional bread is the use of powder from the skin of grape pomace.

This additive is a light brown powder with a moisture content of 9-10% sweet and sour taste, obtained from dried, crushed and sifted grapes, which are waste in the production of wines. The main components of the powder are carbohydrates in the form of mono- and disaccharides, pectin and fiber, as well as vitamins and minerals. An insignificant content of proteins and lipids was noted. The increased content of pectin and fiber in bread when using grape squeezed powder increases their quality and prolongs shelf life.

The use of powder helps to increase the gas and sugar-forming ability of the test and the quality of gluten with an increase in its hydrophilic properties, elasticity and increase in the compressive strain resistance of gluten.

The use of this additive helps to improve physical and chemical parameters in relation to bread from wheat flour without the use of grape squeezed powder. However, the use of non-standard raw materials in the recipe is a consequence of obtaining bread with a dark elastic crumb of a sour-sweet taste.

The use of powder from the skin of grape marc extracts also helps to slow down staling processes, increase the content of mineral and pectin substances in bread, and enrich it with dietary fiber.

2. Main part

In the socio-economic conditions of the Russian Federation at the moment, the formation of a state policy in the field of healthy nutrition is not only relevant, but also a vital task, based on a balanced combination of all the compounds necessary for the functioning of the body.

The main nutritional disorders are the high consumption of animal fats containing high-density lipoproteins (HDL), insufficient dietary fiber intake, and a lack of macro- and micro-minerals and vitamins. The development of new types of food products using non-traditional plant materials rich in useful substances necessary for the human body is an actual direction of the STI of the Russian

Federation "FoodNet". Bread and bakery products occupy a high share in the consumer basket of the population of the Russian Federation, research in the field of designing recipe compositions and technologies for new types of bakery products, including those enriched with physiologically functional ingredients, is essential. [1,2,4,7,14].

Milk thistle and turmeric, Jerusalem artichoke powder with a leaf of lingonberry and carrot were selected as potentially useful additives of plant origin from non-traditional raw materials. In the production of bakery products, properly selected raw materials, an economically sound way of processing them on the principles of affordability, cheap raw materials and technological indicators are an important factor.

The use of raw materials growing in the Stavropol and Krasnodar Territory, reduces the cost of production. An increase in the number of diseases associated with adverse environmental effects necessitates the widespread use of biologically active compounds of hepatoprotective action, which protect the liver from damage of various kinds of external factors. The high concentration of flavolignan silymarin in the milk thistle fruits, which has an antioxidant effect and a pronounced hepatoprotective effect, allows the use of this unconventional raw material as a promising source for innovative selective products. The value of Jerusalem artichoke polysaccharide is inulin, which does not require insulin for biotransformation of substances. Carrot is a source of provitamin A - carotene. In humans and animals, carotene turns into retinol - vitamin A. The biological role of vitamin A, and therefore carotene, lies in the fact that it is part of the retina in the form of rhodopsin [3,2,5,6, 4,7,14].

Milk thistle is known as the most effective natural product for the restoration of liver cells. This is confirmed by many scientific studies. Grass is used in case of jaundice, cirrhosis, fatty degeneration or inflammation of the liver. Milk thistle is also used in case of mushroom poisoning, with psoriasis, gallstones, Alzheimer's disease and diabetes. This product is easily available and inexpensive. The composition of spotted milk thistle includes: silymarin (about 2-3%) - this is the name of the unique complex of flavolignans, consisting of silybin, isosilibine, silydianin, silychristine, taxifolin; phytosterols (campesterol, stigmasterol, beta-sitosterol); flavonoids (quercetin, apigenin, luteolin); oil (20-30%); protein (25-30%). Milk thistle owes its useful properties to the content of silymarin in it (or rather, silybin). This substance, as an effective antioxidant, stimulates protein synthesis and protects cells from various injuries and mutations, a feature of this non-traditional raw material is the absence of contraindications. The use of milk thistle herbs is possible at any age, as part of a separate or complex therapy. If it necessary, milk thistle is used for children.

Also, a rare class of phenolic compounds, flavolignans, is included in the spotted thistle. Almost all of the thistles are spotted thistle - flavonols; dehydrosilibinin and dihydroquercetin (Taxifolin) are flavonones.

The composition also includes essential fatty acids, fat-soluble vitamin K, hepatoprotector betaine, saponins, alkaloids, mucus, resins, as well as other substances that have a synergistic effect of silymarin.

Milk thistle processing products are oil, oilcake and meal and seed powder.

Milk thistle fruit is obtained after pressing the oil from them. It contains a large number of biologically active compounds: dietary fiber, flavonoids, vitamins, macro- and microelements (table 1).

Table 1 - Chemical composition of meal (TU 9141-005-46899394-04)

№	The name of indicators	Value
1	Moisture, %	6.9
2	Protein, %	22.1
3	Fat,%	13.01
4	Fatty acids, % of the total:	
5	linoleic	62.0

6	oleic	21.9
7	arachidonic	1.99
8	linolenic	1.49
9	Water-soluble carbohydrates,%	0.79
10	Dietary fibre,%	27.4
11	Ash,%	6.0
12	Essential oils, %	0.39
13	Vitamins mg /	
14	E	47
15	B1	1.39
16	B2	1.33
17	β - carotene	0.84
18	Flavolignans (in terms of silybin) per 100 g	74 mg

Qualitative indicators of milk thistle powder are shown in table 4.

Table 2 - Qualitative indicators of milk thistle powder

№	The name of indicators	Value
1	Color	Grey-brown
2	Smell	Inherent to milk thistle seed powder, oily, odorless
3	Taste	Inherent to milk thistle seed powder, without extraneous flavors, not sour, not bitter
4	Moisture content,%	8.3
5	Degree of acidity	2.6
6	Presence of signs of mold, insects, pests	Are absent

According to the research of V.S. Agibalova "... the fruits of milk thistle and their processed products are used in the food industry, namely, in the production of butter, soft drinks and coffee drinks, dietary products for therapeutic and preventive purposes ...".

The introduction of milk thistle in the form of a powder composite with the addition of 1% turmeric will enrich the resulting product with biologically active compounds of hepatoprotective, hypoglycemic and antioxidant effects [4,7,8,9,10,11,12,13].

Jerusalem artichoke tubers contain food acids (malic, citric, succinic, etc.), of which citric is predominant. Jerusalem artichoke is characterized by a wide range of physiologically active components, such as macro-, microelements, ascorbic acid, B vitamins and polyphenolic compounds. The biochemical composition of Jerusalem artichoke gives reason to recommend tubers as a raw material for creating functional products.

We propose to use a powdery composite, which includes inulin powder from Jerusalem artichoke and a sheet of lingonberry. The content of lingonberry leaves of arbutin, flavonoids, tannins and tannins causes an anti-inflammatory and immunomodulating effect. The tirpenoids contained in the lingonberry leaf ursolic and oleanic acids have antitumor activity; their content ranges from 100 - 150 mg%. The main constituent in the composition of the powder composite is inulin (the content of 0.5% nanodispersion powder of lingonberry leaf in 100 g of powder enhances the hypoglycemic and diuretic effect). This composite, when used in the formulation of antidiabetic products, will expand the range of therapeutic and prophylactic products. The special value of carrots is explained by its high content of provitamin A - carotene. In humans and animals, carotene is converted into β-from one molecule of β-carotene, two molecules of vitamin A are formed in the body. A person's need for vitamin A is 1 mg of pure vitamin or 2 g of carotene.

Scientists at the Kuban State Technological University have developed a technology for the production of carrot-based sauces, using a composite multifunctional food supplement of a polysaccharide nature, which allows controlling the quality of the resulting products.

In the work of O.V. Perfilova, the rationale for the use of fruit and vegetable powders (carrot, beetroot, etc.) in the production of muffins is given. The resulting products were characterized by low calorie content and a high content (10% or more) of pectin substances, vitamin E, β-carotene, P-active substances, hematopoietic trace element of iron, which allows us to attribute ready-made muffins to functional products [4,7,8,9,10,11,12,13].

D.L. Azin studied the nutritional value of new types of bakery products prepared with the addition of vegetable powders (a combination of apple and carrot powder). [13,14,15,16].

The main biologically active component is (curcumin). Curcumin (Fig. 5), Long's most active component of turmeric ("Zardchoubeh"), is from the ginger family (Zingiberaceae), which is a mixture of 3 different compounds known as curcuminoids (C3 curcumin complex) Fig. 6.

Curcuminoids with respect to oxidative stress radicals: O2 • -, HO •, ROO • and NO • exhibit an antiradical effect similar to quercetin, the most common plant-based flavonoids [13,14,15,16].

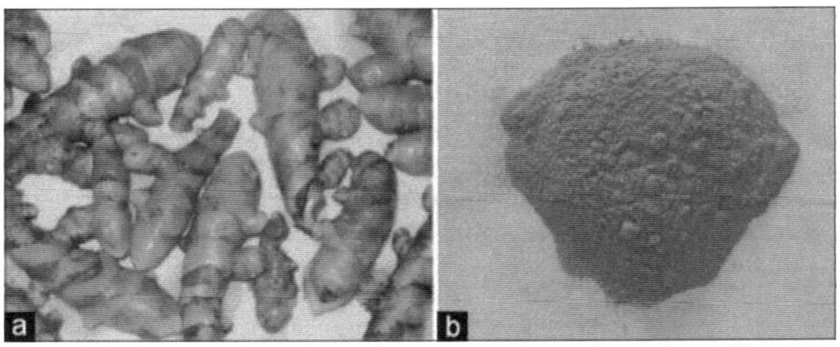

Figure 4. Turmeric Longo and powdered composite

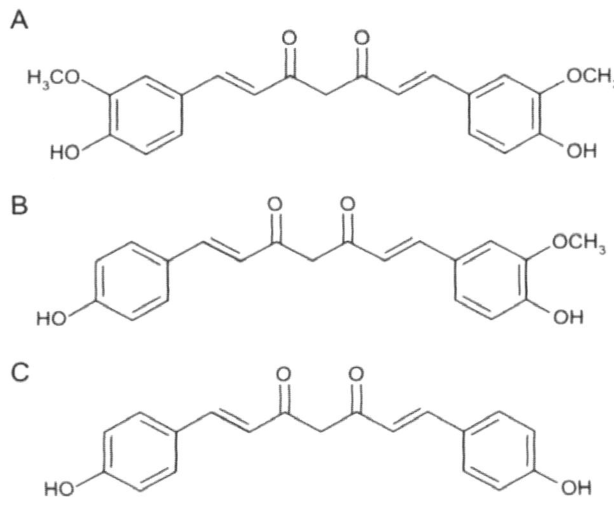

Figure 5. Structure of turmeric curcuminoids [13,14,15,16]

Curcuminoids are able to activate antioxidant enzymes and phase II enzymes of xenobiotic metabolism.

Results and discussion

The effect of powders (milk thistle and turmeric powder; Jerusalem artichoke powder with lingonberry leaf and carrots) on the physicochemical parameters of raw materials, semi-finished products of table. 3

Table 3 - Organoleptic characteristics of flour with non-traditional vegetable dressing

Sample Name	Ash content, %	Organoleptic indicators			Point grade
		Smell, Taste	Crunch	Color	
Control (MAKFA flour, premium)	0.52±0.1	a characteristic crunch for a premium flour	a characteristic crunch for a flour	white	5 5 5
Sample No. 1 (milk thistle 5% + 1% turmeric powder)	0.78±0.2	pronounced aroma, turmeric	a characteristic crunch for a flour	with a yellowish tint	4 5 5
Sample No. 2 (milk thistle 7% + 0.5% turmeric powder)	0.88±0.3	slightly noticeable turmeric flavor	a characteristic crunch for a flour	grayish yellow	4 4 4
Sample No. 3 (milk thistle 10% + 0.5% turmeric powder)	0.95±0.1	slightly noticeable turmeric flavor	a characteristic crunch for a flour	grayish yellow	4 3 3
Sample No. 4 (powder-like composite 5% Jerusalem artichoke + lingonberry leaf 1% carrot powder)	0.76±0.3	slightly noticeable grassy smell	non-characteristic crunch for a flour	taupe	4 5 4
Sample No. 5 (powder-like composite 7% Jerusalem artichoke + lingonberry leaf 1% carrot powder)	0.91±0.3	pronounced grassy odor	non-characteristic crunch for a flour	taupe with orange blotches	3 2 3
Sample No. 6 (powder-like composite 10% Jerusalem artichoke + lingonberry leaf 1% carrot powder)	1.25±0.2	pronounced grassy odor	non-characteristic crunch for a flour	taupe with orange blotches	3 2 3

An analysis of the experimental data showed that when phyto-enrichment agents from non-traditional plant materials are introduced, the organoleptic characteristics of flour change. The most acceptable according to experts were samples No. 1 and No. 4.

The effect of phyto-enrichment agents on titratable acidity of flour is investigated. The acidity of flour shows the ability of protein compounds of flour to bind a certain amount of alkali. With the breakdown of compounds: phytin, lipids, etc., rancidity of flour occurs and an increased content of organic acids occurs in it.

The results of the influence of phytochemicals on acidity are presented in table 4.

Table 4 - Indicators titratable acidity of flour with unconventional phytochemicals

A portion of flour, g	Name of sample	Amount of 0.1 N alkali used for titration, ml	Degree of acidity	Conclusion
5	Control	1.5	3.0	Meets the standard requirements for flour quality
5	Sample №1	1.6	3.2	With the introduction of powder with milk thistle, a slight increase
5	Sample №2	1.7	3.4	With the introduction of powder with milk thistle, a slight increase
5	Sample №3	1.9	3.8	With the introduction of milk thistle powder, an increase in acidity occurs due to the presence of organic acids in the feed
5	Sample №4	2.0	4.0	With the introduction of powder from Jerusalem artichoke, a significant increase
5	Sample №5	2.1	4.2	With the introduction of powder from Jerusalem artichoke, a significant increase
5	Sample №6	2.2	4.4	With the introduction of powder from Jerusalem artichoke, a significant increase

The analysis of experimental data showed that in sample No. 1 there is a slight increase in the acidity of flour, in samples No. 3-4 the acidity is higher, which must be taken into account in the production of bakery products.

The effect of phytochemicals on the quality of flour was studied using the method of swelling in a solution of acetic acid, the so-called flour strength.

The "strength" of flour affects the quality of bakery products. The results of the analysis are presented in table. 5.

Table 5 - "Strength" of flour, depending on the size of sedimentation sediment in experimental samples

Name of sample	V (sediment volume), cm^3	Analysis results, cm^3	Flour class by "strength"
Control	More than 60	50.9	High protein, excellent gluten, strong
Sample №1	40-59	47.8	High protein, good quality gluten, strong
Sample №2	40	42.4	With medium protein and low gluten quality
Sample №3	31-39	39.1	With medium protein and low gluten quality
Sample №4	31-39	48.6	High protein, good quality gluten, strong
Sample №5	31-39	37.7	With medium protein and low gluten quality
Sample №6	Less than 30	30.3	Low-protein mealy with reduced quality gluten

Thus, the most suitable for the production of bakery products are sample No.1, No.2, No.4.

3. Conclusions

In connection with the above, conducted comprehensive studies on the development of bakery product formulations using non-traditional plant materials, namely flour from powder of the seeds of milk thistle, Jerusalem artichoke and carrots, is relevant, as it will contribute to the expansion of the range of preventive bakery products.

Phyto-fortifiers not only improve the quality of the flour used for the production of bakery products, but also enrich them with biologically active compounds of a functional orientation of selective action: hepatoprotective, antioxidant, contributing to a decrease in blood glucose, expanding the range.

References

[1] Fundamentals of the state policy of the Russian Federation in the field of healthy nutrition for the period up to 2020; approved by order of the Government of the Russian Federation of October 25, 2010 No. 1873-r.

[2] Boryaev V E [et al] Edited by Teplova V I 2005 *Functional food.* Belgorod: Cooperative education, 415 p.

[3] Agibalova V S 2013 The use of carrot powder in the production of bakery products with preventive properties. *Integration of science and practice as a mechanism for the effective development of the agro-industrial complex: international materials. scientific and practical. conf. in the framework of the XXII International specialized exhibition "AgroComplex-2013", March 12-15, 2013* Part II Ufa: Bashkir State Agrarian University 20-21.

[4] Agibalova, V S 2016 *Development of scientifically grounded recipes of bakery products of increased nutritional value with the use of promising phyto-fortifiers:* dissertation ... Candidate of Agricultural Sciences: 05.18.01, Michurin state agrarian. un-t, Voronezh. 203 p.

[5] Nikonovich Yu N, Tarasenko N A 2014 Dietary fibers of plant raw materials and features of their application. *News of High School. Food technology.* **5-6** 6-9.

[6] Altunyan S V, Ivanova E E, Altunyan M K 2014 Improvement of technology for the production of sauce based on Jerusalem artichoke and fish grits. *Science. Technics. Technologies (polytechnic bulletin).* 1 61-63.

[7] Konovalova E V, Tarasenko N A, Krasina I B, Buzunar A B 2012 Method for the production of functional cake. Patent for invention No. 2494629 dated 05.06.2012, Publ. 10.10.2013 Bul No 28.

[8] Konovalova E V, Krasina I B, Tarasenko N A, Buzunar A B, Nikonovich Yu N, 2012 Features of the functional and technological properties of food fibers in flour confectionery. *News of High School. Food technology.* **5-6** 35-37.

[9] Pashchenko L P [et al.] 2005 Characteristics of milk thistle - a promising component of bakery products. *Storage and processing of agricultural raw materials.* **9** 60 p.

[10] Smertina, E S 2011 Application of extracts of wild plants in functional bakery products. *Proceedings of the Far Eastern Federal University. Economics and Management.* V **3**. 61-67.

[11] Semyonkina N G 2010 *Development of the technology of bakery products using the products of processing milk thistle.* Author's abstract of thesis ...candidate of technical sciences, Moscow. 26 p.

[12] Korchagin V I 2001 *Promising enrichment of plant origin in the production of bakery products.* Voronezh: Voronezh. state technol. Acad. 278 p.

[13] Iorgacheva I G 2014 The potential of medicinal, aromatic plants in improving the quality of wheat bread. *East European Journal of Advanced Technologies.* V **2/12 (68).** 101-108.

[14] Hema Suryawanshi, Rupali Naik, Pramod Kumar, and Rolly Gupta 2017 Curcuma longa extract - Haldi: A safe, eco-friendly natural cytoplasmic stain. *J Oral Maxillofac Pathol.* **21 (3)** 340–344

[15] Kengo Kawasaki, Koutarou Muroyama, and Shinji Murosak 2018 Effect of a water extract of Curcuma longa on emotional states in healthy participants. *Biosci Microbiota Food Health.* **37(2):** 25–29.

[16] Ngadino, Setiawan, Koerniasari, Ernawati and S A Sudjarwo 2008 Evaluation of antimycobacterial activity of Curcuma xanthorrhiza ethanolic extract against Mycobacterium tuberculosis H37Rv in vitro. *Vet World.* **11 (3)** 368–372

IOP Conf. Series: Materials Science and Engineering **941** (2020) 012059 doi:10.1088/1757-899X/941/1/012059

The Reduction Of Activity Of Antialimentary Compounds Of Vegetable Raw Materials Under Exposure To Electric Discharges

V T Kazub[1], V N Orobinskaya [2], E V Galdin[2], S A Yemelyanov[3]

[1] Pyatigorsk medical and pharmaceutical Institute (Affiliate), Volgograd state medical University, Russia, Pyatigorsk;
[2] Institute of Service, Tourism and Design (Affiliate), North-Caucasus Federal University, Pyatigorsk;
[3] North-Caucasus Federal University, Stavropol;

e-mail: orobinskaya.val@yandex.ru

Abstract: The article deals with the issues related to the harmful effects on the human body of anti-nutritional components (anti-nutritional factors of natural origin) that can block active enzymes that break down protein compounds. A comparative analysis of anti-nartative components of protein-containing plant origin under the influence of thermal nature and parameters of electric-discharge treatment has been carried out

1. Introduction

The main biochemical component of soy is protein. Soy is one of the most high-protein crops cultivated in the world. According to different authors, the seeds of this culture can accumulate an average of 38-42% of the protein, varying this indicator from 30 to 50%.

Soy proteins are heterogeneous in structure and function. Among them there are substances that are considered to be antialimentary (anti-nutritional) components of food. These are inhibitors of proteolytic enzymes, lectins, urease, lipoxygenase and others. Most of soy protein (about 70%) is made up of storage proteins of the class 7% (β-conglycinins) and 11% (glycinins), which are quite normally absorbed by mammals. Soy flour is the most widely used source of protein in the creation of balanced feed, and a food protein ingredient, however, in the process of obtaining it is necessary to inactivate anti-nutritional (anti-nutritional) components [1,4,6,7,8].

Soy beans and soy products (soy milk, okara, etc.) contain more than 5 trypsin inhibitors (5-10% of the total protein content). Their activity ranges from 7 to 38 mg / g. A distinctive feature of these substances is that, interacting with enzymes designed to break down proteins (proteases, etc.), they form stable complexes, devoid of both inhibitory and enzymatic activity. The result of this blockade is a decrease in the absorption of protein substances. Once in the stomach, some of the inhibitors (30-40%) lose their activity, and the most stable ones reach the duodenum in active form and inhibit the enzymes produced by the pancreas. As a result, the pancreas is forced to produce them more intensively, which can ultimately cause its hypertrophy with further development of disorders of carbohydrate-lipid metabolism. In terms of chemical structure, properties and substrate specificity, soybean inhibitors mainly belong to two families: Kunitz inhibitors - water-soluble proteins with a molecular weight of 20,000-25,000 Da, binding one trypsin molecule; Bauman-Birk inhibitors are alcohol-soluble proteins with a molecular weight of 6000-10000 Da and a small number of disulfide bridges that can inhibit both trypsin and chymotrypsin [2,3].

The most common method for reducing the activity of trypsin inhibitors is thermal treatment in the presence of alkali. But this method is not always effective, about 20% of soybean trypsin inhibitors have thermal stability.

Pamirsky I.E. [1,2] using the Internet server BLAST (Basic local Alignment Searcy Tool) and the Bio Edit 5.0.9 program, the high structural homology of the protein pancreatic protease inhibitor aprotinin and the plant analogue of soybean trypsin inhibitor was calculated (Fig. 1).

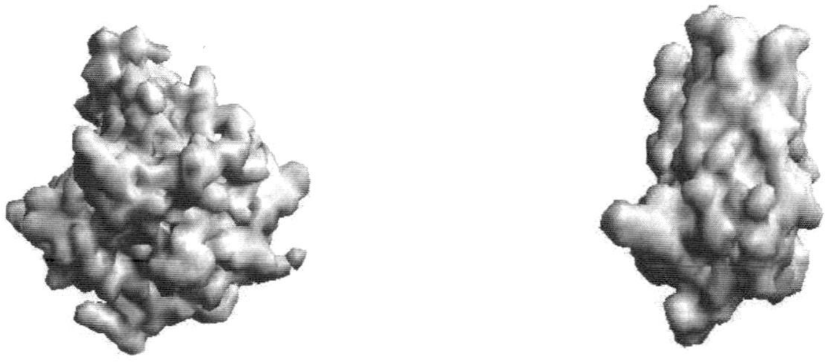

a) soybean trypsin inhibitor b) aprotinin

Figure 1 - Electronic three-dimensional models of the tertiary structures of aprotinin and soybean inhibitor [1,2]

Modern studies of foreign and domestic colleagues in in vitro experiments have shown that aprotinin prevents fibrinolysis, and an inhibitor of soy isolate has a similar effect [3,4,5].

Scientific research of Solomentsev M. showed that the trypsin-inhibiting activity is influenced by the temperature regimes of exposure to table 1 [3,4].

Table 1 - Trypsin-inhibiting activity (Kunitz inhibitor) of thermally exposed soy products, (mg / g) [1]

Sample name	Exposure temperature,C ^{0}		
	60	80	100
Soy milk (soy emulsion)	0.071±0.007	0.0542±0.011	0.0170±0.10
Protein-carbohydrate mass (okara)	1.227±0.01	1.123±0.014	0.740±0.11

Table 1 shows that heat treatment reduces the activity of trypsin inhibitors by almost 98.59% for both soy milk and protein-carbohydrate mass (okara).

Note also that microwave heating of soy protein extracts for 1.5–2 min completely destroys trypsin inhibitors [4].

But the impact of temperatures above 80 C 0, and even more so microwave - heating, is associated with the destruction of protein and other biologically active compounds.

2. Main part

In order to obtain an environmentally friendly target product of soy milk and reduce the activity of anti-nutrients, soybeans and soybean meal were treated with rectangular electric discharges with the following processing parameters: U = 27 kV, C = 0.4 μF, interelectrode gap length 1.5 mm , the ratio "solid: liquid" S: W = 1: 10, the range of change in the number of applied pulses from 150 to 600 imp / l in the electric discharge chamber "sharp - plane".

Table 2 - The yield of protein compounds (in %) and trypsin and antitrypsin (in mU/ml or γ/ml)

Material	Humidity, %	Protein, %	Trypsin, mU/ml	Antitrypsin	
				mU/ml	γ/ml
Soy beans	5.12±0.01	44.7±0.13	5.78±0.02	23.67±0.01	5.29±0.01
Soybean meal	4.46±0.03	50.4±0.014	0.48±0.12	10.24±0.11	2.03±0.03

The protein yield from soybeans changes nonlinearly with an increase in the number of pulses applied (Table 3).

Table 3 - Protein yield from soybeans depending on the number of pulses

Sample number	Number of pulses	Dry residue,%	Protein, mg / ml
1	1500	3.96	11.00
2	3000	6.33	17.70
3	4000	7.99	21.10
4	5000	7.22	31.15
5	6000	7.96	33.15

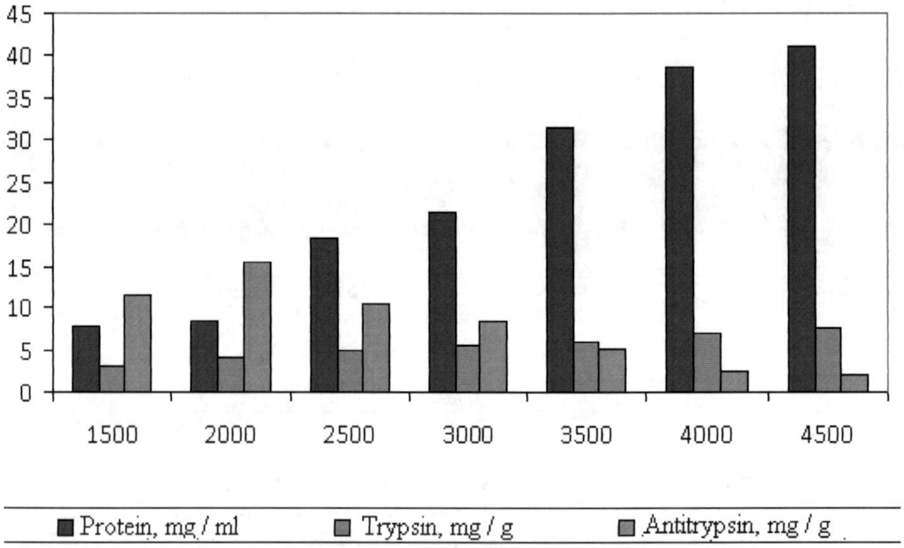

Figure 2. Protein yield mg/ml; trypsin, mg/g, antitrypsin; mg/g, depending on the number of pulses

The extraction efficiency is ~ 90% of the initial content of water-soluble complexes. In the range of 100-300 imp / L, the trypsin concentration increases ~ 8 times, while the protein concentration increases ~ 3 times (Fig. 2).

Analysis of water extracts obtained by electric discharge processing of soybeans showed the presence of the main product - protein, as well as enzymes - trypsin.

To assess the integrity of protein molecules in the extracts, proteins were separated by gel filtration with Toyopearl-55 on a 2.6 x 40 cm column, and standard spectra and spectra of products obtained as a result of electric discharge treatment were compared.

Proteins from the centrifugates of suspensions obtained at 200 and 600 imp / L, as well as products after mechanical treatment of the soy suspension (in an agate mortar), were eluted in 4 stages of 25 ml of Tris-HCl buffer with pH = 7.5. The eluate was collected in 4 ml fractions. The protein in each fraction was determined spectrophotometrically at a wavelength of 280 nm. The wavelength range was determined according to the standard method [11,12].

The spectra of proteins eluted from the column are shown in Fig. 3 a. The abscissa shows the values of the total volumes of the eluate, and the ordinate shows the optical density of the fractions.

For all compared samples, the soy protein eluted from the column in two peaks. The first peak in the region of 20-25 ml corresponds to the molecular mass of the protein of 550 KD, the second peak is found in the region of 50-60 ml and corresponds to the molecular mass of the protein of 200 KD.

Shown in fig. 3b, the dependences are similar in the shape of peaks and the time of protein release, and it can be concluded that the proteins included in the compared samples are identical. In the course of electric discharge processing, the main product - vegetable proteins - retain their structure [7,8,9.10].

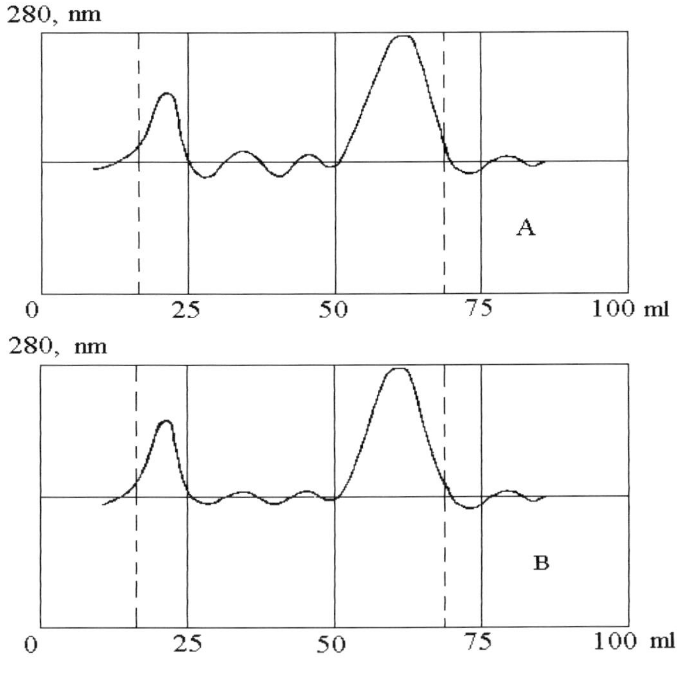

Figure 3. Protein spectrum of soy milk obtained by mechanical method (A), and electric discharge processing (B) - 2000 impulses, (C) - 6000 impulses

The analysis of the activity of inhibitors was determined in accordance with the recommendations of OSP TU 431.1418.990.02-92 and OSK TU 431.1419.990.01-92 in the laboratory of the Pyatigorsk Research Institute of Balneology and Physiotherapy. Samples for analysis were taken during the processing of the soy-water suspension at the hydromodule (S: W = 1: 10) through the number of pulses 100 imp / l, 200 imp / l, 300 imp / l.

In the pulse range of 100-300 imp / l, the antitrypsin content in the liquid phase of the suspension first increased to 10.6 γ/ / ml, and then decreased, reaching 1.65 γ / ml at 300 imp / l, i.e. when more than 300 imp / l is supplied, antitrypsin is cleaved (Kazub V T, Orobinskaya V N, Konovalov D A, 2007).

The phenomenon of the cleavage of antitrypsin, an enzyme that impairs the assimilation of proteins, is of independent scientific interest and requires further study.

For the purity of the experiment, experiments were carried out to reduce the activity of trypsin inhibitors in other types of organic plant raw materials and to obtain protein from new raw materials sources, the burdock variety "Tohako Rio" was used as an experimental object. The study of the chemical composition of this type of organic raw material showed that this variety is rich in proteins, the total protein content in% for air-dry raw materials is 7.55 ± 0.13, shoots of 1 year were used for the study. Comparative analysis of the effect of electrical discharges on the decrease in the activity of anti-nutritional factors: 1) soy isolate trypsin inhibitor; and 2) an antitrypsin inhibitor contained in burdock variety “Tohako Rio” was determined in accordance with the recommendations of OSP TU 431.1418.990.02-92 and OSK TU 431.1419.990.01-92. The soy-water suspension, “burdock extract containing a trypsin inhibitor”, was treated with electric discharges at a hydromodule of 1:10, the antitrypsin content in the liquid phase of the soy isolate suspension first increases to 15.6 mg / g and then decreases to 1.65 mg / g (under the influence, the splitting of antitrypsin occurs).

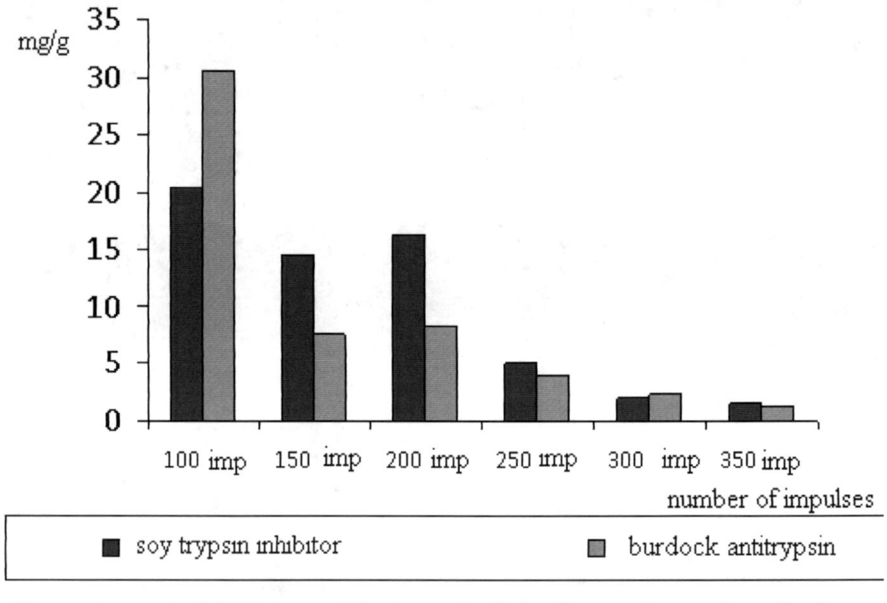

Figure 4. Change in trypsin inhibitory activity in mg / g depending on the number of pulses

A decrease in antitrypsin containing in burdock extract occurs in the range from 300 - 500 imp / l and the residual trypsin inhibitory activity is 1.23 mg / g per air-dry raw material.

The phenomenon of the cleavage of antitrypsin, an enzyme that impairs the assimilation of proteins, is of scientific interest and requires special additional study.

3. Conclusions

When processing soybean suspension with electrical impulse discharges, the yield of protein and trypsin, an enzyme that ensures the assimilation of soy proteins by the body, increases, which improves the quality of the ingredients obtained. With an increase in the number of electrical impulses of 300 imp / l, the Kunitz inhibitors are degraded, which reduces their activity and increases the absorption of protein in the body.

A comparative study of the effect of electric discharge processing on the anti-nutritional factors of soybeans and burdock varieties "Tohako Rio" showed that when exposed to electric discharge processing, protein destruction does not occur, and the activity of trypsin inhibitors and their content decreases, which improves the quality of the products obtained.

References

[1] Pamirski I E, Shtarberg M A, Beloglazova I G, Borodin E A 2007 The influence of soy bean trypsin inhibitor on some biologically relevant proteolitic processes in vitro and in vivo. *Book of Abstract, Commemorating 15 years of Russia-Japan Medical Exchange under the guidance of Japan-Russia Medical Exchange Foundation (1992-2007): Book of Abstract. 21-22 september 2007. Blagoveshchensk.* Blagoveshchensk: ASMA. 82.

[2] Borodin E, Pamirski I, Shtarberg M, Gorin A, Anikin S, Beloglazova I 2007 Proteolysis, protease inhibitors and soya. In: Studies on Drugs Used in Traditional Therapy. *IV Russia and China Medical Forum: Book of Abstract. 10-12 september 2007, Blagoveshchensk.* Blagoveshchensk: ASMA. 31-35.

[3] Solomentsev M V 2009 *X-RAY: a new method for determining the trypsin-inhibiting activity of foods.* M.: Moscow State Open University. 100 p.

[4] Toltoguzov V B 1987 *New forms of protein food (Technological problems and production prospects).* M.: Agropromizdat, 303 p.

[5] Kaishev V G 2011 Food and processing industry in the new millennium. *Food industry.* **1** 9-14

[6] Konarev V G 1981 Resources of vegetable protein and problems of its quality. *Works on applied botany, genetics and breeding.* **V 70** 2 71-84.

[7] Lusas E, Ki Chun Ri 1998 *Soy protein production. Guidelines for the processing and use of soybeans.* Ed. Klyuchkina V V and Domoroshchenkova M L M.: Kolos. 54 p.

[8] Savin V N 2005 Environmental impact on biological objects of electromagnetic fields of extremely low and ultra-low frequencies. *Izvestiya vuzov. Food technology.* **5-6** 126

[9] Kazub V T, Kudimov Yu N, Rudobashta S P, Borisov A G 2004 The role of cavitation and a pulsating vapor-gas cavity in the processes of electric-discharge grinding of raw materials. *Food Industry.* **3** 21-23.

[10] Kazub V T, Kudimov Yu N, Rudobashta S P, Borisov A G 2014 The role of cavitation and a pulsating vapor-gas cavity in the processes of electric-discharge grinding of vegetable raw materials. *Storage and processing of agricultural raw materials.* **9** 21 - 22.

[11] Orobinskaya V N, Kazub V T, Marshalkin M F, Konovalov D A 2009 Influence of electric discharge treatment on lipoxygenase activity of tomato juice. *News of higher educational institutions. Food technology.* **2-3** 112-113.

[12] Orobinskaya V N 2012 *Kinetics and technology of electric discharge extraction of biologically active compounds from organic raw materials.* Author's abstract of thesis … candidate of technical sciences, Tambov. TSTU. 16 p.

[13] Kazub V T, Orobinskaya V N, Konovalov D A 2007 Influence of electrical impulse processing on enzymes of food plant raw materials. *The role of university potential and scientific institutions in the implementation of the Strategy of socio-economic development of the Caucasian Mineral Waters until 2020: mater. VIIIth Int. scientific-practical conf. Antalya, September 6-13, 2007.* 40-43.

IOP Conf. Series: Materials Science and Engineering **941** (2020) 012060 doi:10.1088/1757-899X/941/1/012060

The Use Of Phyto-Enrichment Agents In The Production Of Functional Bakery Products

V N Orobinskaya[1], E N Kholodova[1], E V Galdin[1], S A Yemelyanov[2]

[1]Institute of Service, Tourism and Design (Affiliate), North-Caucasus Federal University, Pyatigorsk;
[2] North-Caucasus Federal University, Stavropol.

e-mail: orobinskaya.val@yandex.ru

Abstract: An anthropogenic pollution, a sedentary lifestyle, a sharp jump in the growth of cardiovascular diseases and oncological diseases led to the need to develop functional products of a new generation that reduce the effects of the harmful factors listed above.

1. Introduction

Currently, factors affecting the ability to work, life expectancy, contributing to the preservation of the gene pool of the nation in the countries of the European Community and the Russian Federation, are elevated to the rank of state policy. "Fundamentals of state policy in the field of healthy nutrition of the population of the Russian Federation for the period until 2020" - the priority task of increasing the production of nutrient-enriched foodstuffs of mass consumption, and especially bakery products [1,2].

Excessive intake of high-density lipoproteins (HDL), low levels of soluble and insoluble dietary fibres (DF) - led to a sharp increase in the number of diseases of the cardiovascular system and endocrine system, cancer. The development of functional products of a selective orientation using non-traditional plant materials containing biologically active compounds, micro and macro elements is in demand.

Plant raw materials and plant cells are the basis of food phytobiotechnology [3]. Bakery products make up a significant share of the consumer basket of the population of the Russian Federation and up to 25% of the total mass of food consumed with daily consumption by men - 200-300 g, women - 180-250 g, in this regard, enrichment with physiologically functional ingredients based on phytobiotechnology is an actual for Foodnet direction of the Russian Federation [3].

According to research by Russian and foreign scientists, "... the bulk of bakery products has low biological and physiological value, but high calorie content ..." [3].

2. Main part

The main goal of the production of bread and bakery products is to increase the biological value and lower calorie content.

The technical regulation of the Customs Union TR TS 021/2011 "On food safety" shows, "...that the food (food) raw materials used in the production of food products must comply with the requirements established by this technical regulation and (or) technical regulations of the Customs Union on certain types of food products, and be traceable ..." [4].

Modern conditions dictate the need to expand the assortment of bakery products and introduce new types of medical-prophylactic and functional products into production [5].

According to the research of V.A. Tutelian's functional bakery products account for approximately 5% of the total output of Fig. 1.

Phyto-enrichment agents of plant origin improve the physicochemical and organoleptic characteristics of bakery products, contribute to the intensification of the technological process and the expansion of the range of products of a functional orientation [7].

The use of aromatic and medicinal raw materials: high elecampane, dioica nettle, medicinal melissa, pharmacy chamomile, etc. improves the quality of bakery products, increasing the biological and nutritional value [8,9].

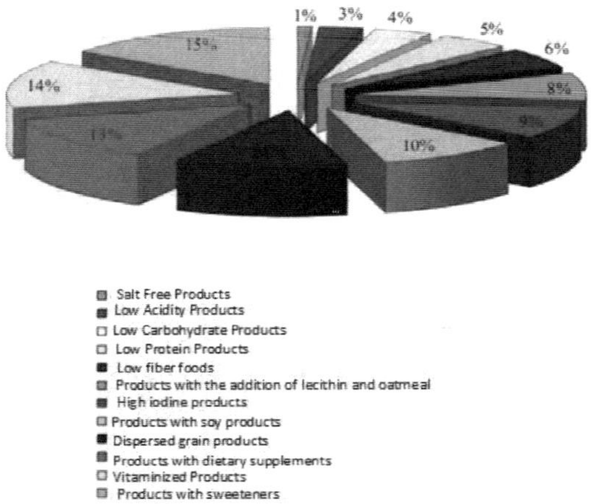

Figure 1. The structure of the production of dietary bakery products by classification groups [6].

For example, the introduction of nettle meal in an amount of 3% by weight of flour improves the colouring of the surface of bakery products, the elasticity of the crumb, due to the formation of oxymethifurfural, the taste and aroma of bread improves; the staling process slows down; increases the content of the group of vitamins C, P and K, β-carotene and minerals [10].

Indian scientists Hooda S. and Jood S. enriched wheat bread with flour from hay fenugreek seeds, enriched with protein (25%), and an essential amino acid - lysine (5.7 g / 16 g), soluble (20%) and insoluble (28%) dietary fibre, as well as calcium, iron and β-carotene [11].

The use of fruit and berry powders leads to an increase in the volume and porosity of products, to obtain products of the correct form with a brightly coloured crust and elastic crumb [12,12].

Pumpkin puree helps to reduce the duration of the fermentation process due to the content of vitamin C and mineral elements, helps to increase the viscosity of the dough, reduces the acidity of bread, extends the shelf life of finished products [14].

The culture of a new generation - mackerel - a hybrid of spinach sorrel - is a valuable plant phyto-enrichment agent, the application of which enriches bakery products with protein, increases the water absorption capacity of the dough, reduces the ability to dilute and spread the dough pieces during fermentation [15].

Hydrobionts of plant origin, for example, brown algae. enrich the target product with nutrients: iodine, polyunsaturated and food acids.

Yu.V. Prikhodko showed that adding ribbed brown seaweed to the bread recipe accelerates the fermentation process; improves the quality of gluten; enriches bakery products: protein, fat, carbohydrates and mineral compounds; increases porosity, crumb with small impregnations of the powder of the fire, improves the physical and chemical and rheological properties of the finished product; reduces the intensity of staling processes during storage [16].

A composite based on mushroom protein: chanterelle mushrooms and champignon mushrooms in the form of a powder increases gas-forming ability, accelerates the fermentation and proofing process of dough; increases porosity, specific volume of bread; reduces the loss of dry matter during baking and storage; slows down the staling process [17].

The functional additive is iodine casein, a composite of iodized milk protein and kelp powder [16] allows to get enriched with protein and iodine bakery products.

Water-alcohol extracts of viburnum, Schisandra chinensis and grapes intensify the fermentation process, increase the fermentative activity of yeast, add a spicy taste and promote the formation of developed porosity, increase the volume of finished products [18].

Untraditional raw materials are used to produce diabetic products. As a promising raw material, chicory roots are used; we propose replacing the scorzonera root [19]. A significant content of fructose and organic acids, the use of puree from chicory roots leads to an increase in the gas-forming ability of flour and accelerate the fermentation activity of yeast. The interaction of amino acids and sugars of the product produces dark-colored products (melanoidins), as a result of the product acquire a more saturated color and a more pronounced taste and aroma. Products obtained using mashed chicory roots differ from bread prepared according to traditional technology with an increased volumetric yield and porosity [19].

One of the promising types of non-traditional plant materials is the powdery silybum thistle composite obtained from milk thistle fruits after squeezing them, rich in compounds such as edible fibers, vitamin flavonoids, macro- and microelements (table 1.) [7,8,9,10].

Table 1 - The chemical composition of milk thistle meal (TU 9141-005-46899394-04)

№	Name of indicators	Value
1	Moisture, %	6.9
2	Protein, %	22.1
3	Fat,%	13.01
4	Fatty acids,% of the total:	
	linoleic	62.0
	oleic	21.9
	arachidonic	1.99
	linolenic	1.49
5	Water-soluble carbohydrates, %	0.79
6	Cellulose,%	27.4
7	Ash %	6.0
8	Essentialoils, %	0.39
9	Vitamins, mg /	
	E	47
	B1	1.39
	B2	1.33
	β - carotene	0.84
10	Flavolignans (in terms of silybin) per 100 g	74 mg

Content in thistle spotted phenolic compounds, incl. 2-3% are flavoigned (silibin, silikristin, silidianin, and also in small quantities - quercetin, silandrin, rosin alcohol, etc.), nitrogen-containing compounds: betanine, resins, up to 0.1% of essential oil and other substances [7,8,9,10].

A distinctive feature of milk thistle is the presence of silymarin flavonoid in its seeds. Most flavonoids are contained in the membrane (up to 7.0%) and a small part in the seed (up to 0.10%) [7,8,9,10] milk thistle substances synergistically enhance the therapeutic effects of silymarin [1,7,8,9,10].

Qualitative indicators of milk thistle powder are shown in table 2.

Table 2. Thistle powder quality indicators

№	Name of indicators	Value
1	Colour	Taupe
2	Smell	Milk thistle seed powder characteristic, oily, without odors
3	Taste	Inherent to milk thistle seed powder, without extraneous flavors,

		not sour, not bitter
4	Moisture content,%	7.9
5	Acidity, degrees	2.4
6	Presence of signs of mold, ins	Are absent

We have developed a powder composite with a water-soluble polysaccharide complex from scorzonera of the Spanish variety "Mamont" - the main component of which is inulin polysaccharide, milk thistle powder, containing antioxidant silybin and turmeric powder - the main biologically active component is curcumin.

Curcumin, the most active ingredient in Long Curcuma ("Zardchoubeh"), is from the Zingiberaceae family and is widely used as a medicinal plant in India and other Asian countries. Rhizomes C. longa are widely used as spices and dyes in the production of dietary food.

C. longa powder is a mixture of 3 different compounds known as curcuminoids (C3 curcumin complex).

The clinical use of curcumin is justified by several pharmacological effects, such as antioxidant, anti-inflammatory, antiparasitic, antimutagenic, chemoprotective, antiviral and antitumor.

Current research on the pharmacological properties of curcumin has proven the activity of this biologically active compound in cancer of the liver and pancreas. Oral curcumin is well tolerated, despite its limited absorption in the gastrointestinal tract and has a pronounced biological activity in some patients with pancreatic and liver cancer.

Pancreatic adenocarcinoma is one of the deadliest types of cancer, most patients die within 1 year. The only compounds that currently have a positive effect are gemcitabine and erlotinib, their effect on patient survival is measured only for weeks [20]

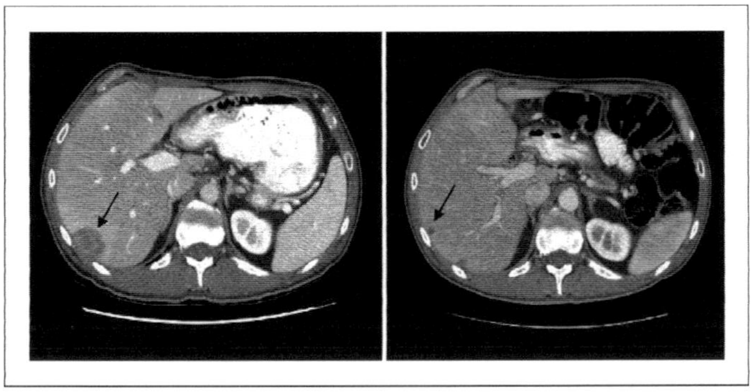

Figure 2. Computed tomography of the abdominal cavity with liver damage in a patient with pancreatic and liver adenocarcinoma. A computed tomographic scan on the left was performed before therapy, and a scan on the right 2 months after the administration of curcumin [20]

There was a general decrease in the size of liver lesions by 73% according to the criteria for evaluating the response in solid tumors.

Studies by Japanese scientists [21,22] showed that C. longa rhizome extracts have atimicrobial activity against Mycobacterium tuberculosis H37Rv tubercle bacillus in experiments in vitro Fig. 3.

IOP Conf. Series: Materials Science and Engineering **941** (2020) 012060 doi:10.1088/1757-899X/941/1/012060

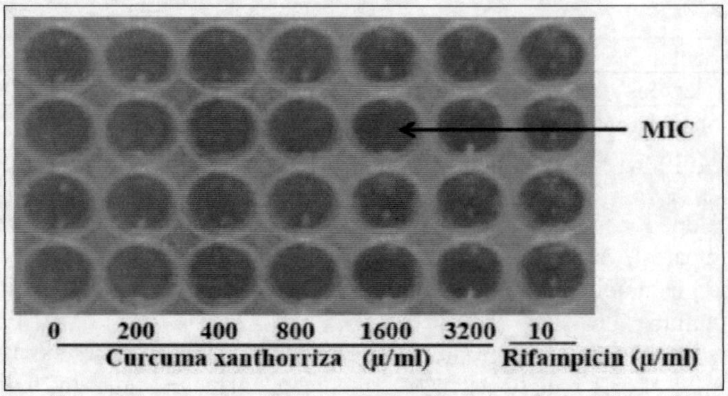

Figure 3. The effect of the extract of C. Longa on the tubercle bacillus of Mycobacterium tuberculosis H37Rv in vitro [21]

The pharmacological effects exhibited by this plant material have made it possible to develop and study the possibility of enriching bakery products with a phytocomposite based on milk thistle, scorzonera and turmeric. The technological parameters affecting the raw materials and the finished product are studied. Experimental studies were carried out in no less than two agility. Deviations did not exceed 3%.

In the work, methods of statistical analysis of the obtained data were used.

Table 3. Options for adding composite

Experience Options	Amount of flour, %	Amount of powder composite, %	NaCl (salt), %	Pressed baker's yeast, %
Premium wheat flour (100%) - control (sample No. 1)	100.0	-	1,50	3.0
Flour 100.0% + powder 1.0 % (sample No. 2)	100,0	1.0	1.50	3.0
Flour 100.0% + powder 3.0 % (sampl eNo. 3)	100.0	3,0	1.50	3.0
Flour 100.0% + powder 5.0 % (sample No. 4)	100.0	5.0	1.50	3.0
Flour 100.0% + powder 7.0 % (sampl eNo. 5)	100.0	7.0	1.50	3.0
Flour 100.0% + powder 10.0 % (sample No. 6)	100.0	10.0	1.50	3.0

An analysis of the experimental data showed that when phyto-enrichment agents from non-traditional plant materials are introduced, the organoleptic characteristics of flour change. The most acceptable according to experts were samples No. 1 and No. 4. The results are shown in table No. 4.

Table 4. Organoleptic characteristics of flour with non-traditional vegetable phyto-enrichment agents

Sample Name	Ash content,%	Organoleptic indicators			Point grade
		Smell	Taste, crunch	Colour	

Control ("MAKFA" flour, premium)	0.52±0.1	Characteristic for premium flour	Characteristic crunch	white	5 4 5
Sample No. 1 (milk thistle 5% + 1% turmeric powder)	0.76±0.2	Pronounced aroma, turmeric	Characteristic crunch	with a yellowish tint	5 5 5
Sample No. 2 (milk thistle 7% + 0.5% turmeric powder)	0.86±0.3	Slightly noticeable turmeric flavor	Characteristic crunch	Grayish yellow	4 4 4
Sample No. 3 (milk thistle 10% + 0.5% turmeric powder)	0.92±0.1	Slightly noticeable turmeric flavor	Characteristic crunch	Grayish yellow	4 3 3
Sample No. 4 (powdery composite 5% scorzonera + lingonberry leaf 1% carrot powder)	0.74±0.3	Grassy odor	Crunch uncharacteristic for flour	Taupe	4 5 4
Sample No. 5 (powder composite 7% scorzonera + lingonberry leaf 1% carrot powder)	0.88±0.3	Pronounced grassy odor	Crunch uncharacteristic for flour	Taupe orange blotches	3 2 3
Sample No. 6 (powdered composite 10% scorzonera + lingonberry leaf 1% carrot powder)	1.22±0.2	Pronounced grassy odor	Crunch uncharacteristic for flour	Taupe orange blotches	3 2 3

Influences of phyto-enrichment agents of unconventional origin on titratable acidity of flour

The acidity of flour shows the ability of protein compounds of flour to bind a certain amount of alkali. With the breakdown of compounds: phytin, lipids, etc., rancidity of flour occurs and an increased content of organic acids occurs in it.

The results of the influence of phyto-enrichment agents on acidity are presented in table 5.

Table 5 - Indicators titratable acidity of flour with non-traditional phytochemicals

A portion of flour, g	Sample name	The amount of 0.1 n alkali, went to the titration, ml	Acidity, degrees	Conclusion
5	Control	1.45	3.0	Meets the standard requirements for flour quality
5	SampleNo. 1	1.58	3.2	With the introduction of milk thistle powder, a slight increase in acidity occurs
5	SampleNo. 2	1.67	3.4	With the introduction of milk thistle powder, a slight increase in acidity occurs
5	SampleNo. 3	1.88	3.8	With the introduction of milk thistle powder, an increase in acidity occurs due to the presence of organic acids in the feed

5	SampleNo. 4	1.97	4.0	With the introduction of powder from scorzonera, a significant increase in acidity occurs.
5	SampleNo. 5	2.1	4.2	When powder is added from scorzonera, a significant increase in acidity occurs.
5	SampleNo. 6	2.2	4.4	When powder is added from scorzonera, a significant increase in acidity occurs.

The analysis of experimental data showed that in sample No. 1 there is a slight increase in the acidity of flour, in samples No. 3-4 the acidity is higher, which must be taken into account in the production of bakery products.

The "strength" of flour affects the quality of bakery products. The results of the analysis are presented in table 6.

Table 6 - "Strength" of flour, depending on the size of sedimentation sediment in experimental samples

Sample Name	V (sediment volume), cm3	Analysis results, cm3	Flour class by "strength"
Control	Moreth an 60	50.9	High-protein, excellent gluten, strong
Sample No. 1	40-59	46.8	High protein, good quality gluten, strong
Sample No. 2	40	43.4	With medium protein and low gluten quality
Sample No. 3	31-39	387.1	With medium protein and low gluten quality
Sample No. 4	31-39	49.6	With medium protein and low gluten quality
Sample No. 5	31-39	38.0	High protein, good quality gluten, strong
Sample No. 6	Less than 30	30.4	With medium protein and low gluten quality

Thus, the most acceptable for the production of bakery products are sample No. 1, No. 2, No. 4

The effect of phyto-additives from unconventional raw materials on the lifting force of baker's yeast

The presence of macro-and micronutrients (Na, K, Ca, Mg, P) in the raw materials of phytobacteria has a special effect on yeast. Na ions activate the synthesis of intracellular substances, stimulating the growth of yeast, Mg stimulates the production of enzymes in yeast cells, and phosphorus is one of the main factors in the activity of yeast cells. Ca - activator of energy processes.

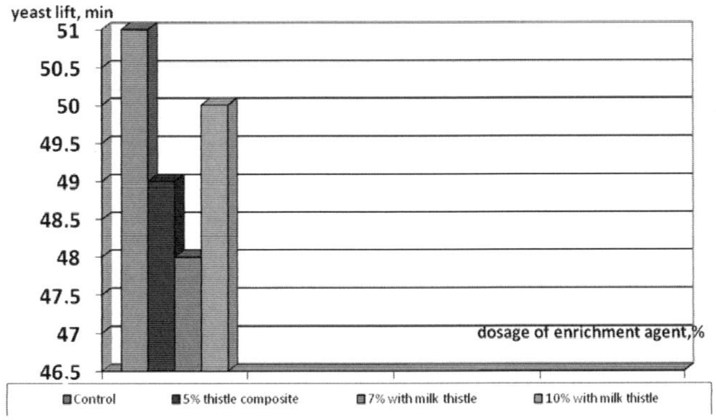

Figure 4. Changes in the lifting force of yeast when making a thistle powder composite

The lifting force of baker's yeast with milk thistle and turmeric is 2 minutes for a sample with a dosage of powdered composite 5%, 3 min - 7% and 1 min with a dosage of powdered composite 10%. Phyto-enricher – is a powdered composite based on milk thistle stimulates the process of alcohol fermentation and the growth of yeast cells due to the microelement composition and vitamin complex.

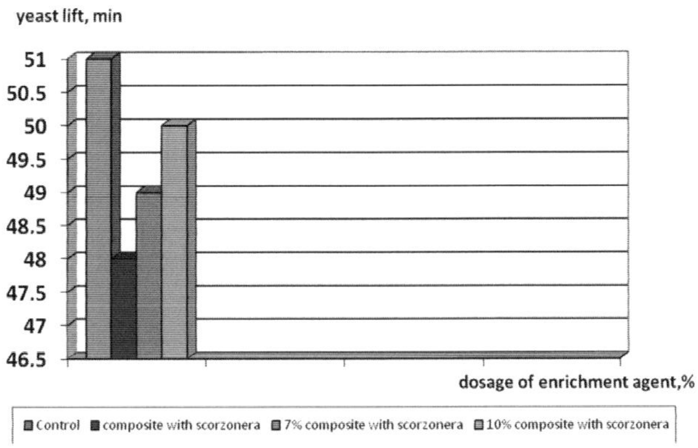

Figure 5. The effect of scorzonera powder with lingonberry leaf on the lifting force of yeast

Bakery yeast with scorzonera and lingonberry leaf (5%) - 3 min, 2 min - 7%, 1 min - 10%.

Thus, with the introduction of 5% milk thistle (sample No. 1, No. 2) and 5% scorzonera (sample No. 4), the lifting force of the yeast increases and the maturation process of the dough is accelerated.

Influence of phyto-fortifiers from non-traditional plant materials on the physicochemical parameters of bakery products.

The results of the effect on the porosity, volume and acidity of finished products are presented in table 7.

Table 7 - The results of the experiment

Sample Name	The porosity of the crumb, %	Volume of bread, cm^3	Acidity, degrees
Control	60.6	115	5.0
Sample No. 1	63.4	125	5.6
Sample No. 2	61.9	116	5.8
Sample No. 3	61.8	110	4.4
Sample No. 4	63.9	130	5.1

Sample No. 5	61.0	123	6.4
Sample No. 6	59.0	100	6.7

Bakery products with the addition of milk thistle powder and scorzonera in an amount of 5% by weight are distinguished by better organoleptic and physicochemical properties compared to the control sample, which is due to

The biologically active compounds contained in the composites form complexes with protein compounds, improving the quality of gluten, increasing the lifting force of yeast. Products turn out to be of greater volume and porosity.

The effect of vegetable dressing from non-traditional raw materials on the organoleptic characteristics of bakery products.

In the study, a total assessment system (integrated) quality assessment of bakery products, i.e. tastes were determined (on a five-point scale) and appearance.

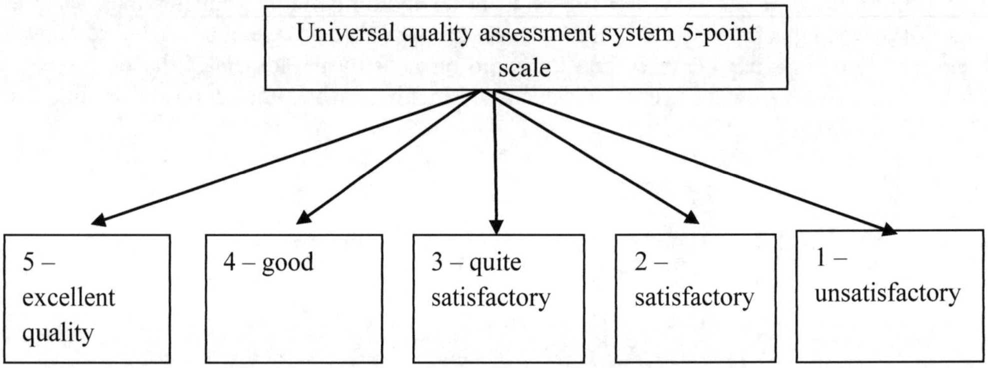

Figure 6. Total quality assessment of bakery products

Bakery products are evaluated according to 6 indicators (the color of the crust, the surface of its condition). The results are presented in table 8, 9.

Table 8 - The coefficient of the indicator of the importance of bakery products

Sample name	Appearance	Taste	Condition of crumb	Condition of crust	Aroma	Freshness	Total significance coefficient
Significance coefficient	4.0	4.0	3.3	3.3	3.3	2.1	20
Control	4.0	3.9	3.2	3.2	3.3	2.0	19.6
Sample No. 1	3.9	4.0	3.3	3.2	3.3	2.0	19.7
Sample No. 2	3.7	3.5	3.0	3.0	3.3	1.7	18.2
Sample No. 3	2.0	3.2	2.1	2.8	3.0	1.3	14.41
Sample No. 4	3.9	4.0	3.3	3.2	3.3	2.0	19.7

Sample No. 5	3.7	3.5	3.0	3.0	3.3	1.7	18.2
Sample No. 6	2.0	3.2	2.1	2.8	3.0	1.3	14.4

The highest significance value in sample No. 1 and sample No. 4, and in the control sample approaches the optimal coefficient of significance for bakery products. Each indicator is evaluated using a five-point system (the total score is 100 points): excellent (100-90 points), very good (89-80 points), good (79-70 points), average (69-55 points), satisfactory (54-40), unsatisfactory (less than 40 points).

Table 9 - Indicators characterizing the quality of the crumb and the appearance of the bread.

Grade, points						Indicators
5	4	3	2	1		
Intense brown with a ruddy tint	Light brown with a golden brown	Yellow gold	Taupe	Pale	Colour	Appearance
Rounded, convex	Medium convex	Weakly convex	Flat	Concave	The form	
Smooth	Flat	Grungy	With small bumps and cracks	Torn edges	Surface	
White with a grayish tint	Intense yellow	Gray yellow or dirty yellow	Grayish Golden	Dark grey	Colour	Crumb
Small, uniform, thin-walled	Small, uneven, thin-walled	Large, uniform	Dense, uneven	Very large, ragged	Porosity	
Easy to restore when pressed, gentle	Delicate, soft structure	With difficulty restores the original form	Crumbles	Haze dup	Elasticity	
Normal bread	Unleavened	Sharply sweet or salty	Yeast	Not peculiar to bread	Normal bread	Taste

According to the results, the highest indicator in samples No. 1 and No. 4.

3. Conclusions

To sum up, a study with a focus on mechanisms of acquired resistance in plants to harmful biotical en Bakery products with the addition of milk thistle powder and scorzonera in an amount of 5% by weight are distinguished by better organoleptic and physico-chemical properties compared to the control sample, which is associated with biologically active compounds contained in the composites form complexes with protein compounds, improving the quality of gluten, increasing the leverage of yeast. Products are obtained with a larger volume and porosity and enrich the functional product with biologically active substances.

References

[1] Mayurnikova L A 2011 Celiac disease. Problems and solutions. *Food industry*. **V 6** 60-63.
[2] Fundamentals of the state policy of the Russian Federation in the field of healthy nutrition for the period up to 2020; approved by order of the Government of the Russian Federation of

October 25, 2010 No. 1873-r.

[3] Pashchenko L P 2006 *Technology of bakery products.* M.: KolosS, 2006 389 p.

[4] The basics of the state policy of the Russian Federation in the field of healthy nutrition for the period up to 2020; approved by order of the Government of the Russian Federation of October 25, 2010 No. 1873-r.

[5] Katznelson Yu M 2015 The quality of our bread. *Newsletter.* V **10** 40-43.

[6] Boryaev V E [et al] Edited by Teplova V I 2005 *Functional food.* Belgorod: Cooperative education, 415 p.

[7] Korchagin V I 2001 *Promising enrichment of plant origin in the production of bakery products.* Voronezh: Voronezh. state technol. Acad. 278 p.

[8] Iorgacheva I G 2014 The potential of medicinal, aromatic plants in improving the quality of wheat bread. *East European Journal of Advanced Technologies.* V **2/12 (68).** 101-108.

[9] Fijalek Z 2003 Determination of some micro- and macroelements in prepared made from peppermint and nettle leaves. *Pharmazie.* V **58** 480-482.

[10] Smertina, E S 2011 Application of extracts of wild plants in functional bakery products. *Proceedings of the Far Eastern Federal University. Economics and Management.* V **3**. 61-67.

[11] Hooda S 2005 Effect of fenugreek flour blending on physical, organoleptic and chemical characteristics of wheat bread. *Nutrition and Food Science*; Bradford.: V **35 3-4** 229-242.

[12] Agibalova V S 2013 The use of carrot powder in the production of bakery products with preventive properties. *Integration of science and practice as a mechanism for the effective development of the agro-industrial complex: international materials. scientific and practical. conf. in the framework of the XXII International specialized exhibition "AgroComplex-2013", March 12-15, 2013* Part II Ufa: Bashkir State Agrarian University 20-21.

[13] Alabushev A V 2000 Unique features of sorghum. *Agriculture.* **3** 20-21.

[14] Gatko N N 2004 The use of herbal additives in the preparation of processed cheese *University proceedings. Food technology.***4** 34-35.

[15] Rakhmetov D B 2011Schnavat: vegetable, fodder, and phytofuel. *Grain.* **3** 62-68.

[16] Prikhodko Yu V 2009 *Scientific and practical justification for the use of raw materials of the Far East as sources for the production of functional food product:* Author's abstract of thesis … doctor technical sciences. Vladivostok, 47-52.

[17] Vlasova M V 2011 The formation of consumer properties and increase the shelf life of bread from wheat flour enriched with mushroom powders: Author's abstract of thesis … candidate technical sciences. Moscow, 89-96.

[18] Smertina E S 2011 Application of extracts of wild plants in bakery products for functional purposes. *Bulletin of the Far Eastern Federal University. Economics and Management.* **3** 61-67.

[19] Orobinskaya V N, Pisarenko O N Kholodova E N 2016 The Dietary Fibers Scorzonerahispanica L. as an Ingredient for the Production of New Foodstuffs. *International journal of advanced biotechnology and research.* **V 7 4** 1705-1711

[20] Glagoleva L E 2004 Formulation optimization of products with natural enterosorbents. *Dairy industry.* **3** 30.

[21] Hema Suryawanshi, Rupali Naik, Pramod Kumar, and Rolly Gupta 2017 Curcuma longa extract - Haldi: A safe, eco-friendly natural cytoplasmic stain. *J Oral Maxillofac Pathol.* **21 (3)** 340–344

[22] Ngadino, Setiawan, Koerniasari, Ernawati and S A Sudjarwo 2008 Evaluation of antimycobacterial activity of Curcuma xanthorrhiza ethanolic extract against Mycobacterium tuberculosis H37Rv in vitro. *Vet World.* **11 (3)** 368–372.

IOP Conf. Series: Materials Science and Engineering **941** (2020) 012061 doi:10.1088/1757-899X/941/1/012061

Future development of transport and environment

S.E. Rudneva

Russian University of Transport (MIIT),
9b9 Obrazcova Ulitca, Moscow, 9127994 Russia

E-mail: tu@miit.ru

Abstract. The reported paper discusses co-existence aspects of an evolving transport infrastructure and the environment. In a long-term perspective, the specific development and upgrade of various transport means shouldn't cause harm to the environment. The paper highlights a necessity of environmental improvement, reduction of noise impact, vibration and hazardous emissions into the atmosphere.

The development of transport systems in human settlements and countries is associated with a necessity to consider probable environmental problems. The transport facilitates the movement of goods and passengers over considerable distances. Increasing velocities of transportation and a level of comfort, enhancing safety and security of all transport modes further the growth of settlements [1], create tourist routes, promote the industry and accelerate production processes.

From the perspective of safety, any means of transportation poses a risk to the environment, remaining a principal source of pollution. When operating cars or busses, damaging substances are emitted into the atmosphere, the smog forms, and the ozone layer tends to depletion. The most poisonous substances produced by modern means of transport include carbon monoxide, dioxin, benzopyrene, nitrogen oxides, and lead compounds. Once they are in the atmosphere, unhealthy substances penetrate into human lungs and blood, provoking the development of diverse diseases.

A system of transport contributes to the depletion of raw resources, carbon, metals, and metallic ores [2]. As a result of washing different transport means water bodies become contaminated. Regular recycling of transport waste products is needed: tires, batteries, metal scrap, plastics, and domestic garbage. Besides the atmosphere, hydrosphere, and lithosphere pollution transport causes the noise pollution.

Railway and sea means of transport are most environmentally friendly. In the pollution of atmosphere a percentage of railway transport amounts to 3.7%, whereas those of sea transport – 2.7%, river transport – 0.9%, air transport – 1.4% and motor vehicles – 91.3%. The use of most energy efficient means of transportation decreases the emission of destructive substances into the environment.

Motor vehicles are most hazardous for the environment since their share is 80 % in total polluting emissions of a transport complex without a pipeline system. A negative effect mobile sources of road transport have on the environment is a rapidly growing number of automobiles [3] and their permanent dispersion, a close proximity of pollution sources to residential areas, higher toxicity in comparison to stationary sources, as well as technically complex safety and pollution prevention equipment. The evaporating gasoline in fuel filling stations remains a source of environmental pollution alongside with stationary sources of road transport, dust formation in air near grounds along highways and transfer of massive land areas to road building companies [4].

A rapidly growing vehicle-to-population ratio [5] intensifies an adverse effect of road transport on the environment, especially in big cities [6]. The exceedance of maximum allowable concentrations along highways, near roads, and noise pollution are reasons for the increasing prevalence. The vehicle fleet is renewed through cars of higher emission standards but the rates total emissions of motor vehicle fall are slow because of the constantly increasing vehicle fleet [7] and insufficient update through cars of higher emission classes.

One of the most environmentally friendly means of transport is a railway line system due to low total emissions of unsafe substances. At the same moment, noise, vibration, air pollution and contaminated sewage have a disastrous effect on the environment. A dangerous influence of stationary sources – locomotive and train sheds, manufacturing and repairing factories of rolling stock, washing and steaming stations – involves atmospheric air pollution with toxic substances, e.g. components of varnishes and paints, products resulting from the incomplete combustion of fuel, water consumption, and contamination of waste waters. The construction of railways, in its turn, causes harm to the environment since it is related to the withdrawal of lands for the building permanent and temporary facilities and installation of utility lines, deforestation, deterioration of hydrogeological conditions of a territory, and genesis of bog soils. However, there are optimistic results reported due to resource saving and nature protection measures being taken by railway companies.

Sea and river transport influence the environment as well, e.g. an excessive use of sea resources, physical transformations of sea species habitat, pollution of water bodies with petroleum products and transported goods, and damage of water and shore ecosystems when dredging river and lake beds.

Pipeline transport is also environmentally unfriendly when its facilities are built, operated, or emergency situations appear. The construction of pipelines leads to the withdrawal of lands and damage of natural landscapes and migration routes of wild animals. When pipelines operated, the atmosphere suffers from pollution caused by gas leakage through cracks. Bulk emissions of oil and gas occur in emergency situations, and the consequences are littering of extended territories, a dramatically high concentration of poisonous substances, death of plants and animals. Making the effort of reducing any probable detrimental effect on the environment, companies maintaining pipelines lower emissions into the atmosphere and dirty discharge, furthermore, they implement the restoration of damaged lands.

An adverse impact on the environment caused by air transport is associated with the noise from running aircraft engines, use of special airport vehicles, auxiliary power plants and contamination of the biosphere with products of aircraft fuel combustion. High-speed aircraft flights at significant heights make products of combustion disappear in upper layers of the atmosphere over vast territories, reducing their catastrophic effect on living organisms.

Transport companies fund the environmental protection and decrease of their destructing effect. Principal solutions to traffic-caused problems are related to the reduction of environmentally unfriendly exhausted gases and use of high-quality clean fuel [8]. There is a need to design principally new types of automobile engines and adapt alternative energy sources. Electrically driven electro mobiles and hybrids have been already offered on the market. Currently, a number of such models are limited, but they are prospective. If a car used, it would be necessary to troubleshoot on time, carry out comprehensive maintenance operations on a constant basis, follow recommendations on allowable loads and appropriate exploitation rules. Undoubtedly, the environment situation will improve provided that purifying and filter equipment is developed and advanced to cut harmful emissions of road transport [9]. Another vital issue is the reconstruction of car engines aimed at the increasing their input-output ratio and lowering of consumed fuel. Moreover, other means of transport are to be used, e.g. trolleybuses and trams [10]. Drivers operating motor cars should utilize them rationally, avoiding any unnecessary negative effect on the environment [11].

Alternative car batteries are designed, e.g. environmentally friendly electrically driven batteries, which don't run on hydrocarbon fuels. They have more capacities and higher input-output ratios than a conventional battery; their production is less complex but still cost-demanding. Once less expensive materials are found, electrical batteries would be a product of large-scale manufacturing. To date, thousands of cars all over the world are equipped with an electric motor. Such cars are four times lighter than traditional ones. Hydrogen is the most energy-intensive substance but its use requires the mounting ceramic machine elements because of chemical reactions evolving in the engine; that is the reason for high manufacturing costs. Another option is to use catalyst agents generating electric power when oxidized, but this type of energy is even more expensive. Automobiles with hydrogen-powered batteries have been already developed but less frequently used than electrically driven cars. In addition, it is reported on chemical, pneumatic, and spring-driven engines. A compressed air car is powered from tanks containing compressed air, which releases under pressure and forces a car driven by clean air. Chemical and spring-driven engines are being developed and improved. The Segway which seemed to be a space innovation is used at present in many countries even by government officials. Its operation principle is the movement under control of a changing spatial body position. When moving, on each wheel autonomous engines run.

IOP Conf. Series: Materials Science and Engineering **941** (2020) 012061 doi:10.1088/1757-899X/941/1/012061

It is recharged when going down a slope or from an electric socket. A self-balancing unicycle is similar to a push bicycle equipped with a powerful engine.

To reduce the environmental pollution caused by railway transport nearly one third of Russian Railways trains is planned to be switched from diesel to natural gas. Several urgent measures are to be taken to eliminate impurities accumulated over a prolonged period of time: cleaning and making more environmentally friendly a railroad right-of-way; purchasing up to date trains, modernizing trains available, introducing new engines with better environmental parameters to decrease soil poisoning with oil products and heavy metals; purchasing the rolling stock which prevents from leakage and spillage of hazardous commodities; using environmentally friendly materials for the construction and repair of railroads; environmental remedial actions related to prior economic activities. A principal goal of JSC "Russian Railways" in the field of the environmental protection is a focus on the nature well-being and human health. To achieve this goal, it is needed to make a negative impact on the environment less serious, reducing it by 70% in 2030; introduce efficient resource saving nature protection technologies and environmentally friendly materials; use nature resources rationally; lower the power consumption of transportation: decrease a specific energy input for haulage of trains by 14.4%, and fuel by 9.1%; advance ecological safety and social responsibility of companies for their economic activities.

To decrease harmful emissions related to engine operation, airline carriers adapt fuel additives, fuel spraying, enrich a mixture in the combustion zone, reduce a number of running motors in steering, when an amount of waste is three to eight times lower. Following strategies of the environment protection and preserving water resources, airline companies constantly measure a volume of sewage supplied to purification and treatment facilities or a special water body provided for the company use. Quality and quantity related parameters of waste waters and performance efficiency of purifying facilities are monitored; residues of anti-icing fluid are collected and recycled after aircraft treatment.

To protect from the electromagnetic environmental pollution, a source of electromagnetic emission is shielded or removed from a near-field region; technical characteristics of the equipment are improved to diminish electromagnetic disturbances and a total input and emission power of the equipment. A period of time operators or civilians are in a zone affected by electromagnetic interferences is limited and monitored. Levels of electromagnetic interferences should be regulated by sanitary and electrocommunication inspection authorities and labor safety engineers at enterprises.

Engines are updated with the purpose to reduce a specific volume of toxic substances in exhaust gases. Besides the improvement of existing turbine engines new ones are being developed with new designs of combustion chambers, better ignition and mixing of a composition supplied to the chamber and its full combustion. A total amount of consumed fuel and emission of noxious substances can be decreased via better maintenance of aircraft, more reasonable landing distances of aircraft in air fields, when they are drawn by hauling units to the start position, bus or conveyor system transportation of passengers from aircraft into the airport building, so an airplane would be in a parking position maximally close to landing and takeoff grounds. Fundamental and applied research works ace focused on issues air transport may confront in future. Researchers are looking for flying apparatus with better aerodynamic characteristics and higher weight coefficient, new, more economical types of engines and environmentally friendly energy carriers.

The policy of airline companies in the field of the environment protection is concentrated on the improvement of energy and ecology efficiency of passengers, luggage, post, and commodities transportation. The aircraft fleet is being modernized while outdated energy-intensive models are replaced by fuel efficient ones. The energy consumption related to operation activities is reduced due to new resource saving processes and technologies. A destination network is optimized and new piloting control techniques are adapted to decrease noise pollution and emission of hazardous substances from aircraft engines into the atmosphere.

Environment protection represents a major concern for sea transport, since 70 % Earth surface is covered by the world ocean, which dominates a great number of processes on the planet, world climate, and contains enormous resources of seafood. Active and purpose-oriented measures in the form of international conventions and codes of laws have been taken to protect the nature from pollution caused by ships. Efforts are made to solve the issue of sea pollution in several ways: preventive actions during marine operations via appropriate designing, equipping and maintaining fleet; reducing a number of sea accidents owing to the introduction of strict and compulsory global standards and navigation procedures with the purpose of sea transport safety improvement; control of polluted areas in emergency situations; assistance in the application of nature protective standards and requirements.

For the elimination of water way impurities garbage and oil film on the water surface are collected mechanically, oil films are treated by chemicals or decomposed biologically. When mechanical cleaning, disinfection is carried out at shore stations. A physical and chemical method is being developed for removing oil spills from the water surface. Special powder or liquid chemical reagents – absorbents are sprayed over oil spills. Absorbing oil, these substances form new, sometimes, toxic chemical compounds, which remain in water and can poison it. To neutralize petroleum products in water, plants are used to block pollutants. This method is adapted for biological neutralization of ballast water on a harbor territory. Research is initiated to explore living species capable for blocking and recycling water pollutants, i.a. carbon hydrates. Biologists suggest that mussels passing daily up to 70 liters of water through their bodies can filter water in the process of their life activities. Anaerobic bacteria propagating quickly in rivers or seas are also considered as a method of water treatment since they consume carbon hydrates and transform pollutants into useful or neutral substances.

In the use of pipelines, it is necessary to control timely any deviation from the design position, develop recommendations for compiling checklists of repair and overhaul directives to ensure their operational reliability and inspect stress states in pipes and their insulation cases. In accidents, measures are to be taken to prevent oil penetration in near located rivers and other water bodies. Once a potential place of fractures is detected, it is necessary to find immediately a right flow direction of a product to do less harm to the environment. Well-timed repair, reconstruction, technical and environmental monitoring improve the reliability and ecological safety of pipeline transportation. For this purpose science intensive and up to date technologies and equipment are required to reduce hazardous emissions and ensure leak resistance of systems.

A pipeline network is directly related to the environment due to the extended geography of pipeline transport and significant length of gas and oil pipelines crossing all natural and climatic zones, underground location of a linear part, as well as the location of pumping and compressor plant stations in diverse natural conditions. On this territory some man-made facilities have a complex relationship with the environment. In most cases, the interdependence of pipeline complexes and the environment acquires a negative character. All man-induced impacts are to be minimized when building and maintaining pipelines, as well as adverse effects of organic components are to be less significant for their reliability and safety.

A pipeline network is frequently considered the most environmentally friendly means of transportation under conditions gas and oil lines are designed, built and operated according to up to date technical and technological standards and meet ecological requirements. Practically oriented programs concerned with the protection of raw resources from any man-induced impacts when building and operating pipelines are conceptualized, studied and adopted. For instance, industrial systems of production and environment monitoring are developed. The mankind has realized we will experience a global ecological disaster soon. The optimization of structure-related rational limitations of construction works and maintenance has gained in importance from the perspective of a minimal effect on the natural landscape, in particular, on specially protected territories.

The construction and operation of oil and gas extraction facilities lead to catastrophic geo-ecological consequences in emergency and regular situations. A green cover is completely killed on about 15% of the total exploration area when planning pipeline networks. The most severe destruction is seen on land surfaces along northern mainstream pipelines built in unstable grounds. In the North natural processes along mainstream gas lines can revitalize fully only in 15-16 years. Damaged territories in the eternal frost areas can be reconstructed successfully via technical re-cultivation and engineering and biological stabilization, which block degradation processes in tundra. A construction process would lack in environmental preparation unless research works and budgets are elaborated in details.

To sum up, all spheres of human activities are confronting with ecological problems [12]. The transport has always played an important role in life of humans; however, it appears to be a major reason for the environmental pollution. We considered perspectives of transport and environment coexistence and revealed a potential to solve urgent issues. A principal aim of designers, constructors and operators seems to be a necessity to manufacture and use environmentally friendly vehicles without any damaging effect on the environment and normal state of the nature. Another radical action of reducing man-made impact of transport on the environment is related to the total saving of energy resources, application of science-intensive technologies and equipment capable to lower hazardous emissions and ensure leak resistance of systems. The experience and knowledge available support solutions to the eliminating consequences of transport and environment coexistence and balancing it. A strict progressive legal framework, an up to date concept of technical diagnostics of transport means, their timely repair, reconstruction, technical and environmental monitoring make transport more reliable and environmentally friendly.

The ecology of transport sets an impact scale on the nature and strategies of environment protection. Each type of transport possesses its advantages; however, they can be environmentally harmful. The reduction of hazardous emissions is a burning issue. Special research and experimental works are carried out with the focus on the developing and advancing technologies of transport power efficiency. The search for alternatives of transportation, means of transport and fuel is supposed to be ongoing.

References

[1] Iodo I.A., Potaev G.A. City development and physical planning. Rostov-on-Don: Fenix, 2008. p. 8-9. (in Russian)
[2] Turabov V.I. Problem of big cities. M.: Publishing house "Sozial'no-politicheskaya mysl'", 2004. p. 22. (in Russian)
[3] Vuchik V. Transport in livable cities. M.: ID "Territoriya budushchego", 2011. p. 267. (in Russian)
[4] Ambartsumyan V.V., Nosov V.B. Environmental safety of road transport. M.: Nauchtekhlitizdat, 1999. (in Russian)
[5] Khorev B.S. Problems of cities. M., 1971. (in Russian)
[6] Guzenko A.V. Development of city passenger transport in a metropolitan city: problems and perspectives // Tomsk State University Journal. 2009. p. 135. (in Russian)
[7] Simpson B. Planning of city development and public transport in Great Britain, France and Germany / translated from English. M.: Transport, 1990. p. 86. (in Russian)
[8] Ecological problems related to road transport development. M., 1997. (in Russian)
[9] Danilov-Danilyan V.I., Losev K.S. Ecological challenge and sustainable development. M.: Progress-Traditsiya, 2000. (in Russian)
[10] Kasyanov P.V. Transition to sustainable development: environmental and economic preconditions. M.: Publishing NUMTS Goskomekologii Rossii, 1998. (in Russian)
[11] Komarov M.A., Melekhin E.S., Kimel'man S.A. Problems of environmental economics development. Kaluga: VIEMS, 1999. (in Russian)
[12] Glasychev V.L. Urbanistics. P. 1. M.: Evropa, 2008. C. 45-46. (in Russian)

Research of the wear mechanism the rubber lining mining trucks

V G Kopchenkov, A G Babich

North-Caucasus Federal University, ul. Pushkina, 1, Stavropol, 355009 Russia

E-mail kopchenkov2@rambler.ru

Abstract. The rubber lining used to protect the bodies of mining trucks is characterized by impact at an angle to the surface of the lining.
Described experimental installation, allowing to model the impact of single solid parti-cles at an angle to the surface of the rubber. It is established that the basis of the mecha-nism of the fracture of rubbers under oblique impact is cracking. The development direc-tion of cracks is approximately equal to the angle of attack, i. e. the direction of the force of impact. The destruction is cyclical in nature and follows the scheme: the tear of the surface — rip crack — turn cracks and separation of the fragment. Factors determining the length and depth of the development of the initial crack set.

Keywords: elastomers, failure mechanism, failure evolution, solid particle impact at an angle to the surface

1. Introduction

The wear of materials by a flux of solid abrasive particles is characteristic for a large group of machines and equipment. Including lining of bodies of mining trucks. These wear modes are specific due to a complex of factors that affect the process of surface deterioration, including the dynamic (impact) character, local effect, dependence of the normal to tangential load ratio on the angle of attack, the small friction path or its absence, short-term contact, etc. [1, 2].

For certain operating conditions, the most effective way to reduce wear is considered the method of gumming [1]. To improve the wear resistance of rubber coatings, it is important to know how rubbers fail under the dynamic action. The investigations conducted today with these aims are still insufficient to formulate a generally accepted viewpoint on the wear mechanism and stages of the deterioration of elastomers in a flow of solid particles. To get answers on these important questions, it is proposed to subdivide the whole complex of phenomena observed during wear and the factors responsible for the failure behavior into simpler constituents and study failure regularities of rubbers by a single particle impact at an angle to the surface.

The most probable trajectory of cracking was determined in the course of investigations of the stress-strain state of the rubber surface layers. It follows from solving Flaman's problem [4, 5] for an elastic half-space subjected to the force action at an angle to the surface [5] that the tensile stresses appear in the backside behind the sliding indenter. They look like a set of radial arcs that originate at the point of contact tangential to the force line. The crack is anticipated to arise perpendicular to the tensile stress; so far, it originates at 90° to the surface and adheres to the arc up to the zero line that separates the regions of compression and tension, thus forming an angle equal to that of the attack, i.e., to the angle of inclination of the force that affects the surface. A characteristic feature of rubbers is their low elasticity modulus, which results in considerable deformations under the shearing force. Therefore, the crack in the near-surface region that appeared at the strongest strain perpendicular to the surface would obtain a certain inclination upon the removal of the load. It follows that the crack is a result of the interference of two processes that occur simultaneously on the surface. This crack finally acquires a direction close to the angle of attack when a particle impacts the rubber surface.

2. Materials and methods

To verify the regularities established previously theoretically, the experimental investigations were fulfilled using a specially designed facility. This facility is able to model a single multiple repeated oblique impact of a solid particle against the elastomeric surface. Its parameters closely approach those of the real impact while providing the invariable position of the contact spot upon each impact.

The above problem is solved using an oscillating crank-and-slot lever mechanism with a rocking stone in the link. On one end of the link, there is an indenter that moves over a preset trajectory. By displacing the point where the stone is fixed on the link in vertical and horizontal directions, it is possible to impart a trajectory to the indenter that repeats the motion of the solid particle upon impacting the rubber surface at an angle in the range of 0° - 80°. This makes possible to keep to a physical similarity of the collision process in what concerns the relative motion mode of the indenter and the sample material in regard to the friction path length, time of penetration, impact force of the particle, and some other parameters.

A general scheme of the facility is shown in Fig. 1. The motor shaft has a fixed crank 1 that sets in motion link 2. The rocking stone 3 of the link serves as another fulcrum for the link able to turn round its axis in the plane of the facility. The adjusting screw mechanism 4 enables to shift the point of fixation of the stone in vertical direction while the screw mechanism 5 in the horizontal direction. This assists in setting the required trajectory of motion of indenter 6 in relation to rubber sample 7. The sample rests on a supplementary rubber spacer 8 and metallic disc supported by a spring 9. The spring is prestressed with the help of disc, screw 10, and a set of adjustable washers.

By displacing the rocking slider *3* vertically using the screw mechanism *4,* the friction path *l* is set (Fig. 1b) and determined by the angle of attack α_1 The screw mechanism assists in setting the given angle of penetration α_1 and of recoiling α_2 (Fig. 1c).

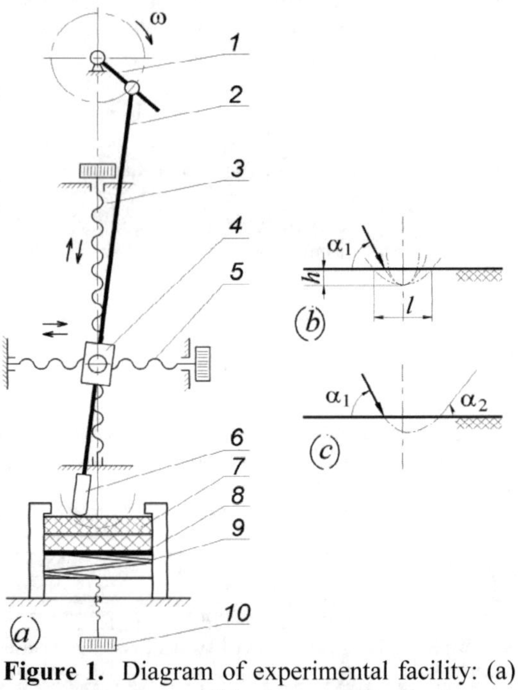

Figure 1. Diagram of experimental facility: (a) structural scheme; (b) change of traj ectory at vertical displacement of rocking stone of the link; (c) at horizontal displacement.

It was established earlier [7] that it is peculiar for the relative motion of the particle and the surface material to shear together initially further followed by the particle sliding. The ratio of these phases of their motion depends on the angle of attack, When the angle of attack is small (0° to 30°—35°) the particle impact results in the elastic shear and relative sliding of the surfaces. In the case when the angle of attack is large (50°—90°), only the elastic shear occurs. In the interval of 35°—45°, there is a transient zone. The kinematics of the relative motion of a particle and the surface determines the surface failure mechanism on the contact spot as well. In this connection, the inves¬tigations of the wear mechanism were conducted for three characteristic angles of attack, namely: 70°, 40°, and 20°. Angles 70° and 20° are the mean in their zone, while 40° represents the processes that occur on the transient boundary portions.

The present facility models an impact of a steel ball 4 mm in diameter, which means that the experiment involves strongly rounded abrasive particles.

The test rubbers were based on isoprene (I2 and I4) and butadiene styrene (C2 and C4) rubbers (Table 1). The physicomechanical properties were controlled by varying the filling degrees of the rubber used by technical carbon of grade P-514. The filling was 20 and 40%, correspondingly. The curing agent was sulfur. Index 2 for the rubbers means a lower modulus and resistance to tear but higher elasticity. The regularities of surface failure are little dependent on the rubber type. Their elastic and strength characteristics are of greater importance for the materials.

Table 1. Main physicomechanical properties

Unit	I2	I4	C2	C4
Modulus at 300%, MPa	4.5	12.5	5	14.0
Ultimate strength, MPa	13	18.1	5.2	15.3
Resistance to tear, KN/m	37.0	37.3	16.1	36.3
Shore hardness, A	50	70	48	68
Elasticity, %	74	51	57	39

3. Results and discussion

Figure 2 presents the photos of the contact spots on the elastomer at the angles of attack 20°, 40°, and 70° for the state before breakage. The evolution of the contact spot wear on the rubber sample at different angles of attack shows no principal differences. The difference consists in the number, direction and depth of the cracks.

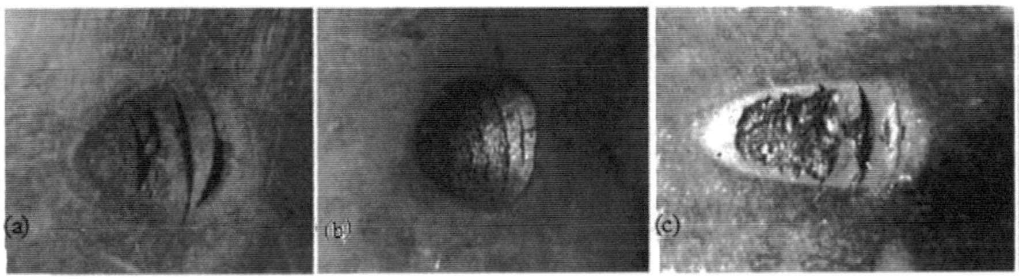

Figure 2. View of a contact spot at different angles of attack during the second stage of breakage (*8): (a) 70°; (b) 40°; (c) 20°.

The general schemes of the wear process were formulated based on the analysis of experimental data for different angles of attack (Fig. 3). The following three stages can be identified: run-in, onset and crack propagation, scoring, and the separation of the fractured rubber fragments.

Frayed patches initially appear and the surface changes its color. In the last third of the contact spot, a small abraded patch is formed. At the center at a slight shift in the direction of the velocity vector, a nonabraded projection appears like the one formed at the direct angle of attack. Afterwards, this projection is subjected to the mechanical destruction. The frayed patch surface is worn-out by the slippage of the indenter and rubber during sliding with formation of microscoring sites. These microscoring sites, which are about 0.01 - 0.05 mm in size, are shown in Fig. 4. This zone migrates from the center to the periphery of the contact spot and back as the indenter penetrates and exits the sample. The rubber with rather high elasticity modulus does not show very high flexibility ($S < 40\%$). Therefore, at the initial stage, residual deformation arises in response to indentation.

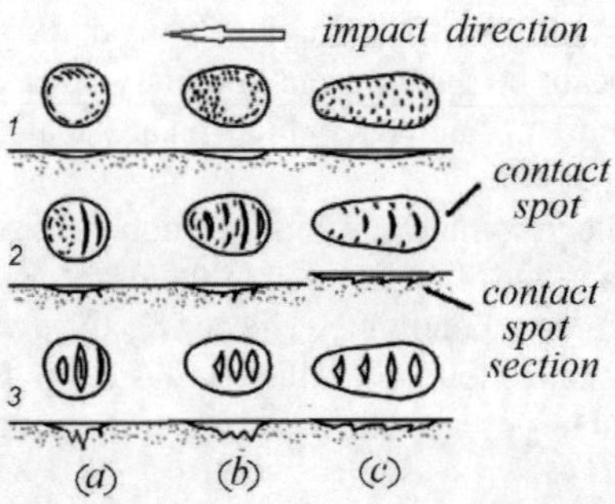

Figure 3. Scheme of surface failure evolution modeled on a facility for impacting at the angles: (a) 70°; (b) 40°; (c) 20°.

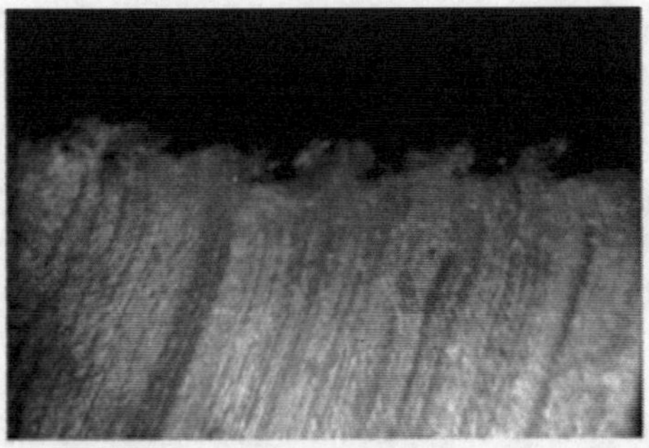

Figure 4. Surface wear of the contact spot with microscore formation (x120).

When the impact energy is not large, the wear mechanism can stabilize during the first stage, so the amount of wear will depend mainly on the number of loading cycles.

During the second stage, the fatigue processes impair the surface layer strength. At a certain fatigue level the sample undergoes tear in response to the impact force, crack formation and propagation (Fig. 2). The surface cracks are arc wise with their ear directed towards the motion. This is because the deformation under the indenter center bears the highest tangential shear in contrast to its edges. The arc-form proves that cracking is initiated in the moment of the maximal shear of the surface by the friction forces. Upon the load removal, the rubber returns into its initial condition. This fact supports the hypothesis on the mechanism of cracking at tension of the surface layer. The angle of the crack onset is close to that of the attack [6], as is shown in Fig. 5 on the contact spot sections.

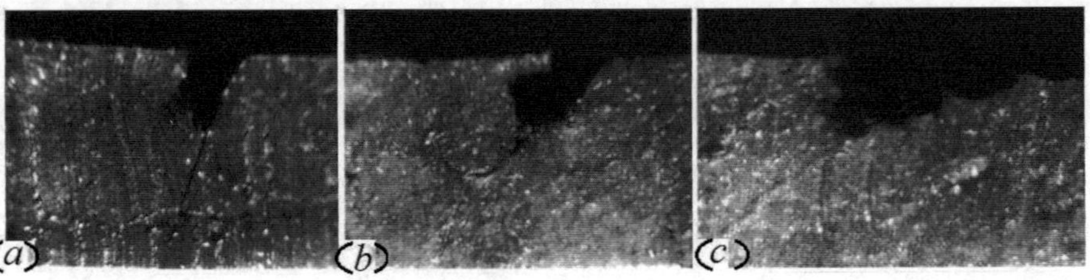

Figure 5. Effect of the angle of attack on crack propagation (x lQQ): (a) $\alpha = 70°$; (b) $\alpha = 40°$; (c) $\alpha = 20°$.

During the third stage, the sample undergoes scoring and spalling accompanied by cracking. The process of scoring includes turn of the initial crack by 90° to the direction of its development, the scoring of the fragment, and its tear away (Fig. 6). As follows from Figs. 5 and 6, the wear fragments approach in their shape a triangular prism. The turn of the crack can be explained by two reasons. Firstly, the fatigue weakening of the surface layer with crack deepening is impeded thus producing insufficient energy for the further crack propagation. Besides, the crack approaches so-called zeroth vector that detaches the tensile stresses from the compressive ones [6].The wear of the surface results in the formation of a specific undulated structure. In general, it resembles Shallamach's picture.

The kinetics of the cracking and rate of transfer from one stage to another depends on the type of rubber, the physicomechanical properties of the rubbers, and the values of the angle of attack and impact energy.

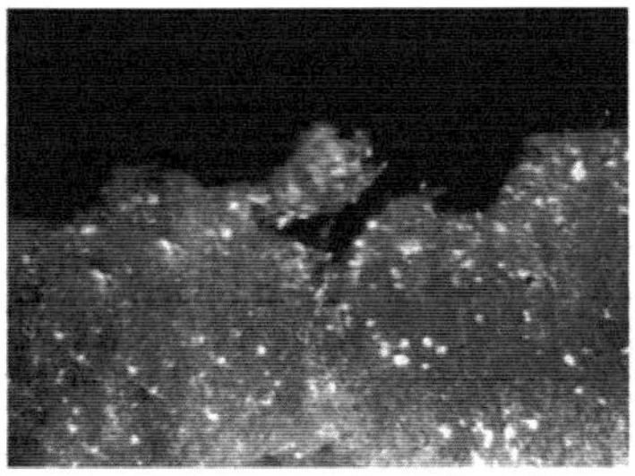

Figure 6. Separation of a fractured surface layer fragment (×80)

Table 2 shows a general principal tendency to abruptly decrease (four times) of the number of cycles up to the formation of a crack with a decreasing angle of attack. However, different types of rubber have their own peculiarities.

Table 2. Effect of the angle of attack on the number of loading cycles on the surface up to its failure under an impact pressure of 1.7/2.4 MPa.

Rubber code	Angle of attack 20°	Angle of attack 20°	Angle of attack 20°
I2	150/110	500/300	2700/2600
C2	200/120	400/350	3000/2800
I4	1100/200	1600/1100	3100/2800
C4	800/300	1800/1000	5300/4100

For instance, for the isoprene-based rubber I4 the time till crack origination is lasts longer at both angles 70° and 20°. Highly elastic rubbers I2 and C2 have a shorter run-in since the friction force is similar to I4 and C4 rubbers, while the rupture strength is 1.5 times less. One of the major factors that determine the rate of crack origination and the failure is the impact energy.

The table shows the effect of the impact energy on the incubation period, i.e., the initiation of rubber decomposition. The impact energy is expressed via the specific pressure. The stronger rubbers I4 and C4 are less dependent on the angle of attack and are more durable than the weaker rubbers. Weaker rubbers I2 and C2 show a low rupture strength and resistance to tear, their impact energy at small angles of attack only slightly affects their durability. At larger angles of attack their durability is comparable to that of stronger materials. This can be attributed to the effect of fatigue properties in addition to the strength ones. The fatigue properties are closely connected with the value of mechanical losses, i.e., with elasticity, which is about 60% for C2 and I2 rubbers.

The wear rate depends not only on the crack formation velocity, but also on that of fragmentation of the worn-out surface layer. The larger the inclination angle of the crack, the more favorable the conditions created for scoring.

Figure 7 illustrates the dependence of the crack depth on the load and physicomechanical properties. For the present investigations, the impact force corresponded to the energy of a 4 mm in diameter steel ball moving at a velocity of 7, 11, 15, and 19 m/s.

The defining effect at 20° is exerted by the friction force, while the fatigue properties are more important at 70°, so the curve can be seen to be flattened. The deepest cracks are characteristic for 70° angles of attack and the most shallow cracks are formed at 20°. Nevertheless, the weaker rubbers, I2 and C2, have the same crack depth at 20°and 70°. It should be noted that

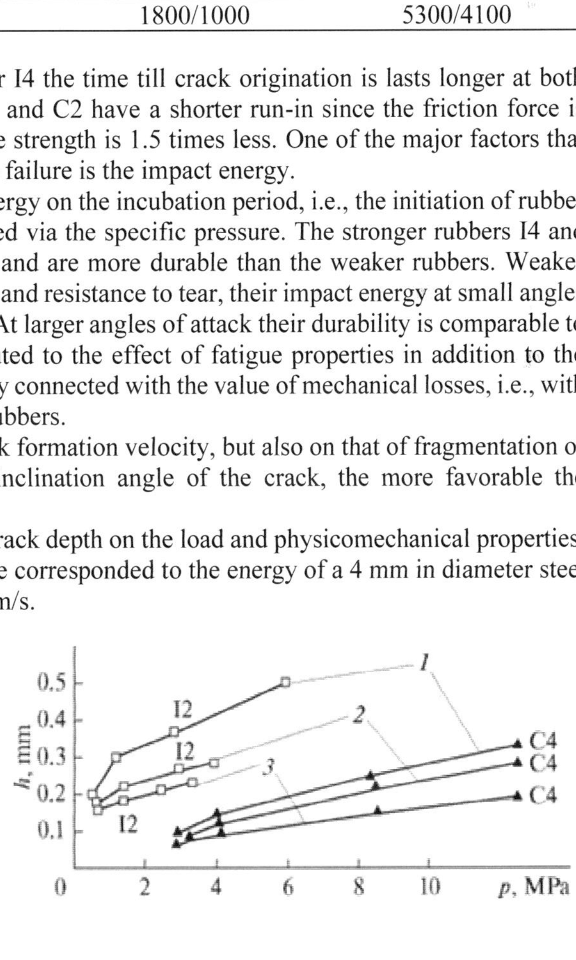

the scoring of the cracks begins at very similar depth values of cracking for all angles. For an angle of inclination of the crack of 20°, we have 0.25 - 0.40 mm, correspondingly, for both stronger and weaker rubbers.

Figure 7. Effect of specific pressure and angle of attack on the crack depth: (1) $a_0 = 70°$; (2) $a_0 = 40°$; *(3)* $a_0 = 20°$.

If to estimate the crack depth h_{cr} in fractions to the penetration depth *h,* it follows that for the weak rubbers ($c_r < 12$ MPa) $h_{cr}/h = 0.15 - 0.20$, and for the stronger rubbers ($c_r < 12$ MPa) is $h_{cr}/h = 0.10 - 0.12$. Consequently, the worn layer thickness makes up on an average of 0.1 - 0.2 of the penetration depth of the impacting particle.

There are investigations in corresponding scientific literature supporting the idea that the principle of cracking lies in the base of the wear mechanism of elastomers [7].

4. Conclusions

Present studies of the failure of the rubber surface by particle impacts at an angle to the surface have led to the following conclusions.

(1) Crack formation on the surface of the contact spot is the cause of elastomer fracture. In the case when the loads are not large, the micro- and submicrocracks are generated, which are by one to two orders less.

(2) It has been experimentally established that the angle of crack propagation relative to the surface coincides with the angle of the indenter action on the surface.

(3) The experiments have shown that with a decreasing angle of attack durability of the rubbers drops abruptly. At small angles of attack, rubbers with high-strength characteristics are the most durable. When the angle of attack is large, rubbers with higher elastic and fatigue properties (resistance to crack generation and propagation) turn out to be the most durable.

(4) The cracks are deeper at larger angles of attack, although the surface breakage rate is greater at smaller angles of attack.

(5) Scoring and the tear of the material begins as the crack reaches a certain length and depth depending on the physicomechanical characteristics of the rubber, as well as the values of the specific load and angle of attack.

(6) The wear mechanism of rubbers takes place in three main stages, namely, crack formation, propagation in depth, and the scoring and tearing of the rubber. The impact-induced wear mechanism of the studied rubbers at an angle to the surface can be defined as scoring.

The above-mentioned results may serve the base for the further development of wear-resistant formulas of rubbers that proceed from the established main causes of fracture in the interrelation with the physico - mechanical properties of rubbers.

References

[1] Penkin N S, Kopchenkov V G, Serbin V M et al 2013 *Gummirovannye detali mashin* (Moscow: Mashinostroenie) p 245
[2] Sare I R, Mardel J I and Hill A J 2001 *Wear* **250** 1–10
[3] Jones L C 2011 *Wear* **271** 1411–17
[4] Aleksandrov AV and Potapov V D 1990 *Osnovy teorii uprugosti i plastichnosti* (Moscow: Vysshaya Shkola) p 400
[5] Samul V I 1982 *Osnovy teorii uprugosti i plastichnosti* (Moscow: Vysshaya Shkola) p 288
[6] Kopchenkov V G 2002 *J. Friction Wear* **23** 623–28
[7] Arnold J C and Hutchings I M 1990*Wear* **138** 33-46
[8] Kopchenkov V G and Penkin N S 2001 *J. Friction Wear* **22** 46–51

IOP Conf. Series: Materials Science and Engineering **941** (2020) 012063 doi:10.1088/1757-899X/941/1/012063

Experimental comparison of smooth movement of an agricultural tractor with balancer suspension and torsion bar suspension

I P Troyanovskaya[1,2], V V Krasnokutsky[1] and M A Rusanov[1,2]

[1] South Ural State University, 76, Lenin Avenue, Chelyabinsk 454080, Russia
[2] South Ural State Agrarian University, 13, Gagarina street, Troitsk 457100, Russia
e-mail: tripav63@mail.ru

Abstract. One of the reasons for the increased fatigue of tracked agricultural tractor operators is vibration. Reducing overall vibration is possible due to a change in suspension. All agricultural tracked tractors are equipped with an elastic balancing suspension, as a rule. Kharkov Tractor Plant has released an experimental batch of tractors with torsion bar suspension. The aim of this study was to determine the effect of the type of suspension on the smooth movement of the tractor. The experiment considered the real modes of operation of the tractor: plowing the ground in first and fourth gears, as well as the transport mode of movement. As a result of the tests, it was determined that the tractor with torsion suspension provides better ride comfort in transport mode. The torsion bar suspension has an advantage in the frequency range of 0.5–4 Hz, while the balancer suspension has an advantage in the high frequency range of 4–12 Hz when plowing the ground. The level of vibration acceleration of the floor of the tractor cab does not depend on the type of suspension in the frequency range above 12 Hz.

1. Introduction

Vibration acts on the driver of an arbitrary vehicle while driving [1-5]. It has a harmful effect on his body. The long-term effect of vibration leads to the destruction of the musculoskeletal system, functional disorders of the vessels, nervous system and vestibular apparatus [6-9]. Vibration issues are most relevant for agricultural tractor units [10-15].

Reducing vibration improves the working conditions of machine operators, reduces their fatigue and occupational morbidity. One of the ways to reduce the vibrational tension of tractors is to change its suspension [16-20].

Most agricultural tracked tractors have an elastic balance suspension [21]. Kharkov Tractor Plant released an experimental batch of crawler tractors HTZ-181 with torsion bar suspension. There was a need to evaluate the impact of the new torsion bar suspension on the vibration of the driver.

2. Research goal

The research goal is to determine the effect of the type of suspension of a caterpillar tractor on the ride when working as part of an arable unit.

3. Objects and methods of experimental research

Two KhTZ-181 caterpillar tractors were taken as objects of experimental research (Figure 1):

- serial tractor with balancer suspension,
- experimental tractor with torsion bar suspension.

a) b)

Figure 1. The KhTZ-181 tractor manufactured by the Kharkov Tractor Plant (a) with balancer suspension and (b) torsion bar suspension

The experiment was conducted in the farm "Veselo-Podolyanskoe" Semenov district of Poltava region of Ukraine. Caterpillar tractors plowed the ground in first (speed 6.5 km / h) and fourth (speed 10 km / h) gears. The vertical components of the acceleration were measured on the cab floor at the place of attachment of the driver's seat during operation (Figure 2) [22-24].

Figure 2. Location of vibration measurements on the floor of the tractor cab

Vibration registration was carried out using vibration measuring equipment manufactured in Denmark (Figure 3), consisting of: piezoelectric vibration transducer, signal amplifier, registrar, narrowband analyzer and real-time recorder.

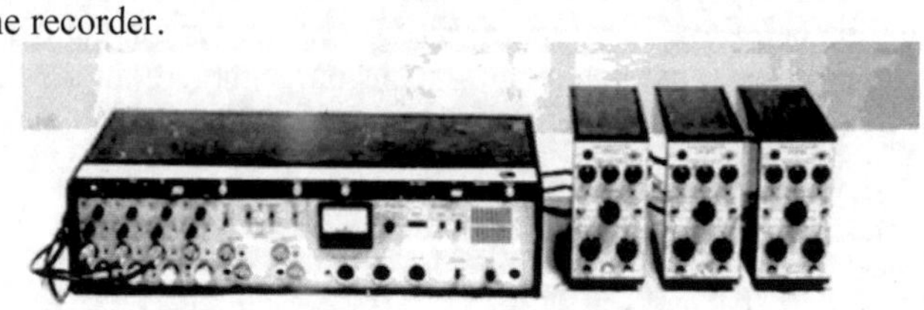

Figure 3. Registration equipment of the company "Bruhl & Kjepr"

The equipment was selected according to the requirements of GOST 12.4.012-83 [25-27]. The signals of the vibration transducer were synchronously recorded by a magnetic recorder with a sampling interval of 0.0039 seconds. This corresponded to 1024 measurements in 4 seconds. The total recording time in each experiment was at least 50 seconds.

4. Processing of the experiment

The spectral composition of vibrational accelerations was determined after testing by distributing the mean square acceleration of the elementary harmonics of the oscillations over frequencies in the range from 0.2 to 100 Hz. Vibration acceleration values were averaged every 0.25 Hz.

$$\breve{a}(f) = \sqrt{\frac{1}{T}\int_0^T a_{\Delta f}^2(t)dt}, \quad (1)$$

where $a_{\Delta f}(t)$ is the measured vibration acceleration in time t in the frequency band with medium non-geometric frequencies f, m / s^2.

The logarithmic levels of vibration acceleration of elementary harmonics were calculated:

$$l_a = 20lg\left(\frac{\breve{a}}{a_0}\right), \quad (2)$$

where $a_0 = 3 \cdot 10^{-4}$ m / s^2 is the reference (zero) value of the mean square value of the vibration acceleration.

5. Research results

The experimental results are presented in the form of envelopes of linear spectra of the logarithmic levels of the vertical components of vibration acceleration of the floor of the KhTZ-181 tractor.

Figure 4 shows the vibration acceleration of the floor of the tractor cab when plowing in first gear. The frequency range of 0.5–4 Hz has the acceleration of a tractor with a torsion bar suspension 2–7 dB lower than that of a tractor with a balancer suspension. The range of 4–12 Hz has vibration acceleration of a tractor with a torsion bar suspension 2–7 dB higher than that of a tractor with a balancer suspension.

Figure 5 shows the vibration acceleration of the floor of the tractor cab when plowing in fourth gear. The vibration acceleration of the cab floor of both tractors has approximately the same values in the range often of 0.5–1 Hz and 9–20 Hz. The range often 1-5 Hz has vibration acceleration lower for a tractor with torsion bar suspension. The range often 6-9 Hz has vibration acceleration lower for a tractor with a balanced suspension.

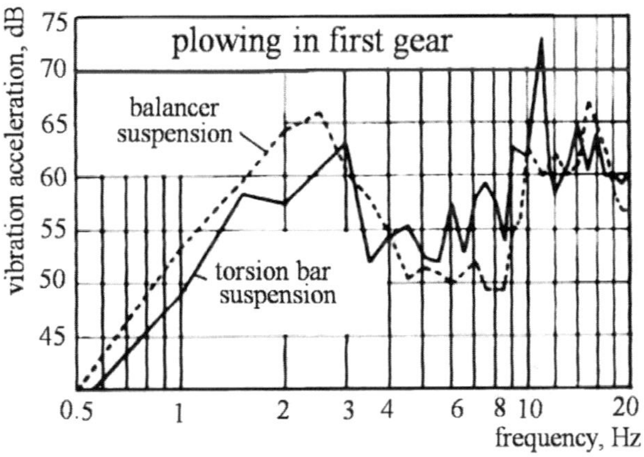

Figure 4. Logarithmic levels of vertical vibration acceleration of the tractor cab floor during plowing at a speed of 6.5 km / h

IOP Conf. Series: Materials Science and Engineering **941** (2020) 012063 doi:10.1088/1757-899X/941/1/012063

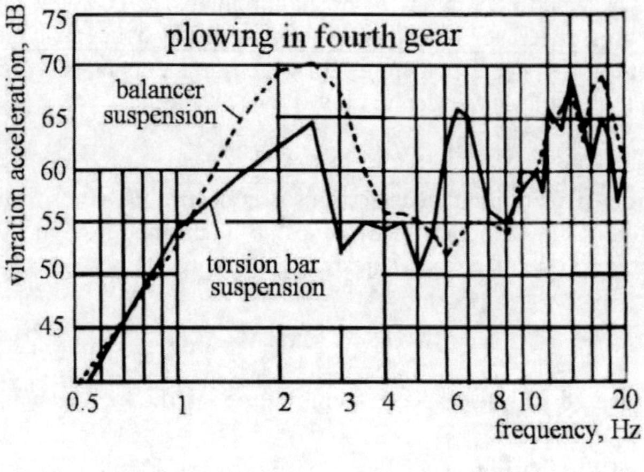

Figure 5. Logarithmic levels of vertical vibration acceleration of the tractor cab floor during plowing at a speed of 10 km / h

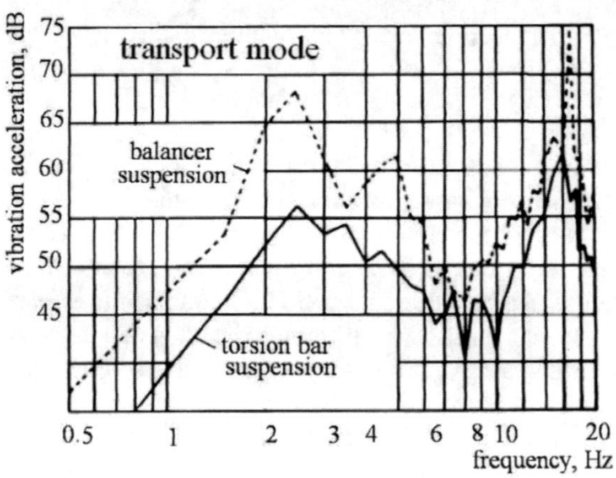

Figure 6. Logarithmic levels of vertical vibration acceleration of the tractor cab floor in the transport mode

Figure 6 shows the vibration acceleration of the floor of the tractor cab in transport mode along the field road. The speed in movement was 10 km / h. The acceleration of vibration of the cab floor in the entire frequency range is 5–10 dB less for a tractor with torsion bar suspension. This indicates that the torsion bar suspension improves tractor performance. Acceleration peaks are in the frequency range 17–19 Hz. They show the transition of the rink from one track link to another.

6. Conclusion

The vibration acceleration of the floor of the tractor cab with torsion bar suspension is 5–10 dB less than that of a tractor with a balancer suspension in transport mode over the entire frequency range. XTZ-181 tractor with torsion bar suspension provides the best ride smoothness in transport mode.

Torsion bar suspension reduces the level of low-frequency vibration acceleration of the cab floor by 2–7 dB in the frequency range of 0.5–4 Hz with plowing. Balanced suspension has an advantage in the high frequency range of 4-12 Hz.

The level of vibration acceleration of the floor of the tractor cab does not depend on the type of suspension in the frequency range above 12 Hz.

References

[1] Cutini M, Costa C and Bisaglia C 2016 Development of a simplified method for evaluating agricultural tractor's operator whole body vibration. *Journal of Terramechanics* **63** 23-32.

[2] Singh A, Singh LP, Singh S and Singh H 2019 Whole-body vibration exposure among tractor drivers. *International Journal of Vehicle Performance* **5(3)** 286–299042

[3] Shismuchi M 1970 Basic Study on the Vibration of the Tractor Used in Forestry Works (I) Vibration of the Crawler-type Tractor. *The Journal of the Japanese Forestry Society* **54(12)** 399-407

[4] Gorshkov YG, Starunova IN, Kalugin AA and Troyanovskaya IP 2019 Investigation of the Slope Angle Influence on the Loading Imbalance of the Wheeled Vehicle Sides and the Change in the Center of Gravity Vector Direction. *Journal of Physics: Conference Series* **1177** 012005

[5] Ahmadi I 2013 Performance assessment of a tractor driver whole body vibration measuring system. *World Applied Sciences Journal*. **21(11)** 1583-6

[6] Ahmadi I 2013 Health hazard assessment of tractor driver whole-body vibration utilizing the ISO 2631 standard. *Agriculturae Conspectus Scientificus* **78(1)** 71-8

[7] Porter JM and Gyi DE 2002 The prevalence of musculoskeletal troubles among car drivers. *Occupational Medicine* **52(1)** 4–12

[8] Jiao K, Li Z, Chen M, Wang C and Qi S 2004 Effect of different vibration frequencies on heart rate variability and driving fatigue in healthy drivers. *International Archives of Occupational and Environmental Health* **77(3)** 205-12

[9] Ljungberg JK and Neely G 2007 Stress, subjective experience and cognitive performance during exposure to noise and vibration. *Journal of Environmental Psychology* **27(1)** 44-54

[10] Gao Z, Xu L, Li Y, Wang Y and Sun P 2017 Vibration measure and analysis of crawler-type rice and wheat combine harvester in field harvesting condition. *Nongye Gongcheng Xuebao/Transactions of the Chinese Society of Agricultural Engineering* **33(20)** 48-55

[11] Cvetanovic B and Zlatkovic D 2013 Evaluation of whole-body vibration risk in agricultural tractor drivers. *Bulgarian Journal of Agricultural Science* **19(5)** 1155-60

[12] Peretti A, Deivecchio S, Bonomini F, Di Bisceglie AP and Colosio C 2013 Vibration on agricultural tractors. *Giornale Italiano di Medicina del Lavoro ed Ergonomia* **35(4)** 297-302

[13] Cutini M, Brambilla M and Bisaglia C 2019 Assessment of a ride comfort number for agricultural tractors: A simplified approach. *Biosystems Engineering* 185 35-44

[14] Singh A, Nawayseh N, Singh LP, Singh S and Singh H 2018 Whole body vibration exposure during rotary soil tillage operation: The relative importance of tractor velocity, draft and soil tillage depth. *International Journal of Automotive and Mechanical Engineering* **15(4)** 5927-40

[15] Dominoni S, Gobbi M, Mastinu G and Previati G 2013 Experimental assessment of the ride comfort of farm tractors. *Proceedings of the ASME Design Engineering Technic Conferencea* **1** V001T01A013

[16] Abdel-Hady S and Abouel-Seoud 2019 Control of driver whole-body vibration ride comfort in agricultural tractor. *Agricultural Engineering International: CIGR Journal* **21(2)** 40-51

[17] Zhang X, Sun D, Song Y and Yan B 2012 Analysis on vibration reduction characteristics of viscoelastic suspension system based on nonlinear behavior. *Nongye Gongcheng Xuebao/Transactions of the Chinese Society of Agricultural Engineering* **28(9)** 47-51

[18] Kim JH, Dennerlein JT and Johnson PW 2018 The effect of a multi-axis suspension on whole body vibration exposures and physical stress in the neck and low back in agricultural tractor applications. *Applied Ergonomics* **68** 80-9

[19] Zehsaz M, Sadeghi MH, Ettefagh MM and Shams F 2011 Tractor cabin's passive suspension parameters optimization via experimental and numerical methods. *Journal of Terramechanics* **48(6)** 439-50

[20] Deprez K, Moshou D, Anthonis J, De Baerdemaeker J and Ramon H 2005 Improvement of vibrational comfort on agricultural vehicles by passive and semi-active cabin suspensions. *Computers and Electronics in Agriculture*. **49(3)** 431-440

[21] Schick FA 1968 *Crawler Tractors* (SAE Technical Papers)

IOP Conf. Series: Materials Science and Engineering **941** (2020) 012063 doi:10.1088/1757-899X/941/1/012063

[22] Xu L, Li Y, Sun P and Pang J 2014 Vibration measurement and analysis of tracked-whole feeding rice combine harvester. *Nongye Gongcheng Xuebao/Transactions of the Chinese Society of Agricultural Engineering*. **30(8)** 49-55

[23] Servadio P, Marsili A and Belfiore NP 2007 Analysis of driving seat vibrations in high forward speed tractors. *Biosystems Engineering* **97(2)** 171-80

[24] Wang F, Cao X, Guo W, Ma G and He C 2007 Research on vibration strength and frequency structure of main driver seat of the wheat combine. *Nongye Jixie Xuebao/Transactions of the Chinese Society of Agricultural Machinery*. **38(4)** 62-5

[25] GOST ISO 10816–1–97. *Mechanical Vibration. Evaluation of Machine Vibration by Measurements on Non-Rotating parts. Part 1. General Guidelines* (Moskow)

[26] De la Hoz-Torres ML, Aguilar-Aguilera AJ, Martínez-Aires MD and Ruiz DP 2019 A comparison of ISO 2631-5:2004 and ISO 2631-5:2018 standards for whole-body vibrations exposure: A case study. *Studies in Systems, Decision and Control* **202** 711-19

[27] Nelson CM and Brereton PF 2005 The European vibration directive. *Industrial Health* **43(3)** 472-9

IOP Conf. Series: Materials Science and Engineering **941** (2020) 012064 doi:10.1088/1757-899X/941/1/012064

Mathematical model of wedge-shaped sliding support with low melting metal coated guide for partially filled working gap

M A Mukutadze, A N Opatskikh, A V Morozova and N S Zadorozhnaya

Rostov State Transport University, 2, Rostovskogo Strelkovogo Polka Narodnogo Opolcheniya sq., Rostov-on-Don, 344038, Russia

e-mail: opatskih@yandex.ru.

Abstract. The reported study presents an asymptotic series solution to a system of differential equations in powers of a small parameter K depending on melting and a mechanical energy dissipation rate. Further, a self-similar solution has been found, as a consequence, a profile of velocities, pressure, principal operational characteristics, load capability, and friction force have been determined for a partially grease-filled working gap.

1. Introduction

An increase in loads designed for friction units of new machines and mechanisms is a key factor of their efficiency, total running time, and reliability [1-10]; it depends by far on the structure and quality of bearing joints. A serious drawback all currently-available calculation methods of sliding bearings suffer from [11-20] is that they hardly take into account the partial grease filling (a pre-failure condition).

Therefore, the development of a mathematical model for the case of partially grease-filled working gap and a guide surface coated with a light-melting metal alloy represents an advanced field of the theoretical tribology research, whose findings might be of a high importance.

2. Problem Statement

The study is focused on a wedge-shaped sliding support to consist of a slider block and a guide. A space between a slider block and a guide is supposed to be partially grease-filled; a slide block is stable, and a guide coated with a light-melting metal alloy goes around axis Oy at a velocity of u^*.

3. Input equations and boundary conditions

In the Cartesian coordinates $x'o'y'$ (Fig. 1) a linear path equation of a sliding block and melted surface is written as follow:

$$y' = h_0 + x'tg\alpha, \quad y' = -\Phi(x), \tag{1}$$

where h_0 – thickness of a grease film in the initial cross section;

Φ_0 – thickness of the melted layer in the initial cross section;

α – inclination angle of a sliding block linear path to axis ox';

$\Phi(x)$ – function characterizing a melt thickness of the guide surface.

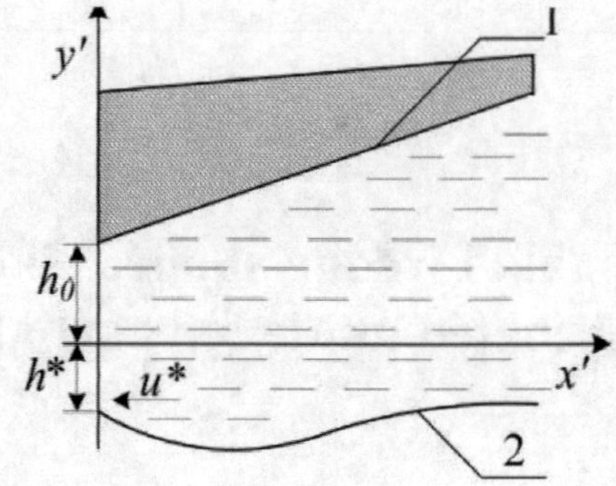

Figure 1. Calculation model

Input base equations are a non-dimensional equation of a viscous uncompressible liquid in approximation of "a thin layer", a continuity equation, and an equation to determine a melted guide edge in terms of mechanical energy dissipation rate. In Cartesian coordinates this system of equations with boundary conditions is written as follow:

$$\frac{\partial v}{\partial y}=0,\quad \frac{\partial^2 v}{\partial y^2}=\frac{dP}{dx},\quad \frac{\partial v}{\partial x}+\frac{\partial u}{\partial y}=0,\quad \frac{d\Phi(x)}{dx}=K\int\limits_{-\Phi(x)}^{h(x)}\left(\frac{\partial v}{\partial y}\right)^2 dy \tag{2}$$

$$v=0, u=0 \text{ for } y=h(x)=1+\eta x,\ v=-1, u=0 \text{ for } y=-\Phi(x),$$

$$\Phi=-\frac{h^*}{l} \text{ for } x=0,\ P(x_1)=P(x_2)=0,\ \Phi(0)=0 \tag{3}$$

where u, v – vector components of a grease medium velocity;

P – hydrodynamic pressure in a greasing layer;

$K=\frac{2\mu u^*}{h_0 L'}$ – parameter to depend on melting and mechanical energy dissipation rate;

L' – specific heat for a volume unit;

x_1 and x_2 – angular coordinates of free surface ends in the grease, respectively.

The relation between non-dimensional and dimensional values complies with formulae:

$$x'=lx,\quad y'=h_0 y,\quad v_{x'}=u^* v,\quad v_{y'}=u^*\frac{h_0}{l}u,\quad P'=P^*P,\quad P^*=\frac{\mu u^* l}{h_0^2} \tag{4}$$

The boundary conditions for u and v on the contour line $y=-\Phi(x)$ are formulated as follow:

$$\begin{gathered} v(0-H(x))=v(0)-\left.\left(\frac{\partial v}{\partial y}\right)\right|_{y=0}\cdot H(x)-\left.\left(\frac{\partial^2 v}{\partial y^2}\right)\right|_{y=0}\cdot H^2(x)+\ldots=-1 \\ u(0-H(x))=u(0)-\left.\left(\frac{\partial u}{\partial y}\right)\right|_{y=0}\cdot H(x)-\left.\left(\frac{\partial^2 u}{\partial y^2}\right)\right|_{y=0}\cdot H^2(x)+\ldots=0 \end{gathered} \tag{5}$$

An analytical solution to a system of differential equations (2) with regard to boundary conditions (3) and (5) is found in series in powers of a small parameter K:

$$\begin{gathered} v(x,y)=v_0(x,y)+Kv_1(x,y)+K^2v_2(x,y)+\ldots \\ u(x,y)=u_0(x,y)+Ku_1(x,y)+K^2u_2(x,y)+\ldots \\ P(x)=P_0(x)+KP_1(x)+K^2P_2(x)+\ldots \\ \Phi(x)=0-K\Phi_1(x)-K^2\Phi_2(x)-K^3\Phi_3(x)-\ldots \end{gathered} \tag{6}$$

Substituting (6) into the system of differential equations (2) ant taking into account boundary conditions (3), following equations are obtained with boundary conditions:

- for zero approximation:

$$\frac{\partial^2 v_0}{\partial y^2}=\frac{dP_0}{dx},\quad \frac{\partial v_0}{\partial x}+\frac{\partial u_0}{\partial y}=0 \tag{7}$$

$$v_0=0, u_0=0 \text{ for } y=1+\eta x, v_0=-1, u_0=0 \text{ for } y=0,\ P_0(x_1)=P_0(x_2)=0,\quad \Phi_0=\frac{h_0^*}{l} \tag{8}$$

- for first approximation:

$$\frac{\partial^2 v_1}{\partial y^2}=\frac{dP_1}{dx},\quad \frac{\partial v_1}{\partial x}+\frac{\partial u_1}{\partial y}=0,\ \frac{d\Phi_1(x)}{dx}=K\int\limits_0^{h(x)}\left(\frac{\partial v_0}{\partial y}\right)^2 dy, u_1=0, v_1=0 \quad \text{при } h(x)=1+\eta x, \tag{9}$$

$$v_1(x)=\left(\frac{\partial v_0}{\partial y}\right)\Bigg|_{y=0}\cdot\Phi_1(x),\quad u_1(x)=\left(\frac{\partial u_0}{\partial y}\right)\Bigg|_{y=0}\cdot\Phi_1(x), P_1(x_1)=P_1(x_2)=0 \tag{10}$$

A solution to the problem in zero approximation is sought as follow:

$$v_0(x,y)=\frac{\partial\psi_0(x,y)}{\partial y}+V_0(x,y),\quad u_0(x,y)=-\frac{\partial\psi_0(x,y)}{\partial x}+V_0(x,y),$$

$$\psi_0(x,y)=\tilde{\psi}_0(\xi),\quad \xi=\frac{y}{h(x)}, V_0(x,y)=\tilde{v}_0(\xi);\quad V_0(x,y)=\tilde{u}_0(\xi)\cdot h'(x) \tag{11}$$

Substituting (11) into the system of differential equations (7) with regard to boundary conditions (8), a system of differential equations is obtained:

$$\psi_0'''(\xi)=\tilde{c}_2;\quad \tilde{v}_0''(\xi)=\tilde{c}_1;\quad \tilde{u}_0'(\xi)-\xi\tilde{v}_0(\xi)=0, \frac{dP_0}{dx}=\frac{\tilde{c}_1}{h^2(x)}+\frac{\tilde{c}_2}{h^2(x)} \tag{12}$$

and boundary conditions:

$$\tilde{\psi}_0'(0)=0, \tilde{\psi}_0'(1)=0, \tilde{u}_0(1)=0, \tilde{v}_0(1)=0, \tilde{u}_0(0)=0,\quad \tilde{v}_0(0)=-1, \int\limits_0^1\tilde{v}_0(\xi)d\xi=0, P_0(x_1)=P_0(x_2)=0 \tag{13}$$

Taking integral of the equation (12) in terms of (13), the result is:

$$\psi_0'(\xi)=\frac{\tilde{c}_2}{2}\left(\xi^2-\xi\right),\quad \tilde{v}_0(\xi)=\tilde{c}_1\frac{\xi^2}{2}+\left(1-\frac{\tilde{c}_1}{2}\right)\xi-1,\quad \tilde{c}_1=-6 \tag{14}$$

Following the formula $P_0(x_1)=P_0(x_2)=0$ within the accuracy of the second order of smallness $o\left(\eta^2\right)$ for $\tilde{c}_2$, it is obtained:

$$\tilde{c}_2=6\left(1+\frac{\eta}{2}(x_2-x_1)\right) \tag{15}$$

in terms of (15) for P_0 it is written:

$$P_0=3\eta\left((x-x_1)^2-(x-x_1)(x_2-x_1)\right) \tag{16}$$

To determine $\Phi_1(x)$ in terms of (14), an equation is formulated:

$$\frac{d\Phi_1(x)}{dx}=h(x)\int\limits_0^1\left(\frac{\tilde{\psi}_0''(\xi)}{h^2(x)}+\frac{\tilde{v}_0(\xi)}{h(x)}\right)^2 d\xi \tag{17}$$

Taking integral of the equation (17), the result is:

$$\Phi_1(x)=\int\limits_0^x\frac{\Delta_1 dx}{h^3(x)}+\int\limits_0^x\frac{\Delta_2 dx}{h^2(x)}+\int\limits_0^x\frac{\Delta_3 dx}{h(x)} \tag{18}$$

$$\text{where } \Delta_1 = \int_0^1 \left(\tilde{\psi}''(\xi)\right)^2 d\xi = \frac{\tilde{c}_2^2}{12};\quad \Delta_2 = \int_0^1 2\psi_0''(\xi)\cdot \tilde{v}_2'(\xi) d\xi = \tilde{c}_2,\quad \Delta_3 = \int_0^1 \left(\tilde{v}_0'(\xi)\right)^2 d\xi = 4 \tag{19}$$

Seeking for a solution to the equation (18)-(19) within the accuracy of the second order of smallness $o\left(\eta^2\right)$, it is obtained:

$$\Phi_1(x) = 13x - \frac{25}{2}\eta x^2 + 9\eta\left(x_2 - x_1\right) \tag{20}$$

Taking into account (20) and boundary conditions (10), it is written as follow:

$$v_1 = 0, u_1 = 0 \text{ for } y = 1 + \eta x, u_1 = \left.\left(\frac{\partial u_0}{\partial y}\right)\right|_{y=0} = 0,$$

$$v_1(x)\big|_{y=0} = \frac{-39}{2}\eta x\left(2x - x_1 - x_2\right) - \frac{51}{2}\eta x^2 - 9\eta x\left(x_2 - x_1\right) + 13x,\ P_1(x_1) = P_1(x_2) = 0. \tag{21}$$

A solution to the equation of a viscous uncompressible liquid for the first approximation (9) and boundary conditions (21) brings about:

$$v_1(x) = \frac{dP_1}{dx}\cdot\frac{y^2}{2} - \frac{\frac{-39}{2}\eta x\left(2x - x_1 - x_2\right) - \frac{51}{2}\eta x^2 - 9\eta x\left(x_2 - x_1\right) + 13x}{1 + \eta x}\cdot y - \\ - \frac{dP_1}{dx}\cdot\frac{1+\eta x}{2}\cdot y - \frac{39}{2}\eta x\left(2x - x_1 - x_2\right) - \frac{51}{2}\eta x^2 - 9\eta x\left(x_2 - x_1\right) + 13x \tag{22}$$

To estimate the hydrodynamic pressure a Leibniz formula vs. parameters is used

$$\frac{\partial}{\partial x}\int_0^{1+\eta x} v_1(x)dy = \frac{\partial}{\partial x}\left[-\frac{1}{12}\frac{dP_1}{dx}\left(1+\eta x\right)^3 - \frac{39}{2}\eta x\left(2x - x_1 - x_2\right) - \frac{51}{2}\eta x^2 - 9\eta x\left(x_2 - x_1\right) + \frac{13}{2}x\left(1+\eta x\right)\right] = 0 \tag{23}$$

Solving an equation (23) for the hydrodynamic equation and boundary conditions (10), it is obtained:

$$P_1 = 12\left[\frac{39}{4}\eta\left(x_2^2 - x^2\right)\left(x_1 + x_2\right) - \frac{155}{6}\eta\left(x_2^3 - x^3\right) - \frac{9}{2}\eta\left(x_2^2 - x^2\right)\left(x_2 - x_1\right) + \frac{13}{4}\left(x_2^2 - x^2\right)\right] \tag{24}$$

4. Results and Discussions

In terms of (7), (8), (16) and (24) for bearing capacity and friction force it is obtained:

$$w = P^*\cdot l\int_0^1\left(P_0 + KP_1\right)dx = \frac{\mu u^* l^2}{h_0^2}\left[3\eta\left(x_1 - x_2\right)^3 + 12\left(x_1 - x_2\right)\left(\eta\left(\frac{147}{24}x_2^3 + \frac{81}{2}x_1^3 + \frac{589}{24}x_1x_2^2 + \frac{13}{2}x_1^2x_2\right) + \frac{13}{12}\left(2x_2^2 - x_1x_2 - x_1^2\right)\right)\right]$$

$$L_{mp} = \mu\int_{x_1}^{x_2}\left[\left.\frac{\partial v_2}{\partial y}\right|_{y=0} + K\left.\frac{\partial v_1}{\partial y}\right|_{y=0}\right]dy = \frac{\mu u^* l^2}{h_0}\left(x_1 - x_2\right)\times$$

$$= \times\left[\frac{7\eta}{2}x_1 - \frac{\eta}{2}x_2 - 1 + 6K\left[\left(-\frac{611}{24}x_2^3 + \frac{13}{4}x_1^3 - \frac{147}{24}x_2^2x_1 + \frac{39}{4}x_1^2x_2 + \frac{3}{2}x_2 - 6x_1 + \frac{7}{12}x_2^2 - \frac{32}{24}x_1^2 + \frac{157}{12}x_2x_1 - \frac{9}{2}x_2 + \frac{155}{4}\right)\eta - \\ - \frac{26}{12}x_2^2 + \frac{13}{6}x_1x_2 + \frac{13}{12}x_1^2 + \frac{13}{2}x_2 + \frac{13}{2}x_1\right]\right] \tag{25}$$

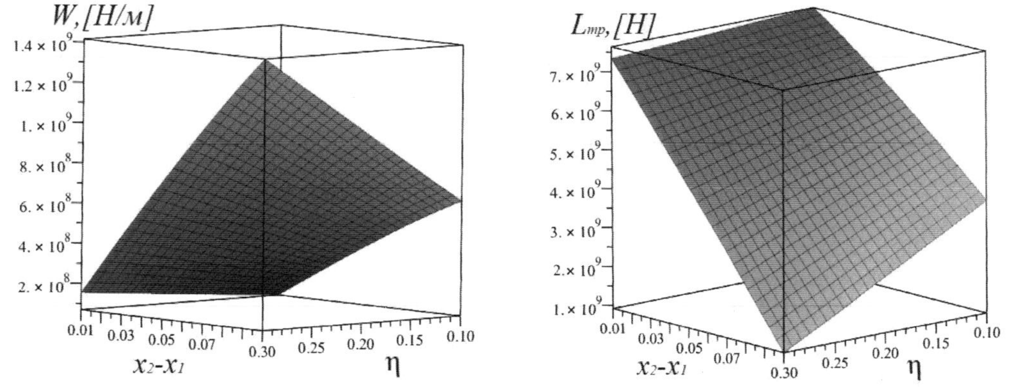

Figure 2. Bearing capacity and friction force vs. length of a loaded zone and eccentricity η

5. Experimental research

Using calculation models developed in the theoretical part, an experimental study is carried out. The most important outcome to emerge from the study is a future application sphere of the designed tribosystem outlined in the form of loading and velocity conditions, as well as principal tribotechnical characteristics.

The experimental study was focused on a sliding bearing with a light-melting metal coating for the case of a partially grease-filled gap (a pre-failure condition).

The experiment findings were used to reveal a period of a hydrodynamic friction mode. A series of experiments has confirmed the validity of developed theoretical calculation models and data on their numerical analysis within the scope of design and performance parameters of tribosystems coated with a light-melting metal, so theoretical and experimental results demonstrate their convergence.

An analysis of the correlations and graphs obtained allows the following conclusions:

1. An expansion of the loaded zone $(x_2 - x_1)$ increases significantly the bearing capacity, whereas the friction falls; this fact is in line with the parameters under consideration, outcomes of other researchers, verifying the efficiency of designed models, as a consequence.
2. If a guide surface is coated with a light-melting metal alloy, the tribotechnical testing carried out by an end-type friction machine has shown the considerable reduction (to 27%) of a wear spot and the longer stability of a greasing film (to 48%) in comparison with a traditional grease.

Conclusion

More specified calculation models of wedge-shaped sliding supports make it possible to change a ratio between its bearing capacity and friction coefficient via varying a light-melting alloy coating. The study has established the acceptable convergence of theoretical and experimental results, supporting therefore theoretical conclusions.

References

[1] I Levanov, A Doykin, E Zadorozhnaya, R Novikov 2017 Investigation of Antiwear Properties of Lubricants with the Geo-Modifiers of Friction *Tribology in Industry* **39 (3)** 302-306.

[2] I V Mukchortov, K A Pochkaylo, E A Zadorozhnaya 2016The Influence of Anti-Wear Additives on the Bearings Hydro-Mechanical Characteristics *Procedia Engineering* **150** 607-611

IOP Conf. Series: Materials Science and Engineering **941** (2020) 012064 doi:10.1088/1757-899X/941/1/012064

[3] E Zadorozhnaya 2013 The Research of Non-Newtonian Properties and Rheology of Thin Lubricant Layers in Hydrodynamic *Journal Bearings, Society of Tribologists and Lubrication Engineers Annual Meeting and Exhibition* 95-97

[4] V V Vasilenko, E O Lagunova, M A Mukutadze, V M Prikhodko 2017 *International Journal of Applied Engineering Research* **12 (19)** 9138-9148

[5] M A Mukutadze, E O Lagunova, A N Garmonina, V V Vasilenko Radial 2018 Slip Bearing with a Pliable Supporting Surface *Russian Engineering Research* **38 (3)** 166-171

[6] M A Mukutadze 2016 Radial Bearing with Porous Elements *Procedia Engineering* **150** 559-570

[7] M A Mukutadze, E O Lagunova, V V Vasilenko 2018 Development of the Design Model of a Hydrodynamic Lubricating Material Formed during Melting of the Axial Bearing, in the Presence of Forced Lubrication *Journal of Machinery Manufacture and Reliability* **47** 271-277

[8] K S Akhverdiev, E O Lagunova, M A Mukutadze 2018 Calculated Model of Wedge-Shaped Sliding Supports in Turbulent Friction Regime *Advances in Engineering Research* (AER), **157** 346-353

[9] K S Akhverdiev, E O Lagunova, M A Mukutadze 2018 Calculated Model of Wedge-Shaped Sliding Supports Taking into Account Rheological Properties of Viscoelastic Lubricant *Advances in Engineering Research* **158** 246-253.

[10] E O Lagunova, M A Mukutadze 2018 Radial Friction Bearings Conditioned by Melt *Proceedings of the 4th International Conference on Industrial Engineering, Lecture Notes in Mechanical Engineering*. 897 – 910

[11] E O Lagunova, M A Mukutadze 2019 Calculation of a Radial Slider Bearing with a Fusible Coating *Journal of Friction and Wear* **40** (1) 88-94.

[12] M A Mukutadze, A M Mukutadze, V V Vasilenko Simulation 2019 Model of Thrust Bearing with a Free-Melting and Porous Coating of Guide and Slide Surfaces *IOP Conf. Series: Materials Science and Engineering* **560** DOI:10.1088/1757-899X/560/1/012031

[13] A N Opatskikh, M A Mukutadze, A M Mukutadze 2019 V-Shaped Journal bearings Using Micropolar Lubricants Caused by a Melt Accounting for the Dependence of Lubricant Viscosity and Porous Layer Permeability on Pressure *IOP Publishing Journal of Physics: Conference Series* **1353** 012025 DOI:10.1088/1742-6596/1353/1/012025

[14] M A Mukutadze, V V Vasilenko, A M Mukutadze, A N Opatskikh 2019 Mathematical Model of a Plain Bearer Lubricated with Molten Metal *IOP Conf. Series: Earth and Environmental Science* **378** 012021 IOP Publishing DOI:10.1088/1755-1315/378/1/012021

[15] K S Akhverdiev, E V Alexandrova, E E Kruchinina, M A Mukutadze 2010 Stratified Course Of Two-Layer Lubrication In The Gap Of A Threshold Bearing, With An Increased Bearing Ability *Vestnik DGTU* **Vol 10 2 (45)** 217-223

[16] K S Akhverdiev, E E Aleksandrova, M A Mukutadze, B E Kopotun 2009 Stratified Course Of Two-Layer Lubrication In The Radial Bearing Clearance With An Increased Bearing Capacity And Damping Properties *Vestnik RGURS* **4 (36)** 133-139

[17] K S Akhverdiev, E E Aleksandrova, M A Mukutadze 2010 Stratified Two-Layer Lubrication Current In The Clearance Of A Complexly Loaded Radial Bearing Of A Final Length With An Increased Loading Capacity *Vestnik RGURS* **1 (37)** 132-137

[18] K S Akhverdiev, A Yu Vovk, M A Mukutadze, M A Savenkova 2008 Analytical Method For Forecasting Values Of Micropolar Lubrication Criteria For A Stable Mode Of Operation Of A Radial Sliding Bearing *Friction and wear* **29 (2)** 184-191.

[19] Akhverdiev K S, M A Mukutadze, V A Zamshin, I S Semenko 2009 Hydrodynamic Calculation Of A Radial Sliding Bearing Operating In A Turbulent Friction Mode When The Viscoelastic Grease Is Not Completely Filled *Vestnik of mechanical engineering* **7** 11-17.

[20] V A Kohanovskij, M A Mukutadze 2001 Matrix materials of antifriction composites *Vestnik of DSTU* **1(2)** 51

Assessing the forcing level of air cooled diesel engine considering parameter spread of boosting system

A M Lartsev, E A Salykin and E A Fedyanov
Volgograd state technical university (VSTU), Lenin Avenue, 28, Volgograd, 400005, Russia
E-mail: esalykin@gmail.com

Abstract. Spread of parameter values of piston internal combustion engine boosting parts is normally due to extreme deviations (spread, error margin) of original parts and quality of assembly. Spread of relation of boosting pressure to counter pressure at outlet of turbo compressors P_k/P_t of the same type affects not only the engine parameters, but also thermal expansion of cylinder-piston group parts and cylinder head. At stable relation value of P_k/P_t, boosting pressure and maximal air flow, turbo compressor allows us to reduce the thermal factor of combustion chamber and cylinder head due to higher values of air excess ratios. This is very important, first of all for air cooled engine as in case of dramatic surge of its thermal factor we face a risk of cracks occurring in inter valve and inter valve- nozzle connectors of cylinder head fire bottom, cracking of setting bevels at valves of gas distribution mechanism. We analyzed the impact of technological spread of boosting details parameter on the value of achieved forcing limit. Impact analysis of technological spread of turbo compressor parameters on diesel engine performance was done on the basis of test results of ten B-400 engines, fitted with randomly selected pairs of TKP8,5-C7 turbo compressors out of a lot of 20 pieces. For each turbo compressor we determined preliminarily characteristics and efficiency factors $\eta_к$, of turbine $\eta_т$ and of a unit as a whole $\eta_{тк}$.

1. Introduction

Power forcing of air cooled diesel engines is limited by maximum allowed thermal condition of parts in cylinder-piston group and cylinder head. Analysis of cylinder temperature fields, piston and cylinder head of air cooled engine at various levels of forcing showed that [1 - 8] maximal temperature gradients of cylinder head are almost three times as high as maximal temperature gradients of cylinder or piston. Based on calculations and test results of B-400 diesel engines it was established that with sufficient reserve of rigidness and hardess of framework and bearing capacity of crankshaft, the factor, limiting permitted forcing level of air cooled engine is thermal condition of cylinder head, having most non-uniform temperature field and levels of thermal stress as compared to parts of cylinder-piston group.

As shown by experience, thermo-cyclic destruction of cylinder head begins either in the point of maximal temperature or close to it as long as the maximal temperature exceeds for a long time the permitted one for the material used. Spread of thermo-cyclic crack is due to spread of material microstructure of cylinder head. Thermometering results of the surface of cylinder head fire bottom of B-400 diesel engine prove that the maximal temperature on the cylinder head bottom seriously affects the maximal temperature gradient. Based on the above, the maximal temperature of cylinder head fire bottom of air cooled diesel engine can be a criteria for finding a forcing limit of such engine.

IOP Conf. Series: Materials Science and Engineering **941** (2020) 012065 doi:10.1088/1757-899X/941/1/012065

Higher size of air cooled diesel engine aggravates the problem of maintaining permitted thermal factor of the structure, as the cooling area of cylinder head, limiting heat removal is rising in proportion to a cube of the cylinder diameter, whereas as heat release increases in proportion to the diameter cube.

For each particular structure of diesel engine, establishing limits of possible forcing is an important practical task, which, if not solved, prevents creation of a feasible engine with maximal high running parameters. Especially acute is the problem of availability and so determining permitted forcing levels for air cooled diesel engines, operating in harsh climatic conditions becomes a critical issue.

We proposed methods of determining the ultimate forcing level of air cooled diesel engine of certain design by boosting. The methods were verified at B-400 engine, having the largest size of all serially produced and being produced now air cooled diesel engines.

2. Object of research

The object of research is a diesel air cooled engine B-400. Other technical characteristics of engine B-400 are given in table 1.

Table 1. Key technical characteristics of engine B-400.

Description of parameter	Parameter value
Stroke	4 strokes
Number and position of cylinders	8, V-90°
Diameter of cylinder, mm	150
Piston stroke, mm	160
Liter capacity, l	22.6
Power, kW	307+15
Rotation frequency, min-1	1700±20
Specific fuel consumption g/kW*h, not more than	210
Boosting pressure, MPa	0.19-0.22

3. Ensuring reliable operation of air cooled engine at forcing limit

When determining limits of possible forcing by boosting of the diesel engine we note high sensitivity of thermal condition of the parts of cylinder- piston group and cylinder head to relation of boosting pressure Pk and pressure Pt of off-gases upstream of turbine [9]. With the increase in relation Pk/Pt at preset forcing level, the excess air factor goes up, the mean integral temperature of the working medium goes down and so does accordingly the temperature of combustion chamber surfaces.

In serial production there is always some spread of indices in a lot of boosting parts, installed in an engine. As an example you can see on Figure 1 an experimentally produced spread of values of relation Pk/Pt for the right and left rows of cylinders in B-400 diesel engine with installed on it a ten pairs of 8,5C-7 turbo compressor, selected randomly from a lot of this type turbo compressors.

It can be assumed that to ensure reliable operation, the limit of air cooled diesel engine forcing should be determined based on smallest values of relation P_k/P_t. As you can see, the maximal deviation of relation P_k/P_t from the relative mean value towards an adverse side, i.e. towards reduction of the relation is 3.5%.

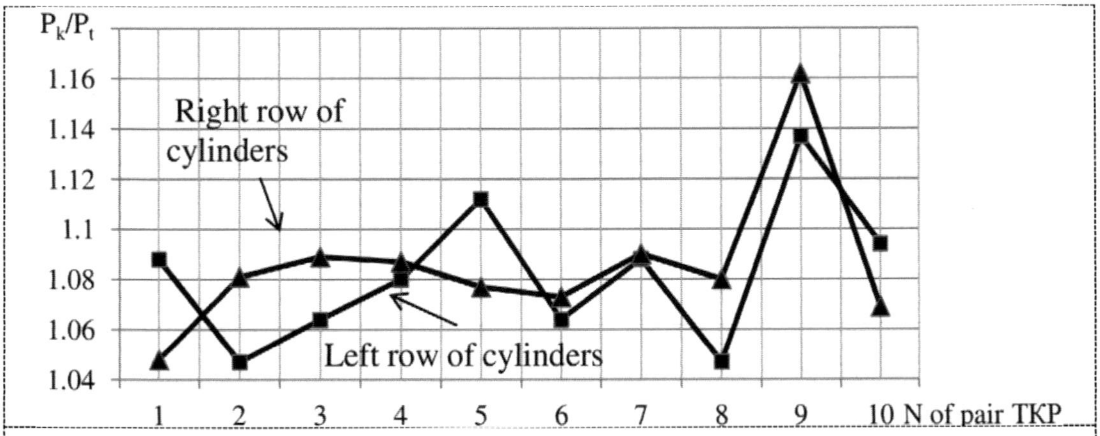

Figure 1. Spread of relation P_k/P_t of diesel engine B-400. N of pair TKP – No. of turbo compressor pair, installed on the engine.

Figure 2-5 below show processing results of experimental data on testing B-400 diesel engine, equipped with ten different pairs of turbo compressors TKP8,5C-7.

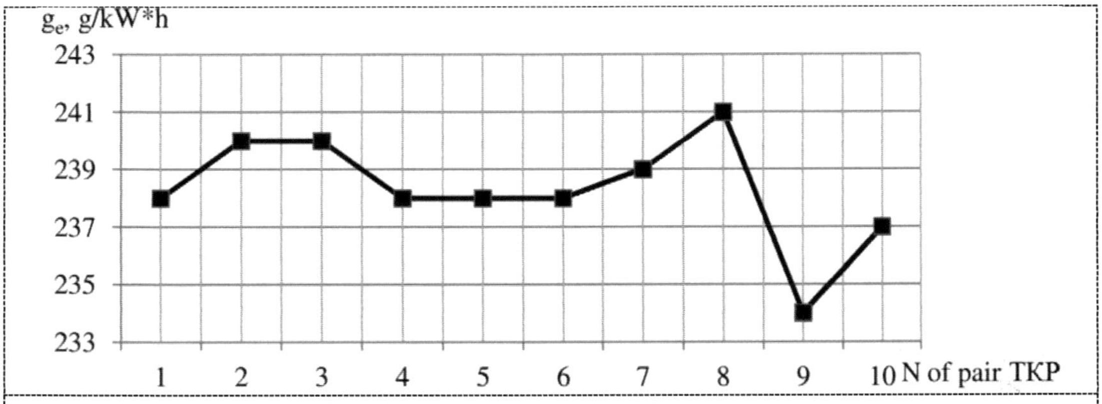

Figure 2. Spread of excess air factor α.

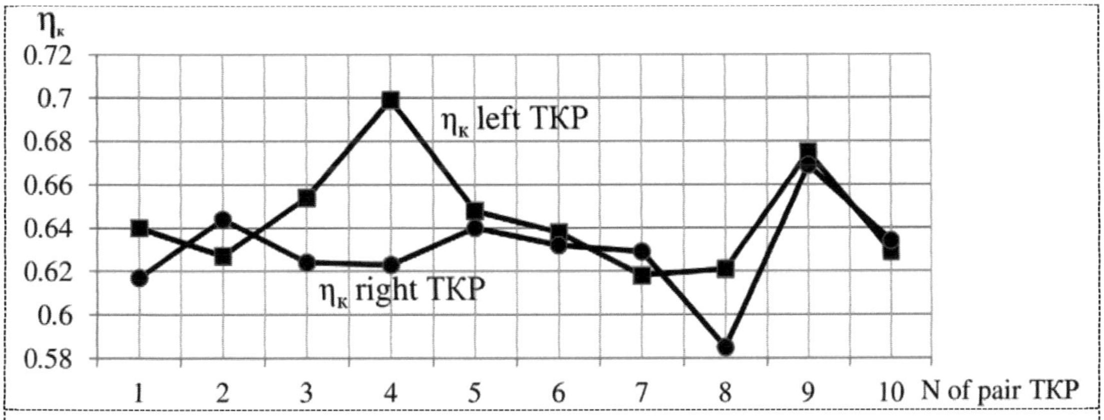

Figure 3. Efficiency factor of used compressors TKP8,5-C7.

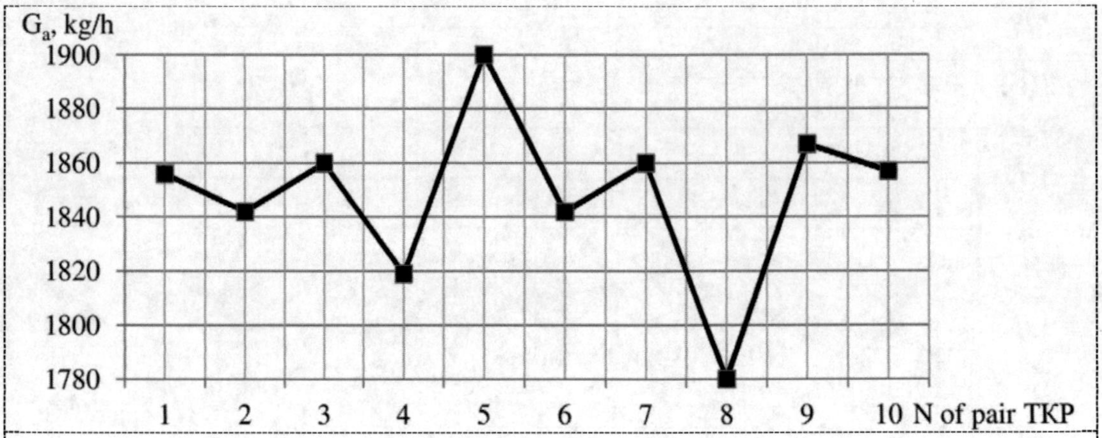

Figure 4. Spread of air consumption per hour by engine G_a.

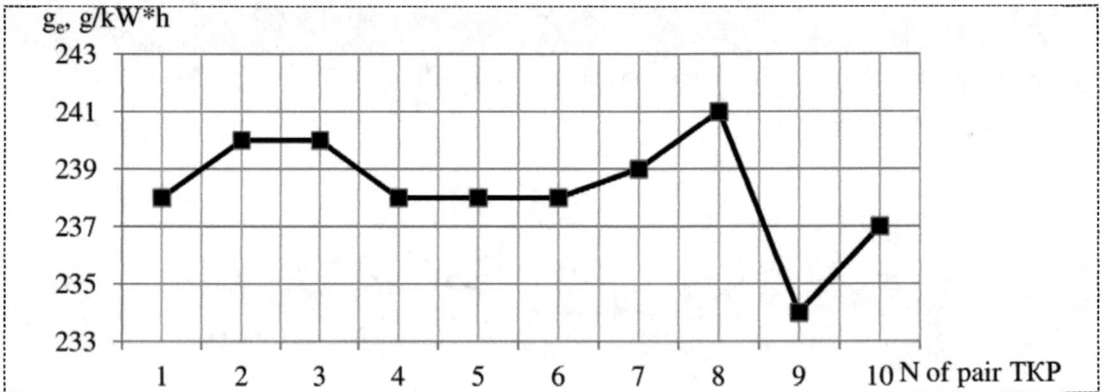

Figure 5. Spread of specific fuel consumption g_e.

We have determined that air cooled [9], diesel engines are especially sensitive to relation Pk/Pt. This means that the engine is at the limit of forcing by thermal condition, technological tolerance of this relation towards an adverse side may have a serious impact on increasing the chance of engine failure due to thermo- cyclic destruction of the cylinder head. In view of the above we made a calculated assessment of technological spread of influence Pk/Pt on thermal condition of the cylinder head. Results of calculations are given in table 2 and in Figure 6.

Table 2. Calculation results of diesel engine B-400 at different relations of value P_k/P_t.

N_e[a] (kW)	N[b] (min^{-1})	P_k/P_t	α	G_a (kg/hour)	P_z[c] (MPa)	T_{wch}[d] (K)	T_{max}[e] (K)
275	1700	0.950	1.74	1609	9.29	488	590
275	1700	1.162	1.86	1693	9.47	479	576

[a]Effective power of engine.

[b]Rated rotation frequency of crankshaft.

[c]Maximal pressure of burning.

[d]Weighted average temperature of cylinder head fire bottom.

[e]Maximal temperature of cylinder head fire bottom.

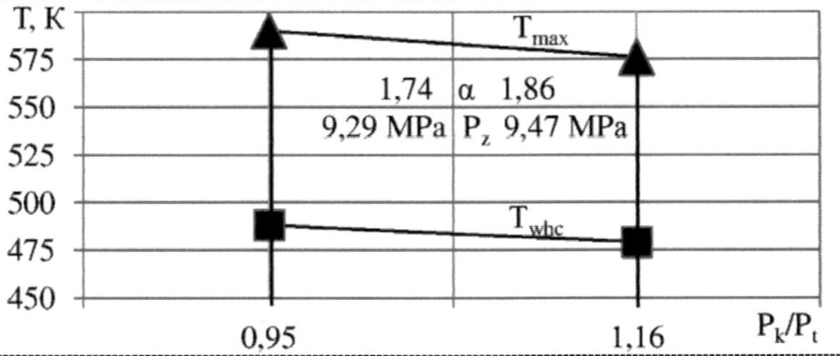

Figure 6. Temperature of cylinder neck bottom at variation of relation P_k/P_t.

4. Conclusion

When using turbo compressors of the same model we can witness due to technological spread of parameters an obvious variation of cylinder head temperature field. Technological spread of parameters TKP8.5C-7 shows as an example significant (within 6%) effect of relation P_k/P_t on maximal temperature of cylinder head fire bottom.

If an air cooled engine is at the ultimate level of forcing, we should consider it and for reliable operation the forcing limit must be established based on minimal values of relation P_k/P_t.

References

[1] Krstic B, Rasuo B, Trifkovic D, Radisavljevic I, Rajic Z and Dinulovic M 2013 *Engineering Failure Analysis* **34** 335-49

[2] Mărdărescu V G, Ispas N and Năstăsoiu M 2014 *INMATEH - Agricultural Engineering* **42** 137-44

[3] Shchinnikov P A, Mikhaylenko A I and Sinelnikov D S 2017 *Proceedings - 2016 11th International Forum on Strategic Technology, IFOST 2016* **2017** 7884233

[4] Tholen P and Killmann I 2017 *Proceedings - 2016 11th* Int Congr on Combust Engines **12** 505-43

[5] Mitianiec W and Buczek K 2007 *Journal of KONES Powertrain and Transport* **14** 3

[6] Honda R&D Co 2014 *SAE International Journal of Engines* **8** 120-34

[7] Ji F Z, Du F R, Xu B and Yang S-C 2014 *Neiranji Gongcheng/Chinese Internal Combustion Engine Engineering* **35** 35-40

[8] Yan T, Yobby J and Vundavilli R 2014 *ASME International Mechanical Engineering Congress and Exposition, Proceedings (IMECE)* **6A** 111737

[9] Vasiliev A V, Lartsev A M and Fedyanov E A 2019 *Lecture Notes in Mechanical Engineering* **2020** 159-65

IOP Conf. Series: Materials Science and Engineering **941** (2020) 012066 doi:10.1088/1757-899X/941/1/012066

Tribological tests of frictional carbon/carbon composite in braking mode

V V Alisin

Blagonravov Institute of Mechanical Engineering, Russian Academy of Sciences Russia

E-mail: vva-imash@yandex.ru

Abstract. The article discusses the issues of friction and wear testing of a carbon-carbon composite of tribological purpose. The experiments were performed on a bench, the work of which is based on the absorption of kinetic energy of rotating masses with materials intended for use in brake devices. The friction coefficient and wear rate of the friction carbon composite in braking mode are investigated. Ring samples are divided into 6 sectors, in each of which, after testing, the parameters of the roughness of the friction surfaces and wear during one braking are determined. The value of linear wear was determined as the average value over 6 sectors for rotating and stationary samples. Changes in the structure of friction surfaces are investigated. Metrological support of the experiments was carried out by statistical processing of the results. The errors in measuring the coefficient of friction and wear rate are determined. High resistance to energy absorption and thermal shock of the studied frictional material is established, which open up the prospects for the mass use of C/C in transport brake devices.

1 Introduction

Current trends in the development of vehicles are based on an increase in speed and traffic safety, which determines the need for improving brake devices. To increase braking efficiency, new materials are developed and existing ones are improved.

Promising friction materials include carbon-carbon composite materials (C/C) because they have unique specific strength characteristics. Much attention is paid to improving tribological properties and test methods. The wear of the friction material and its effect on tribological characteristics were examined in study [1]. To study the tribological characteristics of friction materials, laboratory friction machines operating according to the disk-finger scheme are used [2 — 4].

The structure of the friction layer of carbon-carbon composites after braking tests was studied in [5]. It has been found that carbon fiber tends to wear out into small fragments along the fiber axis. Nanosized carbon particles were discovered in the surface layer, which during friction acted as a lubricant. The use of short carbon fibers to improve the tribological properties of carbon composites was tested [6]. The study of mechanical and tribological properties showed that a composite with a fiber content of 30% demonstrates the highest bending strength (201 MPa) and shear strength (116 MPa) and has the highest wear resistance. It was established in [7] that the effect of carbon fiber on increasing toughness and thermal conductivity is more noticeable than that of aramid fiber. In [8], the influence of the surface roughness of the friction material on the tribological characteristics of disc brakes is shown.

The influence of braking pressure and braking speed on the tribological properties (C / C) of composites was studied in [9]. At higher braking pressure or braking speed, wear increases due to scaly damage to the friction film and high oxidation losses. In [10], the effect of moisture on the friction coefficient of a composite disk was experimentally established. Evidence of a possible transition to a hydrodynamic regime of friction is given. In [11], the braking characteristics of carbon composites, which are promising for high-speed trains and cars, were investigated. It was found that the introduction of a plastic phase into the brake pads favorably reduced the friction coefficient to an acceptable range and effectively increased wear resistance. In [12], a study was made of friction, wear, thermal properties, and temperature fields of three C / C composites with different matrix-carbon textures.

In [13], the tribological properties of carbon composites under dry friction and in an aqueous medium were studied. It was found that carbon composites meet the technical conditions for use in aircraft brakes. In [14], the mechanism of brake disc wear was investigated. Recommendations on hardening discs are proposed to reduce abrasive wear. In [15], studies were performed by the spectroscopic method, which made it possible to identify the physicochemical mechanisms of the degradation of composites. An important stage in the creation of new friction materials is tribological testing. More reliable and accurate tests are carried out on models as close as possible to real braking devices, especially in terms of modeling thermal fields.

The practical need for the results of tribological tests in braking mode determines the relevance of the work.

2. Purpose of work

Determine the friction-wear characteristics of the carbon composite and their compliance with the requirements of the technical conditions for materials for use in brake devices of vehicles.

3.Materials and equipment.

The samples are made of a carbon-carbon composite material in the form of ring disks with an outer diameter of 75 mm and an inner 53 mm separated by radial grooves into 6 sectors with a total of 3 friction pairs of one material.

Friction model tests were carried out on an IM-58 friction machine, which makes it possible to simulate on the samples the temperature field formed in materials during braking. The operation of a friction machine is based on the absorption of kinetic energy of rotating masses by friction materials of brake devices in braking mode. The kinetic energy of the masses is selected from the condition that the load-speed parameters approach the typical values of the friction pairs of real brakes. A feature of the braking mode is the rapid change in temperature in the surface layer of materials. The test procedure is based on the twist reduction of rotating masses with an inertia moment of 0.505 kg.m2 up to 6000 rpm, then the drive is turned off and the friction pair is loaded with an axial force of 160 kgf. Previously, a friction pair is wearing-in by a 3-fold recycling of the braking. The number of wearing-in brakes is 4, the number of valid brakes is 15 brakes.

Roughness measurement according to GOST 27964-88, GOST 2789-7 3 on the profilometer model 130.

4. Results and Discussion

The tribological characteristics of friction pairs depend on the roughness of the friction surfaces. Figure 1 shows a typical profilogram of the wearing-in friction surface.

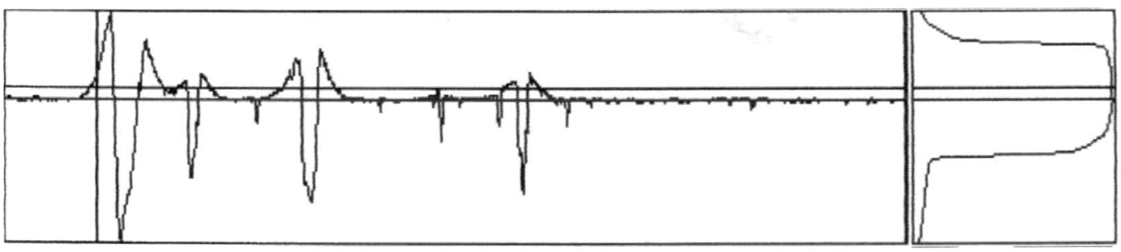

Figure 1. Profilogram of the C/C composite surface in sector 1 and the type of the reference curve.

On the profilogram, individual grooves are clearly visible, the formation of which is associated with the formation of wear products in the contact surfaces. Analysis of the state of the friction surface in all sectors (table 1) according to the accepted roughness parameters shows that the parameters are very close and the process of friction and wear is the same over the entire friction surface.

IOP Conf. Series: Materials Science and Engineering **941** (2020) 012066 doi:10.1088/1757-899X/941/1/012066

Table 1. Roughness parameters of the wearing-in surfaces

letter designation	parameter	dimension	Sector 1	Sector 3	Sector 5
	Surface roughness class				
Ra	Deviations of the arithmetic means	micrometr	2.37	2.36	2.48
Rz	Roughness height by 10 points	micrometr	32.3	34.9	46.1
Rmax	Maximum roughness height	micrometr	51.0	50.1	81.6
Sm	Mean pitch of roughness	micrometr	26.0	26.2	23.1
S	Mean pitch of local peak of profile	micrometr	5.78	5.92	5.59
Rpk	Height of surface asperity	micrometr	15.6	17.1	9.13
Rvk	Depth of surface valley	micrometr	28.0	30.3	63.2
tp1	Top of bearing surface	%	10.1	14.6	7.17
tp2	Bottom of bearing surface	%	92.8	89.2	94.0
Rp	Height of maximum surface asperity	micrometr	19.4	18.9	14.0

In the structure of the working surface of the samples both in the initial state (Fig. 2) and after friction (Fig. 3), there are no visible pore inclusions in the volume of this carbon-carbon composite material, which indicates good wettability of the carbon fibers by the binder and good adhesion bond at the interface between the matrix and C/C filler. This structure provides high stability and reproducibility of test results.

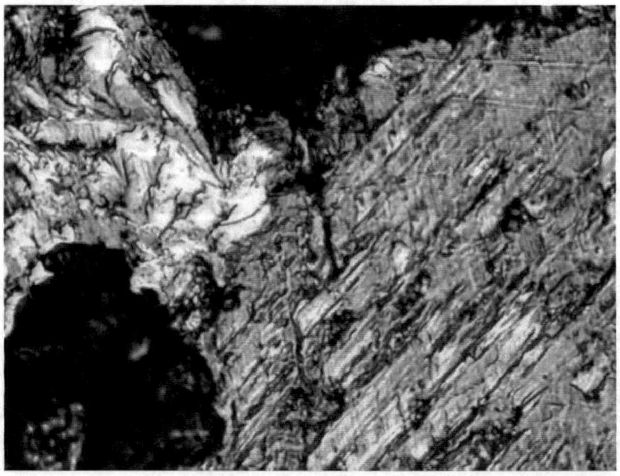

Figure 2. A picture (×50) of the working surface of the disks before friction tests (rotating sample)

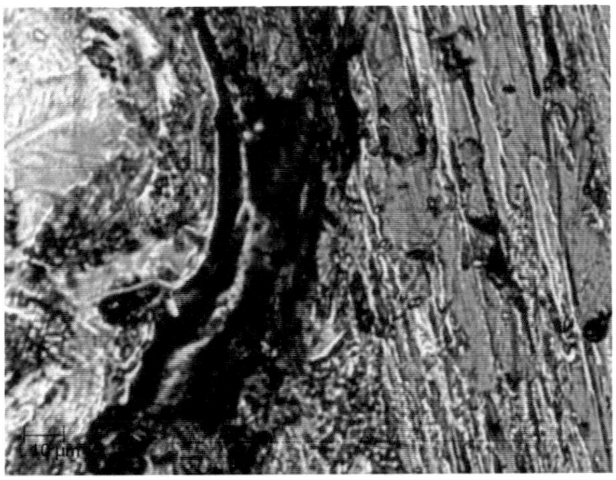

Figure 3. A picture (×50) of the working surface of the discs after friction tests (rotating sample)

During braking tests, temperature was constantly monitored. In between each test, the temperature of the cooling samples was measured (not more than 50 ° C). After each control test on the samples, thickness measurements were carried out in each of the 6 sectors (table 2). The linear wear value was determined as the average value over 6 sectors for rotating and stationary samples. The linear wear of the sample was determined by the difference in thickness of the sample before the valid braking and after the test.

Table 2. The results of a single measurement of linear wear (mm) for 6 sectors and contain note [a)]

№	view	1	2	3	4	5	6	the average
1	n[b)]	13.905	13.915	13.912	13.918	13.916	13.904	13.912
	v[c)]	13.905	13.915	13.912	13.917	13.915	13.902	13.911
	n	13.924	13.921	13.914	13.918	13.941	13.969	13.931
	в	13.924	13.920	13.914	13.918	13.941	13.969	13.911
2	n	13.905	13.917	13.912	13.916	13.914	13.902	13.911
	v	13.926	13.919	13.913	13.918	13.941	13.979	13.933
3	n	13.906	13.917	13.912	13.916	13.914	13.902	13.911
	v	13.924	13.918	13.912	13.916	13.941	13.977	13.931
4	n	13.905	13.917	13.914	13.914	13.914	13.902	13.911
	v	13.924	13.918	13.911	13.914	13.941	13.970	13.930
5	n	13.905	13.915	13.914	13.914	13.912	13.902	13.910
	v	13.922	13.917	13.911	13.914	13.932	13.969	13.928
6	n	13.902	13.913	13.914	13.914	13.912	13.902	13.910
	v	13.922	13.917	13.912	13.914	13.934	13.968	13.928
7	n	13.902	13.916	13.914	13.915	13.910	13.901	13.910
	v	13.922	13.917	13.912	13.916	13.937	13.971	13.929
8	н	13.901	13.915	13.913	13.915	13.909	13.901	13.909
	v	13.922	13.917	13.911	13.915	13.936	13.975	13.929
9	n	13.901	13.915	13.913	13.915	13.907	13.901	13.909
	v	13.922	13.916	13.911	13.915	13.935	13.973	13.929
10	n	13.901	13.915	13.913	13.914	13.909	13.901	13.909

	v	13.922	13.914	13.910	13.914	13.935	13.971	13.928
11	n	13.901	13.915	13.915	13.914	13.909	13.900	13.909
	v	13.922	13.914	13.909	13.914	13.935	13.970	13.927
12	n	13,900	13.915	13.914	13.914	13.909	13.900	13.909
	v	13.922	13.914	13.909	13.913	13.935	13.970	13.927
13	n	13.900	13.915	13.914	13.914	13.909	13.900	13.909
	v	13.922	13.914	13.909	13.913	13.935	13.960	13.926
14	n	13.898	13.912	13.914	13.914	13.909	13.900	13.908
	v	13.921	13.914	13.909	13.913	13.935	13.960	13.925
15	n	13.898	13.913	13.914	13.914	13.908	13.900	13.908
	v	13.920	13.914	13.907	13.913	13.933	13.959	13.924

[b)] n - stationary sample
[c)] v - rotating sample

Metrological provision is given in table 3, where the following symbols are accepted:
{Xi} - average linear size for a non-rotating sample, micrometr;
Sn is the standard deviation of an individual measurement, micrometr;
S is the standard deviation of the arithmetic mean, micrometr;
Δ is the random error for the accepted probability P = 0.95, the Student's coefficient is 2.6 (for n = 6), micrometr;
Δx is the accumulated error of an individual measurement (absolute error), micrometr;
μ - Relative error of an individual measurement of linear dimensions,%.

Table 3. The results of statistical measurement process data of wear of friction surfaces and contain note [a)]

Parameter	Braking 1		Braking 5		Braking 10		Braking 15	
	[b)]1n	[c)]1v	5n	5v	10n	10v	15n	15v
$\{X_i\}$	13.911	13.932	13.910	13.928	13,909	13.928	13.908	13.924
S_n	0.0061	0.0212	0.0055	0.0216	0.0064	0.023	0.0072	0,0185
S	0.0025	0.0087	0.0022	0.0088	0,0026	0.0094	0.0029	0.0075
Δ	0.0065	0.0225	0.0058	0.0229	0.0068	0.0244	0.0076	0.0196
Δ_x	0.021	0.021	0.0208	0.0304	0,0211	0.032	0.0214	0.028
μ	0.15	0.15	0.149	0.218	0.152	0.227	0.154	0.201

[b)]1n – stationary sample №1.
[c)]1v – rotating sample №2.

During each braking, the moment of friction was recorded. Based on the results of planimetric recording of the change in the moment of friction in time, the coefficient of friction was determined. It was found that the arithmetic mean value of the braking time t = 38.9; arithmetic mean value of the friction moment Mtr = 12.7; arithmetic mean value of the coefficient of friction f_{cp} = 0.248. Relative and absolute errors are given in table 4.

Table 4. Evaluation of the measurement error of the coefficient of friction

Sample material	Relative error, δE_f	Absolute error, ΔE_f	ΔF_x H	ΔF_z H	ΔR_{mp} мм	$\Delta(\Delta F_x)$, H	$\Delta(\Delta F_z)$ H	$\Delta(\Delta R_{mp})$ мм
FM 1.1	0.133	0.032	396.9	1600	32	0.05	0.5	0.05

ΔM_{tr}= 12,7 (Нм) - arithmetic mean value of the friction moment;
ΔF_x = 396.9 (H) – arithmetic mean of friction force;
ΔF_z = 1600 (H) - arithmetic mean of the normal load;
ΔR_{mp} =; 0,032 - friction groove radius.

5. Conclusion

Control tests of a friction carbon-carbon composite material on a bench simulating the distribution of heat fluxes in a real braking device showed high stability of tribological characteristics during braking. Testing the frictional heat resistance of the C/C composite showed a good reproduction of the friction coefficient, which is 0.248 ± 0.032. The C/C composite wear resistance was established with an intensity of 0.27 ± 0.02 (μm / brake.) for a non-rotating sample, 0.47 ± 0.03 (μm / brake.) for a rotating sample, which allows to increase the service life of the brakes. High resistance to energy absorption and thermal shock determines the prospects for mass application of C/C composite in transport brake devices. The reliability of the work is confirmed by the results of statistical processing of experimental data obtained in the braking mode

References

[1] Wei L, Choy Y S, Cheung C S and Jin D 2020 Wear 448–449 203215
[2] Federici M, Gialanella S, Leonardi M, Perricone G and Straffelini G 2018 Wear 410–411 202-9
[3] Mutlu I, Eldogan O and Findik F 2006 Tribology International 39(4) 317-25
[4] Österle W, Deutsch C, Gradt T, Orts-Gil G and Dmitriev A I 2014 Tribology International 73 148-55
[5] Wu S, Liu Y, Ge Y, Ran L and Yi M 2016 Tribology International 102 497-506
[6] Tang C, Dang A, Li T, Zhao T and Jiao S 2019 Tribology International 136 404-11
[7] Ahmadijokani F, Shojaei A, Dordanihaghighi S, Jafarpour E and Arjmand M 2020 Wear 452 – 453 203280
[8] Wang D W, Mo J L, Liu M Q, Li J X and Zhou Z R 2017 Wear 376–377 B15 1586-600
[9] Deng H, Li K, Li H, Wang P and Zhang L 2010 Wear 270(1) 95-103
[10] Bian G and Wu H 2015 Tribology International 92 1-11
[11] Ma X, Fan S, Sun H, Luan C and Cheng L 2020 Tribology International 142 105981
[12] Xu H, Huang B, Yi M, Xiong X and Lei B 2011 Tribology International 44(1) 18-24
[13] Xu X, Fan S, Zhang L, Du Y and Cheng L 2014 Tribology International 77 7-14
[14] Cai R, Zhang J, Nie X, Tjong J and Matthews D T A 2020 Wear 452–453 203283
[15] Marin E, Daimon E, Boschetto F, Rondinella A and Pezzotti G 2019 Wear 432–433 202969

IOP Conf. Series: Materials Science and Engineering **941** (2020) 012067 doi:10.1088/1757-899X/941/1/012067

Increasing the efficiency of agricultural transport and technological machines in the operation

G E Kokieva[1] and Y A Shaposhnikov[2]

[1]Department of Energy Supply in Agroindustrial Complex Yakut State Agricultural Academy, 3 Sergelyakhskoye Highway, Yakutsk, Republic of Sakha (Yakutia),677007, Russia

[2]Department of Automobiles and Automobile Economy Polzunov Altai State Technical University, 46 Lenin Ave, Barnaul, 656049, Russian Federation

E-mail: u_shaposhnikov@mail.ru, kokievagalia@mail.ru

Abstract. Production of commercial agricultural products is impossible without transport and technological machines and their technical-readiness, through repair and maintenance. The relevance of this direction is due to the agricultural producers' lack of necessary material and technical base, required production structures, taking into account multistructurality, and various forms of ownership. The use of transport and technological complexes and maintenance tool for their setting and adjustment allows to reduce labor costs and improve the quality of mechanized works. With the increase of commercial agricultural production technical capacity, some issues have emerged: increasing the costs of machines, increasing expenses of production per unit, and also the complicated operation system of mechanisms. As a result, the indicators of their use are not improved. Reformation of management methodsin agriculture is aimed to solve the issues of technical and economicprovision of agricultural products producers with the necessary technical resources. To solve this problem, differential standards of resource requirements should become essential, taking into account the production volume. It is necessary to increase the technical-readiness of transport and technological machines and thus the cost-effectiveness of their use.

1. Introduction

At present, the volume of works on the maintenance of agricultural machines are conducted by 60... 80%, on diagnosticsby 25%. The main reasons for this state - the farms' lack of necessary material and technical base, the poor organization of works performance, the engineers and technical workers' low level of training. The main reasons for this location are the lack of adequate material and technical base in the farms, imperfect organization of these works, and technical workers' weak level of training.

Issues of machines durability and reliability are interested not only scientists and machine engineers, but also consumers who operate machines, arrange their maintenance and repair. It can be explained by the fact that modern machines use structural members, assemblage, regulation, coloring, oiling, and other factors estimated at the different operating time, and requiring replacement or updating in contrast to the low-speed machines with a large strength margin. Unfortunately, many researchers involved in issues of reliability and durability overlook this important feature of modern technology, whereby it makes their recommendations ineffective. Moreover, the indicator of machine durability, like any other compound object, should take into account operating time of elements included in its, but also their share in the total volume and their responsibility for the machine operating time. The latter circumstances show the comparison of two machines with the same operating time, one of them often requires replacement of only small and cheap parts, and the other one requires replacement of large and expensive parts. The characteristics of machine durability and operating time can coincide only in theory, as a rule in practice they don't, machine elements have the same operating time. [1]

2. Methods of research

In agro-industrial production, machines and equipment have different operating time, for example, in crop production - seasonal, in cattlebreeding - year-round, which requires different approaches of machine organization and maintenance. Based on this, it develops a repair plan, calculates annual volume, and identifies the need for replacement parts, units, and materials. Farm specialists or service centers workers conduct maintenance of machines and equipment.

It found that the inputs efficiency to the use of one level of equipmentis different. Thus, operating C and transferred $C + E_n K$ costs reduce to a certain value and further they increase, when inputs $E_n K$ given their norm coefficient E_n in AD-100A and DAS-2B milking parlors(Figure 1).

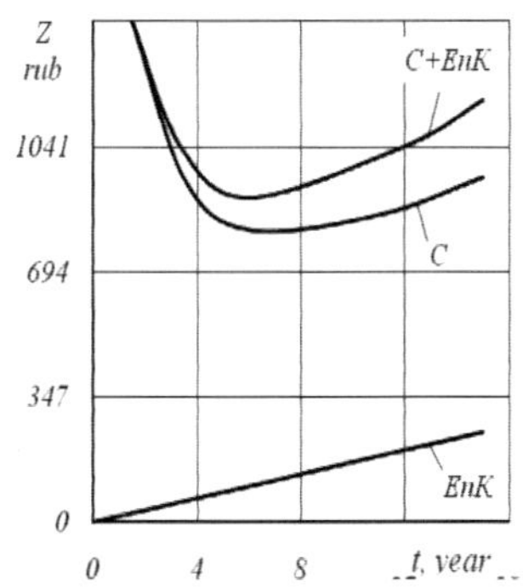

Figure 1. Change of component costs Z- current expenses C andinputs K -on linear installations depending on their operating time t.

Such dependencies are characterized by the fact that the higher production inputs, the better their performance standards, and the lower the production cost. The annual reduction trend in the share of operating costs when using a particular machine is legitimate in cases when the machine is used in production and the conditions for the constancy of forms and method of organizing its operation according to the performance projectionsare maintained. Thus, the operating expenses structure changes by the increase of animal husbandry machines operating time [2]. The changes regularity of operating expenses component on the example of the TKP-70 store-loader for hay towers depending on the operating time is shown in Figure 2.

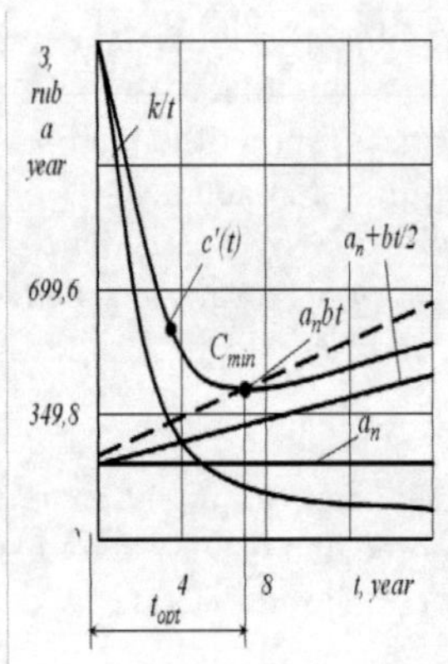

Figure 2. Depreciation charges change (K/t), valid ( ), average

( ), annual total C'(t), operation costs of the TKP-70 haylage towers depending on the operating time t (K – inputs, rub.,– constant costs, rub., b – costs of maintenance and repair).

An important reserve of reducing operating costs and increasing the efficiency of using machines and equipment in animal husbandry [3] is the introduction of a routine-preventive system for maintenance and repair. The idea of the system is that regardless of the machine technical state after some time (scientifically based) according to the maintenance, farms that have a low coverage rates, the maintenance cost per one service quickly increases, and then it slows downwith increasing λ. This connection is characterized by hyperbolic dependence:

$$y_\lambda = 25.3 + 7.3/\lambda \ \text{ if } \ 0.17 \le \lambda \le 0.9 \quad (1)$$

where – the cost of one periodic maintenance of line milking plants, rub.

The obtained value of correlation coefficient (r = 0.75) indicates a high linkage between studied traits [4]. The average cost of one periodic maintenance of all milking parlors deviated from the maintenance average cost by 800 rubles. This difference is because a significant gap in the maintenance conducting leads to premature equipment wear and parts damage, and an increase of remedial measures cost. With the increase of the λ coefficient, the technical readiness coefficient Rr of machines and equipment increases, and consequently, the effectiveness indicator — labor intensity (an hour per 100 kg) of producing milk. This dependence is described by the following equation:

$$y_x = 14.03 - 4.94\lambda - 6.82R_r \quad (2)$$

We determined the R coefficient of multiple correlations, which was 0.967 for identifying the effectiveness indicator tightness with both factors. To determine the influence on studied indicator (λ or Rr), we found the private elasticity coefficient *E*, which characterizes the percentage of average changes of the analyzed indicator with a change of 1% of each factor at the fixed position of others:

E_λ = 0.44; $E_{\lambda r}$ = 0.34. It follows that the λ coefficient has the greatest influence on dairy production labor intensity.

An important requirement for improving the efficiency of technical maintenance of machines and equipment - planning their maintenance and repair [5, 6].Therefore, before the planned period, the farms' engineering service draws up an annual plan of animal husbandry machines maintenance and repair, in which they consider all machines of the farm. The machine technical condition at the beginning of the planned year is the initial data for the annual plan development(machine operation time from the last repair or the first operation), the planned operation time of the machine by quarters or months, and the frequency of maintenance and repairs [7]. Data on machine operating time is determined by chronometer or by time detector N-30 or N-348. It can be used the electric counters for the SCHR-TS1 operation time or another type of counter, which automatically starts and stops along with the machines and equipment. Sometimes single-phase power counter is used together with an incandescent light switched to the machine electric drive; the counter corresponds to the machine's operation time in hours. You can also use the calculation method based on the required volume of work and the machines output per shift. Accounting of machine operating time is necessary to control the equipment use in production lines during the day, which allows to properly organize maintenance. For example, knowing machines operating time, it is possible to timely conduct all operations of maintenance and control their frequency for equipment used in a unit technological complex (the maintenance frequency must coincide or be a short periodicity of the entire complex) [8]. At the same

time, the frequency of - the machines maintenanceshould be lower than - the operating time average value at least than the value - deviation from its average values. Therefore, it is necessary to observe the maintenance precautionary conditions:

$$T_{mm} \leq T_{mm} - \Delta T_o$$

It is possible to determine the need for each machine maintenance and repair by analysis, nomograms, and graphic methods knowing the MC frequency and the machines annual time. In the first case, the need for each machine maintenance and repair:

$$N = (B_o + B_p) / T - N_p \tag{3}$$

where N - the number of types of maintenance and repairs; B_o -machinetime from the last MC or repair, h; B_p - planned machine time for the estimated year, h; T — periodicity execution of the MC or repair for which the calculation is issued, h; N_p - the number of all types of MC or ongoing repairs with a high frequency of the MC type or repairs for which the calculation is estimated.

The calculation is conducted in such sequence: firstly, it determines the required number of repairs, then the number of MC-2 and MC-1. When calculating the required number of repairs, the value is N_p=0.The N value is always rounded to the nearest whole number, mostly to downward, despite the fractional part. The value B_o is remained as balance obtained by dividing the time from the last repair or the beginning of operation by the periodicity of the type of maintenance or repair on which the calculation is conducted. Thus, if refrigeration units and equipment of primary milk processing from the beginning of operation worked 2190 h, in this case, B_o = 30 h.

For the efficient operation of livestock farms equipment under the maintenance routine-preventive system, it is needed the appropriate production base, the basis of which is the maintenance centers (MC). Under the conditions of unified engineering service, the MC of farms are branches of district maintenance centers of livestock machines.

However, projects of livestock farms reconstruction do not always provide constant maintenance centers. Therefore, it is important to determine what reconstructed farms and complexes are needed such points. We found that the criteria of the maintenance center replacement at farms and complexes are the minimum of the spent and transport costs:

$$C_{\Sigma} = C_E + E_n K_c + C_{tr} \to \min \tag{4}$$

where C_{Σ} - total costs for maintenance, rub.; C_E - expenses for machines maintenance and repair, rub.; K_c - capital investment in the construction of maintenance center, rub.; C_{tr} - transport expenses, rub.

The dependence of expenses level on the machines maintenance and repair is linked to the size of livestock farms and expressed by the following equation:

$$C_E = a_o + a_1/N_{an} \tag{5}$$

where a_o and a_1 – coefficient of reserve and cost improvement limits on maintenance; N_{an} - animal numbers on the farm.

Capital investments can be reduced due to cost cuts on construction and mounting works, and equipment installation in the maintenance center. In general:

$$K_c = A/N_{an} + B, \tag{6}$$

where A and B are the coefficient of reserve and capital investment reduction limits.

Transportation costs:

$$C_{tr} + a' + b\,'R, \tag{7}$$

where a' и b' - parameters of the line; R - radius of service a livestock complexes, km.

If we know the cost of maintenance center and repair per animal C_1 (rub.) and N_{an} number animals on the farm, then the total cost of maintenance center and repair per shift on one farm: $a = C_1 N_{an}$.

3. The main part

Projects of livestock farms reconstruction do not always provide permanent maintenance centers. Thus, it is important to determine what types of reconstructed farms and complexes such points are needed. We find the radius (R) of maintenance of livestock farms and complexes according to the formula:

$$R = \sqrt[3]{3C_1 N_{an}/\pi\delta\eta_t'\eta_d' S_{tr}} \tag{8}$$

where δ - the density of farms in the zone calculated per 1km²; η_t' - the coefficient of curve road routes; η_d' - the coefficient of paved roads; S_{tr} - transport costs, rubles per 1 km.

Substituting the expressions (2), (3), (4), (6) in (1) and differentiating by N_{an} we get:

$$\frac{d}{dN_{an}}\left[a_o + a_1\frac{d}{N_{an}} + E_n\left(\frac{A}{N_{an}} + B\right) + a' + b\sqrt[3]{\frac{3C_1 N_{an}}{\pi\delta\eta_t'\eta_d' S_{tr}}}\right] = 0 \tag{9}$$

After transformations:

$$N_{an} = \sqrt[4]{\frac{9(a_{1+E_n}A)^3\pi\delta\eta_t'\eta_d' S_{tr}}{(b')^3 C_1}} \tag{10}$$

It is possible to determine the size of the reconstructed livestock complex for which it needs to build a maintenance center. Thus, if for dairy farms a_1=4249.76; A=15887.6; a_1=11.69 rubles [9];

δ=0.01; S_t=0.06; η'_t=1 (for territories that have an ellipse shape with the ratio of half-axes, equal to 0.8); η'_d=0.5; b'=0.264; E_n=0.15, livestock farm size N_{an}=327 animals.

The size of livestock complexes, for which it is necessary to build a maintenance center, is significantly affected by their density (δ) in the zone. Calculations on the computer [10] showed that with growth δ the size of farms increases. For example, δ=0.05; δ=0.1; 0.2; 0.3; 0.4 and 0.5 (with unchanged values of other indicators) farm sizes are 489, 582, 692, 766, 823 and 870 animals respectively.

4. Conclusion

The routine-preventive system efficiency of the livestock machines maintenance significantly increases due to the specialization of engineering and technical service. The engineering service in animal husbandry is divided into two separate sections of technical operation, repair, and installation. The first section is responsible for all installed equipment efficient use, and the second section is responsible for installation, adjustment, and repair.

Farmers that have small livestock farms and for those full set of expensive equipment is cost-ineffective, they can use them based on cooperation. The supply service provides a centralized procurement of replacement parts, exchange fund of assemblies and units, new equipment. The farms established production departments, which included chief engineers (managers of departments), accountants, fixers, and repair workers. The economy of repair and maintenance work is paid at cost with the accrual of savings in the amount of 5%.

References

[1] Dong Z, Zhao J, Duan J, Wang M and Wang H 2018 Research on Agricultural Machinery Fault Diagnosis System Based on Expert System 2nd IEEE Advanced Information Management,Communicates,Electronic and Automation Control Conf. (IMCEC) (Xi'an: IEEE) pp 2057-60 doi: 10.1109/IMCEC.2018.8469579.

[2] Kastakaev N M and Zhomluhova A Zh 2007 Organization of a system for introducing innovative developments in the agricultural sector Herald of science of the Kazakh State Agrotechnical University. S. Seifullina 1 228-33.

[3] Zelepukin V S 2010 *Cattle* (Moscow: AquariumPrint) p 461.

[4] 1986 Guidelines for the Use of Complexity Units for Planning the Costs of Maintenance and Repair of Machinery in Livestock (Minsk) p 104.

[5] 1988 System of Preventive Maintenance and Maintenance of Livestock Machinery and Equipment (Moscow: Gosagroprom of the USSR) p 114.

[6] Peifeng S. 2011 Structural Analysis of the Remote Performance Monitoring System Used in Modern Agricultural Machinery Fourth Int. Conf. on Intelligent Computation Technology and Automation vol 2 (Shenzhen, Guangdong: IEEE) pp 189-93 doi: 10.1109/ICICTA.2011.338.

[7] Figueredo V, Reis A V, Garcia F and Araújo F. C. 2019 Virtual Reality for Agribusiness in the Development of a Maintenance Simulator for Agricultural Machinery for SenarGoiás 21st Symposium on Virtual and Augmented Reality (SVR) (Rio de Janeiro : IEEE) pp. 17-9 doi: 10.1109/SVR.2019.00019.

[8] Li D, Zheng Y and Zhao W 2019 Fault Analysis System for Agricultural Machinery Based on Big Data IEEE Access 7 99136-51 doi: 10.1109/ACCESS.2019.2928973

[9] Morozov I Yu 2008 Economic Efficiency of Technical Service of Machines in Dairy Farming (Moscow) p. 175.

[10] Krishna K L, Silver O, Malende W F and Anuradha K 2017 Internet of Things application for implementation of smart agriculture system International Conference on I-SMAC (IoT in Social, Mobile, Analytics and Cloud) (I-SMAC) (Palladam: IEEE) pp 54-59 doi: 10.1109/I-SMAC.2017.8058236.

Evaluation of the main indicators of the reliability of the power transmission of a combine harvester John Deere 9660

V V Vakhrushev[1], A E Nemtsev[1], N M Ivanov[1] and A A Melbert[2]

[1] Siberian Federal Scientific Centre of Agro-BioTechnologies of the Russian Academy of Sciences, Novosibirsk region, Novosibirsk district, Krasnoobsk village, Russia, 633501
[2]Polzunov Altai State Technical University, Lenina avenue, 46, Barnaul, Altai region, Russian Federation, 656038

E-mail: my-disk21@yandex.ru

Abstact. The failure-free performance of the components determines the technical and economic efficiency of the use of the combine harvester. A low level of reliability leads to an increase in the cost of harvesting, the need for unscheduled repairs due to failures of parts, assemblies and assembly units. Studies have shown that a significant proportion of failures occur in power transmission units. In this regard, the aim of the work is to assess the reliability of the power transmission of the John Deere 9660 combine harvester. The average time between failures of the power transmission units was 80 hours, which does not meet the regulatory requirements (according to STO AIST 8.22 - 2010 - at least 100 hours). Failures of all powertrain assemblies are caused by a combination of failures: wear and fatigue failure of the bore holes for the bearings of the unit housings, and gears, bearings from dynamic loads.

1 Introduction

The power train of the John Deere 9660 combine harvester consists of a drive axle, final drives and a range box. The importance of assessing the reliability of power transmission is confirmed by a significant number of studies aimed at improving the reliability of agricultural machines [1,2].

The longevity of the elements of combine harvesters is of great importance, since the yield and grade of cultivated crops largely depends on the implementation of the relevant work within the agrotechnical time frame. It should be noted that only recorded and taken into account machine downtime for harvesting crop products account for up to 50% of the total working time of the combine harvester. This indicates the lack of reliability of agricultural harvesting equipment and maintaining it at the proper level during operation and repair. In addition, downtime is associated with insufficient provision of machinery and equipment with spare parts, and their rational use [3].

The power train is one of the most unreliable, systematically failing combine harvester systems. As practice shows, in terms of failure frequency, the power transmission takes the third place among all units and assemblies of the John Deere 9660 combine harvester. The average time to eliminate the failure of the drive axle, range box, final drives, in relation to other units of the John Deere 9600 combine is the longest. The most complete use of the technical resources of the components and assemblies of the power transmission, the timely execution of maintenance work with the minimum labor and materials costs, largely depends on how well the operations on replacing and restoring resource elements are worked out in the machine.

2 Materials and methods

The acquisition of new equipment, including foreign production, is impossible without an analysis of the initial indicators of reliability, durability, maintainability and the rate of change during operation.

Such researches make it possible to obtain general patterns of change in characteristics, to establish their quantitative indicators, on the basis of which to identify and disclose design flaws.

An analysis of these patterns allows us to develop specific recommendations for improving the efficiency of using existing machines.

With this in mind, the objectives of the study are:

1. Determination of failure-free performance indicators of power transmission units of a John Deere 9660 combine harvester;
2. A quantitative assessment of the reliability indicators of units and power transmission units.

At the stages of experimental testing, testing and operation, as a rule, the role of reliability indicators is performed by statistical methods for assessing the corresponding probabilistic characteristics. All reliability indicators, in accordance with GOST 27.002-2017 [4], are defined as probabilistic characteristics. In the work, the failure of the object is considered as a random event, that is, the given structure of the object and the conditions of its operation do not determine exactly the moment and place of occurrence of the failure. The adoption of this more widespread concept predetermines the widespread use of probability theory [1,2,4,5,8].

Reliability indicators are regulated by observation plans. To assess the reliability of power transmission reliability, the observation plan [N, R, T] is the most acceptable. Its use is justified by the minimum time spent on the experiment. The observation plan [N, R, T] provides for the assessment of the following reliability indicators:

- average time to failure;
- mean time between failures;
- average resource.

To calculate the parameters of the distribution law to failure, the first two types of operating time can be used equally if the products are replaced with new ones, as well as with the full restoration of all operating parameters after repair. Information on developments of the third type can also be partially used.

The procedure for collecting information on the reliability of facilities, the forms for recording and accumulating data on malfunctions and failures are defined in GOST 27.502-83 and RD 50 - 204 - 87 “Collection and processing of information on the reliability of products in operation”.

Data collection about power transmission failures was carried out on the basis of: “Technical act on damage and malfunction of an assembly or unit”, “Combine maintenance record book”, on-board computer fault codes.

A group of 37 vehicles was placed under surveillance. During operation, we tracked the time between failures, recorded the time of the case of the failure, the time to restore the working state, the costs of restoring the working state (direct and indirect), the causes of failures, etc.

A special algorithm for fixing faults was developed for each series of combines [9, 10].

The algorithm consisted of a digital code, including: the node where the failure occurred, the failed part, the type of failure. The digital code generated in this way was recorded in a database and processed using a computer.

3 Results and discussion

According to the adopted observation plan [N, R, T], the following indicators are accepted as indicators: average resource, time to failure, time between failures [11,12].Figures 1-6 show function graphs illustrating the distribution of statistical failure indicators of power transmission elements.

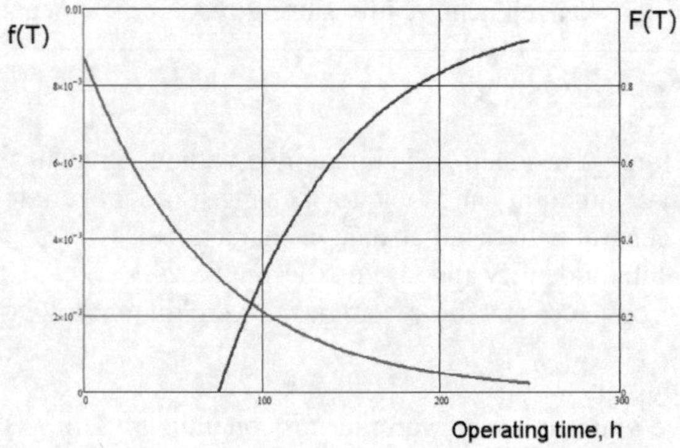

Tcp_1 – mean time to failure;
F(T) - curve of accumulated experimental probabilities of operating time to failure;
f(T) - time-to-failure density functions $f(T) = 3.0 \cdot 10^{-3} \cdot e^{-\left(\frac{T-75}{70}\right)}$,
$$F(T) = 1 - e^{-\left(\frac{T-75}{70}\right)}.$$

Figure 1. Graph of the density distribution function of running hours to failure and a curve of accumulated experimental probabilities of running time to failure of the drive axle of the John Deere 9660 combine harvester

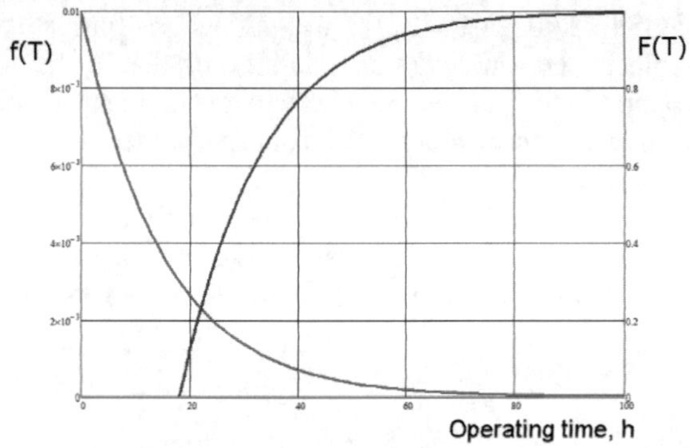

Tcp_0 – mean time between failures ;
F(T) - curve of accumulated experimental probabilities of running between failures;
f(T) - the density function of operation time between failures
$$f(T) = 0.77 \cdot 10^{-3} \cdot e^{-\left(\frac{T-18}{15}\right)},$$
$$F(T) = 1 - e^{-\left(\frac{T-18}{15}\right)}.$$

Figure 2. Graph of the density function of the distribution of running hours between failures and a curve of accumulated experimental probabilities of running time between failures of the axle drive axle of a John Deere 9660 combine harvester

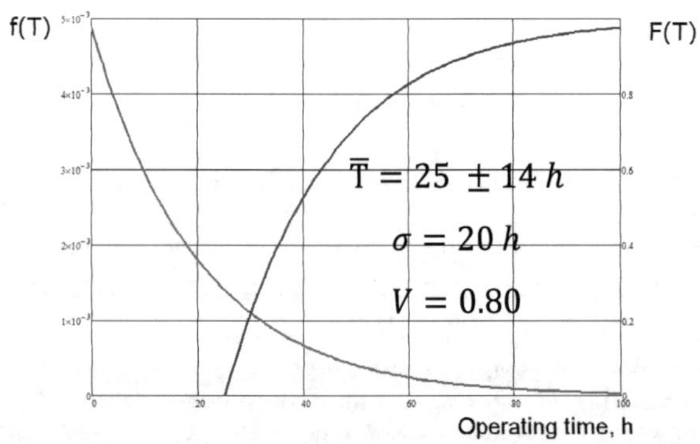

Tcp_1 – mean time to failure;
F(T) - curve of accumulated experimental probabilities of operating time to failure;
f(T) - time-to-failure density
$$f(T) = 1.4 \cdot 10^{-3} \cdot e^{-\left(\frac{T-25}{20}\right)},$$
$$F(T) = 1 - e^{-\left(\frac{T-25}{20}\right)}.$$

Figure 3. Graph of the density distribution function of running hours to failure and a curve of accumulated experimental probabilities of running time to failure of the range box of a John Deere 9660 combine harvester

IOP Conf. Series: Materials Science and Engineering **941** (2020) 012068 doi:10.1088/1757-899X/941/1/012068

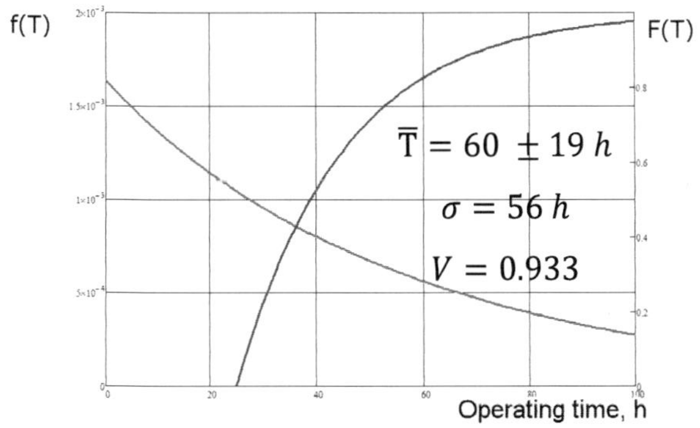

Tcp$_0$ – mean time between failures ;
F(T) - curve of accumulated experimental probabilities of running between failures;
f(T) - the density function of operation time between failures

$$f(T) = 0.56 \cdot 10^{-3} \cdot e^{-\left(\frac{T-60}{56}\right)},$$

$$F(T) = 1 - e^{-\left(\frac{T-60}{56}\right)}.$$

Figure 4. Graph of the density function of the distribution of running hours between failures and the curve of the accumulated experimental probabilities of running between failures ranges of combine harvester John Deere 9660

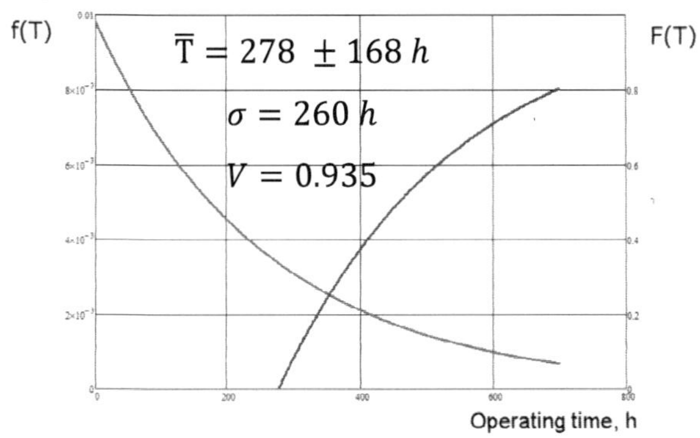

Tcp$_1$ – mean time to failure;
F(T) - curve of accumulated experimental probabilities of operating time to failure;
f(T) - time-to-failure density

$$f(T) = 3.37 \cdot 10^{-3} \cdot e^{-\left(\frac{T-278}{260}\right)},$$

$$F(T) = 1 - e^{-\left(\frac{T-278}{260}\right)}.$$

Figure 5. Graph of the density distribution function of running hours to failure and the curve of the accumulated experimental probabilities of running to failure, final drives of the John Deere 9660 combine harvester

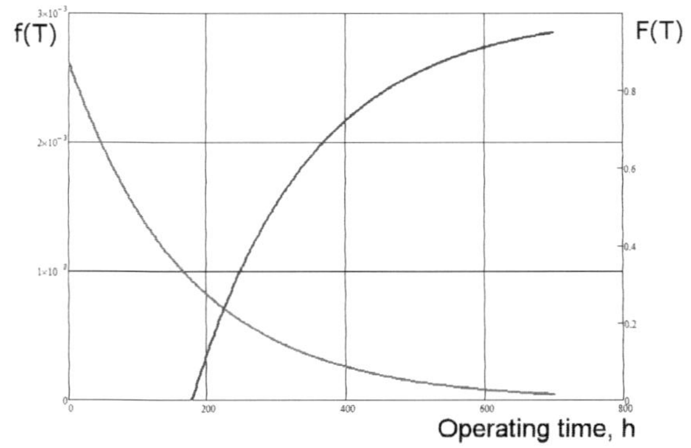

Tcp$_0$ – mean time between failures ;
F(T) - curve of accumulated experimental probabilities of running between failures;
f(T) - the density function of operation time between failures

$$f(T) = 0.93 \cdot 10^{-3} \cdot e^{-\left(\frac{T-179}{173}\right)},$$

$$F(T) = 1 - e^{-\left(\frac{T-179}{173}\right)}.$$

Figure 6. Graph of the density function of distribution of running hours between failures and a curve of accumulated experimental probabilities of running between failures of final drives of a John Deere 9660 combine harvester

Failures of all powertrain assemblies are caused by a combination of failures: wear and fatigue fracture of gears, bearings from dynamic loads.

Thus, when operating a John Deere 9660 combine harvester, the low range box life should be considered. In addition to the main repair actions, it is necessary to plan additionally the maintenance and overhaul of this unit.

4 Conclusions

1. The work has developed and proposed an algorithm for recording failures in the power transmission of a John Deere 9660 combine harvester in the form of a digital code that allows identifying and classifying failures of power transmission units.

2. In this work, we obtained the functions of distributing the operating time between failures, the operating time to failure of the components and assemblies of the power transmission of the John Deere 9660 combine harvester.

3. Failures of all power transmission units are caused by a combination of failures: wear and fatigue failure of gears, bearings from dynamic loads.

Acknowledgments

The authors are grateful to Professor Egorov AV and Associate Professor Zubova EV for the assistance in writing the article.

References

[1] John Deere. Tekhnicheskoye rukovodstvo. Remont kombaynov modeley 9400, 9500, 9600, 1997 *John Deere Harvester Works*, TM1634, 2005 – 1108 r.Nishi S and Seiki T Planning and design of underground space use *Mem. Sch. Eng. Nagoya Univ* **1**

[2] S V Subochev, A Ye Nemtsev, I V Kopteva 2014 Vliyaniye osnovnykh pokazateley nadozhnosti na effektivnost' zernouborochnykh kombaynov, *Vestnik NGAU*, **2 (31)** 157-164

[3] R V Kugel' Nadezhnost' mashin massovogo proizvodstva 1981 (Moscow: Mashinostroyeniye) 244 p.

[4] Kuznetsov Ye.S. 1982 Upravleniye tekhnicheskoy ekspluatatsiyey avtomobiley, (Moscow:Transport) 272 p.

[5] Ushakov I A 1980 Spravochnik po raschetu nadezhnosti (Moscow: Mashinostroyeniye) 606 p.

[6] GOST 27.503 - 81. 1982 Metody otsenki pokazateley nadozhnosti, (Moscow: Gosudarstvennyy komitet SSSR po Standartam) 56 p.

[7] Ye S Venttsel' 1966 Teoriya veroyatnosti (Moscow: Vysshaya shkola) 576 p.

[8] Lipman L N 1966 Posobiye dlya opredeleniya chisla ob"yektov nablyudeniy (ispytaniy), (Moscow: Vysshaya shkola) 200 pp.

[9] Gnedenko B V 1988 Kurs teorii veroyatnostey (Moscow: Vysshaya shkola) 488 p.

[10] Polozheniye o tekhnicheskom obsluzhivanii i remonte podvizhnogo sostava avtomobil'nogo transporta Minavtotransa RSFSR 1986 (Moscow: Transport) 73 p.

[11] A A Pitukhin 2010 Nadozhnost' lesozagotovitel'nykh mashin i oborudovaniya: Uchebnoye posobiye, - (SPb.: Izdatel'stvo «Lan'») 288 p.

[12] A S Grunichev 1974 Tablitsy dlya raschetov nadezhnosti pri raspredelenii Veybulla (Moscow: Standart) 327 p.

Assessment of the service properties of commodity motor oils by hydrogen indicator

V V Vakhrushev, A E Nemtsev and N M Ivanov

Siberian Federal Scientific Centre of Agro-BioTechnologies of the Russian Academy of Sciences, Novosibirsk region, Novosibirsk district, Krasnoobsk village, Russia, 633501

E-mail: my-disk21@yandex.ru

Abstract. Basic oils produced from oil or obtained by the synthesis of monomers are generally chemically neutral. To improve the service properties of base motor oils, additive packages and active ingredients are introduced into it. The presence of anti-seize, anti-wear, and dispersing-stabilizing additives in a composition of commercial oil of a certain concentration of compounds of alkali metal salts changes its chemical activity [1].
Based on this, as a measure of service properties (anti-seize, anti-wear and dispersion-stabilizing properties (DSP)) of commercial motor oils can be used hydrogen indicator - pH.
In the work, as an invariant estimate of the concentration of resource additives and service properties of commercial motor oils, the use of pH is considered. Correlation dependences of pH on BSP and viscosity of the studied motor oil are given.

1Introduction. The quality of fresh commercial motor oil (quality, performance of oil) is characterized by properties that determine its suitability for a particular internal combustion engine [1]. The set of qualities of motor oil is characterized by operational and service properties.

The performance features of fresh commercial motor oil characterize the ability of the oil to work under various conditions and operating models of the internal combustion engine [1-3].

The operational properties of fresh commercial motor oil include:
- physical and chemical properties of the base oil;
- indicators of the composition of the base oil;
- viscous properties;
- antioxidant properties;
- antifoam properties;
- anticorrosion properties.

The operational properties of fresh commercial motor oil are formed, set and strictly standardized by the developer and the manufacturer. Basically, these properties are controlled by the manufacturer on the basis of laboratory chemical research and do not change during the operation of motor oil.

Service properties of fresh commercial motor oil are properties that determine the working capacity (resource) of both the oil itself and the durability of tribomechanical systems of an internal combustion engine [1,2].

The service properties of fresh commercial motor oil include:
- dispersing - stabilizing properties;
- extreme pressure and antiwear properties;

IOP Conf. Series: Materials Science and Engineering **941** (2020) 012069 doi:10.1088/1757-899X/941/1/012069

- antifriction properties;
- washing properties.

Service properties are formed by a specific additive package, as a rule, such packages are low-resource, that is, after their exhaustion, the engine oil must be replaced. Based on the foregoing, the stability of the properties of engine oil and the long-term trouble-free operation of an internal combustion engine should be ensured by a package of additives related to service properties.

To reduce and prevent the formation of carbon deposits, special surfactants called detergents are introduced into motor oils. Among them are ash-less compounds, in particular succinimides. It should be noted that the term “detergents” is, to a certain extent, conditional, since these additives mainly prevent the formation of deposits on heated surfaces, and do not have a washing effect in the literal sense of the term.

The detergent action of additives, in itself, does not ensure the effective operation of the oil, since the products formed have low stability and gradually precipitate, increasing the contamination of engine components and parts. On this basis, the control of detergents does not require consideration.

In this regard, from the energy point of view, the processes occurring at the interface between the dispersing - stabilizing additives and the solid phase between molecules are of great importance [3,4].

Based on this, a hydrogen pH value can be used as an indicator of service properties (by anti-seize, anti-wear, and dispersion-stabilizing properties) of commercial motor oils [1,2]. Accounting for pH in this case allows us to ensure the invariance of the assessment of the concentration of resource additives.

2 Materials and methods. Synthetic and mineral commodity motor oils with service properties according to API SM, SN, SN / CF with viscosity according to SAE 5W30, 5W40, 10W40 were used as the test material.

Actual pH values for each oil sample were measured with a WTW-315i / Set analytical pH meter equipped with a special SenTix electrode for measuring pH in fatty environments with an accuracy of 0.001 pH.

Before measurement, the instrument was calibrated with a buffer solution with a pH value of 7.0.

The pH was measured at a constant controlled temperature close to the calibration temperature of the device, for which the oil samples were thermostatic until the temperature of each of them became constant and equal to the ambient temperature. The control was a sample with a minimum temperature. A thermometer was placed in a container with a control sample, according to the readings of which they monitored the achievement of a given temperature.

The sample temperature was not controlled since the pH meter is equipped with a temperature compensator.

Before the measurement, the electrodes were thoroughly rinsed with a buffer solution, removing the remnants of the previous oil sample, blotted with filter paper, lowered into the analyzed sample and after 1 - 3 minutes. (after establishing a constant value) recorded the readings of the device. The measurement was performed at least two times, repeated measurement after 1 min.

Between measurements, the electrodes were left in a buffer solution.

The pH value was taken as readings from the instrument panel. The measurement result is the arithmetic average of two results of parallel measurements.

Assessment of the activity of the DSP additive included the introduction of an oil of 3.5 ml ashless pollutant in a volume of 2.1 ml into the test sample (distilled water was used as the pollutant) and active mixing of the sample for 0.5 min. Then, by the color, height of the sediment, and the volume of precipitated water, the actual activity of the DSP was evaluated [4].

Evaluation of extreme pressure and anti-wear properties was carried out on a MAST - 1 machine according to a standard procedure [1].

3 Results and discussion. Figure 1 shows the correlation curves obtained by comparing the results of measuring the pH and DSP.

IOP Conf. Series: Materials Science and Engineering **941** (2020) 012069 doi:10.1088/1757-899X/941/1/012069

Analysis of the graphs indicates that the evaluation of the DSP of a commodity is small cannot be carried out by the hydrogen indicator, since the correlation is negligible. Therefore, the use of an ash-less pollution may be the main means of assessing DSP.

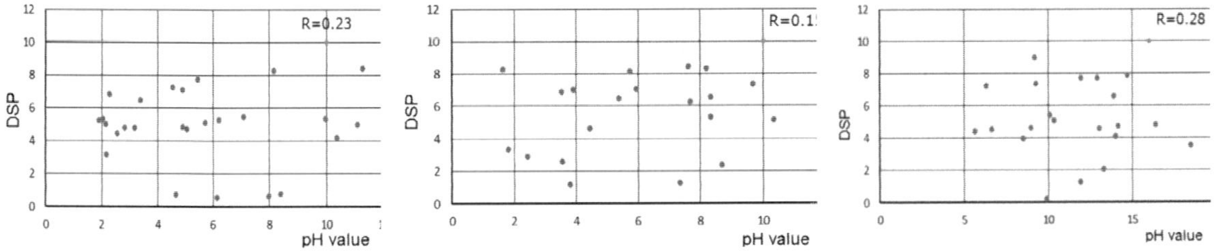

API SM SAE 5W40 API SN SAE 5W40 API SN/CF SAE 5W30

Figure 1. Pair correlation between the signs of the DSP and the hydrogen index

Studies have shown that the average DSP value for API SM SAE 5W40 oils is 6.3 points, API SN SAE 5W40 is 8.2 points, API SN / CF SAE 5W30 is 9.6 points. Based on this, it should be considered that the engine oil resource depends on the activity of the DSP. In addition, the DSP score allows you to compare the performance characteristics of oils in accordance with their classification by groups of operational properties.

Figure 2 shows the pair correlation between the hydrogen index and the friction moment.

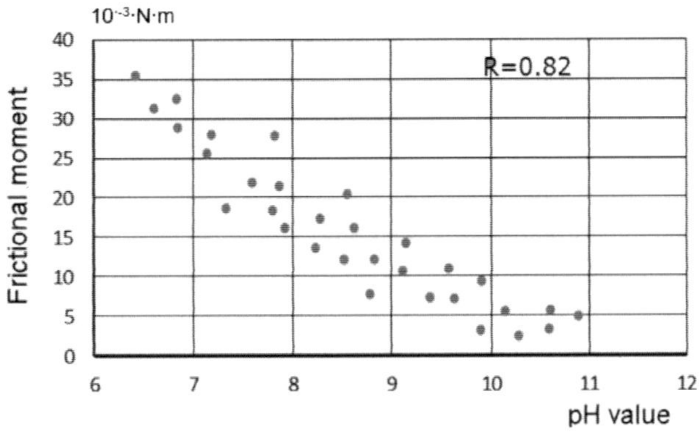

API SM SAE 5W40

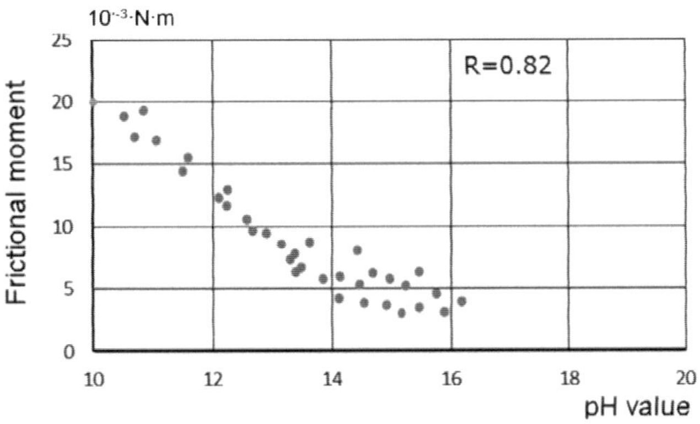

API SN SAE 5W40

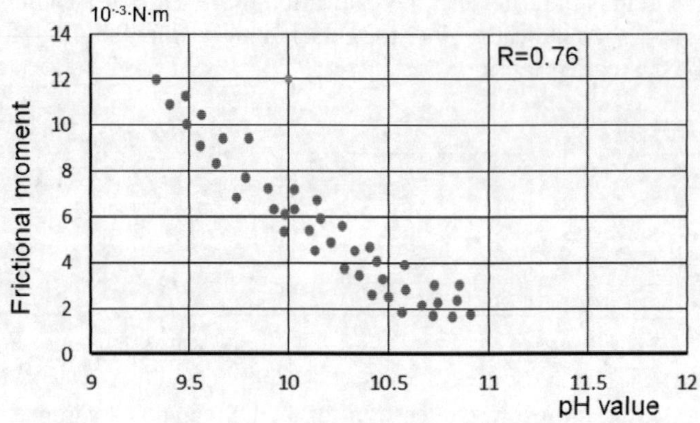

API SN/CF SAE 5W30

Figure 2. Pair correlation between signs of friction moment and hydrogen index

On the contrary, the use of a hydrogen index in assessing extreme pressure and antiwear properties is acceptable, since a close correlation between the friction moment and the hydrogen index is found.

The analysis of the graphs in Figure 2 shows that the average value of the friction moment for API SM SAE 5W40 oils is $27 \cdot 10^{-3} N \cdot m$), for API SN SAE 5W40 it is $12{,}7 \cdot 10^{-3} N \cdot m$, for API SN / CF SAE 5W30 - $6{,}6 \cdot 10^{-3} N \cdot m$. In addition, the graphs (Figure 2) provide an indicative assessment of operational properties by the content of extreme pressure and antiwear additives in terms of pH for the choice of motor oil for trouble-free operation of an internal combustion engine.

4 Conclusions.

1. Measuring instruments and a method for studying the service properties of commercial motor oils according to the DSP and the pH value are proposed;
2. Based on experimental studies, it was found that the average DSP value for fresh marketable oils API SM SAE 5W40 is 6.3 points, API SN SAE 5W40 8.2 points, API SN / CF SAE 5W30 9.6 points;
3. A ballistic assessment of the DSP allows you to compare the performance characteristics of the oils in accordance with their classification by groups of operational properties;
4. The extreme pressure and anti-wear properties of commercial motor oils can be estimated by the pH indicator, since a close correlation between the friction moment and the hydrogen index is found.

References

[1] Motornyye masla. Proizvodstvo. Svoystva. Klassifikatsiya. Primeneniye / *Baltenas R., Safonov A.S., Ushakov A.I, Shergalis V. Nauchno-tekhnicheskoye izdaniye 2000* (Moscow: Al'fa-Lab) 272 p.

[2] *Financial Times Oil Gas.* International Year Book 1995. Longman Group UK Ltd. Harlow, Essex 504 p.

[3] D.K. Klamman (Esso AG) *Lubricants and Related Products*, vol. 15, pp. 423 – 518

[4] Gur'yanov YU.A. (2010) K voprosu ob otsenke neytralizuyushchikh svoystv motornykh masel [On the assessment of the neutralizing properties of engine oils] *Materialy XLIX mezhdunarodnoy nauchno-tekhnicheskoy konferentsii «Dostizheniya nauki – agropromyshlennomu proizvodstvu»* vol. 2, pp. 172–178.

IOP Conf. Series: Materials Science and Engineering **941** (2020) 012070 doi:10.1088/1757-899X/941/1/012070

Application of the photoelasticity method for studying the stress-strain state of the tribomechanical system "ring-bearing housing"

V V Vakhrushev[1], A E Nemtsev[1], N M Ivanov[1] and AV Shindelov[2]

[1] Siberian Federal Scientific Centre of Agro-BioTechnologies of the Russian Academy of Sciences, Novosibirsk region, Novosibirsk district, Krasnoobsk village, Russia, 633501
[2]Novosibirsk State Agricultural University, 160, st. Dobrolyubova, Novosibirsk, Russia, 630039

E-mail: my-disk21@yandex.ru

Abstract. In this paper, we consider the use of the polarization-optical method for assessing the stress-strain state of films of polymeric material in the restored tribomechanical system "bearing ring - housing" and the dependences for determining the magnitude of the principal stresses are proposed.
When studying the stress-strain state, compounds with a diameter of the contact surface of 40 mm from an optically sensitive material imitating an adhesive layer were used as models. The photoelasticity method showed that the maximum stresses occurred on the axis of symmetry of the model, and the contact arc was 180 degrees. A qualitative picture of the strips indicates the identity of the stress-strain state of the adhesive joint model both for the uniform gap model and for the eccentricity models.

1 Introduction. In the practice of repair production, to restore the operability of the tribomechanical system "bearing ring - housing", polymeric materials based on polyfunctional acrylic compounds are used [1]. A feature of the application of such materials is their low elastic modulus [2, 6, 7]. As a consequence of this, there is a need to reasonably assign a minimum level of mechanical properties depending on the degree of wear and geometry of the joint, which in turn requires investigation of the stress state and assessment of the strength of materials in contact with the bearing ring and the housing. This task is complicated by variable loads and operating temperatures, which can change the state of the polymer from elastic to plastic. In this regard, apparently, the polarization-optical method (photoelasticity method) is the only possible method for studying the stress state of thin polymer interlayers in contact problems.

2 Materials and methods. The polarization-optical method is an experimental method for analyzing the stress-strain state of objects, based on the anisotropy of optically sensitive materials under loading.

Photoelasticity allows interpretation of the pattern of bands across the field, to evaluate the nominal values of stresses and gradients, to conduct quantitative measurements. In particular, it is possible to determine the directions of the principal stresses at all points of the photoelastic model, the magnitude and sign of the tangential stresses along the free boundaries and in all areas where the stress state is uniaxial; in the case of a plane stress state, the magnitude and sign of the difference between the principal stresses at selected points of the object under study [3].

The photoelasticity method [1,8] is based on the study of not the parts themselves, but their models.

To study the stress-strain state, models with an affine similarity (that is, with increased thickness) were used as models of a tribomechanical system “bearing ring - housing”, since the thickness of real layers of polymer material in the restored tribomechanical system “bearing ring - housing” does not exceed as a rule, 300 microns [4].

As models, compounds with a contact surface diameter of 40 mm were used. The model consisted of housing K and optically sensitive material C (Figure 1). The model of polymer interlayer C was made of evacuated epoxy resin ED-20, cured at room temperature using PEPA (polyethylene polyamine), modified TEG (triethylene glycol) [5].

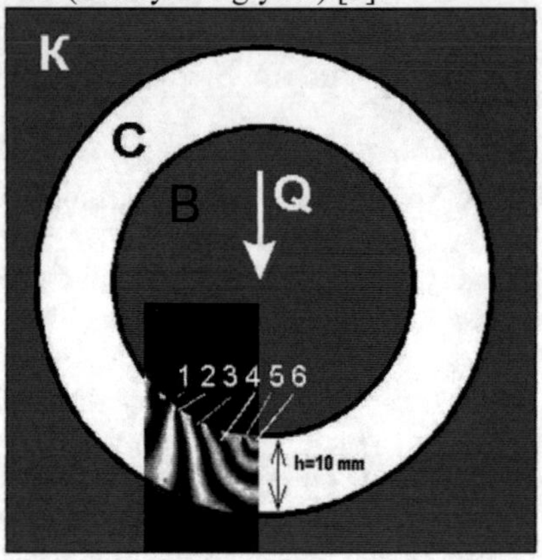

K - hosing
C - is a polymer layer,
B – bearing
Q is the radial load;
1,2,3,4,5,6 - the order of the bands in the model.

Figure 1. Qualitative distribution of the difference of the main stresses in the model

Models made of optically sensitive material (adhesive bonding immitant) were made for uniform clearance between the bearing ring and the housing, and with an eccentricity styling wear of the rolling bearing bore in the gearbox housing.

The wall thickness of the model (Figure 2, a) h of an optically sensitive material for a uniform annular gap (Figure 2, b) was - 10 mm, for a model with an eccentricity - 20 mm (Figure 2, c).

The model was loaded using a ball bearing 203 with a radial load of 300 N. The value of the load was selected from the condition of the elastic behavior of the optically sensitive material.

A Zeiss-300 circular polariscope was used to observe the patterns of the bands; the patterns were recorded using a digital camera [9].

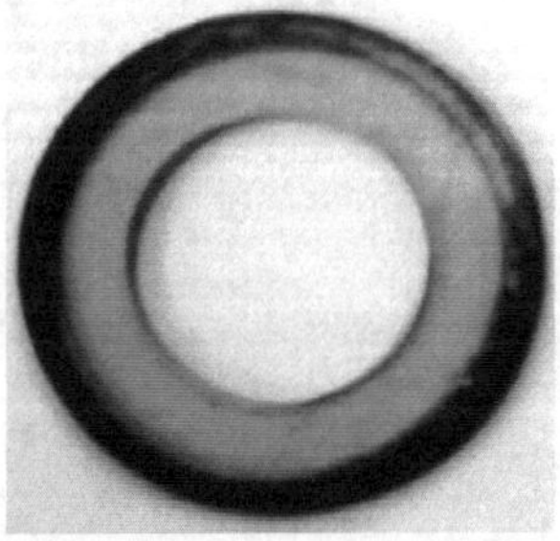

b

a

c

Figure 2. Models for the study of stress-strain state

The polarization-optical method for assessing the stress-strain state allows us to determine the difference between the principal $\sigma_r - \sigma_\theta$ and tangential stresses $\tau_{r\theta}$ for any point in the model.

The equations for determining $\sigma_r - \sigma_\theta$and$\tau_{r\theta}$have the following forms [5]:

$$\sigma_r - \sigma_\theta = \frac{\sigma_0^{(1)} \cdot m}{t} \cos\left(2(\theta - \alpha)\right) \quad (1)$$

$$\tau_{r\theta} = \frac{\sigma_0^{(1)} \cdot m}{2t} \cos\left(2(\theta - \alpha)\right) \quad (2)$$

where m-is the strip order at the considered point of the model;

$\sigma_0^{(1)}$ - is the optical constant of the model material;

t - is the thickness of the flat model;

α - is the band parameter;

θ - is the angular coordinate of the model point.

The optical constant of the model material was determined in accordance with the method described in [8] and amounted to $\cdot 0{,}046 \frac{\mathbf{N}}{\mathbf{mm}}$ band.

Dependencies (1) and (2) do not allow determining the stress components $\sigma_r$and σ_θ. Except that on the free contour of the polymer interlayer, where $\sigma_r = 0$, it is possible to determine the stress values σ_θ from dependence (1).

Differentiation of equations (1) and (2) in polar coordinates we get:

$$\begin{cases} \dfrac{d\sigma_r}{dr} + \dfrac{1}{r} \cdot \dfrac{d\tau_{r\theta}}{d\theta} + \dfrac{\sigma_r - \sigma_\theta}{r} = 0 \\ \dfrac{d\sigma_{r\theta}}{dr} + \dfrac{1}{r} \cdot \dfrac{d\sigma_\theta}{d\theta} + \dfrac{2\tau_{r\theta}}{r} = 0 \end{cases} \quad (3)$$

The second equation of system (3) in a graph-analytical way allows us to determine the values of normal stresses σ_θ from the contact arc of the polymer layer [5].

To determine σ_θ, we transform equation (3) into the form:

$$\sigma_{\theta_n} = \sigma_{\theta_0} - \int_{\theta_0}^{\theta_n} \left(r \frac{d\tau_{r\theta}}{dr} + 2\tau_{r\theta} \right) d\theta \quad (4)$$

Denote the expression $\left(r \frac{d\tau_{r\theta}}{dr} + 2\tau_{r\theta} \right)$ by $f(\theta)$, then:

$$\sigma_{\theta_n} = \sigma_{\theta_0} - \sum_{i=1}^{n} \frac{f(\theta_{i-1}) + f(\theta_i)}{2} (\theta_i - \theta_{i-1}) \quad (5)$$

Thus, equations (2) and (5) make it possible to determine the values of normal stresses σ_θ at the points of the contact surface of the adhesive joint.

The ordinates of the functions f(θ) under the sum sign in dependence (5) are determined by the addition of the products $\left(\frac{d\tau_{r\theta}}{dr}\right)_i r_i$ with the corresponding doubled values of the tangential stresses $\tau_{r\theta}$.

3 Results and discussion. The experiments show that on the internal unloaded contours of the models, one can choose such points for which $\sigma_\theta = 0$, or determine by expression (1). Thus, the unloaded contour of the glue line in the tribomechanical system makes it possible to find the boundary values of the quantity σ_{θ_0} involved in equation (5).

Figure 3 shows the qualitative picture of the distribution of the main stresses in the model.

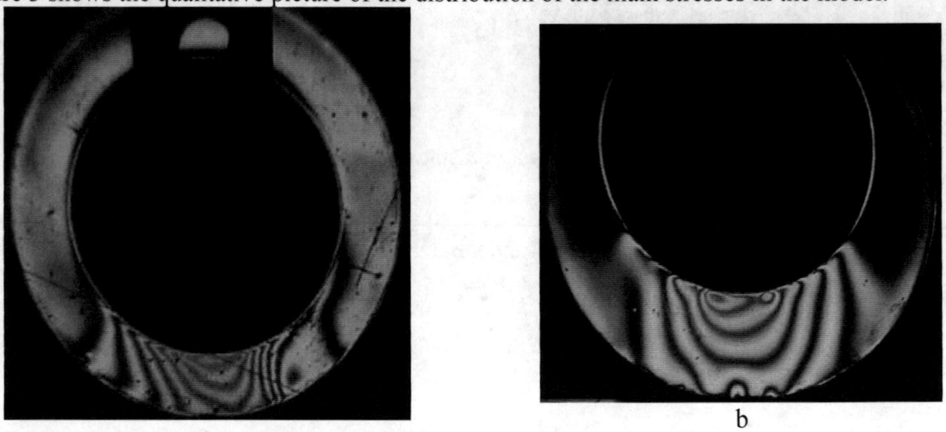

a b

a - model with a radial annular gap of 10 mm;
b - model with eccentricity - 20 mm (external load on the sample 300 N)

Figure 3. Qualitative distribution of the difference of the main stresses in the model

Observation of the picture of isoclines (trajectories of the main stresses) showed that the main sites are located along the radius and circumference. Maximum stresses occurred on the axis of symmetry of the model. The contact arc was 180 degrees. A qualitative picture of the strips indicates the identity of the stress-strain state of the adhesive joint model, both for the uniform gap model and for the eccentricity models. In this case, the same amount and shape of isoclines are observed in different models. The number of isoclines increases with increasing load in all types of models.

Obviously, at zero clearance, the contact area "outer bearing ring - polymer layer" does not change, and therefore the stress value at any point in the model is proportional to the load (linear problem).

Experimental studies show that maximum stresses occur on the axis of symmetry, while the contact arc is 180 degrees.

4 Conclusions.

1. A calculation-experimental method is proposed for assessing the stress-strain state of a polymer film in a tribomechanical system "bearing ring - housing".
2. The study of the stress-strain state of polymer films by photoelasticity showed that the maximum stresses occurred on the axis of symmetry of the model, the contact arc was 180 degrees;
3. A qualitative picture of the strips indicates the identity of the stress-strain state of the adhesive joint model, both for the uniform gap model and for the eccentricity models.
4. The number of isoclines for all models is identical and increases with increasing load on the model.

References

[1] Zhilkin V. A., Vakhrushev V. V. (2008) Obosnovaniye tekhnologii remonta posadochnykh otverstiy pod podshipniki kacheniya [Justification of the technology of repair of bore holes for

rolling bearings] *Vestnik Chelyabinskogo gosudarstvennogo agroinzhenernogo universiteta*, vol. 53, pp. 47-68

[2] Vakhrushev V.V., Egorov A.V., Zubova E.V. (2016) Obespecheniye rabotosposobnosti tribomekhanicheskoy sistemy «kol'tso podshipnika – korpus» v korobkakh peredach transportnykh i tekhnologicheskikh mashin agropromyshlennogo kompleksa primeneniyem polimernykh materialov akrilovogo ryada [Ensuring the operability of the tribomechanical system "bearing ring - housing" in gearboxes of transport and technological machines of the agro-industrial complex using polymeric materials of the acrylic series] *APK Rossii* **75 (1)** pp. 65-70.

[3] Fink K 1961 Izmereniye napryazheniy i deformatsiy (Moscow: Mashinostroyeniye) 352 p.

[4] Gerasimov S. I. (2004) Primeneniye metoda fotouprugosti dlya analiza ostatochnykh napryazheniy v kompakt-diskakh [Application of the photoelasticity method for the analysis of residual stresses in compact discs] *Prikladnaya mekhanika i tekhnicheskaya fizika* **45(3)** pp. 176 – 180

[5] Timoshenko S. P., Gud'yer Dzh 1979 Teoriya uprugosti (Moscow: Nauka) 560 p.

[6] Malysheva G. V 2005 Skleivaniye v mashinostroyenii (Moscow: Nauka i tekhnologii) 2(1) 545 s.

[7] A. F. Nikolayev, V. K. Kryzhanovskiy, V. V. Burlov (2008) Tekhnologiya polimernykh materialov (Saint Petersburg: Professiya) 544 p.

[8] Aleksandrov A. YA., Akhmetzyanov M. KH. Polyarizatsionno-opticheskiye metody mekhaniki deformiruyemogo tela. (Moscow: Nauka) 576 p.

[9] Sharafutdinov G.Z. (1995) Ob osnovakh metoda integral'noy fotouprugosti [On the basics of the integral photoelasticity method] *Vestnik Moskovskogo universiteta. Seriya 1. Matematika i mekhanika* vol. 5 pp. 70–79.

IOP Conf. Series: Materials Science and Engineering **941** (2020) 012071 doi:10.1088/1757-899X/941/1/012071

The rational repair and maintenance cycle of the KamAZ car gearbox taking into account the pre-repair operating time

V V Vakhrushev[1], A E Nemtsev[1], N M Ivanov[1] and V I Poddubny[2]

[1]Siberian Federal Scientific Centre of Agro-BioTechnologies of the Russian Academy of Sciences, Novosibirsk, Russia, 633501
[2]Polzunov Altai State Technical University, Lenina avenue, 46, Barnaul, Altai region, Russian Federation, 656038

E-mail: my-disk21@yandex.ru

Abstract. A significant proportion of the costs and downtime for repairs are on transmission units, and especially on the gearbox (GB). Based on the research results presented in the work, up to 76% of all transmission failures fall on the gearbox of a KamAZ 6540 automobile under ordinary operating conditions. The purpose of the study is to justify the repair and maintenance cycle (RMC) of the gearbox of the KamAZ 6540 automobile, taking into account the pre-repair operating time. The proposed methodology for assessing the pre-repair resource allows you to select the resource elements of the gearbox, taking into account the value of their resource and coefficient of variation. The structure of the RMC proposed in the work provides for: the first current repair during running hours - 125940 km; first overhaul during running hours - 251880 km; second current repair during running hours - 377820 km; second overhaul during running hours - 503760 km. Thus, changing the structure of the gearbox REC allows increasing the gearbox resource from 380 to 500 thousand km.

1. Introduction. Automobiles manufactured by the Kama Automobile Plant are widely used in all industries. The KamAZ cars of the sixth series, in a structural respect, differ significantly from their predecessors. Due to the increased load capacity and speed of the car, Stayer bridges with a final drive ratio of 6.88 were used. According to the manufacturer, the resource of the engine and bridges is from 0.5 to 2 million kilometers [1]. In the basic configuration, a ten-speed gearbox is installed on the chassis of the KamAZ 6540 car. In connection with the above, the study of reliability indicators of this gearbox model is relevant. The change in the technical condition of the units during operation is characterized by significant dispersion, which indicates the advisability of using the statistical method to determine the pre-repair operating time.

2. Materials and methods. The current structure of the repair and maintenance cycle (RMC) of units due to current and overhauls is not rational, since it does not take into account the actual changes in the technical condition during operation and rational forms of repair. Improving the structure of the REC is advisable by calculating the rational periodicity of repair work, taking into account changes in the technical condition during operation. With this in mind, the aim of the study is to determine the pre-repair resource of resource elements and the rationale for a rational RMC GB.

In the study of various random processes using statistical analysis of the selective method. It consists in establishing the generalized characteristics of the totality of observation not for everyone, but only for a part of its constituent elements, taken by random selection. The legitimacy of the application of this method is based on the law of large numbers. According to this law, the average value of the attribute, measured in a large number of tests, approaches the mathematical expectation [2,3,4].

When determining the pre-repair time using the statistical method, the selection of the actual pre-repair time should be made from among the machines of a certain age composition, operated in real conditions.

To find the value of the pre-repair operating time of the GB, its mathematical value is determined, which in the case of a normal distribution has the form [5]:

(1)

Taking into account formula 1, the value of the pre-repair operating time of GB will be determined from the formula:

(2)

where σ_x- is a quarter of the area under the normal distribution curve and corresponds to the abscissa $t = \frac{2}{3} \approx 0,67$ or $x = \frac{2}{3}\sigma_x$.

Then, by the nomogram, the power distribution indicator m and the value of the pre-repair operating time corresponding to the segment on the abscissa, which cuts off 75% of the area of the distribution curve of the pre-repair operating time of the gearbox, are determined (Figure 1) [8,9,10].

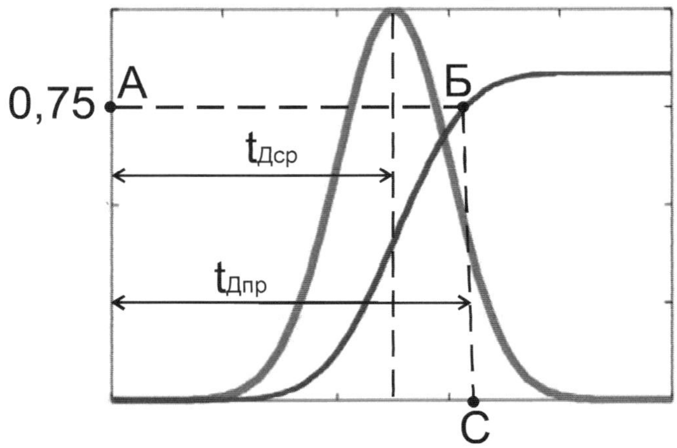

Figure 1. Scheme for determining pre-repair running hours graphoanalytical method

3Results and discussion. The observations were carried out for 23 vehicles that were operated in the Ural-Siberian region from 2007 to 2018.

In order to identify the least durable parts of the gearbox, limiting the amount of pre-repair running hours, data on the number and probability of occurrence of failures are analyzed.

The graphs of the distribution density function of the pre-repair resource and the integral failure curve of parts and assembly units are shown in Figures 2-5.

After the necessary calculations, the function of the distribution density of the pre-repair resource and the integral failure curve of the gearbox of the KamAZ 6540 automobile were obtained:

$$f(T) = 2{,}7 \cdot 10^{-3} \cdot \left(\frac{T - 125940}{40860}\right)^{0,2} \cdot e^{-\left(\frac{T-125940}{40860}\right)^{1,2}}. \quad (3)$$

$$F(T) = 1 - e^{-\left(\frac{T-125940}{40860}\right)^{1,2}} \quad (4)$$

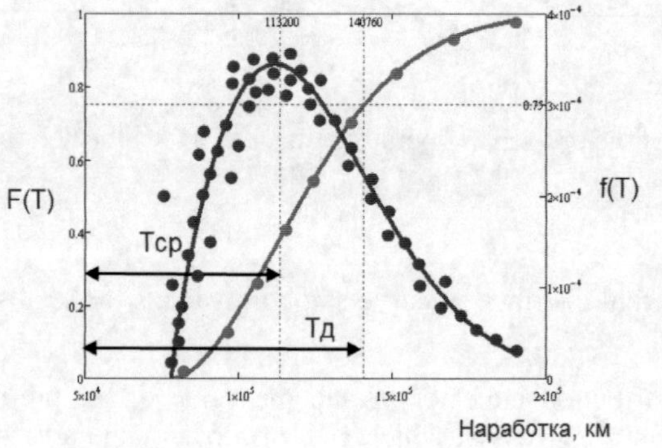

$$f(T) = 0{,}77 \cdot 10^{-3} \cdot \left(\frac{T-113200}{52330}\right)^{0{,}8} \cdot e^{-\left(\frac{T-113200}{52300}\right)^{1{,}8}}$$

$$F(T) = 1 - e^{-\left(\frac{T-113200}{52300}\right)^{1{,}8}}$$

Tср –average operating time;
Tд –allowable operating time;
F(T) - integral failure curve of parts and assembly units;
f(T) - distribution density function of the pre-repair resource
Figure 2.Graphs of the distribution density function of the pre-repair resource and the integral curve of the gear 5 gear failure

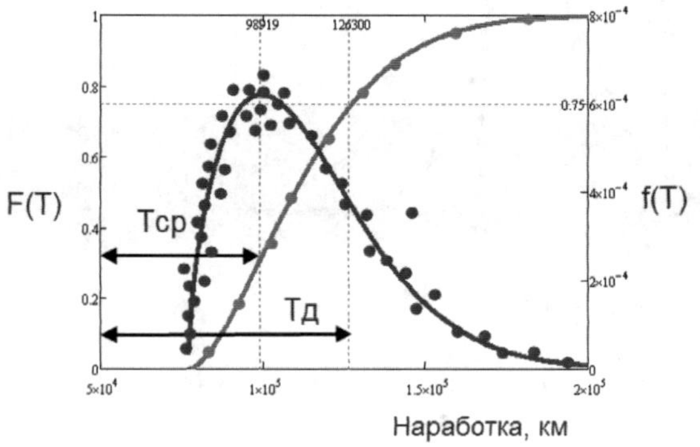

$$f(T) = 1{,}3 \cdot 10^{-3} \cdot \left(\frac{T-98919}{40860}\right)^{0{,}8} \cdot e^{-\left(\frac{T-98919}{40860}\right)^{1{,}8}}$$

$$F(T) = 1 - e^{-\left(\frac{T-98919}{40860}\right)^{1{,}8}}$$

Figure 3. Graphs of the distribution density function of the pre-repair resource and the integral gear failure curve of the 4th gear gear

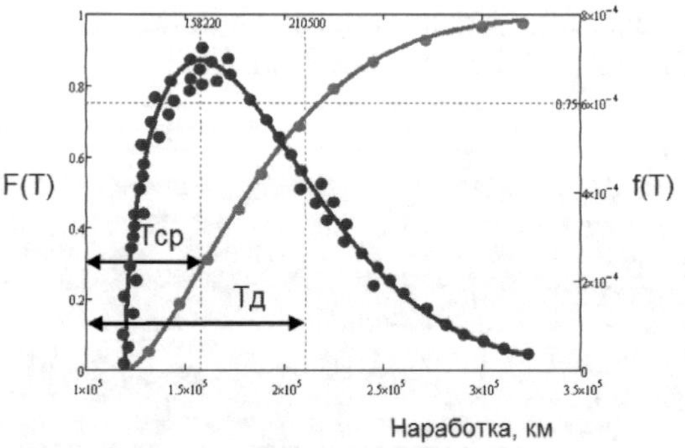

$$f(T) = 1{,}4 \cdot 10^{-3} \cdot \left(\frac{T-158220}{78000}\right)^{0{,}7} \cdot e^{-\left(\frac{T-158220}{78000}\right)^{1{,}7}}$$

$$F(T) = 1 - e^{-\left(\frac{T-158220}{78000}\right)^{1{,}8}}$$

Figure 4. Graphs of the distribution density function of the pre-repair resource and the integral failure curve of the rear driven shaft ball bearing

IOP Conf. Series: Materials Science and Engineering **941** (2020) 012071 doi:10.1088/1757-899X/941/1/012071

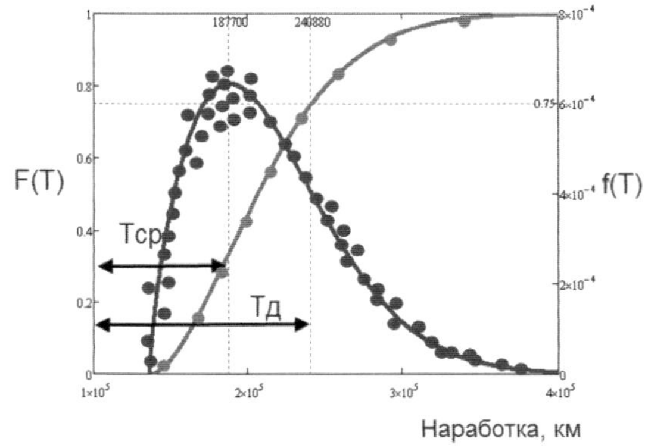

$$f(T) = 1{,}4 \cdot 10^{-3} \cdot \left(\frac{T - 187700}{86310}\right)^{0{,}7} \cdot e^{-\left(\frac{T-187700}{86310}\right)^{1{,}7}}$$

$$F(T) = 1 - e^{-\left(\frac{T-158220}{86310}\right)^{1{,}8}}$$

Figure 5. Graphs of the distribution density function of the pre-repair resource and the integral front roller bearing failure curve

Function graphs are demonstrated in Figure 6.

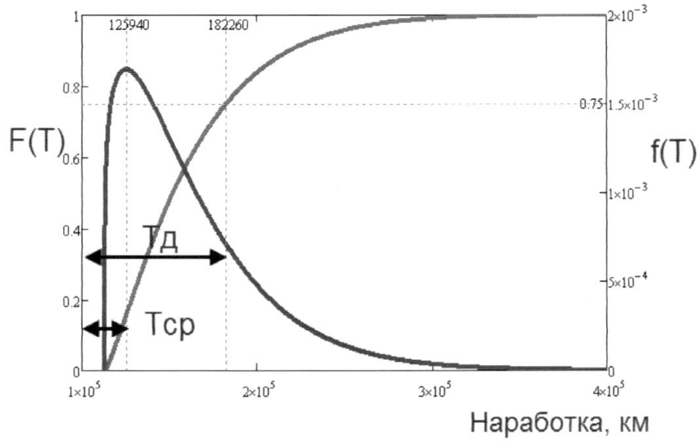

Figure 6. Graphs of the distribution density function of the pre-repair resource and the integral curve of the KamAZ 6540 gearbox failure

To enlarge the KP's repair life in the repair and maintenance cycle, it is necessary to provide for the replacement of a number of unreliable parts and assembly units, in particular: gears 5 gears, gears 4 gears, ball bearing of the rear driven shaft, front roller bearing.

The resource after overhaul of the control box is within 25-72% of the pre-repair resource. The resource of the GB significantly depends on the serial number of the repair. Differentiation by the repair number showed the following dynamics of the GB resource: after the 1st repair - 72%, 2 - 50%, 3 - 32%, 4 - 28% of the pre-repair. This is due to the mutual influence of transmission units, leading to an increase in dynamic loads in the transmission during operation. With more repairs, the service life of the control unit is reduced by 20% and the cost of repairs increases by 25%. From such a statement, it becomes obvious that a change in the gearbox RMC in the conditions of a repair enterprise is justified.

At present, the RMC structure has been formed, consisting of three overhauls and elimination of failures during the operation of the gearbox. As practice shows, such a RMC structure, after the third overhaul, leads to decommissioning of the unit due to its low after-repair resource.

It was recommended that the Kama Automobile Plant carry out overhaul of the gearbox while the vehicle was running for about 300,000 km, however, as our studies show, the actual average pre-repair running time is 125940 km. In the conditions of the repair enterprise, as a rule, without considering the reasons and conditions for the occurrence of a failure, overhaul of the control gear is carried out.

The proposed RMC structure provides for:
- the first current repair during running hours - 125940 km;
- first overhaul during running hours - 251880 km;
- second running repair during running hours - 377820 km;
- second overhaul during running hours - 503760 km.

Maintenance work No. 1 and No. 2 includes the replacement of low-resource parts and assembly units.

Thus, studies of the present work show that the inclusion of two current repairs in the structure of the RMC significantly increases the overhaul and overhaul life of the gearbox.

The transmission RMC for the proposed structure, taking into account the second overhaul, will be 503,760 km.

4 Conclusions.

1. Studies have shown that the average pre-repair running time of the gearbox of the KamAZ 6540 is 182,260 km.
2. The rational scheme of the repair and maintenance cycle of the KamAZ vehicle gearbox was proposed and recommended.

Acknowledgments
The authors are grateful to Professor Egorov AV and Associate Professor Zubova EV for the assistance in writing the article.

[1] V.N.Barun, R.A.Azamatov i dr. Avtomobili KamAZ: Konstruktsiya, tekhnicheskoye obsluzhivaniye i remont, M: Transport 2008 – 670 s.
[2] Avdon'kin F.N. Teoreticheskiye osnovy tekhnicheskoy ekspluatatsii avtomobiley, M.: Transport, 1985 – 215 s.
[3] Kuznetsov Ye.S. Upravleniye tekhnicheskoy ekspluatatsiyey avtomobiley, M.: Transport, 1982 – 272 s.
[4] Ushakov I. A Spravochnik po raschetu nadezhnosti M.: Mashinostroyeniye, 1980 - 606 s. GOST R 52567 – 2006. Avtomobili. Metody ispytaniy, M.: Izdatel'stvo standartov, 2006 – 29 s.
[5] Venttsel' Ye.S. Teoriya veroyatnosti M.: Vysshaya shkola, 1966 – 576 s.
[6] Lipman L.N. Posobiye dlya opredeleniya chisla ob"yektov nablyudeniy (ispytaniy), M.: Vysshaya shkola, 1966 – 200 s.
[7] Kazartsev V.I. Metodika opredeleniya optimal'nogo mezhremontnogo sroka sluzhby detaley mashin. // Vestnik mashinostroyeniya, 1963, №1
[8] Gnedenko B.V. Kurs teorii veroyatnostey, M.: Vysshaya shkola, 1988 – 488 s.
[9] Polozheniye o tekhnicheskom obsluzhivanii i remonte podvizhnogo sostava avtomobil'nogo transporta Minavtotransa RSFSR. M: Transport, 1986 – 73 s.
[10] Perspektivy sovershenstvovaniya tekhnicheskoy ekspluatatsii avtomobiley. M: TSBNTI Minavtotransa RSFSR, 1983 – 56 s.

IOP Conf. Series: Materials Science and Engineering **941** (2020) 012072 doi:10.1088/1757-899X/941/1/012072

Modern approaches of vehicle suspension durability evaluation at early stages of development

V A Kulagin and A I Bokarev

Central Scientific Research Automobile and Automotive Engines Institute, 2 Avtomotornaya Street, Moscow, 125438, Russia

E-mail: viktor.kulagin@nami.ru

Abstract. Ensuring demanded service life of a vehicle is a relevant engineering problem. Currently, resolving this problem is heavily complicated. Modern tendencies of automotive industry assume shortened development cycle for new vehicles and sustaining wide lineup of cars for manufacturers. Performing long-term durability testing of vehicle subsystems is becoming limited. Meanwhile, methods of laboratory testing, which provide higher acceleration degree of damage accumulation in comparison with testing on proving ground, are progressing. Virtual simulation methods, which allow analyzing and optimizing a new design at early development stages, are becoming widespread as well. Modern durability analysis methods, based on laboratory testing and virtual analysis, and load cycle generation methods, used for testing and simulations, are considered. Some points are noted, such as an extensive use of described methods of durability research among foreign car manufacturers, and the relevance of developing this kind of durability evaluation methodology for the Russian automotive industry.

Introduction

Ensuring failure free service life of a vehicle, like any machine, is a relevant engineering problem. In recent years, resolving this problem is heavily complicated by the factors caused by the world automotive market conditions and carmakers competition.

One of the factors, which determine progress of automotive industry, is shortening a development cycle for new vehicles, in average, from 48 to 25 months. Acceleration of vehicle development is explained by both constant vehicle lineup expansion and shortening the period of model generation change. These tendencies are noticeable on the example of Volkswagen AG. Volkswagen Golf's model cycle decreased from 9 to 4 years generation to generation. Since 1950's, the Volkswagen lineup has been broadening from 3 to, at the lowest estimate, 18 models for today [1]. It worth noting that each car model has a number of configurations, which may vary with powertrain and drivetrain type, and available active or adaptive chassis systems, which have an impact on contact patch forces.

Shortened duration of new vehicles development and sustaining a lineup diversity demand an optimization for each stage of the engineering process. One of the optimization way is wide implementation of means of simulation analysis to ensure targeted service properties of vehicle subsystems and components. Meanwhile, duration of on-road testing program is reducing.

In current conditions ensuring the chassis durability is of a particular interest. The complexity of development and evaluation of suspension is based on the purpose of a suspension in a vehicle. During vehicle's service, a suspension is transmitting loads, arising from contact patches, to a vehicle body, damping sprung and unsprung mass oscillations caused by loads and road profile and conceiving reaction loads from drivetrain components. Fatigue properties of suspension arms are highly important for ensuring preassigned service life. According to this, the article is mostly dedicated to suspension durability analysis. The provisions given in the article may also be applicable to other components and subsystems analysis.

Methods of accelerated durability testing are based on providing damaging effect on vehicle subsystems, which is equal to impact, accumulated during service life. On proving ground, acceleration of damage accumulation is achieved by the driving cycle, scaled to target service life mileage. The number of tracks and single obstacles (events), order and driving mode of events determine the driving cycle. Driving route also includes connection roads for moving between tracks, wherein the severity of damaging impact caused by connection roads is negligible. Other circumstances, which affect on proving ground testing duration, are shift-based operation hours of test drivers and weather conditions.

Russian automotive industry enterprises practice performing this kind of testing for prototypes, which are to be prepared in last engineering stages. Meanwhile, the companies have not developed own principles of durability evaluation at early development stages, in contrast to foreign car manufacturers. All leading carmakers perform strength and fatigue assessment of new design before the first vehicle prototypes being built. Utilized research methods for this kind of assessment ensure demanded information value of detected failures and comparison with supposed service mileage. Implementation of this approach involves development of unique methodologies and use of cutting-edge testing equipment.

Ways of laboratory testing

The most effective way of chassis durability analysis at the design and development stage is performing semi-natural testing on a state-of-the art test rig. Advantages of laboratory testing are:

- Ability to be performed at any vehicle development stage for rapid obtaining accurate results of failure detecting;
- Vehicle units and subsystems might be used as a research object as well as a vehicle assembly;
- Providing control and monitoring of input loads signals and physical system response signals;
- Ability to rapid identification of weak points of vehicle design, damage propagation and components properties degradation;
- Repeatability of testing and statistical data accumulation;
- Providing continuous loading of object during a testing cycle;
- Providing higher degree of load severeness needed to accelerate damage accumulation [2].

With optimized input and continuous test rig operation, duration of laboratory fatigue testing can be up to 10 times less than duration of durability testing on proving ground.

One of the most used approaches of performing accelerated laboratory testing of a vehicle frame structure is use of a hydropulsator N-post rig (figure 1). Input load of this facility determined as a wheel platforms vertical displacement, independent from each other, which imitates load and kinematic factors exposure on a vehicle. Emerging normal wheel loads makes a significant contribution to damage accumulation. However, lack of complicated impact of wheel forces and torques as well as reactive loads produced by drivetrain torques makes this approach imperfect for a complete identification of weakest points in a suspension design. According to this, described type of laboratory testing is suitable only for frame structure fatigue analysis.

Figure 1. Hydropulsator N-post test rig

Independent fatigue testing of each component with uniaxial load has the similar issue. Testing facilities, used for this purpose, are unable to provide such loading conditions with critical cross-sections of a component to be loaded in the same way as during vehicle service. In addition, there is a risk to overload other cross-sections, which can lead to component failure to be not similar to the same one happened in a vehicle assembly.

To avoid mistakes described above, 6 degree-of-freedom multichannel spindle-coupled testing systems (figure 2) were developed for suspension assembly testing. These test rigs allow setting contact patch forces and torques (up to 12 inputs for one axle), reaction loads from drivetrain units, steering and active/adaptive suspension systems operation. Laboratory testing on a spindle-coupled facility can provide high accuracy of damage accumulation and full information about weakest points of design and identify components properties changing during the test.

Figure 2. 6-DOF multichannel spindle-coupled test rig

Methods of suspension module accelerated durability testing with a spindle-coupled rig have been confirming their efficiency since the first implementation of them. Thanks to deserved recognition, these test rigs are widely adopted by the leaders of the world automotive industry, such as Porsche Engineering and Magna. Spindle-coupled test rigs find an application for military and racing cars development as well (figure 3).

Figure 3. HMMWV mounted on a spindle-coupled test-rig

The most advanced testing equipment allow considering a negative impact of environment on vehicle units durability. This impact is simulated via supply of a mix of water and salts to an object periodically during the testing cycle.

Computer-aided engineering methods of durability analysis

As the development cycle of new vehicles is constantly shortening, CAE software becomes more widespread for engineering tasks. Making simulation analysis of vehicle performance properties allows to identify weaknesses of a design before first prototypes to be assembled and tested, that helps to avoid undesirable material and financial resources spending. Additional advantage of using CAE tools is the fact that numerical calculation is the only element of fatigue design process, which provides well-controlled certainty and accuracy [3].

Multibody system models are used for calculation of suspension parts loads regarding the impact of wheel loads. The main advantages of MBS models are functionality of a detailed physical object properties description, simplicity of properties editing and event simulation. An MBS model of a vehicle or a suspension subsystem allows considering suspension kinematics, linear and non-linear properties of spring, dampers and bushings, and the operation of existing active or adaptive systems of suspension control.

Suspension loading simulation can be performed with a numerous approaches. The simplest way is to simulate operation of a suspension on a virtual multichannel test rig. This approach is noticeable with the fact that physical object testing results can be compatible with calculation results that allows an MBS model accuracy improvement. Improvement of a model accuracy can be achieved with an iterative editing of properties of suspension components [4]. If it is necessary, an MBS-model can be sophisticated with a detailed description of testing equipment used for performing of physical object testing (figure 4) [5].

Figure 4. A suspension test rig based on the hexapod concept

Some leading companies of an automotive industry use approaches of simulation of full vehicle assembly driving through proving ground tracks and single obstacles. This way of performing a load calculation is the most advanced, but at the same time has higher demands to be implemented. It is necessary to have a number of virtual copies proving ground tracks, obtained with laser scanning technologies. Quality of responded wheel loads is provided by high-accuracy frequency-dependent tyre model, which is able to perceive loads from the road surface in the whole contact patch, not in one contact point. The FTire model is widely used format of this kind of tyre model [6]. In addition, the driver model used in an MBS software can have limits of driving path choice and maximum longitudinal and lateral accelerations to be achieved. It is a very tricky point of simulation of events, which assumes driving on the grip limits. As a result, vehicle chassis can be underloaded, and the targeted chassis service life could not be checked. In the end, higher computation capacity is demanded for this kind of simulation in comparison to a suspension assembly load simulation. As the number of simulated events is not necessary restricted, automatized event formation is preferable.

Estimation of damage accumulation has several approaches as well. The most accurate way is implementing "flexible" parts (e.g. suspension arms) into an MBS-model after being prepared by methods of finite element modelling. Stress and strain of these parts under load is determined as a sum of linear deformation according to eigen modes of parts. Calculated time-dependent stress-strain state is exported to fatigue analysis software. Approach of estimation of pseudodamage accumulation, where stress-strain state replaced by loads on component, is also acceptable [3].

For car manufacturers and engineering centers, which are working on their method of virtual analysis of chassis durability, development of a detailed MBS-model of suspension modules with flexibility of suspension arms is sufficient.

Performing simulation analyses allows effective optimization of durability evaluation at early stages of vehicle development in terms of testing duration and quality of results. Combined durability research with both laboratory testing and simulation analysis ensure high accuracy of results. Vehicle testing on proving ground remains necessary at further development stages as well as semi-natural testing for more detailed failures research.

Load cycle formation

Additional testing duration reducing in comparison with testing on proving ground tracks is provided by use of shorter load cycles with higher values of loads, which ensure damage value to be equal to damage accumulated during service life of vehicle. Load cycle is a set of load signals, which describe contact patch forces and reaction forces caused by drivetrain loads. Approaches of load cycle formation are:

- Editing previously recorded load signals on tracks based on accelerated durability testing driving cycle;
- Determination of artificial (synthetic) signals based on vehicle properties, targeted service life and service conditions rate.

Initial set of signal could be obtained during a driving cycle of accelerated durability testing on proving ground, performed by vehicle, comparable with a new vehicle under development. Adjustment of recorded loads is conducted by scaling according to weight, size and other properties of a new car.

Obtained signals are necessary to be optimized to ensure the same damage accumulation for less duration. It is possible with excluding of parts of signal, which cause a small amount of damage. Thus, signals of driving on connection roads during the driving cycle can be removed completely. In addition, there a plenty of amplitude-based and frequency-based methods of signal editing. Choice of signal processing method depends on the impact on a vehicle. Amplitude-based methods are used for description the signal as a rainflow cycle counting or range pairs of load. Frequency-based methods allows determining input signal based on vehicle properties as an oscillation system. Special attention is required to wheel load data processing because of their simultaneous influence, and synchronicity of load impact is crucial to be saved [3].

Utilizing a synthetic load cycle allows achieving the highest degree of laboratory testing acceleration and ensure expected value of damage accumulation during service life. Synthetic signal consists of a number of sinusoidal signals with determined frequency and amplitude. Synthetic load cycle must take into account such demanded information as type of a vehicle, service conditions rate, expected service mileage, weight properties, stiffness and damping properties of a suspension, etc. According to the correlation of synthetic load cycle with supposed service conditions of a vehicle, the cycle can be generated with achievement of scaling possible number of important load signals in the limits of linear deformation of parts (Hooke's Law) [3]. It means that artificial load cycle can provide much more severe impact than proving ground tracks, so it allows reducing the time of testing performing. Use of synthetic load cycle may exclude a road load data recording phase with similar cars, but it is necessary to approve a creditability of determined load cycle with a reliable statistic data of vehicle service in different conditions.

Conclusion

Current level of technology development suppose use of a variety of vehicle fatigue evaluation methods at early stages of development. This article reflects generalized concepts of existing methods of accelerated durability laboratory testing, simulation analysis and load cycle formation.

Leading car manufacturers and engineering centers are developing or have already had own methodologies of fatigue evaluation. These methodologies are a commercial secret and cannot be found in a public domain.

FSUE "NAMI" has a broad experience in performing accelerated durability testing of vehicles on roads of "NICIAMT" proving ground. Meanwhile, described approaches of chassis durability analysis at early development stages have not been embodied yet by Russian scientists and researchers. Development the methodology, which reflects service conditions on roads of Russia, can be useful for being utilized by Russian manufacturers and foreign car makers, which want to keep competitive performance on Russian automotive market, as well.

To sum up, development and implementation of accelerated durability testing methodology of a vehicle based on virtual and semi-natural experiment technologies is a relevant and highly sought scientific and engineering problem for Russian automotive industry. In addition, problems, which are to be resolved, are:

- Increasing efficiency of fatigue research results;
- Shortening duration of laboratory testing in comparison with road testing;
- Ensuring thc possibility of road testing replacement on laboratory testing;
- Reduction of material and human resources for preparation physical objects for testing.

Taking into account all the advantages and disadvantages of described approaches, possible methodology of accelerated durability evaluation at early development stages can have a good potential with these methods:

- The method of synthetic load cycle formation for chassis;
- The method of performing virtual experiments with a highly detailed MBS-model of a suspension assembly;
- The method of performing semi-natural testing of a suspension assembly on the multichannel spindle-coupled test rig.

References

[1] Sabadka D, Molnar V and Fedorko G 2019 *TEM Journal* **8 (4)** 1295-301
[2] Brudnak M et al 2016 *SAE Int. J. Commer. Veh.* **10 (1)** 113-21
[3] Johannesson P and Speckert M 2014 *Guide to Load Analysis for Durability in Vehicle Engineering* (Chichester: Wiley) p 458
[4] Numata F and Gattringer O 2018 *JSAE Annual Congress (Spring)* **409** 20185409
[5] Dressler K et al 2009 *Vehicle System Dynamics* **47 (4)** 387-16
[6] Li B, Yang X and Yang J *SAE Int. J. Veh. Dyn., Stab., and NVH* **2 (1)** 71-87

IOP Conf. Series: Materials Science and Engineering **941** (2020) 012073 doi:10.1088/1757-899X/941/1/012073

Assessment of the technical and technological condition of the transport complex

A V Gubin[1], A V Pavlova[1], Yu V Kosolapov[1], D E Morkovkin[1], E S Lineva[1] and A A Gibadullin[2]

[1] Russian University of Transport, 9b9, Obrazcova Street, Moscow, 127994, Russia
[2] State University of Management, 99, Ryazan Avenue, Moscow, 109542, Russia

E-mail: 11117899@mail.ru

Abstract. The presented study is devoted to the assessment of the technical and technological state of the transport complex. The work analyzed indicators reflecting the share of fixed assets of the transport complex as part of national fixed assets and the degree of depreciation of production capacities of the transport industry. The analyzed indicators reflect that about a quarter of all fixed assets are related to the transport complex, and the volume of depreciated fixed assets exceeds 50%, while production capacities are updated to increase fixed assets, and not to ensure a qualitative or quantitative upgrade. The study proposed a model for ensuring the qualitative development of the transport complex, which contains measures to update existing fixed assets and new principles for the management and development of the transport complex.

1. Introduction

The development of various fields of activity in Russia was carried out as part of the Soviet Union, when the main types of activity were formed, new types of production were created, large industrial complexes were built and the existing production potential was formed. In the same period, a transport complex was formed, which included not only road, rail and city transport, but also pipelines were built, sea and air transport and other types of transport were created. The indicated types of transport until today carry out interstate transportation, the potential of which is today used almost unchanged [1].

The creation of new modes of transport was caused by the need to transport people over long distances, moving products and services within the framework of a nation state, in order to provide the necessary resources to all consumers. In the same period, the main transport corridors were created, which were based on the assessment of congestion, demand, deficit and excess of goods and services, as well as the construction of storage facilities and transport and logistics centers were carried out [2-3]. At the same time, the use of equipment, devices, buildings and structures from the Soviet period reduces the quality of the production process and negatively affects the environment [4].

In this regard, we consider it relevant to conduct a study in which the technical and technological condition of the transport complex will be assessed, namely, the operating periods of buildings, structures, machinery and equipment, vehicles are determined, and the volumes of renewal, retirement and depreciation of production capacities in transport complex.

2. Materials and methods

The purpose of this study is to assess the technical and technological state of the transport complex and formulate recommendations on the development of the transport complex of Russia. The following tasks were set in the work:

- Assess the technical and technological condition of the transport complex;
- To propose measures for the development of the transport complex.

The study was based on data that was published in open sources, and statistical, logical, and analytical methods were used as methods and approaches, which made it possible to reveal the goal of the study.

3. Results

The modern industrial structure on a global and national scale has different scales and affects different areas of activity. Not only has the development of various types of activity, but also the stability of the state and population depended on the functioning of the industrial complex [5]. One of the industries that affect the stability of the state and the world community is the transport complex, which provides export and import of various goods, increases the volume of trade between states, carries out the movement of passengers and, in general, ensures the stable functioning of the global industrial complex [6].

To conduct an objective study, it is necessary to consider the composition of fixed assets by type of activity in order to determine the share of the transport complex in fixed assets (figure 1) [7].

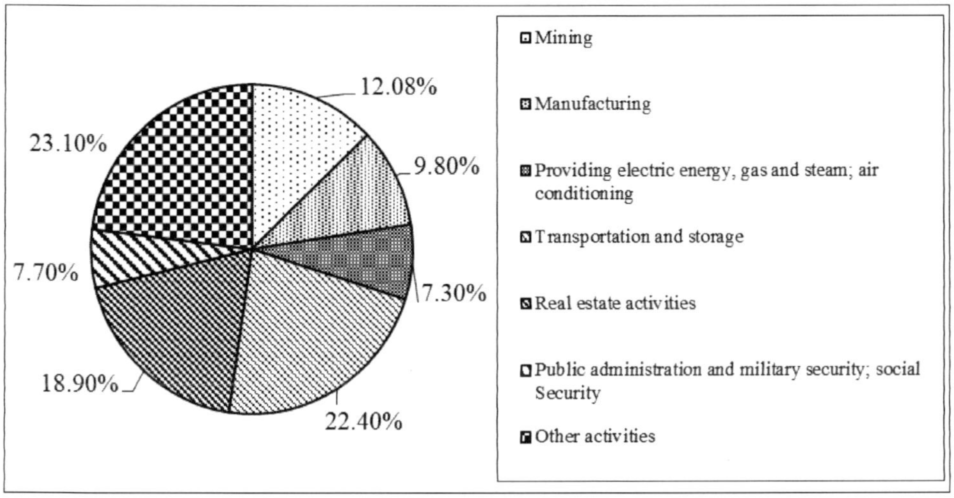

Figure 1. The composition of fixed assets by fields of activity, as a percentage.

From the presented figure, it is clear that 22.4% of fixed assets account for transport and storage, and this indicates that the capital industry in the transportation industry is a leader among other areas of activity.

Further, it is advisable to analyze indicators reflecting the state of fixed assets of the transport complex. Assessment of the technical and technological state of production assets in the modern scientific literature is carried out using various indicators and coefficients. In our opinion, it is advisable to base the assessment of the technical and technological state on the analysis of worn-out capacities, input and output of fixed assets (figure 2) [7].

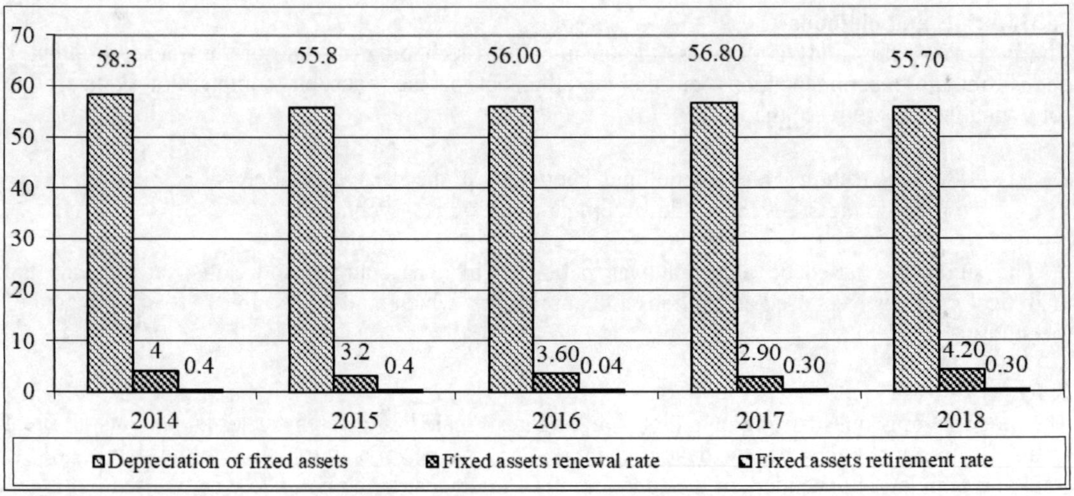

Figure 2. The state of fixed assets, in percent.

The figure indicates that the volume of worn-out capacities is more than 50%, but at the same time, a slight decrease in this coefficient is observed. The coefficient of renewal of fixed assets does not exceed 4%, but in the period under review, this coefficient has variable indicators. The fixed asset retirement rate does not exceed 0.4%, and if we consider this ratio together with the renewal rate, we can conclude that it is not the main production assets of the transport complex that are updated, but their volumes increase.

Consider the volume of completely worn out basic production assets (figure 3) [7].

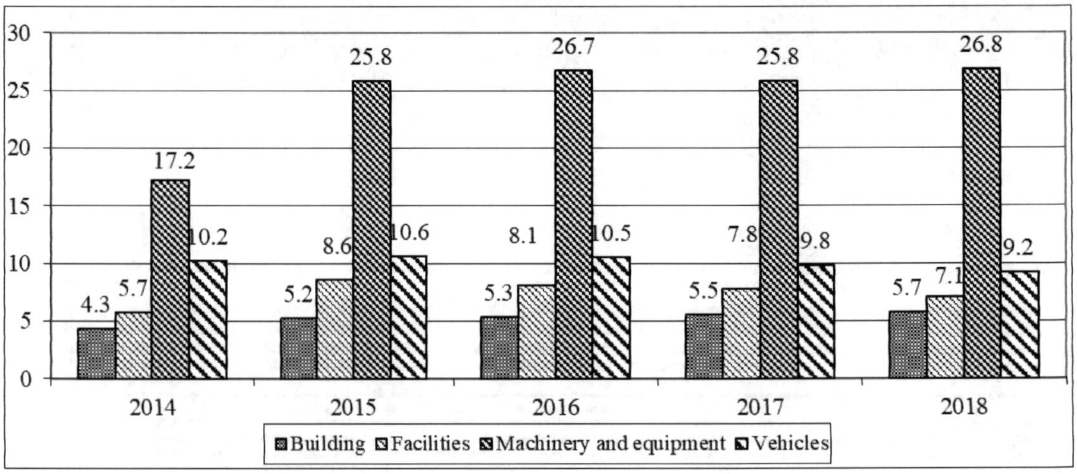

Figure 3. The percentage of completely worn out fixed assets in the transport sector, in percent.

From the presented figure, it can be seen that in the transport complex, "machinery and equipment" and "vehicles" are considered the most worn out. It is worth noting that during the period under review there is no qualitative change in fixed assets, moreover, in terms of "building", "construction" and "machinery and equipment" these indicators increase.

Thus, the analysis showed that the technical and technological condition of fixed assets of the transport complex is deteriorating every year and today it is necessary to carry out activities aimed at the qualitative and quantitative updating of fixed assets [8].

IOP Conf. Series: Materials Science and Engineering **941** (2020) 012073 doi:10.1088/1757-899X/941/1/012073

4. Discussion

Ensuring the development and transition to new technologies, including innovative and digital, requires the transport sector to completely restructure and create certain conditions for this area of activity. Undoubtedly, the transition to innovative technologies that provide an increase in the efficiency of the field of activity will reduce production-operating costs, reduce the consumption of fuel and energy resources and increase the quality characteristics of production [9-12].

In our opinion, the model for ensuring technical and technological development can be represented in the form of the following scheme (figure 4) [13-17].

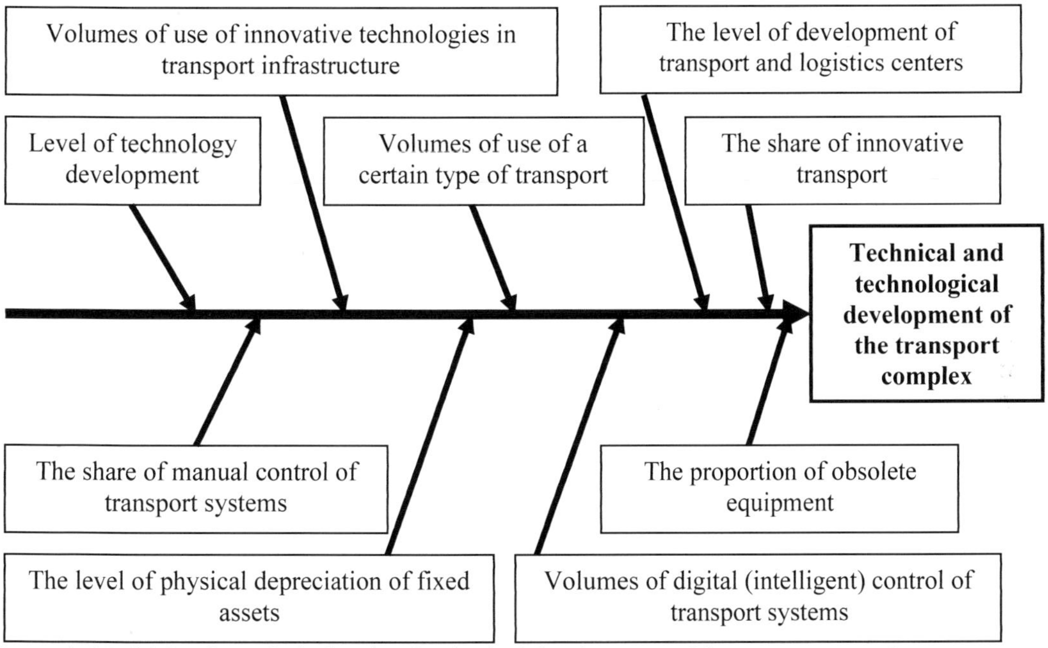

Figure 4. Model for the technical and technological development of the transport complex.

Thus, the presented model will ensure the transition of the transport complex to a qualitatively new path of development; the implementation of the model will allow the introduction of new and updated production capacities and will ensure the transition to innovative and digital (intelligent) transport management systems.

5. Conclusion

The study revealed the problems and prospects for the development of the transport complex. The work analyzed technical and technological indicators that reflect the share of worn-out capacities. The presented analysis showed that the transport complex has a high proportion of worn-out capacities, and at the same time, the capacity is being updated to increase fixed assets, rather than eliminating worn-out facilities. In the work, a model was proposed for ensuring the technical and technological development of the transport complex, which contains mechanisms for updating production capacities, as well as for switching to a qualitatively new path of development and management of the transport complex.

References

[1] Yuryeva A A, Kuternin M I, Gibadullin A A and Romanova Ju A 2019 Formation of mechanisms for the development of innovative activity in the industrial production of the Russian Federation. *Journal of Physics: Conference Series* **1399** 033099

IOP Conf. Series: Materials Science and Engineering **941** (2020) 012073 doi:10.1088/1757-899X/941/1/012073

[2] Ustyuzhanina E, Evsukov S and Komarova I 2018 Network economy as a new economic system. *European Research Studies Journal* **21(3)** 77-89

[3] Gibadullin A A, Pulyaeva V N, Kharitonova E N and Kharitonova N A 2019 Convergence of the Russian power industry in a period of declining reliability and sustainability of the industry. *Journal of Physics: Conference Series* **1399** 033034

[4] Sharipov F F and Timofeev O A 2020 Coal mining and import in China in 2019: prospects for russian exporters. *Ugol' – Russian Coal Journal* **4** 68-71

[5] Rudaya I L 2019 Issues of strategic management of Russian innovative organizations' intellectual property. *Upravlenie* **7 1** 50-9

[6] Sadriddinov M I, Mezina T V, Morkovkin D E, Romanova Ju A and Gibadullin A A 2020 Assessment of technological development and economic sustainability of domestic industry in modern conditions. *IOP Conference Series: Materials Science and Engineering* **734** 012051

[7] *Russian statistical yearbook* 2018 (Moscow: Rosstat) 694

[8] Morkovkin D E, Mamychev A Y, Yakovenko N V, Komov I V, Derevyagina M V and Didenko O V 2016 Factors and material conditions for space-intensive economic development of region. *International Review of Management and Marketing* **6 S1** 67-72

[9] Kolesnikov A V, Zernova L E, Degtyareva V V, Panko Iu V and Sigidov Yu I 2020 Global trends of the digital economy development. *Opción* **26** 523-40

[10] Gureev P M, Degtyareva V V and Prokhorova I S 2020 National Features of Forming a Digital Economy in Russia. *Advances in Intelligent Systems and Computing* **1100** 13-20

[11] Gibadullin A A, Sadriddinov M I, Kurbonova Z M, Shedko Yu N and Shamraeva V V 2020 Assessment of factors ensuring sustainable development of the electric power industry in the context of transition to renewable energy sources of the national economy. *IOP Conference Series: Earth and Environmental Science* **421** 032051

[12] Davnis V V, Tinyakova V I, Blinov A O and Volodin Yu V 2019 Combined Modeling of Projected Evaluation of the Regional Socio-economic Development *International Journal of Economics & Business Administration (IJEBA)* **0** 1 pp 348-54

[13] Shumaev V A, Morkovkin D E, Nikonorova A V, Nezamaikin V N, Yurzinova I L 2018 Innovative aspects of agritourism project management. *SERVE 2017, Financial and Economic Tools Used in the World Hospitality Industry - Proceedings of the 5th International Conference on Management and Technology in Knowledge, Service, Tourism and Hospitality* 241-8

[14] Tolkachev S A et al 2020 *IOP Conf. Ser.: Earth Environ. Sci.* **421** 032041

[15] Romanova I N, Morkovkin D E, Nezamaikin V N, Gibadullin A A and Ivanova M A 2020 Formation of a policy to ensure environmental safety in modern economic conditions. *IOP Conference Series: Materials Science and Engineering* **734(1)** 012166

[16] Bryukhovetskaya S V, Artamonova K A, Gibadullin A A, Ilminskaya S A and Kurbonova Z M 2020 Management of digital technology development in the national economy. *IOP Conference Series: Earth and Environmental Science* **421** 042018

[17] Plotnikov V, Makarov I, Shamrina I and Shirokova O 2010 Transport development as a factor in the economic security of regions and cities. *E3S Web of Conferences* **91** 05032

IOP Conf. Series: Materials Science and Engineering **941** (2020) 012074 doi:10.1088/1757-899X/941/1/012074

Formation of a digital transformation system for the transport complex

S A Tolkachev[1], D E Morkovkin[1], P S Shcherbachenko[1], A A Gibadullin[2] and A A Bykov[3]

[1] Financial University under the Government of the Russian Federation, 49, Leningradsky avenue, Moscow, 125993, Russia
[2] State University of Management, 99, Ryazan Avenue, Moscow, 109542, Russia
[3] Belarus State Economic University, 26, Partizansky avenue, Minsk, 220070, Belarus

E-mail: 11117899@mail.ru

Abstract. The presented study is devoted to assessing the level of digital development of organizations and the formation of a digital transformation system for the transport complex. The digital transformation of the transport complex is considered in the framework of the concept of the cyclic sequence of the spread of basic technologies in the economy, in accordance with which the full-scale deployment of the "transport" technological structure is predicted within the framework of the third mega technological cycle in the middle of the 21st century. The study analyzed indicators reflecting the level of digital development of organizations, transport companies and the amount of data protection tools used. Because of the analysis, it was revealed that in general, organizations have a high share of digitalization, but they are significantly lower in organizations of the transport complex, while about a third of organizations do not use data protection technologies. The study proposed a model for providing digital transformation of the transport complex, containing the types of activities that should be involved in digital transformation, the potential used for digital transformation and tools for digital transformation of the transport complex. In conclusion, conclusions were drawn from the results of the study.

1. Introduction

In recent years, the issues of the transition of industrial complexes to innovative, digital and intelligent technologies in all areas of activity, which will ensure a qualitative transition to a new level of existence and development of activities, have been updated. On a global industrial scale, the stages of transition to a new development path are associated with the introduction of artificial intelligence in the production process, but in Russia, it is based on the need to ensure innovative and digital development. The production potential of Russia was created back in the Soviet period, which today is used almost unchanged and requires the speedy resolution of issues related to the introduction of new technologies in all areas of activity.

The transport complex was also created in the Soviet period, namely during this period transport routes were built, new modes of transport were created, the main transport hubs were built and a large number of transport infrastructure was created. However, during the formation of these types of activities, the task was solved to ensure the transportation of goods and passengers in order to reduce operating and capital costs. To date, the indicated infrastructure and modes of transport do not always meet modern requirements for environmental safety, trends in innovative and digital development [1].

The significance of the study of the laws of digitalization of the transport complex is theoretically justified in the concept of the industry sequence of the spread of basic technologies. In accordance with this concept, each industrial revolution is a mega-technological cycle, consisting of three technological structures that cyclically replace each other - production, transport, and information and communication. The latter plays the role of a transitional link between the industrial revolutions, preparing the mass renewal of production assets in industry and transport.

In the middle of the XIX century, within the framework of the first megacycle, a transport system based on a steam engine developed - steamboats, steam locomotives. In the middle of the twentieth century, based on advances in basic technologies of electricity and an internal combustion engine, the transport system in the form of motor vehicles, electric trains, and aviation received an impetus for growth [2].

Finally, in our time, in the process of deploying the technologies of the fourth industrial revolution in the field of production (digitalization of the manufacturing industry), a technological basis is being prepared for the future transport system, which will enter a phase of rapid growth in the middle of the 21st century. We give it the conditional name "seventh transport technological structure." The fundamental technological basis for it will be well-established developments in the field of the Internet of things, artificial intelligence.

At present, we are already seeing the first breakthrough pioneer developments that are in the phases of a pilot test or first commercial application. In the field of automobile transport, the massive use of “drones” will continue. Very promising in the near future will look and investments in the production of individual vehicles: electric scooters, electro street etc.

In the field of air transport, drones already have a number of important functions: monitoring weather conditions and territories, quickly delivering the necessary cargo, conducting military operations without the need to include human units, etc. However, it is not too long now, when, probably, everyday use will become a reality passenger drones [3-5].

In the field of railway transport, a shift to magnetic repulsion technologies is planned to increase the composition and ensure the movement of the train. Deep technological modernization of rail transport is already taking place in line with the digitalization of its basic technical content.

Finally, low-pressure technologies can become a revolutionary breakthrough based on the use of new physical principles in the field of transport - Hyperloop is the fifth mode of transport after a train, plane, car and ship.

In all areas of activity there is a gradual process of introducing innovative technologies, issues are being considered to create new technologies for production activities and introduce stand-alone devices for managing various processes in enterprises and organizations [6]. The transport complex is also subject to transformation, as a result of which various digital devices for tracking goods and modes of transport are introduced, technologies in the field of intelligent control of transport systems are developed, new modes of transport are created, and much more. At the same time, not all activities of the transport industry have undergone digital development, which is why, we believe, it is necessary to assess the level of digital transformation of the transport complex and form a system of digital and intellectual transition of the transport complex to a new development path.

2. Materials and methods

The purpose of this study is the formation of a digital transformation system for the transport complex. The following tasks were set in the work:

- Assess the level of digital transformation of organizations and the transport complex;
- Form a digital transformation system for the transport complex.

The study used common methods and approaches that made it possible to reveal the purpose of the study, and information published in open sources and on the Internet were used as a statistical base.

3. Results

The digital transformation in Russia began after 2016, when regulatory acts were adopted aimed at ensuring the transition of all areas of activity, including the state apparatus, to innovative and digital technologies. Similar requirements are contained in industry documents and corporate development programs, which are aimed at improving the efficiency of production activities, reducing operating and capital costs, improving the quality of products, reducing negative environmental impact, and much more [7].

Digital transformation is aimed at improving the quality of managerial decisions, monitoring production activities, reducing human participation in production, ensuring remote management, collection, processing, accumulation and storage of various information. In general, for an objective determination of the level of digital development, it is necessary to analyze organizations that use information technology and information and telecommunication networks in their activities (figure 1) [8].

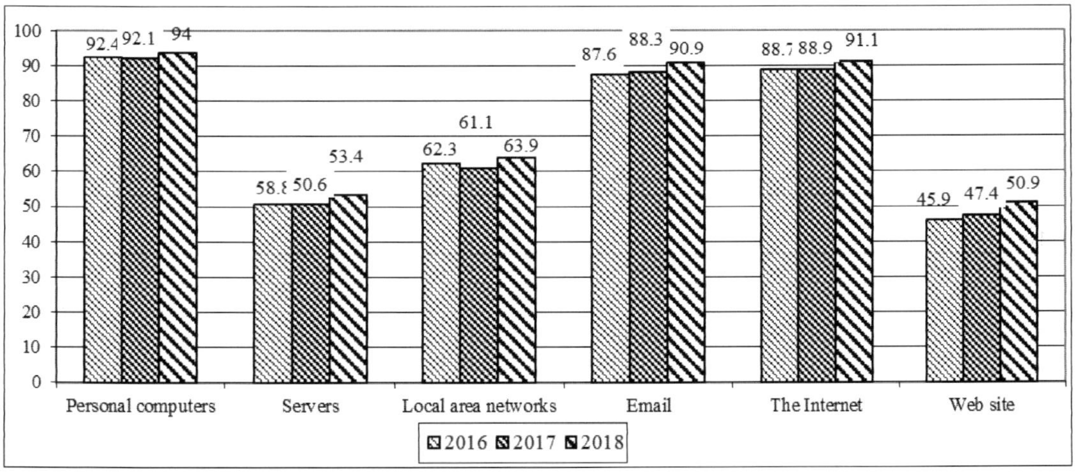

Figure 1. Organizations using information technology and information and telecommunication networks, as a percentage of the total number of organizations.

From the presented figure it can be seen that over the considered period of time, all indicators for information and telecommunication technologies increase. At the same time, it can be noted that almost all organizations use personal computers, the Internet and e-mail, but only half of the organizations use servers, local area networks and websites.

Next, we consider the volume of use of information technology and information and telecommunication networks in the field of "transport and storage" (figure 2) [8].

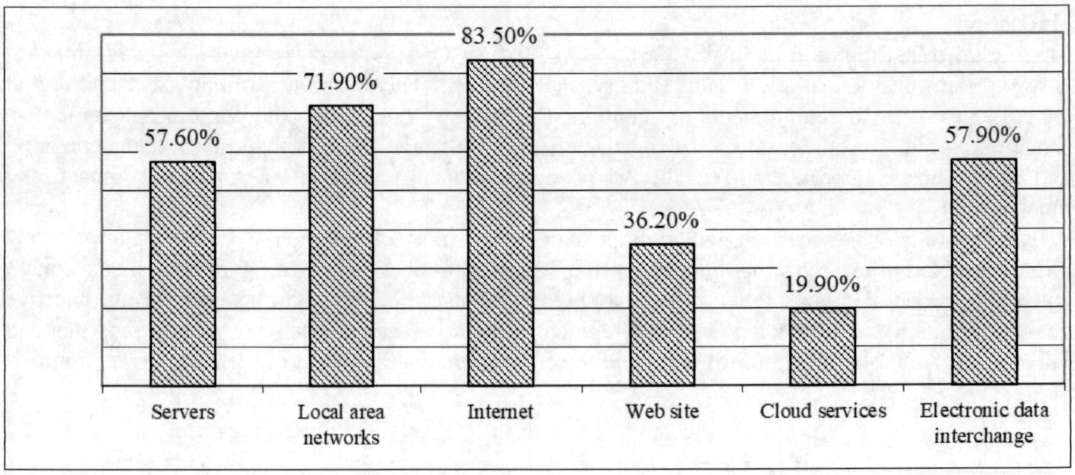

Figure 2. Transport complex organizations using information technologies and information and telecommunication networks, as a percentage of the total number of organizations.

The figure shows that 83.5% of organizations use the Internet, 71.9% of transport companies uses local area networks, about 58% use servers and electronic data exchange, and less than a third of organizations use websites and cloud servers. Thus, we can conclude that the level of digital development in the transport sector is lower than in the economy as a whole.

Consider the number of organizations in the transport complex that use information security tools (figure 3) [8].

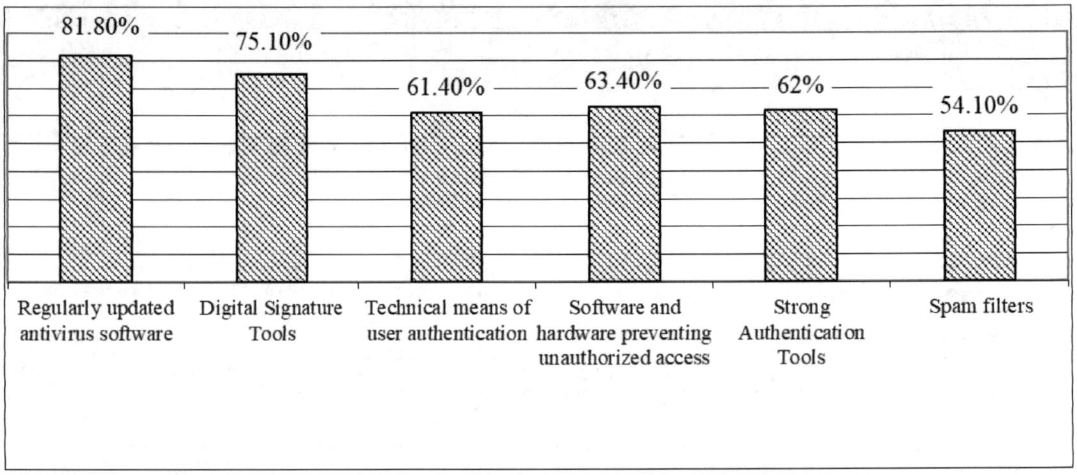

Figure 3. Organizations of the transport complex using information protection tools, as a percentage of the total number of organizations.

The presented graph shows that, in general, the transport complex uses technologies to protect information and data, however, we see that about a third of organizations do not use such means of protection.

Thus, the analysis showed that the digitalization level of the transport complex is lower than the average for all economic organizations. Of course, this situation indicates the need to create a system for digital transformation of the transport complex, which would include measures to introduce common information technologies in production and organizational activities, activities aimed at collecting, processing, accumulating and storing data, as well as protecting them from external interventions and impacts [9-11].

4. Discussion

The formation of a digital transformation system for the transport complex should be based on a set of measures that will ensure a high-quality transition of the industry to new technologies. Among the main areas providing digital transformation, it is advisable to provide for the following activities [12-15]:

- The development of Internet of things technologies in order to ensure continuous interaction of various types of activities within the framework of a single transport and logistics system;
- Implementation of the "intelligent transport" system, which will allow for intelligent decision making and management of the entire transport system;
- Development of an artificial intelligence system that allows to evaluate and control the functioning of the transport-logical system;
- Implementation of GPS - systems for all freight and passenger vehicles;
- Development of a system for protecting data, information and the safe digital functioning of the transport complex.

To implement the tasks set, it is necessary to propose a model that will allow achieving the goals and objectives of the digital transformation of the transport complex (figure 4) [16-19].

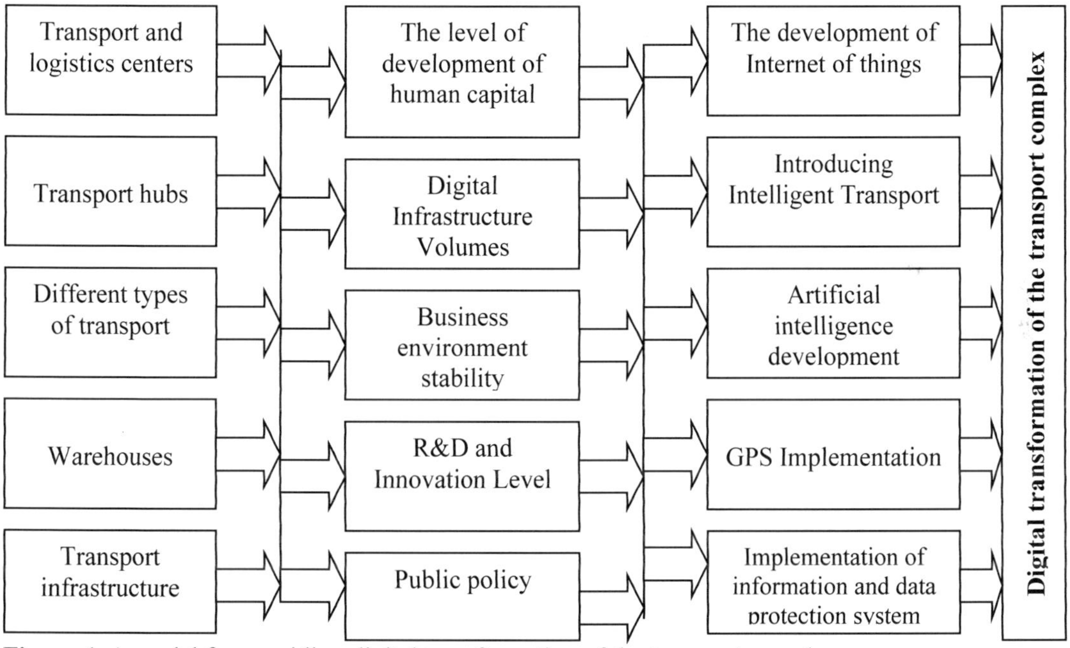

Figure 4. A model for providing digital transformation of the transport complex.

The presented model for ensuring the digital transformation of the transport complex should contain:

- Activities that should be digitalized;
- The necessary potential that will enable the digital transformation;
- Tools that enable digital transformation.

Thus, we can conclude that the activities in the field of digital transformation of the transport complex should contain the necessary potential for its implementation and the tools by which the digital transformation process will be carried out.

5. Conclusion

Thus, in the framework of the presented study, we analyzed the indicators of the level of digital development of economic organizations, the level of digital development of organizations of the transport complex, and the volumes of information protection tools used in the transport industry. Data analysis showed that the level of digital development of economic organizations is at a high level, and in the transport sector, these indicators are much lower, while it was revealed that one third of organizations in the transport industry do not use information security tools. In order to ensure further digital transformation of the transport complex, a model was proposed that reflects the types of activities subject to digital transformation, the presence of the necessary potential and the tools by which digital transformation is carried out. The model presented in the study should ensure a high-quality transition of the transport complex to a digital and innovative development path.

Acknowledgments

The article was prepared as part of the fundamental research work "Cyclical development of world economic structures" according to the state order of the Financial University under the Government of the Russian Federation for 2019-2021.

References

[1] Yuryeva A A, Kuternin M I, Gibadullin A A and Romanova Ju A 2019 Formation of mechanisms for the development of innovative activity in the industrial production of the Russian Federation. *Journal of Physics: Conference Series* **1399** 033099

[2] Tolkachev S A and Teplyakov A Yu 2019 The concept of the cyclic sequence of the spread of basic technologies in the economy and the ontological conditionality of the theories of industrial society. *Economic Revival of Russia* **4(62)** 19-36

[3] Gibadullin A A, Pulyaeva V N, Kharitonova E N and Kharitonova N A 2019 Convergence of the Russian power industry in a period of declining reliability and sustainability of the industry. *Journal of Physics: Conference Series* **1399** 033034

[4] Mitrofanova S V, Demjanchenko N V, Novikov S V, Rudakova O V and Shmanev S V 2017 The role and characteristics of the enterprises working conditions before and after the transition to market relations: A view from macroeconomic perspective. *International Journal of Applied Business and Economic Research* **15(13)** 63-72

[5] Bryukhovetskaya S V, Artamonova K A, Gibadullin A A, Ilminskaya S A and Kurbonova Z M 2020 Management of digital technology development in the national economy. *IOP Conference Series: Earth and Environmental Science* **421** 042018

[6] Tolkachev S A et al 2020 *IOP Conf. Ser.: Earth Environ. Sci.* **421** 032041

[7] Shumaev V A, Morkovkin D E, Nikonorova A V, Nezamaikin V N and Yurzinova I L 2018 Innovative aspects of agritourism project management. *SERVE 2017, Financial and Economic Tools Used in the World Hospitality Industry - Proceedings of the 5th International Conference on Management and Technology in Knowledge, Service, Tourism and Hospitality* 241-8

[8] *Russian statistical yearbook* 2018 (Moscow: Rosstat) 694

[9] Gibadullin A A, Sadriddinov M I, Kurbonova Z M, Shedko Yu N and Shamraeva V V 2020 Assessment of factors ensuring sustainable development of the electric power industry in the context of transition to renewable energy sources of the national economy. *IOP Conference Series: Earth and Environmental Science* **421** 032051

[10] Ustyuzhanina E, Evsukov S and Komarova I 2018 Network economy as a new economic system. *European Research Studies Journal* **21(3)** 77-89

[11] Sadriddinov M I, Mezina T V, Morkovkin D E, Romanova Ju A and Gibadullin A A 2020 Assessment of technological development and economic sustainability of domestic industry in modern conditions. *IOP Conference Series: Materials Science and Engineering* **734** 012051

[12] Ustyuzhanina E V, Evsukov S G, Ustyuzhanin V L and Novikova E S 2019 The distribution of economic rent within global value chains in resource management. *International Journal of Supply Chain Management* **8(6)** 1018-26

[13] Gureev P M, Degtyareva V V and Prokhorova I S 2020 National Features of Forming a Digital Economy in Russia. *Advances in Intelligent Systems and Computing* **1100** 13-20

[14] Davnis V V, Tinyakova V I, Blinov A O and Volodin Yu V 2019 Combined Modeling of Projected Evaluation of the Regional Socio-economic Development *International Journal of Economics & Business Administration (IJEBA)* **0** 1 348-54

[15] Merenkov A O 2018 Digital economy: transport management and intelligent transportation systems. *E-management* **1(1)** 12-8

[16] Romanova I N, Morkovkin D E, Nezamaikin V N, Gibadullin A A and Ivanova M A 2020 Formation of a policy to ensure environmental safety in modern economic conditions. *IOP Conference Series: Materials Science and Engineering* **734(1)** 012166

[17] Kobtseva O N, Novoselova N N, Novoselov S N, Shichiyakh R A, Morkovkin D E and Sidorchukova E V 2017 Organizational and economic features of import substitution formation and realization in the conditions of spatial restrictions. *International Journal of Applied Business and Economic Research* **15(23)** 25-35

[18] Plotnikov V, Makarov I, Shamrina I and Shirokova O 2010 Transport development as a factor in the economic security of regions and cities. *E3S Web of Conferences* **91** 05032

[19] Silverstrov S N, Osipov V S and Zeldner A G 2019 The role of state corporations in redesign of sectoral supply chain management. *International Journal of Supply Chain Management* **8(5)** 314-24

AUTHOR INDEX